2019年四川省重点出版专项资金资助项目
"一带一路"输电线路建设管理专业丛书
OBOR Transmission Line Construction and Management - Book Series

特高压交流输电线路运检专业培训及考核标准

Standard for Professional Training and Assessment
for Operation Maintenance of UHV AC Transmission Line

丛书主编／汤晓青
本册主编／范　宇　朱　康
副 主 编／汤晓青　赵　强　杨　力
编　者／王雯雯　郑和平　廖家俊　陈　立
　　　　　覃　浩　隆　茂　胡永银　全昌前
　　　　　卢国栋　杜　毅　饶建彬　张　杰
　　　　　邱中华　王　鸿　周刘育　郭定海
　　　　　王如鹏　陈　妮
译　者／成都优译信息技术股份有限公司

电子科技大学出版社
University of Electronic Science and Technology of China Press

·成都·

图书在版编目（CIP）数据

特高压交流输电线路运检专业培训及考核标准 / 范宇，朱康主编. -- 成都：电子科技大学出版社，2019.12
ISBN 978-7-5647-7548-3

Ⅰ.①特… Ⅱ.①范… ②朱… Ⅲ.①特高压输电－交流输电－输电线路－电力系统运行－检修－标准 Ⅳ.①TM726.1-65

中国版本图书馆CIP数据核字(2019)第287194号

特高压交流输电线路运检专业培训及考核标准
TEGAOYA JIAOLIU SHUDIAN XIANLU YUNJIAN ZHUANYE PEIXUN JI KAOHE BIAOZHUN

范　宇　朱　康　主编

策划编辑　陈松明　陈　亮　熊晶晶
责任编辑　辜守义

出版发行	电子科技大学出版社 成都市一环路东一段159号电子信息产业大厦　邮编　610051
主　　页	www.uestcp.com.cn
服务电话	028-83203399
邮购电话	028-83201495
印　　刷	成都市火炬印务有限公司
成品尺寸	185mm×260mm
印　　张	47
字　　数	1248千字
版　　次	2019年12月第一版
印　　次	2019年12月第一次印刷
书　　号	ISBN 978-7-5647-7548-3
定　　价	187.00元

版权所有，侵权必究

前　言

"'一带一路'输电线路建设管理专业丛书"共分十册，该丛书全面系统地介绍了国家电网公司在输电线路上的建设、运行及维护知识。每册图书都由中文版和英译文版组成，可作为"一带一路"上的国家和地区的电力职业教育规划教材，职业教育电力技术类专业培训用书。该丛书的出版对于推动"一带一路"的输电线路管理具有重要的意义。

特高压输电是目前世界上最先进的输电技术。中国特高压技术的出现，为构建"全球能源互联网"、落实国家"一带一路"发展战略，提供了强大的基础支撑。

随着多条特高压交流、直流线路的投运，为保障线路正常运行，线路检修、运行和维护工作显得非常重要。国网四川省电力公司技能培训中心在国网四川电力公司大力支持下，联合国网四川省电力公司检修公司编写了本书。国网四川省电力公司技能培训中心长期从事输电线路运检、输电线路带电作业培训和资质认证工作，在教学过程中积累了丰富的培训资源。目前，已经制定了国网四川省电力公司多个输配电线路专业培训和考核标准，培训及考核标准编制经验丰富；四川电力公司检修公司长期从事四川过境特高压线路的运行维护，实践经验丰富，强强联手合力打造，强化、确保了本书的实用性。

本书共分两部分，第一部分为特高压交流输电线路运检专业培训及考核标准总体说明；第二部分为各技能模块培训及考核标准，总计17个模块，由11个带电检修作业、6个停电检修作业组成。各模块均由培训标准和考核标准两部分组成，涵盖了培训和考核的全部元素。

本书的编写得到了国网四川省电力公司人力资源部、技能培训中心及检修公司各位领导及同仁的大力支持和帮助，在此一并致谢。

由于编者水平有限，加之时间仓促，书中难免有不足之处，恳请广大专家读者批评指正。

编　者
2019年9月

Foreword

OBOR Transmission Line Construction and Management - Book Series totally include ten volumes. The books comprehensively and systematically introduce the knowledge of the State Grid in the construction, operation and maintenance of power transmission lines. Each volume of the books consists of Chinese version and English translation version, which can be used as the electric power vocational education planning textbooks for the countries and regions on the "Belt and Road", and the professional training books for vocational education in electric power technologies. The publication of the book series is of great significance for promoting the management of the power transmission lines on the "Belt and Road".

UHV power transmission is most advanced power transmission technique in the world at present. The emergence of UHV technique in China provides strong foundation support for structuring the "Global Energy Internet", and implementing the national "Belt and Road" development strategy.

As multiple UHV AC or DC lines have been put into operation, the line repair, operation and maintenance become very important in order to guarantee the normal operation of the lines. With the great support from the State Grid Sichuan Electric Power Company, the Skill Training Center of State Grid Sichuan Electric Power Company has edited this book jointly with the Maintenance Company of State Grid Sichuan Electric Power Company. The Skill Training Center of State Grid Sichuan Electric Power Company has been engaged in the training and qualification for operation maintenance of power transmission lines, live working on power transmission lines for a long time, with rich training resources accumulated in the process of teaching. Currently, it has formulated the standards for professional training and assessment of multiple power transmission lines under State Grid Sichuan Electric Power Company, and has rich experiences in preparation of training and assessment standards; the Maintenance Company of Sichuan Electric Power Company has been engaged in the operation and maintenance of Sichuan cross-border UHV lines for a long time, and is rich in practical experiences. The win-win co-operation has strengthened and guaranteed the practicability of this book.

This book is divided into two parts: the first part is the general description of the professional

training and assessment standards for UHV AC transmission line operation maintenance; the second part is the standard for training and assessment of totally 17 skill modules, which includes 11 live maintenance modules, and 6 interruption maintenance modules. Each module consists of two parts: training standard and assessment standard, covering all elements of training and assessment.

The strong support and help for the editing of this book from the leaders and colleagues from the Human Resources Department, Skill Training Center and Maintenance Company of State Grid Sichuan Electric Power Company is greatly acknowledged.

Due to the limited level of the editor, as well as the hurry in time, there is inevitably some places that can be improved in the book, and the criticism and correction from experts and readers are appreciated.

Editor
September, 2019

目 录

第一部分　特高压交流输电线路运检专业培训及考核标准 ………………………………1
第二部分　技能模块培训及考核标准 …………………………………………………19
　模块1　带电更换1000kV交流输电线路直线塔Ⅰ型复合绝缘子培训及考核标准 ……20
　模块2　带电更换1000kV交流输电线路直线塔单V型复合绝缘子培训及考核标准 …37
　模块3　带电更换1000kV交流输电线路耐张塔横担侧1～3片玻璃绝缘子培训及考
　　　　　核标准 …………………………………………………………………………54
　模块4　带电更换1000kV交流输电线路耐张塔导线侧1～3片玻璃绝缘子培训及考
　　　　　核标准 …………………………………………………………………………68
　模块5　带电更换1000kV交流输电线路耐张玻璃绝缘子串任意单片绝缘子培训及考
　　　　　核标准 …………………………………………………………………………84
　模块6　带电更换1000kV交流输电线路耐张玻璃绝缘子串任意段绝缘子培训及考核
　　　　　标准 ……………………………………………………………………………100
　模块7　带电更换1000kV交流输电线路导线间隔棒培训及考核标准 …………………115
　模块8　带电更换1000kV交流输电线路导线防振锤培训及考核标准 …………………134
　模块9　带电修补1000kV交流输电线路导线培训及考核标准 …………………………152
　模块10　带电处理1000kV交流输电线路导线引流板发热缺陷培训及考核标准 ………166
　模块11　带电更换1000kV交流输电线路直线塔地线联接金具培训及考核标准 ………180
　模块12　停电更换1000kV交流输电线路直线塔Ⅰ型复合绝缘子培训及考核标准
　　　　　 ……………………………………………………………………………………192
　模块13　停电更换1000kV交流输电线路直线塔单V型复合绝缘子培训及考核标准
　　　　　 ……………………………………………………………………………………203
　模块14　停电更换1000kV交流输电线路耐张整串绝缘子培训及考核标准 ……………214
　模块15　停电修补1000kV交流输电线路架空地线培训及考核标准 ……………………226
　模块16　停电更换1000kV交流输电线路子导线培训及考核标准 ………………………239
　模块17　停电更换1000kV交流输电线路架空地线培训及考核标准 ……………………252

CONTENTS

Part I Standard for Professional Training and Assessment for Operation Maintenance of UHV AC Transmission Line ··265

Part II Skill Module Training and Assessment Standards ······························295

    Module 1 Standards for Training and Assessment on Live Replacement of Type I Composite Insulator for 1000kV AC Transmission Line Tangent Tower 296

    Module 2 Standards for Training and Assessment on Live Replacement of Single-V Composite Insulator for 1000kV AC Transmission Line Tangent Tower 326

    Module 3 Standards for Training and Assessment on Live Replacement of 1 ~ 3 Glass Insulators on the Cross Arm Side of 1000kV AC Transmission Line Resisting-tensile Tower ··357

    Module 4 Standards for Training and Assessment on Live Replacement of 1 ~ 3 Glass Insulators on the Conductor Side of 1000kV AC Transmission Line Resisting-tensile Tower ··381

    Module 5 Standards for Training and Assessment on Live Replacement of Arbitrary Single Insulator of Tensile Glass Insulator String of 1000kV AC Transmission Line ··408

    Module 6 Standards for Training and Assessment on Live Replacement of Arbitrary Section Insulator of Tensile Glass Insulator String of 1000kV AC Transmission Line ··435

    Module 7 Standards for Training and Assessment on Live Replacement of 1000kV AC Transmission Line Conductor Spacer ··463

    Module 8 Standards for Training and Assessment on Live Replacement of 1000kV AC Transmission Line Conductor Vibration Damper ··497

Module 9　Standards for Training and Assessment on Live Repair of 1000kV AC Transmission Line Conductor ……………………………………………531

Module 10　Standards for Training and Assessment on Live Treatment of Heating Defects of Drainage Plate for Conductor of 1000kV AC Transmission Line ……………………………………………………………………………557

Module 11　Training and Assessment Standards for Live Replacement of Ground Wire Connecting Fitting for 1000kV AC Transmission Line Tangent Tower …………………………………………………………………………583

Module 12　Standards for Training and Assessment on Power-cut Replacement of Type I Composite Insulator for 1000kV AC Transmission Line Tangent Tower ………………………………………………………………………607

Module 13　Standards for Training and Assessment on Power-cut Replacement of Type Single V Composite Insulator for 1000kV AC Transmission Line Tangent Tower ………………………………………………………………627

Module 14　Training and Assessment Standard of the Power-cut Replacement of the Whole Strain Insulator String for 1000kV AC Transmission Line ……647

Module 15　Training and Assessment Standard for Power-cut Repair of 1000kV AC Transmission Line Overhead Ground Wire …………………………669

Module 16　Training and Assessment Standard for Power-cut Replacement of 1000kV AC Transmission Line Sub-conductor ……………………………692

Module 17　Training and Assessment Standard for Power-cut Replacement of 1000kV AC Transmission Line Overhead Ground Wire ……………………716

第一部分

特高压交流输电线路运检专业培训及考核标准

一、总则

为贯彻"安全第一、预防为主、综合治理"的方针，规范1000kV交流输电线路运检专业培训和考核工作，全面提升特高压交流输电线路检修人员技能水平，制定本标准。

本标准适用于特高压交流输电线路检修人员专项技能培训考核工作。

二、编制依据

《国家电网公司技能人员岗位能力培训规范 第4部分：输电线路运检（330kV及以上）》（Q/GDW 11372.4—2015）

《国家电网公司技能人员岗位能力培训规范 第7部分：输电带电作业》（Q/GDW 11372.7—2015）

《国家电网公司电力安全工作规程 线路部分》（Q/GDW 1799.2—2013）

三、培训对象

特高压交流输电线路检修人员。

四、培训标准

本标准共涵盖17个培训模块，其中，带电检修作业培训模块11个，停电检修作业模块6个。

（一）模块设置

本标准所设置的培训模块如表1-1所示。

表1-1 培训模块设置

序号	模块名称	学时数
1	带电更换1000kV交流输电线路直线塔Ⅰ型复合绝缘子	21
2	带电更换1000kV交流输电线路直线塔单V型复合绝缘子	21
3	带电更换1000kV交流输电线路耐张塔横担侧1~3片玻璃绝缘子	21
4	带电更换1000kV交流输电线路耐张塔导线侧1~3片玻璃绝缘子	21
5	带电更换1000kV交流输电线路耐张玻璃绝缘子串任意单片绝缘子	21
6	带电更换1000kV交流输电线路耐张玻璃绝缘子串任意段绝缘子	21
7	带电更换1000kV交流输电线路导线间隔棒	21

续表

序号	模块名称	学时数
8	带电更换1000kV交流输电线路导线防振锤	21
9	带电修补1000kV交流输电线路导线	14
10	带电处理1000kV交流输电线路导线引流板发热缺陷	14
11	带电更换1000kV交流输电线路直线塔地线联接金具	14
12	停电更换1000kV交流输电线路直线塔Ⅰ型复合绝缘子	21
13	停电更换1000kV交流输电线路直线塔单V型复合绝缘子	21
14	停电更换1000kV交流输电线路耐张整串绝缘子	28
15	停电修补1000kV交流输电线路架空地线	14
16	停电更换1000kV交流输电线路子导线	21
17	停电更换1000kV交流输电线路架空地线	21

（二）模块描述

本标准各模块培训内容描述如表1-2～表1-18所示。

表1-2　带电更换1000kV交流输电线路直线塔Ⅰ型复合绝缘子培训内容

模块名称	带电更换1000kV交流输电线路直线塔Ⅰ型复合绝缘子
内容描述	学习掌握带电更换1000kV交流输电线路直线塔Ⅰ型复合绝缘子标准化作业流程，分组训练学员通过"吊篮法"进入等电位，采用等电位作业法更换直线塔Ⅰ型复合绝缘子的专项技能，提升检修人员带电更换1000kV交流输电线路直线塔Ⅰ型复合绝缘子项目的作业能力

表1-3　带电更换1000kV交流输电线路直线塔单V型复合绝缘子培训内容

模块名称	带电更换1000kV交流输电线路直线塔单V型复合绝缘子
内容描述	学习掌握带电更换1000kV交流输电线路直线塔单V型复合绝缘子标准化作业流程，分组训练学员通过"吊篮法"进入等电位，采用等电位作业法更换直线塔V型复合绝缘子的专项技能，提升检修人员带电更换1000kV交流输电线路直线塔单V型复合绝缘子项目的作业能力

表1-4　带电更换1000kV交流输电线路耐张塔横担侧1～3片玻璃绝缘子培训内容

模块名称	带电更换1000kV交流输电线路耐张塔横担侧1～3片玻璃绝缘子
内容描述	学习掌握带电更换1000kV交流输电线路耐张塔横担侧1～3片玻璃绝缘子标准化作业流程，分组训练学员采用地电位作业法更换耐张塔横担侧1～3片玻璃绝缘子的专项技能，提升检修人员带电更换1000kV交流输电线路耐张塔横担侧1～3片玻璃绝缘子项目的作业能力

表1-5　带电更换1000kV交流输电线路耐张塔导线侧1~3片玻璃绝缘子培训内容

模块名称	带电更换1000kV交流输电线路耐张塔导线侧1~3片玻璃绝缘子
内容描述	学习掌握带电更换1000kV交流输电线路耐张塔导线侧1~3片玻璃绝缘子标准化作业流程,分组训练学员通过"跨二短三"作业方式沿耐张绝缘子串进入等电位,采用等电位作业法更换耐张塔导线侧1~3片玻璃绝缘子的专项技能,提升检修人员带电更换1000kV交流输电线路耐张导线侧1~3片玻璃绝缘子项目的作业能力

表1-6　带电更换1000kV交流输电线路耐张玻璃绝缘子串任意单片绝缘子培训内容

模块名称	带电更换1000kV交流输电线路耐张玻璃绝缘子串任意单片绝缘子
内容描述	学习掌握带电更换1000kV交流输电线路耐张玻璃绝缘子串任意单片绝缘子标准化作业流程,分组训练学员通过"跨二短三"作业方式沿耐张绝缘子串进入强电场,采用中间电位作业法更换耐张玻璃绝缘子串任意单片绝缘子的专项技能,提升检修人员带电更换1000kV交流输电线路耐张玻璃绝缘子串任意单片绝缘子项目的作业能力

表1-7　带电更换1000kV交流输电线路耐张玻璃绝缘子串任意段绝缘子培训内容

模块名称	带电更换1000kV交流输电线路耐张玻璃绝缘子串任意段绝缘子
内容描述	学习掌握带电更换1000kV交流输电线路耐张玻璃绝缘子串任意段绝缘子标准化作业流程,分组训练学员通过"跨二短三"作业方式沿耐张绝缘子串进入强电场,采用中间电位作业法更换耐张玻璃绝缘子串任意段绝缘子的专项技能,提升检修人员带电更换1000kV交流输电线路耐张玻璃绝缘子串任意段绝缘子项目的作业能力

表1-8　带电更换1000kV交流输电线路导线间隔棒培训内容

模块名称	带电更换1000kV交流输电线路导线间隔棒
内容描述	学习掌握带电更换1000kV交流输电线路导线间隔棒标准化作业流程,分组训练学员通过"跨二短三"作业方式沿耐张绝缘子串进入等电位,采用等电位作业法更换八分裂导线间隔棒的专项技能,提升检修人员带电更换1000kV交流输电线路导线间隔棒项目的作业能力

表1-9　带电更换1000kV交流输电线路导线防振锤培训内容

模块名称	带电更换1000kV交流输电线路导线防振锤
内容描述	学习掌握带电更换1000kV交流输电线路导线防振锤标准化作业流程,分组训练学员通过"跨二短三"作业方式沿耐张绝缘子串进入等电位,采用等电位作业法更换导线防振锤的专项技能,提升检修人员带电更换1000kV交流输电线路导线防振锤项目的作业能力

第一部分
特高压交流输电线路运检专业培训及考核标准

表1-10　带电修补1000kV交流输电线路导线培训内容

模块名称	带电修补1000kV交流输电线路导线
内容描述	学习掌握带电修补1000kV交流输电线路导线标准化作业流程,分组训练学员通过"跨二短三"作业方式沿耐张绝缘子串进入等电位,采用等电位作业法使用预绞丝补修条修补导线的专项技能,提升检修人员带电修补1000kV交流输电线路导线项目的作业能力

表1-11　带电处理1000kV交流输电线路导线引流板发热缺陷培训内容

模块名称	带电处理1000kV交流输电线路导线引流板发热缺陷
内容描述	学习掌握带电处理1000kV交流输电线路导线引流板发热缺陷标准化作业流程,分组训练学员通过"跨二短三"作业方式沿耐张绝缘子串进入等电位,采用等电位作业法对导线引流板连接螺栓进行紧固的专项技能,提升检修人员带电处理1000kV交流输电线路导线引流板发热缺陷项目的作业能力

表1-12　带电更换1000kV交流输电线路直线塔地线联接金具培训内容

模块名称	带电更换1000kV交流输电线路直线塔地线联接金具
内容描述	学习掌握带电更换1000kV交流输电线路直线塔地线联接金具标准化作业流程,分组训练学员采用地电位作业法更换直线塔地线联接金具的专项技能,提升检修人员带电更换1000kV交流输电线路直线塔地线联接金具项目的作业能力

表1-13　停电更换1000kV交流输电线路直线塔Ⅰ型复合绝缘子培训内容

模块名称	停电更换1000kV交流输电线路直线塔Ⅰ型复合绝缘子
内容描述	学习掌握停电更换1000kV交流输电线路直线塔Ⅰ型复合绝缘子标准化作业流程,分组训练学员采用停电作业方式更换直线塔Ⅰ型复合绝缘子的专项技能,提升检修人员停电更换1000kV交流输电线路直线塔Ⅰ型复合绝缘子项目的作业能力

表1-14　停电更换1000kV交流输电线路直线塔单V型复合绝缘子培训内容

模块名称	停电更换1000kV交流输电线路直线塔单V型复合绝缘子
内容描述	学习掌握停电更换1000kV交流输电线路直线塔单V型复合绝缘子标准化作业流程,分组训练学员采用停电作业方式更换直线塔单V型复合绝缘子的专项技能,提升检修人员停电更换1000kV交流输电线路直线塔单V型复合绝缘子项目的作业能力

表1-15　停电更换1000kV交流输电线路耐张整串绝缘子培训内容

模块名称	停电更换1000kV交流输电线路耐张整串绝缘子
内容描述	学习掌握停电更换1000kV交流输电线路耐张整串绝缘子标准化作业流程,分组训练学员采用停电作业方式更换耐张整串绝缘子的专项技能,提升检修人员停电更换1000kV交流输电线路耐张整串绝缘子项目的作业能力

表1-16　停电修补1000kV交流输电线路架空地线培训内容

模块名称	停电修补1000kV交流输电线路架空地线
内容描述	学习掌握停电修补1000kV交流输电线路架空地线标准化作业流程,分组训练学员采用停电作业方式利用飞车到达作业位置,使用预绞丝补修条修补架空地线的专项技能,提升检修人员停电修补1000kV交流输电线路架空地线项目的作业能力

表1-17　停电更换1000kV交流输电线路子导线培训内容

模块名称	停电更换1000kV交流输电线路子导线
内容描述	学习掌握停电更换1000kV交流输电线路子导线标准化作业流程,分组训练学员采用停电作业方式更换子导线的专项技能,提升检修人员停电更换1000kV交流输电线路子导线项目的作业能力

表1-18　停电更换1000kV交流输电线路架空地线培训内容

模块名称	停电更换1000kV交流输电线路架空地线
内容描述	学习掌握停电更换1000kV交流输电线路架空地线标准化作业流程,分组训练学员采用停电作业方式更换架空地线的专项技能,提升检修人员停电更换1000kV交流输电线路架空地线项目的作业能力

（三）教学设计

本标准中各模块培训教学实施过程,分为理论教学、准备工作、作业现场准备、培训师演示、学员分组训练、工作终结6个培训阶段。各培训阶段按照完成工作任务的标准化作业流程来设计,包括培训目标、培训内容、培训学时、培训方法（培训资源）、培训环境和考核评价等内容。

五、考核标准

本标准中的考核模块与培训模块对应,在完成模块培训后可参照相应的考核评分细则实施考核。考核评分细则主要设定了考生信息、考评员信息、考核时间和时长、考核模块、考核对象、考核方式、考核时限、任务描述、工作规范及要求、考核情景准备等内容和范围,按照各模块作业标准化流程，分工作准备阶段、工作实施阶段、工作终结阶段方面，制定了各具体流程段的质量要求、考核分值、扣分标准，并提供考评员填写的扣分原因栏、扣分分值栏、得分分值栏。各模块考核成绩以完成考核评分细则中设定的工作任务全过程，所取得各流程段的评价分值之和。考核模块可以按组进行考核成绩评定，也可以对单个考生进行考核成绩评定，若需对作业中涉及几个角色均进行考核，采用角色轮换方式，完成考生考核成绩评定。本标准所设置的考核模块如表1-19所示。

表1-19 考核模块设置

序号	模块名称	考核方式	考核时限/min	考核角色数
1	带电更换1000kV交流输电线路直线塔Ⅰ型复合绝缘子	操作	120	3
2	带电更换1000kV交流输电线路直线塔单Ⅴ型复合绝缘子	操作	120	3
3	带电更换1000kV交流输电线路耐张塔横担侧1~3片玻璃绝缘子	操作	90	2
4	带电更换1000kV交流输电线路耐张塔导线侧1~3片玻璃绝缘子	操作	90	2
5	带电更换1000kV交流输电线路耐张玻璃绝缘子串任意单片绝缘子	操作	90	2
6	带电更换1000kV交流输电线路耐张玻璃绝缘子串任意段绝缘子	操作	90	3
7	带电更换1000kV交流输电线路导线间隔棒	操作	90	3
8	带电更换1000kV交流输电线路导线防振锤	操作	90	3
9	带电修补1000kV交流输电线路导线	操作	60	3
10	带电处理1000kV交流输电线路导线引流板发热缺陷	操作	60	3
11	带电更换1000kV交流输电线路直线塔地线联接金具	操作	60	2
12	停电更换1000kV交流输电线路直线塔Ⅰ型复合绝缘子	操作	150	3
13	停电更换1000kV交流输电线路直线塔单Ⅴ型复合绝缘子	操作	150	3
14	停电更换1000kV交流输电线路耐张整串绝缘子	操作	360	4
15	停电修补1000kV交流输电线路架空地线	操作	100	3
16	停电更换1000kV交流输电线路子导线	操作	360	4
17	停电更换1000kV交流输电线路架空地线	操作	360	4

六、保障条件

（一）师资配置

本标准中各培训模块师资配置及要求如表1-20所示。

表1-20 培训模块师资配置及要求

模块名称	主讲培训师		技能操作演示培训师		地面辅助人员	
	数量	要求	数量	要求	数量	要求
带电更换1000kV交流输电线路直线塔Ⅰ型复合绝缘子	1	明确带电更换1000kV交流输电线路直线塔Ⅰ型复合绝缘子标准化作业流程,熟悉特高压交流输电线路发展趋势,具备扎实的高电压技术和带电作业理论知识,熟悉作业中的技术关键点和危险点预控,熟悉吊篮法进出强电场和电位转移棒使用方法,能对各作业步骤动作要领、原因和方法进行充分的讲解,能灵活使用合适教学方法,具备良好教学组织能力	5	熟悉带电更换1000kV交流输电线路直线塔Ⅰ型复合绝缘子工作内容和标准化作业流程,多次参与带电更换1000kV交流输电线路直线塔Ⅰ型复合绝缘子作业,具备丰富的作业经验和熟练的作业技能。工作负责人角色能正确组织整个作业,能判定作业现场条件和工作票安全措施是否完备并补充,明确各项安全措施和危险点预控,能组织执行工作票所列安全措施,能有效地监督作业人员按规程执行现场安全措施;塔上电工角色需熟悉工作内容和工作流程,掌握安全措施,明确工作范围和危险点及预控措施,能熟练使用工器具及防护用品	5	明确带电更换1000kV交流输电线路直线塔Ⅰ型复合绝缘子工作内容和标准化作业流程,具备相应的电工操作技能,能及时理解工作负责人角色下达的命令,能正确操作相关工器具,能正确起吊、传递工器具、传递绝缘子等
带电更换1000kV交流输电线路直线塔单V型复合绝缘子	1	明确带电更换1000kV交流输电线路直线塔单V型复合绝缘子标准化作业流程,熟悉特高压交流输电线路发展趋势,具备扎实的高电压技术和带电作业理论知识,熟悉作业中的技术关键点和危险点预控,熟悉吊篮法进出强电场和电位转移棒使用方法,能对各作业步骤动作要领、原因和方法进行充分的讲解,能灵活使用合适教学方法,具备良好教学组织能力	5	熟悉带电更换1000kV交流输电线路直线塔单V型复合绝缘子工作内容和标准化作业流程,多次参与带电更换1000kV交流输电线路直线塔单V型复合绝缘子作业,具备丰富的作业经验和熟练的作业技能。工作负责人角色能正确组织整个作业,能判定作业现场条件和工作票安全措施是否完备并补充,明确各项安全措施和危险点预控,能组织执行工作票所列安全措施,能有效地监督作业人员按规程执行现场安全措施;塔上电工角色需熟悉工作内容和工作流程,掌握安全措施,明确工作范围和危险点及预控措施,能熟练使用工器具及防护用品	6	明确带电更换1000kV交流输电线路直线塔单V型复合绝缘子工作内容和标准化作业流程,具备相应的电工操作技能,能及时理解工作负责人角色下达的命令,能正确操作相关工器具,能正确起吊、传递工器具、传递绝缘子等

续表

模块名称	主讲培训师 数量	主讲培训师 要求	技能操作演示培训师 数量	技能操作演示培训师 要求	地面辅助人员 数量	地面辅助人员 要求
带电更换1000kV交流输电线路耐张塔横担侧1~3片玻璃绝缘子	1	明确带电更换1000kV交流输电线路耐张塔横担侧1~3片玻璃绝缘子标准化作业流程,熟悉特高压交流输电线路发展趋势,具备扎实的高电压技术和带电作业理论知识,熟悉作业中的技术关键点和危险点预控,熟悉单片绝缘子更换方法和质量标准,能对各作业步骤动作要领、原因和方法进行充分的讲解,能灵活使用合适教学方法,具备良好教学组织能力	2	熟悉带电更换1000kV交流输电线路耐张塔横担侧1~3片玻璃绝缘子工作内容和标准化作业流程,多次参与带电带电更换1000kV交流输电线路耐张塔横担侧1~3片玻璃绝缘子作业,具备丰富的作业经验和熟练的作业技能。工作负责人角色能正确组织整个作业,能判定作业现场条件和工作票安全措施是否完备并补充,明确各项安全措施和危险点预控,能组织执行工作票所列安全措施,能有效地监督作业人员按规程执行现场安全措施;塔上电工角色需熟悉工作内容和工作流程,掌握安全措施,明确工作范围和危险点及预控措施,能熟练使用工器具及防护用品	3	明确带电更换1000kV交流输电线路耐张塔横担侧1~3片玻璃绝缘子工作内容和标准化作业流程,具备相应的电工操作技能,能及时理解工作负责人角色下达的命令,能正确操作相关工器具,能正确传递工器具、传递绝缘子等
带电更换1000kV交流输电线路耐张塔导线侧1~3片玻璃绝缘子	1	明确带电更换1000kV交流输电线路耐张塔导线侧1~3片玻璃绝缘子标准化作业流程,熟悉特高压交流输电线路发展趋势,具备扎实的高电压技术和带电作业理论知识,熟悉作业中的技术关键点和危险点预控,熟悉沿绝缘子进出电场作业方式原理和电位转移棒使用方法,熟悉单片绝缘子更换方法和质量标准,能对各作业步骤动作要领、原因和方法进行充分的讲解,能灵活使用合适教学方法,具备良好教学组织能力	2	熟悉带电更换1000kV交流输电线路耐张塔导线侧1~3片玻璃绝缘子工作内容和标准化作业流程,多次参与带电更换1000kV交流输电线路耐张塔导线侧1~3片玻璃绝缘子作业,具备丰富的作业经验和熟练的作业技能。工作负责人角色能正确组织整个作业,能判定作业现场条件和工作票安全措施是否完备并补充,明确各项安全措施和危险点预控,能组织执行工作票所列安全措施,能有效地监督作业人员按规程执行现场安全措施;塔上电工角色需熟悉工作内容和工作流程,掌握安全措施,明确工作范围和危险点及预控措施,能熟练使用工器具及防护用品	3	明确带电更换1000kV交流输电线路耐张塔导线侧1~3片玻璃绝缘子工作内容和标准化作业流程,具备相应的电工操作技能,能及时理解工作负责人角色下达的命令,能正确操作相关工器具,能正确传递工器具、传递绝缘子等

续表

模块名称	主讲培训师		技能操作演示培训师		地面辅助人员	
	数量	要求	数量	要求	数量	要求
带电更换1000kV交流输电线路耐张玻璃绝缘子串任意单片绝缘子	1	明确带电更换1000kV交流输电线路耐张玻璃绝缘子串任意单片绝缘子标准化作业流程,熟悉特高压交流输电线路发展趋势,具备扎实的高电压技术和带电作业理论知识,熟悉作业中的技术关键点和危险点预控,熟悉沿绝缘子进出电场作业方式原理,熟悉单片绝缘子更换方法和质量标准,能对各作业步骤动作要领、原因和方法进行充分的讲解,能灵活使用合适教学方法,具备良好教学组织能力	2	熟悉带电更换1000kV交流输电线路耐张玻璃绝缘子串任意单片绝缘子工作内容和标准化作业流程,多次参与带电更换1000kV交流输电线路耐张玻璃绝缘子串任意单片绝缘子作业,具备丰富的作业经验和熟练的作业技能。工作负责人角色能正确组织整个作业,能判定作业现场条件和工作票安全措施是否完备并补充,明确各项安全措施和危险点预控,能组织执行工作票所列安全措施,能有效地监督作业人员按规程执行现场安全措施;塔上电工角色需熟悉工作内容和工作流程,掌握安全措施,明确工作范围和危险点及预控措施,能熟练使用工器具及防护用品	3	明确带电更换1000kV交流输电线路耐张玻璃绝缘子串任意单片绝缘子工作内容和标准化作业流程,具备相应的电工操作技能,能及时理解工作负责人角色下达的命令,能正确操作相关工器具,能正确传递工器具、传递绝缘子等
带电更换1000kV交流输电线路耐张玻璃绝缘子串任意段绝缘子	1	明确带电更换1000kV交流输电线路耐张玻璃绝缘子串任意段绝缘子标准化作业流程,熟悉特高压交流输电线路发展趋势,具备扎实的高电压技术和带电作业理论知识,熟悉作业中的技术关键点和危险点预控,熟悉沿绝缘子进出电场作业方式原理,能对各作业步骤动作要领、原因和方法进行充分的讲解,能灵活使用合适教学方法,具备良好教学组织能力	5	熟悉带电更换1000kV交流输电线路耐张玻璃绝缘子串任意段绝缘子工作内容和标准化作业流程,开展过带电更换1000kV交流输电线路耐张玻璃绝缘子串任意段绝缘子作业,具备丰富的作业经验和熟练的作业技能。工作负责人角色能正确组织整个作业,能判定作业现场条件和工作票安全措施是否完备并补充,明确各项安全措施和危险点预控,能组织执行工作票所列安全措施,能有效地监督作业人员按规程执行现场安全措施;塔上电工角色需熟悉工作内容和工作流程,掌握安全措施,明确工作范围和危险点及预控措施,能熟练使用工器具及防护用品	6	明确带电更换1000kV交流输电线路耐张玻璃绝缘子串任意段绝缘子工作内容和标准化作业流程,具备相应的电工操作技能,能及时理解工作负责人角色下达的命令,能正确操作相关工器具,能正确传递工器具、传递绝缘子等

第一部分
特高压交流输电线路运检专业培训及考核标准

续表

模块名称	主讲培训师		技能操作演示培训师		地面辅助人员	
	数量	要求	数量	要求	数量	要求
带电更换1000kV交流输电线路导线间隔棒	1	明确带电更换1000kV交流输电线路导线间隔棒标准化作业流程,熟悉特高压交流输电线路发展趋势,具备扎实的高电压技术和带电作业理论知识,熟悉作业中的技术关键点和危险点预控,熟悉沿绝缘子进出电场作业方式原理和电位转移棒使用方法,能对各作业步骤动作要领、原因和方法进行充分的讲解,能灵活使用合适教学方法,具备良好教学组织能力	3	熟悉带电更换1000kV交流输电线路导线间隔棒工作内容和标准化作业流程,多次参与带电更换1000kV交流输电线路导线间隔棒作业,具备丰富的作业经验和熟练的作业技能。工作负责人角色能正确组织整个作业,能判定作业现场条件和工作票安全措施是否完备并补充,明确各项安全措施和危险点预控,能组织执行工作票所列安全措施,能有效地监督作业人员按规程执行现场安全措施;塔上电工角色需熟悉工作内容和工作流程,掌握安全措施,明确工作范围和危险点及预控措施,能熟练使用工器具及防护用品	2	明确带电更换1000kV交流输电线路导线间隔棒工作内容和标准化作业流程,具备相应的电工操作技能,能及时理解工作负责人角色下达的命令,能正确操作相关工器具,能正确传递工器具、传递间隔棒等
带电更换1000kV交流输电线路导线防振锤	1	明确带电更换1000kV交流输电线路导线防振锤标准化作业流程,熟悉特高压交流输电线路发展趋势,具备扎实的高电压技术和带电作业理论知识,熟悉作业中的技术关键点和危险点预控,熟悉沿绝缘子进出电场作业方式原理和电位转移棒使用方法,能对各作业步骤动作要领、原因和方法进行充分的讲解,能灵活使用合适教学方法,具备良好教学组织能力	3	熟悉带电更换1000kV交流输电线路导线防振锤棒工作内容和标准化作业流程,多次参与带电更换1000kV交流输电线路导线防振锤作业,具备丰富的作业经验和熟练的作业技能。工作负责人角色能正确组织整个作业,能判定作业现场条件和工作票安全措施是否完备并补充,明确各项安全措施和危险点预控,能组织执行工作票所列安全措施,能有效地监督作业人员按规程执行现场安全措施;塔上电工角色需熟悉工作内容和工作流程,掌握安全措施,明确工作范围和危险点及预控措施,能熟练使用工器具及防护用品	2	明确带电更换1000kV交流输电线路导线防振锤工作内容和标准化作业流程,具备相应的电工操作技能,能及时理解工作负责人角色下达的命令,能正确操作相关工器具,能正确传递工器具、传递防振锤等

续表

模块名称	主讲培训师		技能操作演示培训师		地面辅助人员	
	数量	要求	数量	要求	数量	要求
带电修补1000kV交流输电线路导线	1	明确带电修补1000kV交流输电线路导线标准化作业流程,熟悉特高压交流输电线路发展趋势,具备扎实的高电压技术和带电作业理论知识,熟悉作业中的技术关键点和危险点预控,熟悉沿绝缘子进出电场作业方式原理和电位转移棒使用方法,熟悉导线修补方法和质量标准能,对各作业步骤动作要领、原因和方法进行充分的讲解,能灵活使用合适教学方法,具备良好教学组织能力	3	熟悉带电修补1000kV交流输电线路导线工作内容和标准化作业流程,多次参与带电修补1000kV交流输电线路导线作业,具备丰富的作业经验和熟练的作业技能。工作负责人角色能正确组织整个作业,能判定作业现场条件和工作票安全措施是否完备并补充,明确各项安全措施和危险点预控,能组织执行工作票所列安全措施,能有效地监督作业人员按规程执行现场安全措施;塔上电工角色需熟悉工作内容和工作流程,掌握安全措施,明确工作范围和危险点及预控措施,能熟练使用工器具及防护用品	2	明确带电修补1000kV交流输电线路导线工作内容和标准化作业流程,具备相应的电工操作技能,能及时理解工作负责人角色下达的命令,能正确操作相关工器具,能正确传递工器具、传递预绞丝等
带电处理1000kV交流输电线路导线引流板发热缺陷	1	明确带电处理1000kV交流输电线路导线引流板发热缺陷标准化作业流程,熟悉特高压交流输电线路发展趋势,具备扎实的高电压技术和带电作业理论知识,熟悉作业中的技术关键点和危险点预控,熟悉沿绝缘子进出电场作业方式原理和电位转移棒使用方法,对各作业步骤动作要领、原因和方法进行充分的讲解,能灵活使用合适教学方法,具备良好教学组织能力	3	熟悉带电处理1000kV交流输电线路导线引流板发热缺陷工作内容和标准化作业流程,多次参与带电处理1000kV交流输电线路导线引流板发热缺陷作业,具备丰富的作业经验和熟练的作业技能。工作负责人角色能正确组织整个作业,能判定作业现场条件和工作票安全措施是否完备并补充,明确各项安全措施和危险点预控,能组织执行工作票所列安全措施,能有效地监督作业人员按规程执行现场安全措施;塔上电工角色需熟悉工作内容和工作流程,掌握安全措施,明确工作范围和危险点及预控措施,能熟练使用工器具及防护用品	2	明确带电处理1000kV交流输电线路导线引流板发热缺陷工作内容和标准化作业流程,具备相应的电工操作技能,能及时理解工作负责人角色下达的命令,能正确操作相关工器具,能正确传递工器具等

第一部分 特高压交流输电线路运检专业培训及考核标准

续表

模块名称	主讲培训师		技能操作演示培训师		地面辅助人员	
	数量	要求	数量	要求	数量	要求
带电更换1000kV交流输电线路直线塔地线联接金具	1	明确带电更换1000kV交流输电线路直线塔地线联接金具标准化作业流程,熟悉特高压交流输电线路发展趋势,具备扎实的高电压技术和带电作业理论知识,熟悉地电位作业基本原理和等效电路图,熟悉地线提线器使用方法,熟悉地线金具更换方法和质量标准,对各作业步骤动作要领、原因和方法进行充分的讲解,能灵活使用合适教学方法,具备良好教学组织能力	2	熟悉带电更换1000kV交流输电线路直线塔地线联接金具工作内容和标准化作业流程,开展过带电更换1000kV交流输电线路直线塔地线联接金具作业,具备丰富的作业经验和熟练的作业技能。工作负责人角色能正确组织整个作业,能判定作业现场条件和工作票安全措施是否完备并补充,明确各项安全措施和危险点预控,能组织执行工作票所列安全措施,能有效地监督作业人员按规程执行现场安全措施;塔上电工角色需熟悉工作内容和工作流程,掌握安全措施,明确工作范围和危险点及预控措施,能熟练使用工器具及防护用品	2	明确带电更换1000kV交流输电线路直线塔地线联接金具工作内容和标准化作业流程,具备相应的电工操作技能,能及时理解工作负责人角色下达的命令,能正确操作相关工器具,能正确传递工器具等
停电更换1000kV交流输电线路直线塔Ⅰ型复合绝缘子	1	明确停电更换1000kV交流输电线路直线塔Ⅰ型复合绝缘子标准化作业流程,熟悉特高压交流输电线路发展趋势,具备特高压交流输电线路结构知识和力学知识,熟悉停电检修流程和工器具使用,熟悉直线塔Ⅰ型复合绝缘子更换方法和质量标准,对各作业步骤动作要领、原因和方法进行充分的讲解,能灵活使用合适教学方法,具备良好教学组织能力	4	熟悉停电更换1000kV交流输电线路直线塔Ⅰ型复合绝缘子工作内容和标准化作业流程,多次参与停电更换1000kV交流输电线路直线塔Ⅰ型复合绝缘子作业,具备丰富的作业经验和熟练的作业技能。工作负责人角色能正确组织整个作业,能判定作业现场条件和工作票安全措施是否完备并补充,明确各项安全措施和危险点预控,能组织执行工作票所列安全措施,能有效地监督作业人员按规程执行现场安全措施;塔上电工角色需熟悉工作内容和工作流程,掌握安全措施,明确工作范围和危险点及预控措施,能熟练使用各类工器具	5	明确停电更换1000kV交流输电线路直线塔Ⅰ型复合绝缘子工作内容和标准化作业流程,具备相应的电工操作技能,能及时理解工作负责人角色下达的命令,能正确操作相关工器具,能正确传递工器具、传递绝缘子等

续表

模块名称	主讲培训师		技能操作演示培训师		地面辅助人员	
	数量	要求	数量	要求	数量	要求
停电更换1000kV交流输电线路直线塔单V型复合绝缘子	1	明确停电更换1000kV交流输电线路直线塔单V型复合绝缘子作业流程,熟悉特高压交流输电线路发展趋势,具备特高压交流输电线路结构知识和力学知识,熟悉停电检修流程和工器具使用,熟悉直线塔单V型复合绝缘子更换方法和质量标准,对各作业步骤动作要领、原因和方法进行充分的讲解,能灵活使用合适教学方法,具备良好教学组织能力	4	熟悉停电更换1000kV交流输电线路直线塔单V型复合绝缘子工作内容和标准化作业流程,多次参与停电更换1000kV交流输电线路直线塔单V型复合绝缘子作业,具备丰富的作业经验和熟练的作业技能。工作负责人角色能正确组织整个作业,能判定作业现场条件和工作票安全措施是否完备并补充,明确各项安全措施和危险点预控,能组织执行工作票所列安全措施,能有效地监督作业人员按规程执行现场安全措施;塔上电工角色需熟悉工作内容和工作流程,掌握安全措施,明确工作范围和危险点及预控措施,能熟练使用各类工器具	5	明确停电更换1000kV交流输电线路直线塔单V型复合绝缘子工作内容和标准化作业流程,具备相应的电工操作技能,能及时理解工作负责人角色下达的命令,能正确操作相关工器具,能正确传递工器具、传递绝缘子等
停电更换1000kV交流输电线路耐张整串绝缘子	1	明确停电更换1000kV交流输电线路耐张整串绝缘子作业流程,熟悉特高压交流输电线路发展趋势,具备特高压交流输电线路结构知识和力学知识,熟悉滑轮组安装及使用,熟悉停电检修流程和工器具使用,对各作业步骤动作要领、原因和方法进行充分的讲解,能灵活使用合适教学方法,具备良好教学组织能力	7	熟悉停电更换1000kV交流输电线路耐张整串绝缘子工作内容和标准化作业流程,开展过停电更换1000kV交流输电线路耐张整串绝缘子作业,具备丰富的作业经验和熟练的作业技能。工作负责人角色能正确组织整个作业,能判定作业现场条件和工作票安全措施是否完备并补充,明确各项安全措施和危险点预控,能组织执行工作票所列安全措施,能有效地监督作业人员按规程执行现场安全措施;塔上电工角色需熟悉工作内容和工作流程,掌握安全措施,明确工作范围和危险点及预控措施,能熟练使用各类工器具	9	明确停电更换1000kV交流输电线路耐张整串绝缘子工作内容和标准化作业流程,具备相应的电工操作技能,能及时理解工作负责人角色下达的命令,能正确操作相关工器具,能正确控制和操作绞磨,能正确传递工器具、绝缘子等

续表

模块名称	主讲培训师		技能操作演示培训师		地面辅助人员	
	数量	要求	数量	要求	数量	要求
停电修补1000kV交流输电线路架空地线	1	明确停电修补1000kV交流输电线路架空地线作业流程,熟悉特高压交流输电线路发展趋势,具备特高压交流输电线路结构知识和力学知识,熟悉飞车结构及使用,熟悉架空地线修补方法和质量标准,对各作业步骤动作要领、原因和方法进行充分的讲解,能灵活使用合适教学方法,具备良好教学组织能力	3	熟悉停电修补1000kV交流输电线路架空地线工作内容和标准化作业流程,多次参与停电修补1000kV交流输电线路架空地线作业,具备丰富的作业经验和熟练的作业技能。工作负责人角色能正确组织整个作业,能判定作业现场条件和工作票安全措施是否完备并补充,明确各项安全措施和危险点预控,能组织执行工作票所列安全措施,能有效地监督作业人员按规程执行现场安全措施;塔上电工角色需熟悉工作内容和工作流程,掌握安全措施,明确工作范围和危险点及预控措施,能熟练使用各类工器具	2	明确停电修补1000kV交流输电线路架空地线工作内容和标准化作业流程,具备相应的电工操作技能,能及时理解工作负责人角色下达的命令,能正确操作相关工器具,能正确传递工器具、预绞丝等
停电更换1000kV交流输电线路子导线	1	明确停电更换1000kV交流输电线路子导线作业流程,熟悉特高压交流输电线路发展趋势,具备特高压交流输电线路结构知识和力学知识,熟悉更换子导线作业现场布置,对各作业步骤动作要领、原因和方法进行充分的讲解,能灵活使用合适教学方法,具备良好教学组织能力	8	熟悉停电更换1000kV交流输电线路子导线工作内容和标准化作业流程,开展过停电更换1000kV交流输电线路子导线作业,具备丰富的作业经验和熟练的作业技能。工作负责人角色能正确组织整个作业,熟悉作业现场布置,能判定作业现场条件和工作票安全措施是否完备并补充,明确各项安全措施和危险点预控,能组织执行工作票所列安全措施,能有效地监督作业人员按规程执行现场安全措施;塔上电工角色需熟悉工作内容和工作流程,掌握安全措施,明确工作范围和危险点及预控措施,能熟练使用各类工器具	14	明确停电更换1000kV交流输电线路子导线工作内容和标准化作业流程,具备相应的电工操作技能,能及时理解工作负责人角色下达的命令,能正确操作相关工器具,能正确控制和操作绞磨,能正确传递工器具等

续表

模块名称	主讲培训师		技能操作演示培训师		地面辅助人员	
	数量	要求	数量	要求	数量	要求
停电更换1000kV交流输电线路架空地线	1	明确停电更换1000kV交流输电线路架空地线作业流程,熟悉特高压交流输电线路发展趋势,具备特高压交流输电线路结构知识和力学知识,熟悉更换架空地线作业现场布置,对各作业步骤动作要领、原因和方法进行充分的讲解,能灵活使用合适教学方法,具备良好教学组织能力	8	熟悉停电更换1000kV交流输电线路架空地线工作内容和标准化作业流程,开展过停电更换1000kV交流输电线路架空地线作业,具备丰富的作业经验和熟练的作业技能。工作负责人角色能正确组织整个作业,熟悉作业现场布置,能判定作业现场条件和工作票安全措施是否完备并补充,明确各项安全措施和危险点预控,能组织执行工作票所列安全措施,能有效地监督作业人员按规程执行现场安全措施;塔上电工角色需熟悉工作内容和工作流程,掌握安全措施,明确工作范围和危险点及预控措施,能熟练使用各类工器具	14	明确停电更换1000kV交流输电线路架空地线工作内容和标准化作业流程,具备相应的电工操作技能,能及时理解工作负责人角色下达的命令,能正确操作相关工器具,能正确控制和操作绞磨,能正确传递工器具等

(二) 实训条件

本标准中各培训模块实训条件如表1-21所示。

表1-21 培训模块实训条件

模块名称	设施设备	工器具及材料
带电更换1000kV交流输电线路直线塔Ⅰ型复合绝缘子	1000kV交流输电线路直线塔,Ⅰ型复合绝缘子	表2-1-3
带电更换1000kV交流输电线路直线塔单V型复合绝缘子	1000kV交流输电线路直线塔,V型复合绝缘子	表2-2-3
带电更换1000kV交流输电线路耐张塔横担侧1~3片玻璃绝缘子	1000kV交流输电线路耐张塔,玻璃绝缘子	表2-3-4
带电更换1000kV交流输电线路耐张塔导线侧1~3片玻璃绝缘子	1000kV交流输电线路耐张塔,玻璃绝缘子	表2-4-4
带电更换1000kV交流输电线路耐张玻璃绝缘子串任意单片绝缘子	1000kV交流输电线路耐张塔,玻璃绝缘子	表2-5-4
带电更换1000kV交流输电线路耐张玻璃绝缘子串任意段绝缘子	1000kV交流输电线路耐张塔,玻璃绝缘子	表2-6-3
带电更换1000kV交流输电线路导线间隔棒	1000kV交流输电线路耐张塔,八分裂导线	表2-7-3

续表

模块名称	设施设备	工器具及材料
带电更换1000kV交流输电线路导线防振锤	1000kV交流输电线路耐张塔	表2-8-3
带电修补1000kV交流输电线路导线	1000kV交流输电线路耐张塔	表2-9-3
带电处理1000kV交流输电线路导线引流板发热缺陷	1000kV交流输电线路耐张塔	表2-10-3
带电更换1000kV交流输电线路直线塔地线联接金具	1000kV交流输电线路直线塔,架空地线型号：JLB20A-150	表2-11-2
停电更换1000kV交流输电线路直线塔Ⅰ型复合绝缘子	1000kV交流输电线路直线塔,Ⅰ型复合绝缘子	表2-12-2
停电更换1000kV交流输电线路直线塔单V型复合绝缘子	1000kV交流输电线路直线塔,V型复合绝缘子	表2-13-2
停电更换1000kV交流输电线路耐张整串绝缘子	1000kV交流输电线路耐张塔	表2-14-2
停电修补1000kV交流输电线路架空地线	1000kV交流输电线路1档,架空地线型号：JLB20A-150	表2-15-2
停电更换1000kV交流输电线路子导线	1000kV交流输电线路1耐张段,导线型号：8×JL/G1A-630/45	表2-16-2
停电更换1000kV交流输电线路架空地线	1000kV交流输电线路1耐张段,架空地线型号：JLB20A-150	表2-17-2

七、编制说明

（一）编制背景

国家电网公司从2004年开始启动特高压电网研究和建设，截至2016年9月，已经投运特高压工程21项，其中特高压交流工程7项，特高压直流工程14项。与此同时，国家电网公司进一步深入研究大电网运行规律和安全机理，科学规划建设合理的网架结构，重点解决特高压"强直弱交"问题，加快形成坚强的特高压骨干网架，不断增强抵御严重故障的能力。随着特高压交流输电线路不断建设和投运，线路运检工作量将大幅增加，特高压交流检修人员技能水平提升以及该类人才的储备迫在眉睫，但目前尚无特高压交流输电线路运检专业培训及考核标准。为此，国家电网四川省电力公司技能培训中心在国家电网四川电力公司大力支持下，联合国家电网四川省电力公司检修公司编写了本标准，归纳出17个特高压交流输电线路检修模块，编制培训标准及其考核评分细则，其中带电检修作业11个、停电检修作业6个。

(二）编制主要原则

本标准的编制思路是以提升特高压交流检修人员完成常用检修作业项目能力为导向，遵循单项实用、自选体系、培评一体的自主培训考核架构，建立多元化培训考核模式。

本标准根据以下原则编制。

1. 以提升人员专项技能水平为重点，遵循"知识够用、技能必备"。
2. 归纳典型作业任务，选择代表项目，采用模块化培训和考核。
3. 实现可单模块进行专项提升培训考核，也可选择其中需要的模块进行系统培训考核。
4. 考核标准与培训标准同步。
5. 实现考核标准适用于某角色、单人、作业组的成绩评定。

(三）标准结构

本标准分两部分，第一部分为特高压交流输电线路运检专业培训及考核标准总体说明，细分为7个部分，由总则、编制依据、培训对象、培训标准、考核标准、保障条件、编制说明组成。

第二部分为各技能模块培训及考核标准，共17个模块，模块1至模块11为带电检修作业，模块12至模块17为停电检修作业，且难易不分先后。各模块均由培训标准和考核标准两部分组成，培训标准细分为4个部分，由培训要求、引用规程规范、培训教学设计、作业流程组成；考核标准为特高压交流输电线路运检技能考核评分细则。

选择多个模块构成系统培训内容时，应去掉重复的理论教学培训内容。

第二部分

技能模块培训及考核标准

模块1 带电更换1000kV交流输电线路直线塔Ⅰ型复合绝缘子培训及考核标准

一、培训标准

(一) 培训要求

模块名称	带电更换1000kV交流输电线路直线塔Ⅰ型复合绝缘子	培训类别	操作类
培训方式	实操培训	培训学时	21学时
培训目标	1.掌握直线塔进、出1000kV强电场时采用"吊篮法"作业方式的电学意义。 2.能完成采用"吊篮法"进入1000kV等电位作业点。 3.能独立完成更换1000kV交流输电线路直线塔Ⅰ型复合绝缘子的操作(等电位作业法)		
培训场地	特高压交流实训线路		
培训内容	等电位和地电位配合,通过"吊篮法"进入等电位作业,采用等电位作业法更换直线塔Ⅰ型复合绝缘子		
适用范围	特高压交流输电线路检修人员		

(二) 引用规程规范

(1)《电工术语》(GB/T 2900.55—2002)。

(2)《带电作业用绝缘滑车》(GB/T 13034—2008)。

(3)《带电作业用绝缘绳索》(GB/T 13035—2008)。

(4)《交流线路带电作业安全距离计算方法》(GB/T 18037—2000)。

(5)《1000kV交流带电作业用屏蔽服装》(GB/T 25726—2010)。

(6)《1000kV架空输电线路设计规范》(GB 50665—2011)。

(7)《1000kV交流输电线路检修规范》(DL/T 209—2008)。

(8)《1000kV交流输电线路运行规范》(DL/T 307—2010)。

(9)《1000kV交流输电线路带电作业技术导则》(DL/T 392—2015)。

(10)《带电作业绝缘配合导则》(DL/T 876—2004)。

(11)《带电作业工器具、装置和设备使用的一般要求》(DL/T 877—2004)。

(12)《带电作业工具、装备和设备预防性试验规程》(DL/T 976—2005)。

(13)《国家电网公司电力安全工作规程(线路部分)》(Q/GDW 1799.2—2013)。

(三)培训教学设计

本设计以完成"带电更换1000kV交流输电线路直线塔Ⅰ型复合绝缘子"为工作任务,按工作任务完成的标准化作业流程来设计各个培训阶段,每个阶段包括了具体的培训目标、培训内容、培训学时、培训方法(培训资源)、培训环境和考核评价等内容,如表2-1-1所示。

表2-1-1 带电更换1000kV交流输电线路直线塔Ⅰ型复合绝缘子培训内容设计

培训流程	培训目标	培训内容	培训学时	培训方法与资源	培训环境	考核评价
1.理论教学	1.初步掌握吊篮法进出1000kV强电场基本方法; 2.熟悉电位转移的方法; 3.熟悉输电线路直线塔Ⅰ型复合绝缘子更换方法	1.吊篮法进出强电场作业方式的电学意义; 2.进、出特高压强电场时电位转移棒的使用方法; 3.输电线路直线塔Ⅰ型复合绝缘子更换方法和质量标准	2	培训方法:讲授法。 培训资源:PPT、相关规程规范	多媒体教室	考勤、课堂提问和作业
2.准备工作	能完成作业前准备工作	1.作业现场查勘; 2.编制培训标准化作业卡; 3.填写培训操作工作票; 4.完成本操作的工器具及材料准备	1	培训方法: 1.现场查勘和工器具及材料清理采用现场实操方法; 2.编写作业卡和填写工作票采用讲授方法; 培训资源: 1.1000kV实训线路; 2.特高压工器具库房; 3.空白工作票	1.特高压输电实训线路 2.多媒体教室	
3.作业现场准备	能完成作业现场准备工作	1.作业现场复勘; 2.工作申请; 3.作业现场布置; 4.班前会; 5.工器具及材料检查	1	培训方法:演示与角色扮演法。 资源:1000kV实训线路	1000kV实训线路	

续表

培训流程	培训目标	培训内容	培训学时	培训方法与资源	培训环境	考核评价
4.培训师演示	通过现场观摩,使学员初步领会本任务操作流程	1.塔上工器具布置安装; 2.等电位电工采用吊篮法进、出强电场; 3.地电位电工与等电位电工相互配合利用荷载转移装置完成Ⅰ型复合绝缘子更换	2	培训方法:演示法。 资源:1000kV实训线路	1000kV实训线路	
5.学员分组训练	1.能完成进、出1000kV强电场操作; 2.能完成1000kV输电线路直线塔Ⅰ型复合绝缘子更换	1.学员分组(11人一组)训练进、出1000kV强电场和直线塔Ⅰ型复合绝缘子更换; 2.培训师对学员操作进行指导和安全监护	14	培训方法:角色扮演法。 资源:1000kV实训线路	1000kV实训线路	采用技能考核评分细则对学员操作评分
6.工作终结	1.使学员进一步辨析操作过程不足之处,便于后期提升; 2.培训学员安全文明生产的工作作风	1.作业现场清理; 2.向调度汇报工作; 3.班后会,对本次工作任务进行点评总结	1	培训方法:讲授和归纳法	1000kV实训线路	

(四)作业流程

1. 工作任务

等电位和地电位配合,采用"吊篮法"进入等电位作业,更换直线塔Ⅰ型复合绝缘子。

2. 天气及作业现场要求

(1)带电更换1000kV交流输电线路直线塔Ⅰ型复合绝缘子应在良好的天气进行。

如遇雷电(听见雷声、看见闪电)、雪、雹、雨、雾等,禁止进行带电作业。风力大于5级,或空气相对湿度大于80%时,不宜进行带电作业;恶劣天气下必须开展带电抢修时,应组织有关人员充分讨论并编制必要的安全措施,经本单位批准后方可进行。

(2)作业人员精神状态良好,熟悉工作中保证安全的组织措施和技术措施;应持有在有效期内的带电作业资质证书。

(3)工作负责人应事先组织相关人员完成现场勘察,根据勘察结果确定本次作业方法和所需工器具,以及应采取的必要措施,并办理带电作业工作票。

(4)作业现场应合理设置围栏,并妥当布置警示标示牌,禁止非工作人员入内。

（5）本项目需停用线路重合闸装置。

（6）工作中安全距离及有效绝缘长度如表2-1-2所示。

表2-1-2　带电更换1000kV交流输电线路直线塔Ⅰ型复合绝缘子的安全距离

海拔高度/m	最小安全距离/m		最小组合间隙/m		绝缘工具最小有效绝缘长度/m	转移电位时人体裸露部分与带电体的最小距离/m
	中相	边相	中相	边相		
$H \leqslant 1000$	6.8	6.0	6.9	6.7	6.8	0.5
$1000 < H \leqslant 2000$	7.4	6.6	7.6	7.3	7.2	0.5

3. 准备工作

3.1　危险点及其预控措施

（1）危险点——触电伤害

预控措施包括下述要点。

①工作前，工作负责人应与值班调控人员联系，停用线路重合闸，并履行许可手续。

②工作负责人必须进行安全措施、技术措施和工作任务交底，塔上地电位作业人员登塔前，必须仔细核对线路命名、杆塔编号、相别，确认无误后方可上塔。

③工作中，如遇线路突然停电，作业人员应视其仍然带电。工作负责人应尽快与调控人员联系，值班调控人员未与工作负责人取得联系前不准强送电。

④绝缘工具及绝缘绳索不得损坏、受潮、变形、失灵，不准使用非绝缘绳索（如棉纱绳、白棕绳、钢丝绳）。

⑤等电位作业人员应穿着阻燃内衣，衣服外面应穿戴全套屏蔽服（包括帽、衣裤、手套、袜和鞋），且各部分应连接良好，全套屏蔽服电阻不大于20Ω。

⑥等电位作业人员在电位转移前，应得到工作负责人的许可，人体裸露部分与带电体的最小距离不小于0.5m；等电位作业人员必须使用电位转移棒进行电位转移；地电位作业人员与带电体的安全距离小于表2-1-2的规定。

⑦用绝缘绳索传递大件金属物品时，地电位作业人员应将金属物品接地后再接触。

⑧工作负责人和专责监护人应对作业人员进行不间断监护，随时纠正其不规范或违章动作。重点关注高处作业人员，使其保持足够的安全距离（符合表2-1-2的规定），禁止同时接触两个非连通的带电体或带电体与接地体。

（2）危险点——高处坠落

预控措施：

①高处作业人员登高前，必须具备符合本项作业要求的身体状况、精神状态和技能素质。

②地电位电工登塔至作业点位后应打好安全绳并检查确认牢固，布置吊篮轨迹绳，等电

位电工人体后备保护绳应合理且可靠锚固。地电位电工使用闭合钩将绝缘吊拉绳与吊篮可靠连接。等电位电工系好绝缘保护绳、主保护绳，对吊篮做冲击试验合格后，进入吊篮。等电位电工系好安全带后，即可开展相关工作。等电位电工在作业点位应系好保护绳。

③监护人员应随时纠正其不规范或违章动作，重点关注作业人员在转位的过程中不得失去安全带或绝缘后备保护绳的保护，严禁低挂高用。

④人员登塔时检查脚钉和塔材的紧固情况，登塔时手抓主材，严禁手抓脚钉。

（3）危险点——高处坠物伤人。

预控措施：

①高处作业人员的个人工具及零星材料应装入工具袋，严禁在高处浮置物件、口中含物。

②地面作业人员必须正确佩戴安全帽，正确使用绳结，与作业点垂直下方距离不得小于坠落半径。

③作业现场设置围栏并挂好警示标示牌。监护人员应随时注意，禁止非工作人员及车辆进入作业区域。

3.2 工器具及材料选择

带电更换1000kV交流输电线路直线塔Ⅰ型复合绝缘子所需工器具及材料见表2-1-3。工器具出库前，应认真核对工器具的使用电压等级和试验周期，并检查确认外观良好、连接牢固、转动灵活，且符合本次工作任务的要求；工器具出库后，应存放在工具袋或工具箱内进行运输，防止脏污、受潮；金属工具和绝缘工器具应分开装运，防止因混装运输导致工器具变形、损伤等现象发生。

表2-1-3 带电更换1000kV交流输电线路直线塔Ⅰ型复合绝缘子所需工器具及材料

序号	工器具名称	规格型号	单位	数量	备注
1	全套屏蔽服	屏蔽效率≥60dB（屏蔽面罩屏蔽效率≥20dB）	套	5	
2	导电鞋	尺码视穿着人员而定	双	5	
3	阻燃内衣	纯桑蚕丝	套	2	
4	双保险安全带	背带式	根	4	
5	安全帽		顶	11	
6	防坠器	与杆塔防坠器装置型号对应	只	4	
7	绝缘保护绳		根	4	
8	八分裂提线器		只	2	
9	液压丝杠		根	2	

续表

序号	工器具名称	规格型号	单位	数量	备注
10	平面丝杠		只	2	
11	专用接头		个	4	
12	机动绞磨	3T	台	1	
13	电位转移棒		根	2	
14	钢丝绳套		根	4	
15	绝缘传递绳	TJS-14	根	3	
16	绝缘轨迹绳	TJS-16	根	1	
17	2-2绝缘滑车	JH20-2	只	2	
18	绝缘滑车	JH10-1	只	6	
19	绝缘吊杆	$\phi 53mm$	组	2	
20	绝缘绳套	1T	根	3	
21	吊篮		套	1	
22	绝缘电阻表	5000V	块	1	
23	温湿度表		块	1	
24	风速风向仪		块	1	
25	对讲机		个	若干	
26	万用表		块	1	
27	防潮苫布		块	4	
28	工作负责人监护服		件	1	
29	专责监护人监护服		件	1	
30	工具袋(箱)		个	若干	
31	安全围栏		套	若干	
32	警示标示牌	"在此工作""从此进出""从此上下"	套	1	
33	复合绝缘子	FXBZ-1000/420	支	1	

3.3 作业人员分工

本任务作业人员分工如表2-1-4所示。

表2-1-4 带电更换1000kV交流输电线路直线塔Ⅰ型复合绝缘子人员分工表

序号	工作岗位	数量(人)	工作职责
1	工作负责人	1	负责本次工作任务的人员分工、工作前的现场查勘、作业方案的制订、工作票的填写、现场复勘、办理工作许可手续、召开工作班前会、落实现场安全措施、负责作业过程中的安全监督、工作中突发情况的处理、工作质量的监督、工作后的总结

续表

序号	工作岗位	数量(人)	工作职责
2	专责监护人	1	负责作业现场的安全把控
3	等电位电工	2	配合地电位电工安装提线系统(平面丝杠、专用接头、绝缘吊杆、液压丝杠),操作液压丝杠转移导线荷载,拆装绝缘子串等
4	地电位电工	2	负责安装吊篮、提线系统(平面丝杠、专用接头、绝缘吊杆、八分裂提线器、液压丝杠)、绝缘磨绳及配合等电位电工进出电位,拆装合成绝缘子串等
5	地面电工	5	负责传递工器具及合成绝缘子串等

4. 工作程序

本任务工作流程如表2-1-5所示。

表2-1-5　带电更换1000kV交流输电线路直线塔Ⅰ型复合绝缘子工作流程表

序号	作业内容	作业步骤及标准	安全措施及注意事项	责任人
1	现场复勘	工作负责人负责完成以下工作: (1)现场核对线路名称、杆塔编号,相别无误;基础及杆塔完好无异常;交叉跨越距离符合安全要求;确认缺陷情况及导地线规格型号等。 (2)检测风速、湿度等现场气象条件符合作业要求。 (3)检查地形环境符合作业要求。 (4)检查工作票所列安全措施与现场实际情况相符,必要时予以补充	(1)正确穿戴安全帽、工作服、工作鞋、劳保手套。 (2)不得在危及作业人员安全的气象条件下作业。 (3)严禁非工作人员、车辆进入作业现场	
2	工作许可	(1)工作负责人负责联系值班调控人员,按工作票内容申请停用线路重合闸。 (2)经值班调控人员许可后,方可开始带电作业工作	不得未经值班调控人员许可即开始工作	
3	现场布置	正确装设安全围栏并悬挂标示牌: (1)安全围栏范围应充分考虑高处坠物,以及对道路交通的影响。 (2)安全围栏出入口设置合理。 (3)妥当布置"从此进出""在此工作""从此上下"等标示	对道路交通安全影响不可控时,应及时联系交通管理部门强化现场交通安全管控	

续表

序号	作业内容	作业步骤及标准	安全措施及注意事项	责任人
4	召开班前会	(1)全体工作成员列队。 (2)工作负责人宣读工作票,明确工作任务及人员分工;讲解工作中的安全措施和技术措施;查(问)全体工作成员精神状态;告知工作中存在的危险点及采取的预控措施。 (3)全体工作成员在工作票上签名确认	(1)工作票填写、签发和许可手续规范,签名完整。 (2)全体工作成员精神状态良好。 (3)全体工作成员明确任务分工、安全措施和技术措施	
5	检查工具	(1)塔上地电位电工和等电位电工正确地穿戴好屏蔽服并检测合格,由负责人监督检查。 (2)正确佩戴个人安全用具(大小合适,锁扣自如),由负责人监督检查。 (3)测量风速风向、湿度,检查绝缘工具的绝缘性能,并做好记录	(1)金属、绝缘工具使用前,应仔细检查其是否损坏、变形、失灵。绝缘工具应使用2500V及以上绝缘电阻表进行分段绝缘检测,阻值应不低于700MΩ,并用清洁干燥的毛巾将其擦拭干净。 (2)用万用表测量屏蔽服衣裤最远端点之间的电阻值不得大于20Ω。工作负责人认真检查作业电工屏蔽服的连接情况。 (3)检查工具组装情况并确认连接可靠。 (4)现场所使用的带电作业工具应放置在防潮帆布上	
6	登塔	(1)核对线路名称、杆塔编号无误后,塔上地电位电工和等电位冲击检查安全带、防坠器受力情况。 (2)塔上地电位电工携带绝缘传递绳登塔,等电位电工随后登塔,两人至横担作业点,选择合适位置系好安全带,塔上地电位电工将绝缘滑车和绝缘传递绳安装在横担合适位置。然后配合地面电工将绝缘传递绳分开作起吊准备	(1)核对线路名称和杆塔编号无误后,方可登塔作业。 (2)登塔过程中应使用塔上安装的防坠装置;杆塔上移动及转位时,不准失去安全保护,作业人员必须攀抓牢固构件。 (3)作业电工必须穿全套合格的屏蔽服,且全套屏蔽服必须连接可靠。在横担进入等电位前,等电位电工要检查确认屏蔽服各个部位连接可靠后方能进行下一步操作	
7	安装滑车及吊篮	(1)地面电工利用绝缘传递绳将吊篮、吊篮轨迹绳、绝缘保护绳、2-2绝缘滑车组以及电位转移棒传至横担。 (2)塔上电工将2-2绝缘滑车组及吊篮安装在横担上平面合适位置,将吊篮轨迹绳安装在横担合适位置	(1)传递时绝缘吊绳要起吊平稳、无磕碰、无缠绕。 (2)吊篮安装好后由塔上电工对吊篮情况进行认真检查核对。 (3)2-2滑车组及吊篮应在横担上合适位置可靠安装	

续表

序号	作业内容	作业步骤及标准	安全措施及注意事项	责任人
8	进入强电场	(1)1号等电位电工系好绝缘保护绳进入吊篮,地面电工缓慢松出2-2绝缘滑车组控制绳、待吊篮距带电导线约2m处放慢速度。 (2)在吊篮向导线继续移动过程中,等电位电工将电位转移棒置于手中面向带电导线,同时向工作负责人申请电位转移,得到同意后,等电位电工待吊篮距导线0.5m时迅速伸出电位转移棒,将其钩在最近的子导线上进行电位转移。 (3)等电位电工进入强电场后系好安全带,并根据作业需要决定是否解除绝缘保护绳,同时等电位电工要控制头部不超过导线侧均压环。 (4)地面电工收紧2-2绝缘滑车组控制绳,将吊篮向上传至横担部位。2号等电位电工系好绝缘保护绳进入吊篮,用同样的方法进入强电场	(1)进入等电位前,等电位电工要再次检查确认屏蔽服各部位、电位转移棒与绝缘屏蔽服连接可靠后方能进行下一步操作。专责监护人负责检查等电位电工屏蔽服连接、屏蔽服与电位转移棒连接可靠。 (2)等电位电工进入电位前必须得到工作负责人的许可。 (3)等电位电工进入吊篮前必须系好保护绳。 (4)地面电工配合等电位电工进入等电位过程中收放滑车组控制绳应平稳。 (5)等电位电工在进入电位过程中与接地体和带电体两部分间隙所组成的组合间隙不得小于6.9m(中相)/6.7m(边相)。 (6)专责监护人负责监护等电位电工进入强电场的安全注意事项。对塔上作业人员的危险、不规范动作应及时提醒,必要时应制止	
9	安装工具并转移导线荷载	(1)地面电工将平面丝杆、绝缘吊杆、八分裂提线器、液压紧线系统等工具传递至工作位置,由等电位电工和地电位电工配合将绝缘子更换工具安装在需更换的复合绝缘子串两侧(顺绝缘子串垂直安装,平面丝杆、液压紧线系统安装在横担侧)。 (2)检查承力工具各部件安装可靠得到工作负责人同意后,地电位电工先收紧平面丝杆,待平面丝杆适当受力后,再收紧液压紧线系统,使绝缘子串松弛。 (3)检查承力工具受力正常得到工作负责人同意后,等电位电工拆开导线侧碗头挂板螺栓。 (4)地面电工将复合绝缘子串控制绳传递给等电位电工,等电位电工将其安装在复合绝缘子串尾部。 (5)地电位电工将绝缘传递绳系在复合绝缘子串上端,然后取出复合绝缘子串与球头挂环连接的锁紧销。地面电工启动机动绞磨,与地电位电工配合脱开复合绝缘子串与球头挂环的连接	(1)上、下作业电工要密切配合,所有作业电工要听从等工作负责人的统一指挥。 (2)地电位电工对带电体、等电位电工对接地体的最小安全距离不得小于6.8m(中相)/6.0m(边相)。绝缘吊杆、绝缘绳索的有效绝缘长度不得小于6.8m。 (3)杆塔上、下传递工具绑扎绳扣应正确可靠,塔上电工不得高空落物。 (4)工具受力后应试冲击检查无误后,报告工作负责人,在得到工作负责人许可后,方可继续作业。 (5)专责监护人对塔上作业人员的危险、不规范动作应及时提醒,必要时应制止	

续表

序号	作业内容	作业步骤及标准	安全措施及注意事项	责任人
10	更换绝缘子串	(1)地面电工控制好复合绝缘子串控制绳,利用机动绞磨缓慢将复合绝缘子串放至地面。注意控制好复合绝缘子串的控制绳,不得碰撞承力工具、导线及杆塔。 (2)地面电工将绝缘传递绳和复合绝缘子串控制绳分别转移到新复合绝缘子上。 (3)地面电工启动机动绞磨,将新复合绝缘子串传递至塔上工作位置。地电位电工恢复新复合绝缘子串与球头挂环的连接,并复位锁紧销。 (4)地面电工缓慢松出机动绞磨使复合绝缘子串自然垂直,地电位电工恢复碗头挂板与金属联板的连接,并装好开口销。	(1)绝缘子串起吊时应注意不要碰撞杆塔,地面电工应拉好绝缘子串尾绳。 (2)绳索不得与杆塔摩擦。绑扎绳扣应正确可靠。 (3)利用绞磨起吊绝缘子串时绞磨应安置平稳,尾绳应由有带电工作经验的电工控制,随时拉紧,不可疏忽放松。 (4)利用机动绞磨起吊绝缘子串时,必须检查绞磨及转向滑车的受力情况,无误后方可进行作业。 (5)专责监护人对塔上作业人员的危险、不规范动作应及时提醒,必要时应制止	
11	拆除工具	(1)经检查复合绝缘子串连接可靠,得到工作负责人同意后,地电位电工松出液压紧线系统和平面丝杆。 (2)经检查复合绝缘子串受力正常得到工作负责人的同意后,地电位电工与等电位电工配合拆除平面丝杆、绝缘吊杆、八分裂提线器、液压紧线系统等,并传至地面	(1)复合绝缘子安装复位后,应详细检查各部位连接正常无误,并得到工作负责人的同意后方可拆除提线工具。 (2)工具在传递过程中不得碰撞杆塔,绑扎绳扣应正确可靠。 (3)专责监护人对塔上作业人员的危险、不规范动作应及时提醒,必要时应制止	
12	退出电位	(1)1号等电位电工系好绝缘保护绳,将电位转移棒的金属端钩在子导线上,一只手握紧绝缘手柄,进入吊篮,然后保持手臂伸直状态使吊篮距子导线0.5m。 (2)等电位电工向工作负责人申请退出电位,得到同意后,等电位电工迅速脱开电位转移棒与子导线的连接,并将电位转移棒收回放置在吊篮中。 (3)地面电工同时迅速收紧2-2绝缘滑车组控制绳,将吊篮向上拉至横担部位停住,然后等电位电工登上横担,并系好安全带。 (4)地面电工利用绝缘传递绳将吊篮传至2号等电位电工处,等电位电工检查导线上无遗留物后进入吊篮,用同样的方法退出电位	(1)上、下作业电工要密切配合,听从工作负责人的指挥。 (2)等电位电工退出电位前必须得到工作负责人的许可。 (3)等电位电工进入吊篮前必须系好保护绳。 (4)地面电工配合等电位电工进入等电位时收放滑车组控制绳应平稳。 (5)等电位电工在退出电位过程中与接地体和带电体两部分间隙所组成的组合间隙不得小于6.9m(中相)/6.7m(边相)。 (6)专责监护人负责监护等电位电工退出强电场的安全注意事项。对塔上作业人员的危险、不规范动作应及时提醒,必要时应制止	

续表

序号	作业内容	作业步骤及标准	安全措施及注意事项	责任人
13	拆除吊篮返回地面	(1)塔上电工配合拆除吊篮轨迹绳、绝缘保护绳、2-2绝缘滑车组及吊篮并传至地面。 (2)塔上电工检查塔上无遗留物后,向工作负责人汇报,得到工作负责人同意后携带绝缘传递绳下塔	(1)工具在传递过程中不得碰撞,绑扎绳扣应正确可靠。 (2)登塔过程中应使用塔上安装的防坠装置;杆塔上移动及转位时,不准失去安全保护,作业人员必须抓牢固构件	
14	工作结束	(1)清理现场及工具,认真检查杆(塔)上有无遗留物,工作负责人全面检查工作完成情况,清点人数,无误后,宣布工作结束,撤离施工现场。 (2)通知调度工作完毕,履行工作票完工手续	不得约时恢复线路重合闸	

二、考核标准

特高压交流输电线路运检技能考核评分细则

考生填写栏	编号:	姓名:	所在岗位:	单位:	日期:	年 月 日		
考评员填写栏	成绩:	考评员:	考评组长:	开始时间:	结束时间:	操作时长:		
考核模块	带电更换1000kV交流输电线路直线塔Ⅰ型复合绝缘子		考核对象	特高压交流输电线路检修人员		考核方式	操作	考核时限	120min
任务描述	带电更换1000kV交流输电线路直线塔Ⅰ型复合绝缘子							
工作规范及要求	1. 带电作业工作应在良好天气下进行。如遇雷、雨、雪、雾天气不得进行带电作业。风力大于5级、湿度大于80%时,一般不宜进行带电作业。 2. 本项作业需工作负责人1名,专责监护人1名,地电位电工2名,地面电工5名,等电位电工2名,采用吊篮摆渡法进入强电场进行绝缘子更换工作。 3. 工作负责人职责:负责本次工作任务的人员分工、工作票的宣读、办理线路停用重合闸、办理工作许可手续、召开工作班前会、工作中突发情况的处理、工作质量的监督、工作后的总结。 4. 专责监护人:负责作业现场的安全把控。 5. 等电位电工职责:配合地电位电工安装提线系统(平面丝杠、专用接头、绝缘吊杆、液压丝杠),操作液压丝杠转移导线荷载,拆装绝缘子串。 6. 塔上地电工职责:负责安装吊篮、提线系统(平面丝杠、专用接头、绝缘吊杆、八分裂提线器、液压丝杠)、绝缘磨绳及配合等电位电工进出电位,拆装合成绝缘子串。 7. 地面电工职责:负责传递工具、材料配合等电位电工进出等电位。 8. 在带电作业中,如遇雷、雨、大风或其他任何情况威胁到工作人员的安全时,工作负责人或监护人可根据情况,临时停止工作。							

第二部分
技能模块培训及考核标准

续表

工作规范及要求	给定条件： 1. 培训基地：特高压交流1000kV实训线路直线塔A相Ⅰ型复合绝缘子，绝缘子型号：FXBZ-1000/420。 2. 工作票已办理，安全措施已经完备（重合闸已停用），工作开始、工作终结时应口头提出申请（调度或考评员）。 3. 安全、正确地使用仪器对绝缘工具进行检测。 4. 必须按工作程序进行操作，工序错误扣除应做项目分值，出现重大人身、器材和操作安全隐患，考评员可下令终止操作（考核）
考核情景准备	1. 线路：特高压交流10000kV实训线路直线塔A相，工作内容：带电更换1000kV交流输电线路直线塔Ⅰ型复合绝缘子，绝缘子型号：FXBZ-1000/420。 2. 所需作业工器具：绝缘传递绳3根(TJS-14)，绝缘轨迹绳1根(TJS-16)，绝缘滑车6个(JH10-1)，2-2绝缘滑车2个(JH20-2)，吊篮1个，液压紧线系统2只，八分裂提线器2只，绝缘吊杆2组，机动绞磨1台，绝缘检测仪，电位转移棒1根，绝缘电阻表（5000V型），屏蔽服（屏蔽效率≥60dB）5套，万用表1块，苫布4块，绝缘测试仪1台，温湿度表、风速仪各1台，纯棉毛巾2条。 3. 作业现场做好监护工作，作业现场安全措施（围栏等）已全部落实；禁止非作业人员进入现场，工作人员进入作业现场必须戴安全帽。 4. 考生自备工作服，阻燃纯棉内衣，安全帽，线手套，安全带（含二保绳）
备注	1. 各项目得分均扣完为止，出现重大人身、器材和操作安全隐患，考评员可下令终止操作。 2. 设备、作业环境、安全带、安全帽、工器具、屏蔽服等不符合作业条件考评员可下令终止操作

序号	项目名称	质量要求	分值	扣分标准	扣分原因	扣分	得分
1	现场复勘	1)工作负责人到作业现场核对线路名称和杆塔编号、现场工作条件、缺陷部位等。 2)检测风速、湿度等现场气象条件符合作业要求。 3)检查工作票填写完整，无涂改，检查是否所列安全措施与现场实际情况相符，必要时予以补充	5	1)未进行核对双重称号扣1分。 2)未核实现场工作条件（气象）、缺陷部位扣1分。 3)工作票填写出现涂改，每项扣0.5分，工作票编号有误，扣1分。工作票填写不完整，扣1.5分			
2	工作许可	1)工作负责人联系值班调控人员，按工作票内容申请停用线路重合闸。 2)汇报内容规范、完整	2	1)未联系调度部门（裁判）停用重合闸扣2分。 2)汇报专业用语不规范或不完整的各扣0.5分			
3	现场布置	正确装设安全围栏并悬挂标示牌： 1)安全围栏范围应充分考虑高处坠物，以及对道路交通的影响。 2)安全围栏出入口设置合理。 3)妥当布置"从此进出""在此工作""从此上下"等标示	3	1)作业现场未装设围栏扣0.5分。 2)未设立警示牌扣0.5分。 3)未悬挂登塔作业标志扣0.5分			

续表

序号	项目名称	质量要求	分值	扣分标准	扣分原因	扣分	得分
4	召开班前会	1)全体工作成员全体人员正确佩戴安全帽、工作服。 2)工作负责人佩戴红色背心,宣读工作票,明确工作任务及人员分工;讲解工作中的安全措施和技术措施;查(问)全体工作成员精神状态;告知工作中存在的危险点及采取的预控措施。 3)全体工作成员在工作票上签名确认	3	1)工作人员着装不整齐扣0.5分,工作人员着装不整齐每人次扣0.5分。 2)未进行分工本项不得分,分工不明扣1分。 3)现场工作负责人未穿佩安全监护背心扣0.5分。 4)工作票上工作班成员未签字或签字不全的扣1分			
5	工器具检查	1)工作人员按要求将工器具放在防防潮苫布上;防潮苫布应清洁、干燥。 2)工器具应按定置管理要求分类摆放;绝缘工器具不能与金属工具、材料混放;对工器具进行外观检查。 3)绝缘工具表面不应磨损、变形损坏,操作应灵活。绝缘工具应使用2500V及以上绝缘电阻表进行分段绝缘检测,阻值应不低于700MΩ,并用清洁干燥的毛巾将其擦拭干净。 4)塔上地电位和登电位人员按要求正确穿戴全套合格的屏蔽服、导电鞋,且各部分连接应良好,屏蔽服内不得贴身穿着化纤类衣服,并系好安全带;工作负责人应认真检查是否穿戴正确。 5)登塔人员再次核对双重名称、杆号、相别并报告	7	1)未使用防潮布并定置摆放工器具扣1分。 2)未检查工器具试验合格标签及外观检查扣每项扣0.5分。 3)未正确使用检测仪器对工器具进行检测每项扣1分。 4)作业人员未正确穿戴屏蔽服且各部位连接良好每人次扣2分。 5)现场工作负责人未对登塔作业人员进安全防护装备进行检查扣1分。 6)登塔人员未核对线路双重名称、杆号、相别每人扣2分。 7)登塔人员未报告核对结果每人扣2分			
6	登塔	1)塔上地电位电工、等电位电工穿好全套合格的屏蔽服,将安全带做冲击试验后,系好安全带后携带绝缘传递绳相继登塔。 2)登塔过程中系好防坠落保护装置,登塔至合适位置,系好安全带,布置好绝缘传递绳,然后配合地面电工将绝缘传递绳分开作起吊准备。	5	1)未系安全带或安全带及后备保护绳未进行冲击试验各扣2分。 2)手抓脚钉扣2分。 3)滑车传递绳悬挂位置不便工具取用扣1分。 4)传递时金属工具难以保证安全距离扣2分;工具绑扎不牢扣2分。 5)传递时高空落物扣2分。			

续表

序号	项目名称	质量要求	分值	扣分标准	扣分原因	扣分	得分
6	登塔	3)登塔过程中应系好防坠落保护装置，匀速登塔，手抓主材，将安全带挂在肩上并与带电体保持6.8m（中相）/6.0m（中相）以上安全距离，工作负责人加强作业监护	5	6)传递过程工具与塔身磕碰扣2分。 7)传递工具绳索打结混乱扣1分。 8)工作负责人监护不到位扣2分。 9)塔上电工操作不正确扣2分			
7	安装滑车组及吊篮	1)传递时绝缘吊绳起吊要平稳、无磕碰、无缠绕。 2)吊篮安装好后由塔上电工对吊篮情况进行认真检查核对。 3)2-2滑车组及吊篮应在横担上合适位置可靠	5	1)2-2滑车组绳子缠绕扣0.5分。 2)轨迹绳安装位置不合理扣1分。 3)绝缘保护绳、长度不合适扣1分。 4)传递工器具不平稳、磕碰扣1分			
8	进入强电场	1)等电位再次检查确认屏蔽服各部位连接可靠后对吊篮进行冲击实验，汇报工作负责人后系好保护绳登上吊篮行。 2)地面电工缓慢松出2-2绝缘滑车组控制绳，当距离导线约2m处放慢速度，等电位电工在距离导线0.5m处向工作负责人申请电位转移，得到同意后，迅速伸出电位转移棒，将其钩在最近的子导线上进行电位转移。 3)等电位电工进入强电场后做好人体后保护，要控制头部不超过导线侧均压环。 4)电位电工在进入电位过程中与接地体和带电体两部分间隙所组成的组合间隙不得小于6.9m（中相）/6.7m（边相），进入强电场必须用电位转移棒进行电位转移	10	1)等电位电工未对吊篮进行冲击扣1分。 2)未系好绝缘保护绳扣1分。 3)地电位电工未检查等电位电工安全措施扣2分。 4)地面电工控制滑车尾绳不平稳扣1分。 5)等电位电工进入强电场前未向工作负责人申请扣2分；申请了但未得同意即开始扣1分。 6)转移电位动作不熟练扣1分，电位转移过程未使用电位转移棒的扣5分。 7)等电位电工进入强电场后安全带未系好扣2分。 8)等电位电工进入强电场后头部超过导线侧均压环扣2分			

续表

序号	项目名称	质量要求	分值	扣分标准	扣分原因	扣分	得分
9	安装工具并转移导线荷载	1)地面电工将平面丝杆、绝缘吊杆、八分裂提线器、液压紧线系统等工具传递至工作位置,由等电位电工和地电位电工配合将绝缘子更换工具安装在需更换的复合绝缘子串两侧。 2)检查承力工具各部件安装可靠得到工作负责人同意后,地电位电工先收紧平面丝杆,待平面丝杆适当受力后,再收紧液压紧线系统,使绝缘子串松弛。 3)检查承力工具受力正常得到工作负责人同意后,等电位电工拆开导线侧碗头挂板螺栓。 4)地面电工将复合绝缘子串控制绳传递给等电位电工,等电位电工将其安装在复合绝缘子串尾部。 5)地电位电工将绝缘传递绳系在复合绝缘子串上端,然后取出复合绝缘子串与球头挂环连接的锁紧销。地面电工启动机动绞磨,与地电位电工配合脱开复合绝缘子串与球头挂环的连接	15	1)工器具传递过程不平稳扣1分;有磕碰每次扣1分。 2)未检查承力工具安装可靠、受力良好扣2分,未汇报并取得工作负责人同意扣2分。 3)地电位电工与等电位电工沟通无效扣1分。 4)拆除绝缘子串前未对承力工具进行检查扣5分,检查结果未汇报扣2分。 5)绝缘绳系复合绝缘子位置不合适扣2分。 6)绝缘子串传递时有磕碰每次扣2分。 7)地面电工机动绞磨控制不到位扣1分			
10	更换绝缘子串	1)地面电工控制好复合绝缘子串控制绳,利用机动绞磨缓慢将复合绝缘子串放至地面。注意控制好复合绝缘子串的控制绳,不得碰撞承力工具、导线及杆塔。 2)地面电工将绝缘传递绳和复合绝缘子串控制绳分别转移到新复合绝缘子上。 3)地面电工启动机动绞磨,将新复合绝缘子串传递至塔上工作位置。地电位电工恢复新复合绝缘子串与球头挂环的连接,并复位锁紧销。 4)地面电工缓慢松出机动绞磨使复合绝缘子串自然垂直,地电位电工恢复碗头挂板与金属联板的连接,并装好开口销	17	1)地面电工未控制好绝缘子尾绳扣1分。 2)绑扎绳扣不合理扣1分。 3)未检查绞磨转向及滑车的受力情况扣2分。 5)绝缘子串传递时有磕碰每次扣2分。 4)绝缘子串安装不到位扣5分。 5)作业人员未检查绝缘子串连扣5分。 6)作业人员未检查销子安装到位扣2分。 7)专责监护人未尽监护职责扣2分			

第二部分 技能模块培训及考核标准

续表

序号	项目名称	质量要求	分值	扣分标准	扣分原因	扣分	得分
11	拆除工具	1)经检查复合绝缘子串连接可靠,得到工作负责人同意后,地电位电工松出液压紧线系统和平面丝杆。 2)经工作负责人的同意后,地电位电工与等电位电工配合拆除平面丝杆、缘吊杆、八分裂提线器、液压紧线系统等,并传至地面	5	1)未详细检查各部位连接正常无误扣1分,未得到工作负责人同意扣1分。 2)工器具传递过程不平稳扣1分;有磕碰每次扣1分			
12	退出电位	1)一名等电位电工系好绝缘保护绳,将电位转移棒的金属端钩在子导线上,一只手握紧绝缘手柄,进入吊篮,然后保持手臂伸直状态使吊篮距子导线0.5m。 2)等电位电工向工作负责人申请退出电位,得到同意后,等电位电工迅速脱开电位转移棒与子导线的连接,并将电位转移棒收回放置在吊篮中。 3)地面电工同时迅速收紧2-2绝缘滑车组控制绳,将吊篮向上拉至横担部位停住,然后等电位电工登上横担,并系好安全带。 4)地面电工利用绝缘传递绳将吊篮传至另一名等电位电工处,等电位电工检查导线上无遗留物后进入吊篮,用同样的方法退出电位	8	1)未报告工作结束扣2分,强电场内有遗留物每件扣1分。 2)等电位未系好绝缘保护绳扣1分。 3)地面电工控制滑车尾绳不平稳扣1分。 4)等电位电工进入强电场前未向工作负责人申请扣2分,申请了但未得同意即开始扣1分。 5)转移电位动作不熟练扣1分,电位转移过程未使用电位转移棒的扣5分			
13	拆除吊篮返回地面	1)塔上电工配合拆除吊篮轨迹绳、绝缘保护绳、2-2绝缘滑车组及吊篮并传至地面。 2)塔上电工检查塔上无遗留物后,向工作负责人汇报,得到工作负责人同意后携带绝缘传递绳下塔	5	1)下塔过程未使用防坠装置扣2分。 2)塔上移位失去安全带保护的扣2分。 3)下塔抓塔钉,每处扣1分。 4)塔上有遗留物的,扣2分			

续表

序号	项目名称	质量要求	分值	扣分标准	扣分原因	扣分	得分
14	工作结束	1）工作负责人组织全体工作成员整理工器具和材料，将工器具清洁后放入专用的箱（袋）中；清理现场，做到"工完料尽场地清"。 2）召开班后会，工作负责人进行工作总结和点评工作。点评本次工作的施工质量；点评全体工作成员的安全措施落实情况。 3）工作负责人向值班调控人员汇报工作结束，申请恢复线路重合闸，终结工作票	10	1）工器具未清理扣2分。 2）工器具有遗漏扣2分。 3）未开班后会不得2分。 4）未拆除围栏扣2分。 5）未向调度汇报不得2分。			
	合计		100				

模块2 带电更换1000kV交流输电线路直线塔单Ⅴ型复合绝缘子培训及考核标准

一、培训标准

(一) 培训要求

模块名称	带电更换1000kV交流输电线路直线塔单Ⅴ型复合绝缘子	培训类别	操作类
培训方式	实操培训	培训学时	21学时
培训目标	1.掌握直线塔进、出1000kV强电场时采用"吊篮法"作业方式的电学意义。 2.能完成采用"吊篮法"进入1000kV等电位作业点。 3.能独立完成更换1000kV交流输电线路直线塔单Ⅴ型复合绝缘子的操作(等电位作业法)		
培训场地	特高压交流实训线路		
培训内容	等电位和地电位配合,通过"吊篮法"进入等电位作业,采用等电位作业法更换直线塔单Ⅴ型复合绝缘子		
适用范围	特高压交流输电线路检修人员		

(二) 引用规程规范

(1)《电工术语》(GB/T 2900.55—2002)。

(2)《带电作业用绝缘滑车》(GB/T 13034—2008)。

(3)《带电作业用绝缘绳索》(GB/T 13035—2008)。

(4)《交流线路带电作业安全距离计算方法》(GB/T 18037—2000)。

(5)《1000kV交流带电作业用屏蔽服装》(GD/T 25726—2010)。

(6)《1000kV架空输电线路设计规范》(GB 50665—2011)。

(7)《1000kV交流输电线路检修规范》(DL/T 209—2008)。

(8)《1000kV交流输电线路运行规范》(DL/T 307—2010)。

(9)《1000kV交流输电线路带电作业技术导则》(DL/T 392—2015)。

(10)《带电作业绝缘配合导则》(DL/T 876—2004)。

(11)《带电作业工器具、装置和设备使用的一般要求》(DL/T 877—2004)。

(12)《带电作业工具、装备和设备预防性试验规程》(DL/T 976—2005)。
(13)《国家电网公司电力安全工作规程(线路部分)》(Q/GDW 1799.2—2013)。

(三)培训教学设计

本设计以完成"带电更换1000kV交流输电线路直线塔单V型复合绝缘子"为工作任务,按工作任务完成的标准化作业流程来设计各个培训阶段,每个阶段包括了具体的培训目标、培训内容、培训学时、培训方法(培训资源)、培训环境和考核评价等内容,如表2-2-1所示。

表2-2-1 带电更换1000kV交流输电线路直线塔单V型复合绝缘子培训内容设计

培训流程	培训目标	培训内容	培训学时	培训方法与资源	培训环境	考核评价
1.理论教学	1.初步掌握吊篮法进出1000kV强电场基本方法。2.熟悉电位转移的方法。3.熟悉输电线路直线塔单V型复合绝缘子更换方法	1.吊篮法进出强电场作业方式的电学意义。2.进、出特高压强电场时电位转移棒的使用方法。3.输电线路直线塔单V型复合绝缘子更换方法和质量标准	2	培训方法:讲授法。培训资源:PPT、相关规程规范	多媒体教室	考勤、课堂提问和作业
2.准备工作	能完成作业前准备工作	1.作业现场查勘。2.编制培训标准化作业卡。3.填写培训操作工作票。4.完成本操作的工器具及材料准备	1	培训方法:1.现场查勘和工器具及材料清理采用现场实操方法。2.编写作业卡和填写工作票采用讲授方法。培训资源:1.1000kV实训线路。2.特高压工器具库房。3.空白工作票	1.特高压输电实训线路;2.多媒体教室	
3.作业现场准备	能完成作业现场准备工作	1.作业现场复勘。2.工作申请。3.作业现场布置。4.班前会。5.工器具及材料检查	1	培训方法:演示与角色扮演法。资源:1000kV实训线路	1000kV实训线路	

续表

培训流程	培训目标	培训内容	培训学时	培训方法与资源	培训环境	考核评价
4.培训师演示	通过现场观摩，使学员初步领会本任务操作流程	1.塔上工器具布置安装。 2.等电位电工采用吊篮法进、出强电场。 3.地电位电工与等电位电工相互配合利用荷载转移装置完成单V型复合绝缘子更换	2	培训方法：演示法。 资源：1000kV实训线路	1000kV实训线路	
5.学员分组训练	1.能完成进、出1000kV强电场操作。 2.能完成1000kV输电线路直线塔单V型复合绝缘子更换	1.学员分组（12人一组）训练进、出1000kV强电场和直线塔单V型复合绝缘子更换。 2.培训师对学员操作进行指导和安全监护	14	培训方法：角色扮演法。 资源：1000kV实训线路	1000kV实训线路	采用技能考核评分细则对学员操作评分
6.工作终结	1.使学员进一步辨析操作过程不足之处，便于后期提升。 2.培训学员安全文明生产的工作作风	1.作业现场清理。 2.向调度汇报工作。 3.班后会，对本次工作任务进行点评总结	1	培训方法：讲授和归纳法	1000kV实训线路	

（四）作业流程

1. 工作任务

等电位和地电位配合，采用"吊篮法"进入等电位作业，更换直线塔单V型复合绝缘子。

2. 天气及作业现场要求

（1）带电更换1000kV交流输电线路直线塔单V型复合绝缘子应在良好的天气进行。

如遇雷电（听见雷声、看见闪电）、雪、雹、雨、雾等，禁止进行带电作业。风力大于5级或空气相对湿度大于80%时，不宜进行带电作业；恶劣天气下必须开展带电抢修时，应组织有关人员充分讨论并编制必要的安全措施，经本单位批准后方可进行。

（2）作业人员精神状态良好，熟悉工作中保证安全的组织措施和技术措施；应持有在有效期内的带电作业资质证书。

（3）工作负责人应事先组织相关人员完成现场勘察，根据勘察结果确定本次作业方法和所需工器具，以及应采取的必要措施，并办理带电作业工作票。

（4）作业现场应合理设置围栏，并妥当布置警示标示牌，禁止非工作人员入内。

(5) 本项目需停用线路重合闸装置。

工作中安全距离及有效绝缘长度如表2-2-2所示。

表2-2-2　带电更换1000kV交流输电线路直线塔单V型复合绝缘子的安全距离

海拔高度/m	最小安全距离/m		最小组合间隙/m		绝缘工具最小有效绝缘长度/m	转移电位时人体裸露部分与带电体的最小距离/m
	中相	边相	中相	边相		
$H \leqslant 1000$	6.8	6.0	6.9	6.7	6.8	0.5
$1000 < H \leqslant 2000$	7.4	6.6	7.6	7.3	6.8	0.5

3. 准备工作

3.1　危险点及其预控措施

（1）危险点——触电伤害

预控措施：

①工作前，工作负责人应与值班调控人员联系，停用线路重合闸，并履行许可手续。

②工作负责人必须进行安全措施、技术措施和工作任务交底，塔上地电位作业人员登塔前，必须仔细核对线路命名、杆塔编号、相别，确认无误后方可上塔。

③工作中，如遇线路突然停电，作业人员应视其仍然带电。工作负责人应尽快与调控人员联系，值班调控人员未与工作负责人取得联系前不准强送电。

④绝缘工具及绝缘绳索不得损坏、受潮、变形、失灵，不准使用非绝缘绳索（如棉纱绳、白棕绳、钢丝绳）。

⑤等电位作业人员应穿着阻燃内衣，衣服外面应穿戴全套屏蔽服（包括帽、衣裤、手套、袜和鞋），且各部分应连接良好，全套屏蔽服电阻不大于20Ω。

⑥等电位作业人员在电位转移前，应得到工作负责人的许可，人体裸露部分与带电体的最小距离不小于0.5m；等电位作业人员必须使用电位转移棒进行电位转移；地电位作业人员与带电体的安全距离小于表2-2-2的规定。

⑦用绝缘绳索传递大件金属物品时，地电位作业人员应将金属物品接地后再接触。

⑧工作负责人和专责监护人应对作业人员进行不间断监护，随时纠正其不规范或违章动作。重点关注高处作业人员，使其保持足够的安全距离（符合表2-2-2的规定），禁止同时接触两个非连通的带电体或带电体与接地体。

（2）危险点——高处坠落

预控措施：

①高处作业人员登高前，必须具备符合本项作业要求的身体状况、精神状态和技能素质。

②地电位电工登塔至作业点位后应打好安全绳并检查确认牢固，布置吊篮轨迹绳，等电

位电工人体后备保护绳应合理且可靠锚固。地电位电工使用闭合钩将绝缘吊拉绳与吊篮可靠连接。等电位电工系好绝缘保护绳、主保护绳，对吊篮做冲击试验合格后，进入吊篮。等电位电工系好安全带后，即可开展相关工作。等电位电工在作业点位应系好保护绳。

③监护人员应随时纠正其不规范或违章动作，重点关注作业人员在转位的过程中不得失去安全带或绝缘后备保护绳的保护，严禁低挂高用。

④人员登塔时检查脚钉和塔材的紧固情况，登塔时手抓主材，严禁手抓脚钉。

（3）危险点——高处坠物伤人。

预控措施：

①高处作业人员的个人工具及零星材料应装入工具袋，严禁在高处浮置物件、口中含物。

②地面作业人员必须正确佩戴安全帽，正确使用绳结，与作业点垂直下方距离不得小于坠落半径。

③作业现场设置围栏并挂好警示标示牌。监护人员应随时注意，禁止非工作人员及车辆进入作业区域。

3.2 工器具及材料选择

带电更换1000kV交流输电线路直线塔Ⅰ型复合绝缘子所需工器具及材料见表2-2-3。工器具出库前，应认真核对工器具的使用电压等级和试验周期，并检查确认外观良好、连接牢固、转动灵活，且符合本次工作任务的要求；工器具出库后，应存放在工具袋或工具箱内进行运输，防止脏污、受潮；金属工具和绝缘工器具应分开装运，防止因混装运输导致工器具变形、损伤等现象发生。

表2-2-3 带电更换1000kV交流输电线路直线塔单V型复合绝缘子所需工器具及材料

序号	工器具名称	规格型号	单位	数量	备注
1	全套屏蔽服	屏蔽效率≥60dB（屏蔽面罩屏蔽效率≥20dB）	套	5	
2	导电鞋	尺码视穿着人员而定	双	5	
3	绝缘拉棒		套	2	
4	张力转移器		台	1	
5	双保险安全带	背带式	根	5	
6	吊篮		个	1	
7	吊篮轨迹绳	TJS-ϕ16mm	根	1	
8	导线后备保护绳	ϕ32mm	根	1	
9	绝缘传递绳	TJS-14	根	2	

续表

序号	工器具名称	规格型号	单位	数量	备注
10	绝缘导轨绳	TJS-16	根	1	
11	绝缘后备保护绳	TJS-16	根	4	
12	绝缘滑车	JH10-1	个	2	
13	绝缘滑车	JH20-2	个	1	
14	2-2绝缘滑车	JH20-2	组	1	
15	电位转移棒		根	1	
16	液压丝杆		根	2	
17	导线翼型卡		个	1	
18	万用表		套	1	
19	风速、温湿度测试仪	HT-8321	只	1	
20	安全围网		套	若干	
21	警示标示牌	"在此工作""从此进出"	套	1	
22	红马甲	"工作负责人"	件	1	
23	防潮苫布	2m×4m	块	5	
24	机动绞磨	1T	台	1	
25	U型环	3T、5T、8T	个	3	
26	绝缘电阻表	5000V	块	1	
27	防坠器	与杆塔防坠器装置型号对应	个	4	
28	专用接头		个	4	
29	绝缘千斤		根	6	
30	钢丝千斤		根	4	
31	对讲机		个	5	
32	个人工器具		套	4	
33	安全帽		顶	12	
34	复合绝缘子	FXBZ-1000/420	支	1	

3.3 作业人员分工

本任务作业人员分工如表2-2-4所示。

表2-2-4 带电更换1000kV交流输电线路直线塔单V型复合绝缘子人员分工表

序号	工作岗位	数量(人)	工作职责
1	工作负责人	1	负责本次工作任务的人员分工、工作前的现场查勘、作业方案的制定、工作票的填写、现场复勘、办理工作许可手续、召开工作班前会、落实现场安全措施、负责作业过程中的安全监督、工作中突发情况的处理、工作质量的监督、工作后的总结
2	专责监护人	1	负责作业现场的安全把控
3	等电位电工	2	配合地电位电工安装提线系统(平面丝杠、专用接头、绝缘吊杆、液压丝杠),操作液压丝杠转移导线荷载,拆装绝缘子串等
4	地电位电工	2	负责安装吊篮、提线系统(平面丝杠、专用接头、绝缘吊杆、液压丝杠)、绝缘磨绳及配合等电位电工进出电位,拆装合成绝缘子串等
5	地面电工	6	负责传递工器具及合成绝缘子串等

4. 工作程序

本任务工作流程如表2-2-5所示。

表2-2-5 带电更换1000kV交流输电线路直线塔单V型复合绝缘子工作流程表

序号	作业内容	作业步骤及标准	安全措施及注意事项	责任人
1	现场复勘	工作负责人负责完成以下工作: (1)现场核对线路名称、杆塔编号、相别无误;基础及杆塔完好无异常;交叉跨越距离符合安全要求;确认缺陷情况及导地线规格型号等。 (2)检测风速、湿度等现场气象条件符合作业要求。 (3)检查地形环境符合作业要求。 (4)检查工作票所列安全措施与现场实际情况相符,必要时予以补充	(1)正确穿戴安全帽、工作服、工作鞋、劳保手套。 (2)不得在危及作业人员安全的气象条件下作业。 (3)严禁非工作人员、车辆进入作业现场	
2	工作许可	(1)工作负责人负责联系值班调控人员,按工作票内容申请停用线路重合闸。 (2)经值班调控人员许可后,方可开始带电作业工作	不得未经值班调控人员许可即开始工作	
3	现场布置	正确装设安全围栏并悬挂标示牌: (1)安全围栏范围应充分考虑高处坠物,以及对道路交通的影响。 (2)安全围栏出入口设置合理。 (3)妥当布置"从此进出""在此工作""从此上下"等标示	对道路交通安全影响不可控时,应及时联系交通管理部门强化现场交通安全管控	

续表

序号	作业内容	作业步骤及标准	安全措施及注意事项	责任人
4	召开班前会	(1)全体工作成员列队。 (2)工作负责人宣读工作票,明确工作任务及人员分工;讲解工作中的安全措施和技术措施;查(问)全体工作成员精神状态;告知工作中存在的危险点及采取的预控措施。 (3)全体工作成员在工作票上签名确认	(1)工作票填写、签发和许可手续规范,签名完整。 (2)全体工作成员精神状态良好。 (3)全体工作成员明确任务分工、安全措施和技术措施	
5	检查工器具	(1)塔上地电位电工和等电位电工正确地穿戴好屏蔽服并检测合格,由负责人监督检查。 (2)正确佩戴个人安全用具(大小合适,锁扣自如),由负责人监督检查。 (3)测量风速风向、湿度,检查绝缘工具的绝缘性能,并做好记录	(1)金属、绝缘工具使用前,应仔细检查其是否损坏、变形、失灵。绝缘工具应使用2500V及以上绝缘电阻表进行分段绝缘检测,阻值应不低于700MΩ,并用清洁干燥的毛巾将其擦拭干净。 (2)用万用表测量屏蔽服衣裤最远端点之间的电阻值不得大于20Ω。工作负责人认真检查作业电工屏蔽服的连接情况。 (3)检查工具组装情况并确认连接可靠。 (4)现场所使用的带电作业工具应放置在防潮帆布上	
6	登塔	(1)核对线路名称、杆塔编号无误后,塔上地电位电工和等电位冲击检查安全带、防坠器受力情况。 (2)塔上地电位电工携带绝缘传递绳登塔,等电位电工随后登塔,两人至横担作业点,选择合适位置系好安全带,塔上地电位电工将绝缘滑车和绝缘传递绳安装在横担合适位置。然后配合地面电工将绝缘传递绳分开作起吊准备	(1)核对线路名称和杆塔编号无误后,方可登塔作业。 (2)登塔过程中应使用塔上安装的防坠装置;杆塔上移动及转位时,不准失去安全保护,作业人员必须攀抓牢固构件。 (3)作业电工必须穿全套合格的屏蔽服,且全套屏蔽服必须连接可靠。在横担进入等电位前,等电位电工要检查确认屏蔽服个部位连接可靠后方能进行下一步操作	
7	安装滑车组、吊篮及磨绳	(1)地面电工利用绝缘传递绳将吊篮、吊篮轨迹绳、绝缘保护绳、2-2绝缘滑车组以及电位转移棒传至横担。 (2)塔上电工将2-2绝缘滑车组及吊篮安装在横担上平面合适位置,将吊篮轨迹绳安装在横担合适位置	(1)传递时绝缘吊绳要起吊平稳、无磕碰、无缠绕。 (2)吊篮安装好后由塔上电工对吊篮情况进行认真检查核对。 (3)2-2滑车组及吊篮应在横担上合适位置可靠安装	

续表

序号	作业内容	作业步骤及标准	安全措施及注意事项	责任人
8	进入强电场	(1)1号等电位电工系好绝缘保护绳进入吊篮,地面电工缓慢松出2-2绝缘滑车组控制绳、待吊篮距带电导线约2m处放慢速度。 (2)在吊篮向导线继续移动过程中、等电位电工将电位转移棒置于手中面向带电导线,同时向工作负责人申请电位转移,得到同意后,等电位电工待吊篮距导线0.5m时迅速伸出电位转移棒,将其钩在最近的子导线上进行电位转移。 (3)等电位电工进入强电场后系好安全带,并根据作业需要决定是否解除绝缘保护绳,同时等电位电工要控制头部不超过导线侧均压环。 (4)地面电工收紧2-2绝缘滑车组控制绳,将吊篮向上传至横担部位。2号等电位电工系好绝缘保护绳进入吊篮,用同样的方法进入强电场	(1)进入等电位前,等电位电工要再次检查确认屏蔽服各部位、电位转移棒与绝缘屏蔽服连接可靠后方能进行下一步操作。专责监护人负责检查等电位电工屏蔽服连接、屏蔽服与电位转移棒连接可靠。 (2)等电位电工进入电位前必须得到工作负责人的许可。 (3)等电位电工进入吊篮前必须系好保护绳。 (4)地面电工配合等电位电工进入等电位过程中收放滑车组时控制绳应平稳。 (5)等电位电工在进入电位过程中与接地体和带电体两部分间隙所组成的组合间隙不得小于6.9m(中相)/6.7m(边相)。 (6)专责监护人负责监护等电位电工进入强电场的安全注意事项。对塔上作业人员的危险、不规范动作应及时提醒,必要时应制止	
9	安装工具并转移导线荷载	(1)地面电工携带绝缘绳至中相导线上方横担作业点,将绝缘滑车和绝缘传递绳安装在合适位置。 (2)地面电工将翼型卡、绝缘拉棒、液压丝杠、导线后备保护绳传递至工作位置,由地电位电工与等电位电工配合将工具正确安装。 (3)检查各部构件可靠后,等电位电工收紧液压丝杠,使之稍受力,检查受力点情况。 (4)工作负责人同意后,等电位电工继续均匀收紧液压丝杠,使复合绝缘子松弛	(1)上、下作业电工要密切配合,所有作业电工要听从等工作负责人的统一指挥。 (2)地电位电工对带电体、等电位电工对接地体的最小安全距离不得小于6.8m(中相)/6.0m(边相)。绝缘吊杆、绝缘绳索的有效绝缘长度不得小于6.8m。 (3)杆塔上、下传递工具绑扎绳扣应正确可靠,塔上电工不得高空落物。 (4)工具受力后应试冲击检查无误后,报告工作负责人,在得到工作负责人许可后,方可继续作业。 (5)专责监护人对塔上作业人员的危险、不规范动作应及时提醒,必要时应制止。 (6)导线后备保护绳必须安装可靠,将8根子导线全部兜住	

续表

序号	作业内容	作业步骤及标准	安全措施及注意事项	责任人
10	拆除绝缘子串	(1)地电位电工收紧张力转移器、拆开平行挂板处的连接螺栓,放松张力转移器约300mm。 (2)等电位电工将绝缘传递绳安装在复合绝缘子尾部。并装好反束滑车,地面电工收紧反束绳。 (3)等电位电工取出碗头挂板螺栓。地面电工配合使绝缘子串自然垂直。 (4)地面电工装好起吊绝缘子串磨绳,启动机动绞磨,配合拆除张力转移器。 (5)地面电工控制好复合绝缘子串尾绳,配合机动绞磨缓慢将复合绝缘子串放至地面	(1)绝缘子串在退出运行时,详细检查受理部件情况,征得工作负责人同意方可拆除。 (2)利用绞磨起吊时应安置平稳,尾绳控制不可疏忽放松。 (3)必须检查绞磨及转向滑车的受力情况,无误后方可进行作业。 (4)绝缘子串尾绳应随时拉好,确保不碰到杆塔	
11	更换新的绝缘子串	(1)地面电工利用机动绞磨将新的复合绝缘子串传递至塔上。地电位电工恢复新的复合绝缘子与张力转移器的连接。 (2)地面电工缓慢松出机动绞磨使复合绝缘子自然垂直。 (3)地面电工收紧复合绝缘子反束绳将复合绝缘子串尾部拉至导线侧地电位电工工作位置。等电位电工恢复碗头挂板与导线联板的连接,装好开口销。 (4)地电位电工收紧张力转移器,恢复平行挂板与新合成绝缘子的连接,并装好开口销	(1)绝缘子串起吊过程中不得与杆塔发生碰撞。 (2)绳索不得与杆塔摩擦,绑扎牢固可靠。 (3)起吊尾绳随时拉紧,不可放松。 (4)必须检查绞磨及转向滑车的受力情况,无误后方可进行作业。 (5)张力转移器和复合绝缘子串连接可靠	
12	拆除工具	(1)经检查复合绝缘子串连接可靠,得到工作负责人同意后,地电位电工松出液压紧线系统和平面丝杆。 (2)经检查复合绝缘子串受力正常得到工作负责人的同意后,地电位电工与等电位电工配合拆除绝缘拉棒、液压丝杆、翼型卡、导线后备保护绳等,并传至地面等,并传至地面	(1)复合绝缘子安装复位后,应详细检查各部位连接正常无误,并得到工作负责人的同意后方可拆除提线工具。 (2)工具在传递过程中不得碰撞杆塔,绑扎绳扣应正确可靠。 (3)专责监护人对塔上作业人员的危险、不规范动作应及时提醒,必要时应制止	

续表

序号	作业内容	作业步骤及标准	安全措施及注意事项	责任人
13	退出电位	(1)1号等电位电工系好绝缘保护绳,将电位转移棒的金属端钩在子导线上,一只手握紧绝缘手柄,进入吊篮,然后保持手臂伸直状态使吊篮距子导线0.5m。 (2)1号等电位电工向工作负责人申请退出电位,得到同意后,等电位电工迅速脱开电位转移棒与子导线的连接,并将电位转移棒收回放置在吊篮中。 (3)地面电工同时迅速收紧2-2绝缘滑车组控制绳,将吊篮向上拉至横担部位停住,然后等电位电工登上横担,并系好安全带。 (4)地面电工利用绝缘传递绳将吊篮传至2号等电位电工处,等电位电工检查导线上无遗留物后进入吊篮,用同样的方法退出电位	(1)上、下作业电工要密切配合,听从工作负责人的指挥。 (2)等电位电工退出电位前必须得到工作负责人的许可。 (3)等电位电工进入吊篮前必须系好保护绳。 (4)地面电工配合等电位电工进入等电位时收放滑车组控制绳应平稳。 (5)等电位电工在退出电位过程中与接地体和带电体两部分间隙所组成的组合间隙不得小于6.9m(中相)/6.7m(边相)。 (6)专责监护人负责监护等电位电工退出强电场的安全注意事项。对塔上作业人员的危险、不规范动作应及时提醒,必要时应制止	
14	拆除吊篮返回地面	(1)塔上电工配合拆除吊篮轨迹绳、绝缘保护绳、2-2绝缘滑车组及吊篮并传至地面。 (2)塔上电工检查塔上无遗留物后,向工作负责人汇报,得到工作负责人同意后携带绝缘传递绳下塔	(1)工具在传递过程中不得碰撞,绑扎绳扣应正确可靠。 (2)登塔过程中应使用塔上安装的防坠装置;杆塔上移动及转位时,不准失去安全保护,作业人员必须抓牢固构件	
15	工作结束	(1)清理现场及工具,认真检查杆(塔)上有无遗留物,工作负责人全面检查工作完成情况,清点人数,无误后宣布工作结束,撤离施工现场。 (2)通知调度工作完毕,履行工作票完工手续	不得约时恢复线路重合闸	

二、考核标准

特高压交流输电线路运检技能考核评分细则

考生填写栏	编号： 姓 名： 所在岗位： 单 位： 日 期： 年 月 日				
考评员填写栏	成绩： 考评员： 考评组长： 开始时间： 结束时间： 操作时长：				
考核模块	带电更换1000kV交流输电线路直线塔单V型复合绝缘子	考核对象	特高压交流输电线路检修人员	考核方式	操作	考核时限	120min
任务描述	带电更换1000kV交流输电线路直线塔单V型复合绝缘子				
工作规范及要求	1. 带电作业工作应在良好天气下进行。如遇雷、雨、雪、雾天气不得进行带电作业。风力大于5级、湿度大于80%时，一般不宜进行带电作业。 2. 本项作业需工作负责人1名，专责监护人1名，地电位电工2名，地面电工6名，等电位电工2名，采用吊篮摆渡法进入强电场进行绝缘子更换工作。 3. 工作负责人职责：负责本次工作任务的人员分工、工作票的宣读、办理线路停用重合闸、办理工作许可手续、召开工作班前会、工作中突发情况的处理、工作质量的监督、工作后的总结。 4. 专责监护人：负责作业现场的安全把控。 5. 等电位电工职责：配合地电位电工安装提线系统（翼形卡、平面丝杠、专用接头、绝缘吊杆、液压丝杠），操作液压丝杠转移导线荷载，拆装绝缘子串。 6. 塔上地电工职责：负责安装吊篮、提线系统（平面丝杠、专用接头、绝缘吊杆、液压丝杠）、绝缘磨绳及配合等电位电工进出电位，拆装合成绝缘子串。 7. 地面电工职责：负责传递工具、材料配合等电位电工进出等电位。 8. 在带电作业中，如遇雷、雨、大风或其他任何情况威胁到工作人员的安全时，工作负责人或监护人可根据情况，临时停止工作。 给定条件： 1. 培训基地：特高压交流1000kV实训线路直线塔B相单V型复合绝缘子，绝缘子型号：FXBZ-1000/420。 2. 工作票已办理，安全措施已经完备（重合闸已停用），工作开始、工作终结时应口头提出申请（调度或考评员）。 3. 安全、正确地使用仪器对绝缘工具进行检测。 4. 必须按工作程序进行操作，工序错误扣除应做项目分值，出现重大人身、器材和操作安全隐患，考评员可下令终止操作（考核）。				
考核情景准备	1. 线路：特高压交流10000kV实训线路直线塔B相，工作内容：带电更换1000kV交流输电线路直线塔I型复合绝缘子，绝缘子型号：FXBZ-1000/420。 2. 所需作业工器具：绝缘传递绳3根（TJS-14），绝缘轨迹绳1根（TJS-16），绝缘滑车6个（JH10-1），2-2绝缘滑车2个（JH20-2），吊篮1个，液压紧线系统2只，绝缘吊杆2组，机动绞磨1台，绝缘检测仪，电位转移棒1根，绝缘电阻表（5000V型），屏蔽服（屏蔽效率≥60dB）5套，万用表1块，苫布4块，绝缘测试仪1台，温湿度表、风速仪各1台，纯棉毛巾2条。 3. 作业现场做好监护工作，作业现场安全措施（围栏等）已全部落实；禁止非作业人员进入现场，工作人员进入作业现场必须戴安全帽。 4. 考生自备工作服，阻燃纯棉内衣，安全帽，线手套，安全带（含二保绳）				
备注	1. 各项目得分均扣完为止，出现重大人身、器材和操作安全隐患，考评员可下令终止操作。 2. 设备、作业环境、安全带、安全帽、工器具、屏蔽服等不符合作业条件考评员可下令终止操作				

第二部分 技能模块培训及考核标准

续表

序号	项目名称	质量要求	分值	扣分标准	扣分原因	扣分	得分
1	现场复勘	1）工作负责人到作业现场核对线路名称和杆塔编号、现场工作条件、缺陷部位等。 2）检测风速、湿度等现场气象条件符合作业要求。 3）检查工作票填写完整，无涂改，检查是否所列安全措施与现场实际情况相符，必要时予以补充	5	1）未进行核对双重称号扣1分。 2）未核实现场工作条件（气象）、缺陷部位扣1分。 3）工作票填写出现涂改，每项扣0.5分，工作票编号有误，扣1分。工作票填写不完整，扣1.5分			
2	工作许可	1）工作负责人联系值班调控人员，按工作票内容申请停用线路重合闸。 2）汇报内容规范、完整	2	1）未联系调度部门（裁判）停用重合闸扣2分。 2）汇报专业用语不规范或不完整的各扣0.5分			
3	现场布置	正确装设安全围栏并悬挂标示牌： 1）安全围栏范围应充分考虑高处坠物，以及对道路交通的影响。 2）安全围栏出入口设置合理。 3）妥当布置"从此进出""在此工作""从此上下"等标示	3	1）作业现场未装设围栏扣0.5分。 2）未设立警示牌扣0.5分。 3）未悬挂登塔作业标志扣0.5分			
4	召开班前会	1）全体工作成员全体人员正确佩戴安全帽、工作服。 2）工作负责人佩戴红色背心，宣读工作票，明确工作任务及人员分工；讲解工作中的安全措施和技术措施；查（问）全体工作成员精神状态；告知工作中存在的危险点及采取的预控措施。 3）全体工作成员在工作票上签名确认	3	1）工作人员着装不整齐扣0.5分，工作人员着装不整齐每人次扣0.5分。 2）未进行分工本项不得分，分工不明扣1分。 3）现场工作负责人未穿佩安全监护背心扣0.5分。 4）工作票上工作班成员未签字或签字不全的扣1分			

续表

序号	项目名称	质量要求	分值	扣分标准	扣分原因	扣分	得分
5	工器具检查	1)工作人员按要求将工器具放在防防潮苫布上;防潮苫布应清洁、干燥。 2)工器具应按定置管理要求分类摆放;绝缘工器具不能与金属工具、材料混放;对工器具进行外观检查。 3)绝缘工具表面不应磨损、变形损坏,操作应灵活。绝缘工具应使用2500V及以上绝缘电阻表进行分段绝缘检测,阻值应不低于700MΩ,并用清洁干燥的毛巾将其擦拭干净。 4)塔上地电位和登电位人员按要求正确穿戴全套合格的屏蔽服、导电鞋,且各部分连接应良好,屏蔽服内不得贴身穿着化纤类衣服,并系好安全带;工作负责人应认真检查是否穿戴正确。 5)登塔人员再次核对双重名称、杆号、相别并报告	7	1)未使用防潮布并定置摆放工器具扣1分。 2)未检查工器具试验合格标签及外观检查扣每项扣0.5分。 3)未正确使用检测仪器对工器具进行检测每项扣1分。 4)作业人员未正确穿戴屏蔽服且各部位连接良好每人次扣2分。 5)现场工作负责人未对登塔作业人员进安全防护装备进行检查扣1分。 6)登塔人员未核对线路双重名称、杆号、相别每人扣2分。 7)登塔人员未报告核对结果每人扣2分。			
6	登塔	1)塔上地电位电工、等电位电工穿好全套合格的屏蔽服,将安全带做冲击试验后,系好安全带后携带绝缘传递绳相继登塔。 2)登塔过程中系好防坠落保护装置,登塔至合适位置,系好安全带,布置好绝缘传递绳,然后配合地面电工将绝缘传递绳分开作起吊准备。 3)登塔过程中应系好防坠落保护装置,匀速登塔,手抓主材,将安全带挂在肩上并与带电体保持6.8(中相)/6.0(中相)m以上安全距离,工作负责人加强作业监护	5	1)未系安全带或安全带及后备保护绳未进行冲击试验各扣2分。 2)手抓脚钉扣2分。 3)滑车传递绳悬挂位置不便工具取用扣1分。 4)传递时金属工具难以保证安全距离扣2分;工具绑扎不牢扣2分。 5)传递时高空落物扣2分。 6)传递过程工具与塔身磕碰扣2分。 7)传递工具绳索打结混乱扣1分。 8)工作负责人监护不到位扣2分。 9)塔上电工操作不正确扣2分			

续表

序号	项目名称	质量要求	分值	扣分标准	扣分原因	扣分	得分
7	安装滑车组及吊篮	1)传递时绝缘吊绳起吊要平稳、无磕碰、无缠绕。 2)吊篮安装好后由塔上电工对吊篮情况进行认真检查核对。 3)2-2滑车组及吊篮应在横担上合适位置可靠	5	1)2-2滑车组绳子缠绕扣0.5分。 2)轨迹绳安装位置不合理扣1分。 3)绝缘保护绳、长度不合适扣1分。 4)传递工器具不平稳、磕碰扣1分。			
8	进入强电场	1)等电位再次检查确认屏蔽服各部位连接可靠后对吊篮进行冲击实验,汇报工作负责人后系好保护绳登上吊篮行。 2)地面电工缓慢松出2-2绝缘滑车组控制绳,当距离导线约2m处放慢速度,等电位电工在距离导线0.5m处向工作负责人申请电位转移,得到同意后,迅速伸出电位转移棒,将其钩在最近的子导线上进行电位转移。 3)等电位电工进入强电场后做好人体后保护,要控制头部不超过导线侧均压环。 4)电位电工在进入电位过程中与接地体和带电体两部分间隙所组成的组合间隙不得小于6.9m(中相)/6.7m(边相),进入强电场必须用电位转移棒进行电位转移	10	1)等电位电工未对吊篮进行冲击扣1分。 2)未系好绝缘保护绳扣1分。 3)地电位电工未检查等电位电工安全措施扣2分。 4)地面电工控制滑车尾绳不平稳扣1分。 5)等电位电工进入强电场前未向工作负责人申请扣2分;申请了但未得同意即开始扣1分。 6)转移电位动作不熟练扣1分,电位转移过程未使用电位转移棒的扣5分。 7)等电位电工进入强电场后安全带未系好扣2分。 8)等电位电工进入强电场后头部超过导线侧均压环扣2分。			
9	安装工具并转移导线荷载	1)地面电工携带绝缘绳至中相导线上方横担作业点,将绝缘滑车和绝缘传递绳安装在合适位置。 2)地面电工将罩型卡、绝缘拉棒、液压丝杠、导线后备保护绳传递至工作位置,由地电位电工与等电位电工配合将工具正确安装。 3)检查各部构件可靠后,等电位电工收紧液压丝杠,使之稍受力,检查受力点情况。 4)工作负责人同意后,等电位电工继续均匀收紧液压丝杠,使复合绝缘子松弛	15	1)工器具传递过程不平稳扣1分;有磕碰每次扣1分。 2)未检查承力工具安装可靠、受力良好扣2分,未汇报并取得工作负责人同意扣2分。 3)地电位电工与等电位电工沟通无效扣1分。 4)拆除绝缘子串前未对承力工具进行检查扣5分,检查结果未汇报扣2分。 5)绝缘绳系复合绝缘子位置不合适扣2分。 6)绝缘子串传递时有磕碰每次扣2分。 7)地面电工机动绞磨控制不到位扣1分。			

续表

序号	项目名称	质量要求	分值	扣分标准	扣分原因	扣分	得分
10	更换绝缘子串	1)地面电工控制好复合绝缘子串控制绳,利用机动绞磨缓慢将复合绝缘子串放至地面。注意控制好复合绝缘子串的控制绳,不得碰撞承力工具、导线及杆塔。2)地面电工将绝缘传递绳和复合绝缘子串控制绳分别转移到新复合绝缘子上。3)地面电工启动机动绞磨,将新复合绝缘子串传递至塔上工作位置。地电位电工恢复新复合绝缘子串与球头挂环的连接,并复位锁紧销。4)地面电工缓慢松出机动绞磨使复合绝缘子串自然垂直,地电位电工恢复碗头挂板与金属联板的连接,并装好开口销	17	1)地面电工未控制好绝缘子尾绳扣1分。2)绑扎绳扣不合理扣1分。3)未检查绞磨转向及滑车的受力情况扣2分。5)绝缘子串传递时有磕碰每次扣2分。4)绝缘子串安装不到位扣5分。5)作业人员未检查绝缘子串连接扣5分。6)作业人员未检查销子安装到位扣2分。7)专责监护人未尽监护职责扣2分			
11	拆除工具	1)经检查复合绝缘子串连接可靠,得到工作负责人同意后,地电位电工松出液压紧线系统和平面丝杆。2)经检查复合绝缘子串受力正常得到工作负责人的同意后,地电位电工与等电位电工配合拆除绝缘拉棒、液压丝杆、翼型卡、导线后备保护绳等,并传至地面等,并传至地面	5	1)未详细检查各部位连接正常无误扣1分,未得到工作负责人同意扣1分。2)工器具传递过程不平稳扣1分;有磕碰每次扣1分			
12	退出电位	1)1号等电位电工系好绝缘保护绳,将电位转移棒的金属端钩在子导线上,一只手握紧绝缘手柄,进入吊篮,然后保持手臂伸直状态使吊篮距子导线0.5m。2)等电位电工向工作负责人申请退出电位,得到同意后,等电位电工迅速脱开电位转移棒与子导线的连接,并将电位转移棒收回放置在吊篮中。3)地面电工同时迅速收紧2-2绝缘滑车组控制绳,将吊篮向上拉至横担部位停住,然后等电位电工登上横担,并系好安全带。4)地面电工利用绝缘传递绳将吊篮传至另一名等电位电工处,等电位电工检查导线上无遗留物后进入吊篮,用同样的方法退出电位	8	1)未报告工作结束扣2分,强电场内有遗留物每件扣1分。2)等电位未系好绝缘保护绳扣1分。3)地面电工控制滑车尾绳不平稳扣1分。4)等电位电工进入强电场前未向工作负责人申请扣2分,申请了但未得同意即开始扣1分。5)转移电位动作不熟练扣1分,电位转移过程未使用电位转移棒的扣5分			

续表

序号	项目名称	质量要求	分值	扣分标准	扣分原因	扣分	得分
13	拆除吊篮返回地面	1)塔上电工配合拆除吊篮轨迹绳、绝缘保护绳、2-2绝缘滑车组及吊篮并传至地面。 2)塔上电工检查塔上无遗留物后，向工作负责人汇报，得到工作负责人同意后携带绝缘传递绳下塔	5	1)下塔过程未使用防坠装置扣2分。 2)塔上移位失去安全带保护的扣2分。 3)下塔抓塔钉，每处扣1分。 4)塔上有遗留物的，扣2分			
14	工作结束	1)工作负责人组织全体工作成员整理工器具和材料，将工器具清洁后放入专用的箱（袋）中；清理现场，做到"工完料尽场地清"。 2)召开班后会，工作负责人进行工作总结和点评工作。点评本次工作的施工质量；点评全体工作成员的安全措施落实情况。 3)工作负责人向值班调控人员汇报工作结束，申请恢复线路重合闸，终结工作票	10	1)工器具未清理扣2分。 2)工器具有遗漏扣2分。 3)未开班后会不得分。 4)未拆除围栏扣2分。 5)未向调度汇报不得2分			
	合计		100				

特高压交流输电线路运检专业培训及考核标准
TEGAOYA JIAOLIU SHUDIAN XIANLU YUNJIAN ZHUANYE PEIXUN JI KAOHE BIAOZHUN

模块3 带电更换1000kV交流输电线路耐张塔横担侧1~3片玻璃绝缘子培训及考核标准

一、培训标准

(一) 培训要求

模块名称	带电更换1000kV交流输电线路耐张塔横担侧1~3片玻璃绝缘子	培训类别	操作类
培训方式	实操培训	培训学时	21学时
培训目标	1.掌握地电位作业法中电磁防护的电学意义。 2.能独立完成带电更换1000kV交流输电线路耐张塔横担侧1~3片玻璃绝缘子的操作(地电位作业法)		
培训场地	特高压交流实训线路		
培训内容	采用地电位作业法带电更换1000kV交流输电线路耐张塔横担侧1~3片玻璃绝缘子的操作		
适用范围	特高压交流输电线路检修人员		

(二) 引用规程规范

(1)《1000kV架空输电线路设计规范》(GB 50665—2011)。

(2)《1000kV交流输电线路检修规范》(DL/T 209—2008)。

(3)《1000kV交流输电线路运行规范》(DL/T 307—2010)。

(4)《1000kV交流输电线路带电作业技术导则》(DL/T 392—2015)。

(5)《交流线路带电作业安全距离计算方法》(GB/T 19185—2008)。

(6)《带电作业用绝缘配合导则》(DL/T 867—2004)。

(7)《架空输电线路带电安装导则及作业工具设备》(DL/T 1007—2006)。

(8)《国家电网公司带电作业工作管理规定(试行)》(国家电网生〔2007〕751号)。

(9)《国家电网公司电力安全工作规程(线路部分)》(Q/GDW1799.2—2013)。

(10)《电工术语架空线路》(GB/T 2900.51—1998)。

(11)《电工术语带电作业》(GB/T 2900.55—2002)。

(12)《带电作业工具设备术语》(GB/T 14286—2002)。

(13)《带电作业用工具、装置和设备使用的一般要求》(DL/T 877—2004)。

(14)《带电作业工具、装置和设备预防性试验规程)》(DL/T 976—2005)。

(15)《带电作业用绝缘滑车》(GB/T 13034—2008)。

(16)《带电作业用绝缘绳索》(GB 13035—2008)。

(17)《1000kV交流带电作业用屏蔽服装》(GB/T 25726—2010)。

(三)培训教学设计

本设计以完成"带电更换1000kV交流输电线路耐张塔横担侧1~3片玻璃绝缘子"为工作任务,按工作任务完成的标准化作业流程来设计各个培训阶段,每个阶段包括了具体的培训目标、培训内容、培训学时、培训方法(培训资源)、培训环境和考核评价等内容,如表2-3-1所示。

表2-3-1 带电更换1000kV交流输电线路耐张塔横担侧1~3片玻璃绝缘子

培训流程	培训目标	培训内容	培训学时	培训方法与资源	培训环境	考核评价
1.理论教学	1.初步掌握地电位作业法中电磁防护的基本方法。 2.熟悉输电线路耐张单片绝缘子更换方法	1.掌握地电位作业法中电磁防护的电学意义。 2.输电线路耐张单片绝缘子更换方法和质量标准	2	培训方法:讲授法。 培训资源:PPT、相关规程规范	多媒体教室	考勤、课堂提问和作业
2.准备工作	能完成作业前准备工作	1.作业现场查勘。 2.编制培训标准化作业卡。 3.填写培训操作工作票。 4.完成本操作的工器具及材料准备	1	培训方法: 1.现场查勘和工器具及材料清理采用现场实操方法。 2.编写作业卡和填写工作票采用讲授方法。 培训资源: 1.1000kV实训线路。 2.特高压工器具库房。 3.空白工作票	1.特高压输电实训线路。 2.多媒体教室	
3.作业现场准备	能完成作业现场准备工作	1.作业现场复勘。 2.工作申请。 3.作业现场布置。 4.班前会。 5.工器具及材料检查	1	培训方法:演示与角色扮演法。 资源:1000kV实训线路	1000kV实训线路	

续表

培训流程	培训目标	培训内容	培训学时	培训方法与资源	培训环境	考核评价
4.培训师演示	通过现场观摩,使学员初步领会本任务操作流程	1.地电位电工组装工器具。 2.地电位电工完成单片玻璃绝缘子的更换工作	2	培训方法:演示法。 资源:1000kV实训线路	1000kV实训线路	
5.学员分组训练	1.能完成横担侧耐张单片玻璃绝缘子的更换工作	1.学员分组(6人一组)训练更换绝缘子技能操作。 2.培训师对学员操作进行指导和安全监护	14	培训方法:角色扮演法。 资源:1000kV实训线路	1000kV实训线路	采用技能考核评分细则对学员操作评分
6.工作终结	1.使学员进一步辨析操作过程不足之处,便于后期提升。 2.培训学员安全文明生产的工作作风	1.作业现场清理。 2.向调度汇报工作终结。 3.班后会,对本次工作任务进行点评总结	1	培训方法:讲授和归纳法	1000kV实训线路	

(四)作业流程

1. 工作任务

带电更换1000kV交流输电线路耐张塔横担侧1~3片玻璃绝缘子。

2. 天气及作业现场要求

(1)带电更换1000kV交流输电线路耐张塔横担侧1~3片玻璃绝缘子应在良好的天气进行。如遇雷电(听见雷声、看见闪电)、雪、雹、雨、雾等,禁止进行带电作业。风力大于5级,不宜进行带电作业;相对湿度大于80%的天气,若需进行带电作业,应采用具有防潮性能的绝缘工具。恶劣天气下必须开展带电抢修时,应组织有关人员充分讨论并编制必要的安全措施,经本单位批准后方可进行。

(2)作业人员精神状态良好,熟悉工作中保证安全的组织措施和技术措施,掌握高处应急救援及触电急救的方法;应持有在有效期内的带电作业资质证书。

(3)工作负责人应事先组织相关人员完成现场勘察,根据勘察结果确定本次作业方法和所需工器具,以及应采取的必要措施,并办理带电作业工作票。

(4)作业现场应合理设置围栏,并妥当布置警示标示牌,禁止非工作人员入内。

(5)本项目须停用线路重合闸装置。

(6)作业方式:地电位作业。

(7)工作中安全距离及有效绝缘长度如表2-3-2所示。

表2-3-2 带电更换1000kV交流输电线路耐张玻璃绝缘子串任意单片玻璃绝缘子的安全距离

海拔高度/m	中间电位作业人员与带电体之间的最小安全距离/m		绝缘工器具的最小有效绝缘长度/m
	中相	边相	
$H \leq 1000$	6.8	6.0	6.8
$1000 < H \leq 2000$	7.4	6.6	7.2

注：表中数值不包括人体占位间隙，作业中需考虑人体占位间隙不得小于0.5 m。

（8）地电位电工进入耐张绝缘子串横担侧时，人体短接绝缘子片数不得多于4片。耐张绝缘子串中扣除人体短接和不良绝缘子片数后，良好绝缘子最少片数应满足表2-3-3的规定。

表2-3-3 最小组合间隙和良好绝缘子的最小片数

海拔高度	单片玻璃绝缘子结构高度（mm）	良好绝缘子串的总长度最小值（m）	良好绝缘子的最少片数
$H \leq 1000$	170	7.2	43
	195		37
	205		36
$1000 < H \leq 2000$	170	8.0	47
	195		41
	205		39

注：表中数值不包括人体占位间隙，作业中需考虑人体占位间隙不得小于0.5 m。

3. 准备工作

3.1 危险点及其预控措施

（1）危险点——触电伤害

预控措施：

①工作前，工作负责人应与值班调控人员联系，停用线路重合闸，并履行许可手续。

②塔上作业人员登塔前，必须仔细核对线路双重命名、杆塔编号、相别，确认无误后方可上塔。

③工作中，如遇线路突然停电，作业人员应视其仍然带电。工作负责人应尽快与调控人员联系，值班调控人员未与工作负责人取得联系前不准强送电。

④绝缘工具及绝缘绳索不得损坏、受潮、变形、失灵，不准使用非绝缘绳索（如棉纱绳、白棕绳、钢丝绳）。

⑤地面电工操作绝缘工具时应戴清洁、干燥的手套，进入作业现场应将使用的带电作业工具放置在防潮的帆布或绝缘垫上，防止绝缘工具在使用中脏污和受潮。

⑥地电位电工应穿着阻燃内衣，衣服外面应穿戴合格全套屏蔽服（包括帽、衣裤、手套、袜和鞋），且各部分应连接良好。

⑦地电位电工进入耐张绝缘子串横担侧时，手与脚的位置必须保持对应一致，且人体和工具短接的绝缘子片数应符合表2-3-3规定。

⑧用绝缘绳索传递大件金属物品时，地电位电工及地面电工应将金属物品接地后再接触。

⑨带电作业过程中，工作负责人（监护人）应对作业人员进行不间断监护，随时纠正其不规范或违章动作。重点关注高处作业人员，使其保持足够的安全距离（符合表2-3-2的规定），禁止同时接触两个非连通的带电体或带电体与接地体。

（2）危险点——高处坠落

预控措施：

①高处作业人员登高前，必须具备符合本项作业要求的身体状况、精神状态和技能素质。

②高处作业人员应使用双保险安全带。上、下塔时，应手抓主材、脚踩脚钉，匀速行进。

③地电位电工作业前应认真检查液压丝杠、闭式卡等，确保承力工具合格；沿绝缘子串移动时，手与脚的位置必须保持对应一致，安全带应系挂在手扶的绝缘子串上，并同步移动；更换绝缘子时，承力工具安装应可靠，荷载转移前、后应做冲击试验判定其可靠性，并及时向工作负责人汇报，得到工作负责人许可后方可实施。

④监护人员应随时纠正其不规范或违章动作，重点关注高处作业人员在转位的过程中不得失去安全带或绝缘后备保护绳的保护，严禁低挂高用。

（3）危险点——高处坠物伤人。

预控措施：

①高处作业人员的个人工具及零星材料应装入工具袋，严禁在高处浮置物件、口中含物。

②地面电工必须正确佩戴安全帽，正确使用绳结，与作业点垂直下方距离不得小于坠落半径。

③作业现场设置围栏并挂好警示标示牌。监护人员应随时注意，禁止非工作人员及车辆进入作业区域。

3.2 工器具及材料选择

带电更换1000kV交流输电线路耐张塔横担侧1~3片玻璃绝缘子所需工器具及材料见表2-3-4。工器具出库前，应认真核对工器具的使用电压等级和试验周期，并检查确认外观良好、连接牢固、转动灵活，且符合本次工作任务的要求；工器具出库后，应存放在工具袋或工具箱内进行运输，防止脏污、受潮；金属工具和绝缘工器具应分开装运，防止因混装运输导致工器具变形、损伤等现象发生。

表2-3-4 带电更换1000kV交流输电线路耐张塔横担侧1~3片玻璃绝缘子所需工器具及材料表

序号	名称	规格型号	单位	数量	备注
1	屏蔽服	Ⅰ型	套	2	个人防护用具
2	导电鞋	尺码视穿着人员而定	双	2	个人防护用具
3	阻燃内衣	纯桑蚕丝	套	2	个人防护用具
4	双保险安全带	背带式	根	2	个人防护用具
5	安全帽		顶	7	个人防护用具
6	护目镜		副	2	个人防护用具
7	绝缘传递绳	$\phi 14mm$,长度与起吊高度匹配	根	1	绝缘工具
8	绝缘后备保护绳	$\phi 16mm$	根	2	绝缘工具
9	绝缘绳套	$\phi 14mm$	根	2	绝缘工具
10	绝缘滑车	1T	个	1	绝缘工具
11	耐张端部卡		个		金属工具
12	液压丝杠		根		金属工具
13	闭式卡（后卡）		个		金属工具
14	绝缘电阻测试仪	2500V,电极宽2cm、极间宽2cm	套	1	其他工具
15	万用表		套	1	其他工具
16	风速、温湿度测试仪		只	1	其他工具
17	安全围网		套	若干	其他工具
18	警示标示牌	"在此工作""从此进出""车辆慢行""车辆绕行"	套	1	其他工具
19	红马甲	"工作负责人""专责监护人"	件	1	其他工具
20	防潮苫布	3m×3m	块	2	其他工具
21	个人工具	扳手、老虎钳	套		其他工具
22	拔销器		把	1	其他工具
23	防坠器	与杆塔防坠落装置型号对应	只	2	其他工具
24	毛巾	棉质	条	1	其他工具
25	绝缘子		片	1	材料

3.3 作业人员分工

本任务作业人员分工如表2-3-5所示。

表2-3-5 带电更换1000kV交流输电线路耐张塔横担侧1～3片玻璃绝缘子人员分工表

序号	工作岗位	数量(人)	工作职责
1	工作负责人	1	负责本次工作任务的人员分工、工作票的宣读、办理线路停用重合闸、办理工作许可手续、召开工作班前会、工作中突发情况的处理、工作质量的监督、工作后的总结
2	专责监护人	1	负责作业现场的安全把控
3	地电位电工	1	负责工器具安装及绝缘子更换工作
4	地面电工	3	负责本次作业过程的地面辅助工作

4. 工作程序

本任务工作流程如表2-3-6所示。

表2-3-6 带电更换1000kV交流输电线路耐张塔横担侧1～3片玻璃绝缘子工作流程表

序号	作业内容	作业步骤及标准	安全措施及注意事项	责任人
1	现场复勘	工作负责人负责完成以下工作： (1)现场核对线路名称、杆塔编号、双重编号无误；基础及杆塔完好无异常；交叉跨越距离符合安全要求；确认缺陷情况及导地线规格型号等。 (2)检测风速、湿度等现场气象条件符合作业要求。 (3)检查地形环境符合作业要求。 (4)检查工作票所列安全措施与现场实际情况相符，必要时予以补充	(1)正确穿戴安全帽、工作服、工作鞋、劳保手套。 (2)不得在危及作业人员安全的气象条件下作业。 (3)严禁非工作人员、车辆进入作业现场	
2	工作许可	(1)工作负责人负责联系值班调控人员，按工作票内容申请停用线路重合闸。 (2)经值班调控人员许可后，方可开始带电作业工作	不得未经值班调控人员许可即开始工作	
3	现场布置	正确装设安全围栏并悬挂标示牌： (1)安全围栏范围应充分考虑高处坠物，以及对道路交通的影响。 (2)安全围栏出入口设置合理。 (3)妥当布置"从此进出""在此工作""车辆慢行"或"车辆绕行"等标示	对道路交通安全影响不可控时，应及时联系交通管理部门强化现场交通安全管控	
4	召开班前会	(1)全体工作成员列队。 (2)工作负责人宣读工作票，明确工作任务及人员分工；讲解工作中的安全措施和技术措施；查(问)全体工作成员精神状态；告知工作中存在的危险点及采取的预控措施。 (3)全体工作成员在工作票上签名确认	(1)工作票填写、签发和许可手续规范，签名完整。 (2)全体工作成员精神状态良好。 (3)全体工作成员明确任务分工、安全措施和技术措施	

续表

序号	作业内容	作业步骤及标准	安全措施及注意事项	责任人
5	检查工器具	(3)在防潮苫布上,将工器具按作业要求准备齐备,并分类定置摆放整齐。检查工器具外观和试验合格证,无遗漏。 (2)使用绝缘电阻测试仪检测绝缘工具及绝缘绳索的表面绝缘电阻值,方法正确,不得低于700MΩ。 (3)将新绝缘子擦拭干净,外观检查完好,不得有锈蚀、裂纹及破损。使用绝缘电阻测试仪测试其绝缘电阻值,方法正确,不得低于500 MΩ。 (4)使用万用表检测全套屏蔽服内阻,方法正确,不得大于20Ω。 (5)检查人员向工作负责人汇报各项检查结果符合作业要求	(3)防潮苫布数量足够,设置位置合理,保持清洁、干燥。 (2)金属、绝缘工器具在使用前,应仔细检查其是否无损伤、受潮、变形、失灵现象,合格证在有效期内。 (3)绝缘工具及绝缘绳索检测合格	
6	登塔	(3)地电位电工再次核对线路双重名称及相别,检查并确认脚钉齐全、牢固,系好安全带、加挂防坠器;对安全带、防坠器做冲击试验,方法正确;工作负责人检查并确认其穿戴的双保险安全带各部件的连接情况良好,包括肩带、胸带、腰带、腿带、后背保护绳、扣和环。 (2)背上工具包,携带绝缘传递绳(含绝缘滑车),方法正确。 (3)将安全带主带和后备保护绳跨肩上。 (4)清洁鞋底,经工作负责人许可后依次登塔。 (5)脚踩脚钉、手抓主材,匀步登塔至横担适当位置,系好安全带,脱离防坠器	(3)安全带、防坠器冲击试验合格。 (2)防止安全带、绝缘传递绳钩挂塔材。 (3)人体与导线保持的最小安全距离应符合表2-3-2的规定。 (4)禁止手抓脚钉。 (5)正确使用防坠器。 (6)转位时,不得失去安全带的保护	
7	安装工具并转移导线张力	(3)地面电工使用绝缘传递绳将闭式卡(前卡)、液压丝杠、耐张端部卡等分别传至地电位电工。起吊过程平稳、无磕碰、无缠绕,正确使用绳结。 (2)地电位电工先在牵引板上安装耐张端部卡,后将闭式卡(后卡)安装在横担侧第3片绝缘子上,并连接好液压丝杠。承力工具各部分安装牢固可靠。 (3)检查并确认承力工具各部分安装情况良好,经工作负责人许可后,操作液压丝杠使其逐渐受力,使需更换的绝缘子松弛。两根液压丝杠的受力应均匀	(3)人体与带电体之间的安全距离不得小于表2-3-2的规定。 (2)防止高处坠物。 (3)扣除劣质绝缘子、人体操作和工具短接的绝缘子后,良好绝缘子片数应符合表2-3-3的规定	

续表

序号	作业内容	作业步骤及标准	安全措施及注意事项	责任人
8	更换绝缘子	(3)地电位电工做冲击试验,检查并确认承力工具受力正常,经工作负责人许可后,用绝缘传递绳系好绝缘子,取出旧绝缘子两端锁紧销,继续操作并收紧液压丝杠,直至拆除旧绝缘子。两根液压丝杠的受力应均匀,操作手柄不得敲击绝缘子。 (2)地面电工用绝缘传递绳的另一端系好新绝缘子,采用旧下、新上的方法,将新绝缘子起吊给地电位电工。起吊过程平稳、无磕碰、无缠绕,正确使用绳结。 (3)地电位电工安装新绝缘子,并复位其两端锁紧销,安装到位	(3)人体与带电体之间的安全距离不得小于表2-3-2的规定。 (2)防止高处坠物	
9	拆除工具	(3)地电位电工检查并确认新绝缘子连接可靠,经工作负责人许可后,操作并松出液压丝杠,使更换的绝缘子逐渐受力。 (2)荷载转移完毕后,地电位电工做冲击试验,检查并确认新绝缘子受力情况良好,经工作负责人许可后,拆除系在绝缘子上的绝缘传递绳,并将其系牢于承力工具适当位置,拆除闭式卡(前卡)、液压丝杠、耐张端部卡等承力工具,在地面电工配合下传递至地面。传递过程平稳、无磕碰、无缠绕,正确使用绳结	(3)人体与带电体之间的安全距离不得小于表2-3-2的规定。 (2)防止高处坠物	
10	撤离杆塔	地电位电工检查塔上无遗留物,拆除绝缘传递绳。经工作负责人许可后,挂好防坠器,解开并整理好安全带,正确携带绝缘传递绳,脚踩脚钉、手抓主材、匀步下塔至地面	(3)人体与带电体之间的安全距离不得小于表2-3-2的规定。 (2)转位时不得失去安全带保护。 (3)防止手滑脱、脚踏空,禁止手抓脚钉。 (4)正确使用防坠器。 (5)防止绝缘传递绳、安全带钩挂塔材或脚钉	
11	工作结束	(3)工作负责人组织全体工作成员整理工器具和材料,将工器具清洁后放入专用的箱(袋)中;清理现场,做到"工完料尽场地清"。 (2)召开班后会,工作负责人进行工作总结和点评工作。点评本次工作的施工质量;点评全体工作成员的安全措施落实情况。 (3)工作负责人向值班调控人员汇报工作结束,并申请恢复线路重合闸,终结工作票		

二、考核标准

特高压交流输电线路运检技能考核评分细则

考 生 填写栏	编号：	姓 名：	所在岗位：		单 位：		日 期：		年 月 日
考评员 填写栏	成绩：	考评员：		考评组长：		开始时间：		结束时间：	操作时长：
考核 模块	带电更换1000kV交流输电线路耐张塔横担侧1～3片玻璃绝缘子		考核 对象	特高压交流输电线路检修人员			考核 方式	操 作	考核 时限	90min
任务 描述	带电更换1000kV交流输电线路耐张塔横担侧1～3片玻璃绝缘子									
工作规范及要求	1. 带电作业工作应在良好天气下进行。如遇雷、雨、雪、雾天气不得进行带电作业。风力大于5级时,不宜进行带电作业。湿度大于80%时,若需进行带电作业,应采用具有防潮性能的绝缘工具。 2. 本项作业需6人,其中工作负责人1名,专责监护人1名,地电位电工1人,地面电工3名。 3. 工作负责(监护)人职责:负责本次工作任务的人员分工、工作票的宣读、办理线路停用重合闸、办理工作许可手续、召开工作班前会、负责作业过程中的安全监督、工作中突发情况的处理、工作质量的监督、工作后的总结。 4. 在带电作业中,遇雷、雨、大风或其他任何情况威胁到工作人员的安全时,工作负责人或监护人可根据情况,临时停止工作 给定条件: 1. 工作票已办理,安全措施已经完备(重合闸已停用),工作开始、工作终结时应口头提出申请(调度或考评员)。 2. 安全、正确地使用仪器对绝缘工具进行检测。 3. 必须按工作程序进行操作,工序错误扣除应做项目分值,出现重大人身、器材和操作安全隐患,考评员可下令终止操作(考核)									
考核情景准备	1. 塔形:1000kV交流耐张塔。 2. 所需作业工器具:安全带(含二保绳)2根,I型屏蔽服2套,防潮布2张,万用表1,绝缘电阻检测仪1个,风速仪、温湿度二合一1台,液压丝杠2根,绝缘绳2根,闭式卡1套,滑车1个,护目镜2个,拔销器1把,手动工具1套。 3. 作业现场做好监护工作,作业现场安全措施(围栏等)已全部落实;禁止非作业人员进入现场,工作人员进入作业现场必须戴安全帽。 4. 考生自备工作服,安全帽,线手套									
备 注	1、各项目得分均扣完为止,出现重大人身、器材和操作安全隐患,考评员可下令终止操作。 2、设备、作业环境、安全带、安全帽、工器具、屏蔽服等不符合作业条件考评员可下令终止操作									

续表

序号	项目名称	质量要求	分值	扣分标准	扣分原因	扣分	得分
1	现场复勘	1)工作负责人到作业现场核对线路名称、杆塔编号、现场工作条件、缺陷部位等无误。2)检测风速、湿度等现场气象条件符合作业要求。3)检查工作票填写完整，无涂改，检查是否所列安全措施与现场实际情况相符，必要时予以补充	5	1)未核对线路名称、杆塔编号、现场工作条件、缺陷部位等，扣1分/项。2)未检测风速、湿度等现场气象条件，扣1分/项。3)工作票填写出现涂改，扣0.5分/处；工作票编号有误，扣1分；工作票填写不完整，扣1.5分			
2	工作许可	1)工作负责人联系值班调控人员(裁判)，按工作票内容申请停用线路重合闸。2)汇报内容规范、完整	2	1)未联系调度部门(裁判)停用重合闸，扣2分。2)汇报专业用语不规范或不完整，扣1分			
3	现场布置	正确装设安全围栏并悬挂标示牌：1)安全围栏范围应充分考虑高处坠物，以及对道路交通的影响。2)安全围栏出入口设置合理。3)妥当布置"从此进出""在此工作""从此上下"等标示	3	1)作业现场未装设围栏，扣1分。2)未设立警示牌，扣1分。3)未悬挂登塔作业标志，扣1分			
4	召开班前会	1)全体工作成员全体人员正确佩戴安全帽、工作服。2)工作负责人佩戴红色背心，宣读工作票，明确工作任务及人员分工；讲解工作中的安全措施和技术措施；查(问)全体工作成员精神状态；告知工作中存在的危险点及采取的预控措施。3)全体工作成员在工作票上签名确认	3	1)工作人员着装不整齐，扣0.5分/人。2)未进行分工，扣3分；分工不明确，扣1分。3)现场工作负责人未穿佩安全监护背心，扣1分。4)工作票上工作班成员未签字或签字不全，扣1分			

续表

序号	项目名称	质量要求	分值	扣分标准	扣分原因	扣分	得分
5	工器具检查	1)在防潮苫布上,将工器具按作业要求准备齐备,并分类定置摆放整齐。检查工器具外观和试验合格证,无遗漏。 2)使用绝缘电阻测试仪检测绝缘工具及绝缘绳索的表面绝缘电阻值,方法正确,不得低于 700MΩ。 3)将新绝缘子擦拭干净,外观检查完好,不得有锈蚀、裂纹及破损。使用绝缘电阻测试仪测试其绝缘电阻值,方法正确,不得低于 500 MΩ。 4)使用万用表检测全套屏蔽服内阻,方法正确,不得大于 20Ω。 5)检查人员向工作负责人汇报各项检查结果符合作业要求	7	1)未使用防潮苫布并定置摆放工器具,扣1分。 2)未检查工器具外观及试验合格证,扣0.5分/项。 3)未正确使用检测仪器对工器具进行检测,扣1分/项。 4)汇报检测结果不规范,扣1分;不完整,扣0.5分/项			
6	登塔	1)地电位电工再次核对线路双重名称及相别,检查并确认脚钉齐全、牢固;系好安全带、加挂防坠器;对双保险安全带、防坠器做冲击试验,方法正确;并进行汇报。工作负责人检查并确认地电位电工穿戴的双保险安全带各部件的连接情况良好,包括肩带、胸带、腰带、腿带、后背保护绳、扣和环。 2)背上工具包,携带绝缘传递绳(含绝缘滑车),方法正确。 3)登塔过程中系好防坠落保护装置,脚踩脚钉、手抓主材、匀步登塔至合适位置,系好安全带,脱离防坠器	5	1)中间电位电工未核对线路双重名称、杆号、相别、塔材情况,扣1分/项;核对完未汇报,扣1分。 2)双保险安全带及防坠器未进行冲击试验,扣2分/项。 3)现场工作负责人未对中间电位电工进行安全防护装备进行检查,扣1分。 4)手抓脚钉,扣0.5分/次。 5)滑车传递绳悬挂位置不合理,扣1分。 6)转位时失去安全带保护,扣5分			

续表

序号	项目名称	质量要求	分值	扣分标准	扣分原因	扣分	得分
7	安装工具	1)地面电工使用绝缘传递绳将闭式卡(前卡)、液压丝杠、耐张端部卡等分别起吊给地电位电工。起吊过程平稳、无磕碰、无缠绕,正确使用绳结。 2)地电位电工先在牵引板上安装耐张端部卡,后将闭式卡(后卡)安装在横担侧第3片绝缘子上,并连接好液压丝杠。承力工具各部分安装牢固可靠。 3)检查并确认承力工具各部分安装情况良好,经工作负责人许可后,操作液压丝杠使其逐渐受力,使需更换的绝缘子松弛。两根液压丝杠的受力应均匀	20	1)起吊过程不平稳,出现磕碰、缠绕,扣1分/次。 2)高处坠物,扣2分/次。 3)卡具安装不正确、固定不到位,扣2分。 4)未检查承力工具安装情况,扣3分;检查了未报告,扣1分;报告了但工作负责人未同意即开始收紧丝杠,扣1分。 5)作业过程短接绝缘子片数超过4片,扣3分/次。 6)安装卡具出现绝缘子碰撞破损,扣2分。 7)未均衡收紧丝杠,扣2分			
8	更换绝缘子	3)地电位电工做冲击试验,检查并确认承力工具受力正常,经工作负责人许可后,用绝缘传递绳系好旧绝缘子,取出旧绝缘子两端锁紧销,继续操作并收紧液压丝杠,直至拆除旧绝缘子。两根液压丝杠的受力应均匀,操作手柄不得敲击绝缘子。 2)地面电工用绝缘传递绳的另一端系好新绝缘子,采用旧下、新上的方法,将新绝缘子起吊给地电位电工。起吊过程平稳、无磕碰、无缠绕,正确使用绳结。 3)地电位电工安装新绝缘子,并复位其两端锁紧销	20	1)未检查承力工具受力情况,扣3分;检查了未报告,扣2分;报告了但工作负责人未同意即取出旧绝缘子两端锁紧销,扣1分。 2)未均衡收紧丝杆,扣2分。 3)操作手柄敲击绝缘子,扣1分/次。 4)绳结错误,扣1分。 5)高处坠物,扣2分/次。 6)新旧绝缘子相互碰撞,扣1分。 7)传递绝缘子与塔身相互碰撞,扣1分/次。 8)绝缘绳缠绕,扣2分			

序号	项目名称	质量要求	分值	扣分标准	扣分原因	扣分	得分
9	拆除工具	1)地电位电工检查新绝缘子连接可靠,经工作负责人许可后,操作并松出液压丝杠,使更换的绝缘子逐渐受力。 2)荷载转移完毕后,等电位电工做冲击试验,检查并确认新绝缘子受力情况良好,经工作负责人许可后,拆除系在绝缘子上的绝缘传递绳,并将其系牢于承力工具适当位置,拆除闭式卡(前卡)、液压丝杠、耐张端部卡等承力工具,在地面电工配合下传递至地面。传递过程平稳、无磕碰、无缠绕,正确使用绳结	20	1)未检查新绝缘子连接情况,扣3分;检查了未报告,扣2分;报告了但工作负责人未同意即松液压丝杠,扣1分。 2)未检查新绝缘子受力情况,扣3分;检查了未报告,扣2分;报告了但工作负责人未同意即拆除液压丝杠,扣1分。 3)捆扎工具时,未正确使用绳结,扣1分。 4)高处坠物,扣2分/次。 5)工器具相互碰撞扣1分/次;工器具与带电体或塔身相互碰撞,扣1分/次;绝缘绳缠绕,扣2分			
10	返回地面	地电位电工检查塔上无遗留物,拆除绝缘传递绳。经工作负责人许可后,挂好防坠器,解开并整理好安全带,正确携带绝缘传递绳,脚踩脚钉、手抓主材、匀步下塔至地面	5	1)下塔过程未使用防坠器,扣5分。 2)塔上移位失去安全带保护,扣5分。 3)下塔手抓脚钉,扣1分/次。 4)塔上有遗留物,扣2分			
11	工作结束	1)工作负责人组织全体工作成员整理工器具和材料,将工器具清洁后放入专用的箱(袋)中;清理现场,做到"工完料尽场地清"。 2)召开班后会,工作负责人进行工作总结和点评工作。点评本次工作的施工质量;点评全体工作成员的安全措施落实情况。 3)工作负责人向值班调控人员汇报工作结束,申请恢复线路重合闸,终结工作票	10	1)工器具未清理,扣2分。 2)工器具有遗漏,扣2分。 3)未开班后会,扣10分。 4)未拆除围栏,扣2分。 5)未向调度汇报,扣2分			
	合计		100				

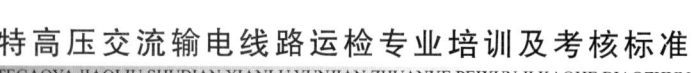

模块4 带电更换1000kV交流输电线路耐张塔导线侧 1~3片玻璃绝缘子培训及考核标准

一、培训标准

(一) 培训要求

模块名称	带电更换1000kV交流输电线路耐张塔导线侧1~3片玻璃绝缘子	培训类别	操作类
培训方式	实操培训	培训学时	21学时
培训目标	1. 掌握沿耐张绝缘子串进、出1000kV强电场时采用"跨二短三"作业方式,以及电位转移的电学意义。 2. 能完成沿耐张绝缘子串进入1000kV等电位作业点。 3. 能独立完成带电更换1000kV交流输电线路耐张塔导线侧1~3片玻璃绝缘子(等电位作业法)。		
培训场地	特高压交流实训线路		
培训内容	采用"跨二短三"作业方式沿耐张绝缘子串进入等电位,采用等电位作业法带电更换1000kV交流输电线路耐张塔导线侧1~3片玻璃绝缘子的操作		
适用范围	特高压交流输电线路检修人员		

(二) 引用规程规范

(1)《1000kV架空输电线路设计规范》(GB 50665—2011)。

(2)《1000kV交流输电线路检修规范》(DL/T 209—2008)。

(3)《1000kV交流输电线路运行规范》(DL/T 307—2010)。

(4)《1000kV交流输电线路带电作业技术导则》(DL/T 392—2015)。

(5)《交流线路带电作业安全距离计算方法》(GB/T 19185—2008)。

(6)《带电作业用绝缘配合导则》(DL/T 867—2004)。

(7)《架空输电线路带电安装导则及作业工具设备》(DL/T 1007—2006)。

(8)《国家电网公司带电作业工作管理规定(试行)》(国家电网生〔2007〕751号)。

(9)《国家电网公司电力安全工作规程(线路部分)》(Q/GDW1799.2—2013)。

(10)《电工术语架空线路》(GB/T 2900.51—1998)。

(11)《电工术语带电作业》(GB/T 2900.55—2002)。

(12)《带电作业工具设备术语》(GB/T 14286—2002)。

(13)《带电作业用工具、装置和设备使用的一般要求》(DL/T 877—2004)。

(14)《带电作业工具、装置和设备预防性试验规程)》(DL/T 976—2005)。

(15)《带电作业用绝缘滑车》(GB/T 13034—2008)。

(16)《带电作业用绝缘绳索》(GB 13035—2008)。

(17)《1000kV 交流带电作业用屏蔽服装》(GB/T 25726—2010)。

(三)培训教学设计

本设计以完成"带电更换1000kV交流输电线路耐张塔导线侧1~3片玻璃绝缘子"为工作任务,按工作任务完成的标准化作业流程来设计各个培训阶段,每个阶段包括了具体的培训目标、培训内容、培训学时、培训方法(培训资源)、培训环境和考核评价等内容,如表2-4-1所示。

表2-4-1 带电更换1000kV交流输电线路耐张塔导线侧1~3片玻璃绝缘子

培训流程	培训目标	培训内容	培训学时	培训方法与资源	培训环境	考核评价
1.理论教学	1.初步掌握沿绝缘子串进出1000kV强电场基本方法。 2.熟悉转移电位的方法。 3.熟悉输电线路耐张单片绝缘子更换方法	1.沿绝缘子进出强电场"跨二短三"作业方式的电学意义。 2.电位转移棒的使用方法。 3.输电线路耐张单片绝缘子更换方法和质量标准	2	培训方法:讲授法。 培训资源:PPT、相关规程规范	多媒体教室	考勤、课堂提问和作业
2.准备工作	能完成作业前准备工作	1.作业现场查勘。 2.编制培训标准化作业卡。 3.填写培训操作工作票。 4.完成本操作的工器具及材料准备	1	培训方法: 1.现场查勘和工器具及材料清理采用现场实操方法。 2.编写作业卡和填写工作票采用讲授方法。 培训资源: 1.1000kV实训线路。 2.特高压工器具库房。 3.空白工作票	1.特高压输电实训线路。 2.多媒体教室	

续表

培训流程	培训目标	培训内容	培训学时	培训方法与资源	培训环境	考核评价
3.作业现场准备	能完成作业现场准备工作	1.作业现场复勘。 2.工作申请。 3.作业现场布置。 4.班前会。 5.工器具及材料检查	1	培训方法： 演示与角色扮演法。 资源： 1000kV实训线路	1000kV实训线路	
4.培训师演示	通过现场观摩，使学员初步领会本任务操作流程	1.等电位电工沿耐张绝缘子串进、出强电场。 2.等电位电工组装工器具。 3.等电位电工完成单片玻璃绝缘子的更换工作	2	培训方法： 演示法。 资源： 1000kV实训线路	1000kV实训线路	
5.学员分组训练	1.能完成进、出1000kV强电场操作。 2.能完成单片玻璃绝缘子的更换工作	1.学员分组（6人一组）训练进、出1000kV强电场和更换绝缘子技能操作。 2.培训师对学员操作进行指导和安全监护	14	培训方法： 角色扮演法。 资源： 1000kV实训线路	1000kV实训线路	采用技能考核评分细则对学员操作评分
6.工作终结	1.使学员进一步辨析操作过程不足之处，便于后期提升。 2.培训学员安全文明生产的工作作风	1.作业现场清理。 2.向调度汇报工作。 3.班后会，对本次工作任务进行点评总结	1	培训方法： 讲授和归纳法	1000kV实训线路	

（四）作业流程

1. 工作任务

带电更换1000kV交流输电线路耐张塔导线侧1~3片玻璃绝缘子。

2. 天气及作业现场要求

（1）带电更换1000kV交流输电线路耐张塔导线侧1~3片玻璃绝缘子应在良好的天气进行。如遇雷电（听见雷声、看见闪电）、雪、雹、雨、雾等，禁止进行带电作业。风力大于5级，不宜进行带电作业；相对湿度大于80%的天气，若需进行带电作业，应采用具有防潮性能的绝缘工具。恶劣天气下必须开展带电抢修时，应组织有关人员充分讨论并编制必要的安全措施，经本单位批准后方可进行。

（2）作业人员精神状态良好，熟悉工作中保证安全的组织措施和技术措施，掌握高处应急救援及触电急救的方法；应持有在有效期内的带电作业资质证书。

（3）工作负责人应事先组织相关人员完成现场勘察，根据勘察结果确定本次作业方法和所需工器具，以及应采取的必要措施，并办理带电作业工作票。

（4）作业现场应合理设置围栏，并妥当布置警示标示牌，禁止非工作人员入内。

（5）本项目须停用线路重合闸装置。

（6）作业方式：等电位作业

（7）工作中安全距离及有效绝缘长度如表2-4-2所示。

表2-4-2　带电更换1000kV交流输电线路耐张塔导线侧1～3片玻璃绝缘子的安全距离

海拔高度/m	等电位电工与接地构架之间的最小安全距离/m		绝缘工器具的最小有效绝缘长度/m	最小组合间隙/m	
	中相	边相		中相	边相
$H \leqslant 1000$	6.8	6.0	6.8	6.9	6.7
$1000 < H \leqslant 2000$	7.4	6.6	7.2	7.6	7.3

注：表中数值不包括人体占位间隙，作业中需考虑人体占位间隙不得小于0.5 m。

（8）等电位电工沿耐张绝缘子串进入等电位时，人体短接绝缘子片数不得多于4片。耐张绝缘子串中扣除人体短接和不良绝缘子片数后，良好绝缘子最少片数应满足表2-4-3的规定。

表2-4-3　最小组合间隙和良好绝缘子的最小片数

海拔高度	单片玻璃绝缘子结构高度（mm）	良好绝缘子串的总长度最小值（m）	良好绝缘子的最少片数
$H \leqslant 1000$	170	7.2	43
	195		37
	205		36
$1000 < H \leqslant 2000$	170	8.0	47
	195		41
	205		39

注：表中数值不包括人体占位间隙，作业中需考虑人体占位间隙不得小于0.5 m。

3. 准备工作

3.1　危险点及其预控措施

（1）危险点——触电伤害

预控措施：

①工作前，工作负责人应与值班调控人员联系，停用线路重合闸，并履行许可手续。

②塔上作业人员登塔前，必须仔细核对线路双重命名、杆塔编号、相别，确认无误后方

可上塔。

③工作中，如遇线路突然停电，作业人员应视其仍然带电。工作负责人应尽快与调控人员联系，值班调控人员未与工作负责人取得联系前不准强送电。

④绝缘工具及绝缘绳索不得损坏、受潮、变形、失灵，不准使用非绝缘绳索（如棉纱绳、白棕绳、钢丝绳）。

⑤地面电工操作绝缘工具时应戴清洁、干燥的手套，进入作业现场应将使用的带电作业工具放置在防潮的帆布或绝缘垫上，防止绝缘工具在使用中脏污和受潮。

⑥等电位电工应穿着阻燃内衣，衣服外面应穿戴合格全套屏蔽服（包括帽、衣裤、手套、袜和鞋），且各部分应连接良好。

⑦等电位电工沿绝缘子串移动时，手与脚的位置必须保持对应一致，且人体和工具短接的绝缘子片数应符合表2-4-3规定。

⑧等电位电工采用"跨二短三"作业方式（也称自由作业法）进入强电场。当作业人员平行移动至距导线侧均压环三片绝缘子处，应停止移动，利用电位转移棒进行电位转移，电位转移棒长度为0.4m，电位转移时，人体面部与带电体距离不得小于0.5m。

⑨用绝缘绳索传递大件金属物品时，地面作业人员应将金属物品接地后再接触。

⑩带电作业过程中，工作负责人（监护人）应对作业人员进行不间断监护，随时纠正其不规范或违章动作。重点关注高处作业人员，使其保持足够的安全距离及组合间隙（符合表2-4-2的规定），禁止同时接触两个非连通的带电体或带电体与接地体。

（2）危险点——高处坠落

预控措施：

①高处作业人员登高前，必须具备符合本项作业要求的身体状况、精神状态和技能素质。

②高处作业人员应使用双保险安全带。上、下塔时，应手抓主材、脚踩脚钉，匀速行进。

③等电位电工作业前应认真检查液压丝杠、闭式卡、导线端部卡等，确保承力工具合格；沿绝缘子串移动时，手与脚的位置必须保持对应一致，安全带应系挂在手扶的绝缘子串上，并同步移动；更换绝缘子时，承力工具安装应可靠，荷载转移前、后应做冲击试验判定其可靠性，并及时向工作负责人汇报，得到工作负责人许可后方可实施。

④监护人员应随时纠正其不规范或违章动作，重点关注高处作业人员在转位的过程中不得失去安全带或绝缘后备保护绳的保护，严禁低挂高用。

（3）危险点——高处坠物伤人。

预控措施：

①高处作业人员的个人工具及零星材料应装入工具袋，严禁在高处浮置物件、口中含物。

②地面作业人员必须正确佩戴安全帽，正确使用绳结，与作业点垂直下方距离不得小于

坠落半径。

③作业现场设置围栏并挂好警示标示牌。监护人员应随时注意，禁止非工作人员及车辆进入作业区域。

3.2 工器具及材料选择

带电更换1000kV交流输电线路耐张塔导线侧1~3片玻璃绝缘子所需工器具及材料见表2-4-4。工器具出库前，应认真核对工器具的使用电压等级和试验周期，并检查确认外观良好、连接牢固、转动灵活，且符合本次工作任务的要求；工器具出库后，应存放在工具袋或工具箱内进行运输，防止脏污、受潮；金属工具和绝缘工器具应分开装运，防止因混装运输导致工器具变形、损伤等现象发生。

表2-4-4 带电更换1000kV交流输电线路耐张塔导线侧1~3片玻璃绝缘子所需工器具及材料表

序号	名称	规格型号	单位	数量	备注
1	屏蔽服	Ⅰ型	套	2	个人防护用具
2	导电鞋	尺码视穿着人员而定	双	2	个人防护用具
3	阻燃内衣	纯桑蚕丝	套	2	个人防护用具
4	双保险安全带	背带式	根	2	个人防护用具
5	安全帽		顶	7	个人防护用具
6	护目镜		副	2	个人防护用具
7	绝缘传递绳	ϕ14mm，长度与起吊高度匹配	根	1	绝缘工具
8	绝缘后备保护绳	ϕ16mm	根	3	绝缘工具
9	绝缘绳套	ϕ14 mm	根	2	绝缘工具
10	绝缘滑车	1T	个	1	绝缘工具
11	导线端部卡		个	1	金属工具
12	液压丝杠	8T	根	2	金属工具
13	闭式卡	Tc4	套	1	金属工具
14	电位转移棒		根	2	其他工具
15	拔销器		把	1	其他工具
16	绝缘电阻测试仪	2500V，电极宽2cm、极间宽2cm	套	1	其他工具
17	万用表		套	1	其他工具
18	风速、温湿度测试仪		只	1	其他工具
19	安全围网		套	若干	其他工具
20	警示标示牌	"在此工作""从此进出""车辆慢行""车辆绕行"	套	1	其他工具
21	红马甲	"工作负责人""专责监护人"	件	1	其他工具

续表

序号	名称	规格型号	单位	数量	备注
22	防潮苫布	3m×3m	块	2	其他工具
23	个人工具	扳手、老虎钳	套	1	其他工具
24	防坠器	与杆塔防坠落装置型号对应	只	2	其他工具
25	毛巾	棉质	条	1	其他工具
26	绝缘子		片	1	材料

3.3 作业人员分工

本任务作业人员分工如表2-4-5所示。

表2-4-5 带电更换1000kV交流输电线路耐张塔导线侧1~3片玻璃绝缘子人员分工表

序号	工作岗位	数量(人)	工作职责
1	工作负责人	1	负责本次工作任务的人员分工、工作票的宣读、办理线路停用重合闸、办理工作许可手续、召开工作班前会、工作中突发情况的处理、工作质量的监督、工作后的总结
2	专责监护人	1	负责作业现场的安全把控
3	等电位电工	1	负责工器具安装及绝缘子更换工作
4	地面电工	3	负责本次作业过程的地面辅助工作

4. 工作程序

本任务工作流程如表2-4-6所示。

表2-4-6 带电更换1000kV交流输电线路耐张塔导线侧1~3片玻璃绝缘子工作流程表

序号	作业内容	作业步骤及标准	安全措施及注意事项	责任人
1	现场复勘	工作负责人负责完成以下工作： (1)现场核对线路名称、杆塔编号,双重编号无误;基础及杆塔完好无异常;交叉跨越距离符合安全要求;确认缺陷情况及导地线规格型号等。 (2)检测风速、湿度等现场气象条件符合作业要求。 (3)检查地形环境符合作业要求。 (4)检查工作票所列安全措施与现场实际情况相符,必要时予以补充	(1)正确穿戴安全帽、工作服、工作鞋、劳保手套。 (2)不得在危及作业人员安全的气象条件下作业。 (3)严禁非工作人员、车辆进入作业现场	
2	工作许可	(1)工作负责人负责联系值班调控人员,按工作票内容申请停用线路重合闸。 (2)经值班调控人员许可后,方可开始带电作业工作	不得未经值班调控人员许可即开始工作	

续表

序号	作业内容	作业步骤及标准	安全措施及注意事项	责任人
3	现场布置	正确装设安全围栏并悬挂标示牌： (3)安全围栏范围应充分考虑高处坠物，以及对道路交通的影响。 (2)安全围栏出入口设置合理。 (3)妥当布置"从此进出""在此工作""车辆慢行"或"车辆绕行"等标示	对道路交通安全影响不可控时，应及时联系交通管理部门强化现场交通安全管控	
4	召开班前会	(3)全体工作成员列队。 (2)工作负责人宣读工作票，明确工作任务及人员分工；讲解工作中的安全措施和技术措施；查(问)全体工作成员精神状态；告知工作中存在的危险点及采取的预控措施。 (3)全体工作成员在工作票上签名确认	(3)工作票填写、签发和许可手续规范，签名完整。 (2)全体工作成员精神状态良好。 (3)全体工作成员明确任务分工、安全措施和技术措施	
5	检查工器具	(3)在防潮苫布上，将工器具按作业要求准备齐备，并分类定置摆放整齐。检查工器具外观和试验合格证，无遗漏。 (2)使用绝缘电阻测试仪检测绝缘工具及绝缘绳索的表面绝缘电阻值，方法正确，不得低于700MΩ。 (3)将新绝缘子擦拭干净，外观检查完好，不得有锈蚀、裂纹及破损。使用绝缘电阻测试仪测试其绝缘电阻值，方法正确，不得低于500 MΩ。 (4)使用万用表检测全套屏蔽服内阻，方法正确，不得大于20Ω。 (5)检查人员向工作负责人汇报各项检查结果符合作业要求	(3)防潮苫布数量足够，设置位置合理，保持清洁、干燥。 (2)金属、绝缘工器具在使用前，应仔细检查其是否无损伤、受潮、变形、失灵现象，合格证在有效期内。 (3)绝缘工具及绝缘绳索检测合格	
6	登塔	(3)等电位电工再次核对线路双重名称及相别，检查并确认脚钉齐全、牢固；系好安全带、加挂防坠器；对安全带、防坠器做冲击试验，方法正确；工作负责人检查并确认等电位电工穿戴的双保险安全带各部件的连接情况良好，包括肩带、胸带、腰带、腿带、后背保护绳、扣和环。 (2)背上工具包，携带绝缘传递绳(含绝缘滑车)，方法正确。 (3)将安全带主带和后备保护绳斜跨肩上。 (4)清洁鞋底，经工作负责人许可后登塔。 (5)脚踩脚钉、手抓主材、匀步登塔至横担适当位置，系好安全带，脱离防坠器	(3)安全带、防坠器冲击试验合格。 (2)防止安全带、绝缘传递绳钩挂塔材。 (3)人体与导线保持的最小安全距离应符合表2-4-2的规定。 (4)禁止手抓脚钉。 (5)正确使用防坠器。 (6)转位时，不得失去安全带的保护	

续表

序号	作业内容	作业步骤及标准	安全措施及注意事项	责任人
7	进入强电场	(3)等电位电工携带绝缘传递绳,转位至作业相耐张绝缘子串挂点处,将安全带主带系挂在绝缘子串连接金具上,将安全带后备保护绳系留在横担适当位置。 (2)再次检查并确认屏蔽服各部分连接良好、绝缘子串连接良好、及故障绝缘子位置,经工作负责人许可后,双手抓扶一串,双脚踩另一串,采用"跨二短三"作业方式,沿绝缘子串进入强电场。当作业人员到达导线侧均压环外三片绝缘子处时,应停止移动,使用电位转移棒进行电位转移,实现作业人员与导线等电位。 (3)转移安全带主带,系挂在导线适当位置,越过均压环到达导线上作业点,在导线适当位置布置绝缘传递绳,牢固可靠,方便作业	(3)防止安全带、绝缘传递绳钩挂塔材。 (2)转位时,不得失去安全带的保护。 (3)人体与接地体之间的安全距离、人体与接地体和带电体间的组合间隙不得小于表2-4-2的规定。 (4)转移电位时,人体面部与带电体距离不得小于0.5m	
8	安装承力工具	(3)地面电工使用绝缘传递绳将闭式卡(前卡)、液压丝杠、导线端部卡等分别起吊给等电位电工。起吊过程平稳、无磕碰、无缠绕,正确使用绳结。 (2)等电位电工在导线侧合适位置上安装导线端部卡,将闭式卡(前卡)安装在导线侧第3片绝缘子上,连接好液压丝杠。承力工具各部分安装牢固可靠。 (3)检查并确认承力工具各部分安装情况良好,经工作负责人许可后,操作液压丝杠使其逐渐受力,使需更换的绝缘子松弛。两根液压丝杠的受力应均匀	(3)人体与接地体之间的安全距离不得小于表2-4-2的规定。 (2)防止高处坠物。 (3)扣除劣质绝缘子、人体操作和工具短接的绝缘子后,良好绝缘子片数应符合表2-4-3的规定	
9	更换绝缘子	(3)等电位电工做冲击试验,检查并确认承力工具受力正常,经工作负责人许可后,用绝缘传递绳系牢旧绝缘子,取出旧绝缘子两端锁紧销,继续操作并收紧液压丝杠,直至拆除旧绝缘子。两根液压丝杠的受力应均匀,操作手柄不得敲击绝缘子。 (2)地面电工用绝缘传递绳的另一端系牢新绝缘子,采用旧下、新上的方法,将新绝缘子传给中间电位电工。起吊过程平稳、无磕碰、无缠绕,正确使用绳结。 (3)安装新绝缘子,并复位其两端锁紧销	(3)人体与接地体之间的安全距离不得小于表2-4-2的规定。 (2)防止高处坠物	

续表

序号	作业内容	作业步骤及标准	安全措施及注意事项	责任人
10	拆除工具	(3)等电位电工检查新绝缘子连接可靠,经工作负责人许可后,操作并松出液压丝杠,使更换的绝缘子逐渐受力。 (2)荷载转移完毕后,等电位电工做冲击试验,检查并确认新绝缘子受力情况良好,经工作负责人许可后,拆除系在绝缘子上的绝缘传递绳,并将其系牢于承力工具适当位置,拆除液压丝杠、闭式卡、导线端部卡等承力工具,在地面电工配合下传递至地面。传递过程平稳、无磕碰、无缠绕,正确使用绳结	(3)人体与接地体之间的安全距离不得小于表2-4-2的规定。 (2)防止高处坠物	
11	退出强电场	(3)等电位电工检查作业部位无遗留物后,拆除并整理绝缘传递绳。 (2)转移安全带主带,系挂在绝缘子串适当位置,越过均压环回到绝缘子串上,将电位转移棒钩紧均压环适当位置,沿绝缘子串向横担侧移动到均压环外三片绝缘子时,停止移动;一只手抓紧绝缘子,另一只手握紧电位转移棒,利用电位转移棒快速脱离等电位。 (3)按照"跨二短三"作业方式沿绝缘子串到达横担	(3)人体与接地体和带电体间的组合间隙不得小于表2-4-2的规定。 (2)转位时,不得失去安全带的保护	
12	撤离杆塔	等电位电工检查塔上无遗留物,经工作负责人许可后,挂好防坠器,解开并整理好安全带,正确携带绝缘传递绳,脚踩脚钉、手抓主材,匀步下塔至地面	(3)转位时不得失去安全带保护。 (2)防止手滑脱、脚踏空,禁止手抓脚钉。 (3)正确使用防坠器。 (4)防止绝缘传递绳、安全带钩挂塔材或脚钉	
13	工作结束	(3)工作负责人组织全体工作成员整理工器具和材料,将工器具清洁后放入专用的箱(袋)中;清理现场,做到"工完料尽场地清"。 (2)召开班后会,工作负责人进行工作总结和点评工作。点评本次工作的施工质量;点评全体工作成员的安全措施落实情况。 (3)工作负责人向值班调控人员汇报工作结束,终结工作票		

二、考核标准

特高压交流输电线路运检技能考核评分细则

考生填写栏	编号：	姓名：	所在岗位：	单位：	日期：	年 月 日			
考评员填写栏	成绩：	考评员：	考评组长：	开始时间：	结束时间：	操作时长：			
考核模块	带电更换1000kV交流输电线路耐张塔导线侧1~3片玻璃绝缘子		考核对象	特高压交流输电线路检修人员	考核方式	操作	考核时限	90min	
任务描述	带电更换1000kV交流输电线路耐张塔导线侧1~3片玻璃绝缘子								
工作规范及要求	1. 带电作业工作应在良好天气下进行。如遇雷、雨、雪、雾天气不得进行带电作业。风力大于5级时，不宜进行带电作业。湿度大于80%时，若需进行带电作业，应采用具有防潮性能的绝缘工具。 2. 本项作业需6人，其中工作负责人1名，专责监护人1名，等电位电工1人，地面电工4名。 3. 工作负责（监护）人职责：负责本次工作任务的人员分工、工作票的宣读、办理线路停用重合闸、办理工作许可手续、召开工作班前会、负责作业过程中的安全监督、工作中突发情况的处理、工作质量的监督、工作后的总结。 4. 在带电作业中，遇雷、雨、大风或其他任何情况威胁到工作人员的安全时，工作负责人或监护人可根据情况，临时停止工作 给定条件： 1. 工作票已办理，安全措施已经完备（重合闸已停用），工作开始、工作终结时应口头提出申请（调度或考评员）。 2. 安全、正确地使用仪器对绝缘工具进行检测。 3. 必须按工作程序进行操作，工序错误扣除应做项目分值，出现重大人身、器材和操作安全隐患，考评员可下令终止操作（考核）								
考核情景准备	1. 塔形：1000kV交流耐张塔； 2. 所需作业工器具：安全带（含二保绳）2根，Ⅰ屏蔽服2套，防潮布2张，万用表1，绝缘电阻检测仪1个，风速仪、温湿度二合一1台，液压丝杠2根，绝缘绳2根，闭式卡1套，滑车1个，电位转移棒2把，护目镜2个，拔销器1把，手动工具1套。 3. 作业现场做好监护工作，作业现场安全措施（围栏等）已全部落实；禁止非作业人员进入现场，工作人员进入作业现场必须戴安全帽。 4. 考生自备工作服，安全帽，线手套								
备注	1. 各项目得分均扣完为止，出现重大人身、器材和操作安全隐患，考评员可下令终止操作。 2. 设备、作业环境、安全带、安全帽、工器具、屏蔽服等不符合作业条件考评员可下令终止操作								

续表

序号	项目名称	质量要求	分值	扣分标准	扣分原因	扣分	得分
1	现场复勘	1)工作负责人到作业现场核对线路名称、杆塔编号、现场工作条件、缺陷部位等无误。 2)检测风速、湿度等现场气象条件符合作业要求。 3)检查工作票填写完整,无涂改,检查是否所列安全措施与现场实际情况相符,必要时予以补充	5	1)未核对线路名称、杆塔编号、现场工作条件、缺陷部位等,扣1分/项。 2)未检测风速、湿度等现场气象条件,扣1分/项。 3)工作票填写出现涂改,扣0.5分/处;工作票编号有误,扣1分;工作票填写不完整,扣1.5分			
2	工作许可	1)工作负责人联系值班调控人员(裁判),按工作票内容申请停用线路重合闸。 2)汇报内容规范、完整	2	1)未联系调度部门(裁判)停用重合闸,扣2分。 2)汇报专业用语不规范或不完整,扣1分			
3	现场布置	正确装设安全围栏并悬挂标示牌: 1)安全围栏范围应充分考虑高处坠物,以及对道路交通的影响。 2)安全围栏出入口设置合理。 3)妥当布置"从此进出""在此工作""从此上下"等标示	3	1)作业现场未装设围栏,扣1分。 2)未设立警示牌,扣1分。 3)未悬挂登塔作业标志,扣1分			
4	召开班前会	1)全体工作成员全体人员正确佩戴安全帽、工作服。 2)工作负责人穿红色背心,宣读工作票,明确工作任务及人员分工;讲解工作中的安全措施和技术措施;查(问)全体工作成员精神状态;告知工作中存在的危险点及采取的预控措施。 3)全体工作成员在工作票上签名确认	3	1)工作人员着装不整齐,扣0.5分/人。 2)未进行分工,扣3分;分工不明确,扣1分。 3)现场工作负责人未穿安全监护背心,扣1分。 4)工作票上工作班成员未签字或签字不全,扣1分			

续表

序号	项目名称	质量要求	分值	扣分标准	扣分原因	扣分	得分
5	工器具检查	1)在防潮苫布上,将工器具按作业要求准备齐备,并分类定置摆放整齐。检查工器具外观和试验合格证,无遗漏。 2)使用绝缘电阻测试仪检测绝缘工具及绝缘绳索的表面绝缘电阻值,方法正确,不得低于700MΩ。 3)将新绝缘子擦拭干净,外观检查完好,不得有锈蚀、裂纹及破损。使用绝缘电阻测试仪测试其绝缘电阻值,方法正确,不得低于500MΩ。 4)使用万用表检测全套屏蔽服内阻,方法正确,不得大于20Ω。 5)检查人员向工作负责人汇报各项检查结果符合作业要求	7	1)未使用防潮苫布并定置摆放工器具,扣1分。 2)未检查工器具外观及试验合格证,扣0.5分/项。 3)未正确使用检测仪器对工器具进行检测,扣1分/项。 4)汇报检测结果不规范,扣1分;不完整,扣0.5分/项			
6	登塔	1)等电位电工再次核对线路双重名称及相别,检查并确认脚钉齐全、牢固;系好安全带、加挂防坠器;对双保险安全带、防坠器做冲击试验,方法正确;并进行汇报。工作负责人检查并确认等电位电工穿戴的双保险安全带各部件的连接情况良好,包括肩带、胸带、腰带、腿带、后背保护绳、扣和环。 2)背上工具包,携带绝缘传递绳(含绝缘滑车),方法正确。 3)登塔过程中系好防坠落保护装置;脚踩脚钉、手抓主材、匀步登塔至合适位置,系好安全带,脱离防坠器	5	1)等电位电工未核对线路双重名称、杆号、相别、塔材情况,扣1分/项;核对完未汇报,扣1分。 2)双保险安全带及防坠器未进行冲击试验,扣2分/项。 3)现场工作负责人未对中间电位电工进行安全防护装备进行检查,扣1分。 4)手抓脚钉,扣0.5分/次。 5)滑车传递绳悬挂位置不合理,扣1分。 6)转位时失去安全带保护,扣5分			

续表

序号	项目名称	质量要求	分值	扣分标准	扣分原因	扣分	得分
7	进入强电场	1)等电位电工携带绝缘传递绳,转位至作业相耐张绝缘子串挂点处,将安全带主带系挂在绝缘子串连接金具上,将安全带后备保护绳系留在横担适当位置。 2)再次检查并确认屏蔽服各部分连接良好、绝缘子串连接良好及故障绝缘子位置,经工作负责人许可后,双手抓扶一串,双脚踩另一串,采用"跨二短三"作业方式,沿绝缘子串进入强电场。当作业人员到达导线侧均压环外三片绝缘子处时,应停止移动,使用电位转移棒进行电位转移,实现作业人员与导线等电位。 3)转移安全带主带,系挂在导线适当位置,越过均压环到达导线上作业点,在导线适当位置布置绝缘传递绳,牢固可靠,方便作业	5	1)安全带后背保护绳系留位置不合理、使用不规范,扣2分。 2)等电位电工未检查屏蔽服连接情况、绝缘子串连接情况及故障绝缘子位置,扣1分/项。 3)未得到工作负责人许可就进入强电场,扣5分。 4)等电位电工进入等电位动作不正确,反复放电,扣2分/次。 5)电位转移动作不正确,扣3分;未汇报,扣2分,汇报了工作负责人未同意即进行电位转移,扣1分。 6)绝缘传递绳安装位置不合理,扣1分。 7)高处坠物,扣2分/次。 8)转位时失去安全带保护,扣5分			
8	安装工具	1)地面电工使用绝缘传递绳将闭式卡(前卡)、液压丝杠、导线端部卡等分别传至等电位电工。起吊过程平稳、无磕碰、无缠绕,正确使用绳结。 2)等电位电工先在导线侧合适位置上安装导线端部卡,后将闭式卡(前卡)安装在导线侧第3片绝缘子上,并连接好液压丝杠。承力工具各部分安装牢固可靠。 3)检查并确认承力工具各部分安装情况良好,经工作负责人许可后,操作液压丝杠使其逐渐受力,使需更换的绝缘子松弛。两根液压丝杠的受力应均匀	15	1)起吊过程不平稳,出现磕碰、缠绕,扣1分/次。 2)高处坠物,扣2分/次。 3)卡具安装不正确、固定不到位,扣2分。 4)未检查承力工具安装情况,扣3分;检查了未报告,扣1分;报告了但工作负责人未同意即开始收紧丝杠,扣1分。 5)作业过程短接绝缘子片数超过4片,扣3分/次。 6)安装卡具出现绝缘子碰撞破损,扣2分。 7)未均衡收紧丝杠,扣2分			

续表

序号	项目名称	质量要求	分值	扣分标准	扣分原因	扣分	得分
9	更换绝缘子	1)等电位电工做冲击试验,检查并确认承力工具受力正常,经工作负责人许可后,用绝缘传递绳系好旧绝缘子,取出旧绝缘子两端锁紧销,继续操作并收紧液压丝杠,直至拆除旧绝缘子。两根液压丝杠的受力应均匀,操作手柄不得敲击绝缘子。 2)地面电工用绝缘传递绳的另一端系好新绝缘子,采用旧下、新上的方法,将新绝缘子传给中间电位电工。起吊过程平稳、无磕碰、无缠绕,正确使用绳结。 3)安装新绝缘子,并复位其两端锁紧销	20	1)未检查承力工具受力情况,扣3分;检查了未报告,扣2分;报告了但工作负责人未同意即取出旧绝缘子两端锁紧销,扣1分。 2)未均衡收紧丝杆,扣2分。 3)操作手柄敲击绝缘子,扣1分/次。 4)绳结错误,扣1分。 5)高处坠物,扣2分/次。 6)新旧绝缘子相互碰撞,扣1分。 7)传递绝缘子与塔身相互碰撞,扣1分/次。 8)绝缘绳缠绕,扣2分			
10	拆除工具	1)等电位电工检查新绝缘子连接可靠,经工作负责人许可后,操作并松出液压丝杠,使更换的绝缘子逐渐受力。 2)荷载转移完毕后,等电位电工做冲击试验,检查并确认新绝缘子受力情况良好,经工作负责人许可后,拆除系在绝缘子上的绝缘传递绳,并将其系牢于承力工具适当位置,拆除液压丝杠、闭式卡、导线端部卡等承力工具,在地面电工配合下传递至地面。传递过程平稳、无磕碰、无缠绕,正确使用绳结	15	1)未检查新绝缘子连接情况,扣3分;检查了未报告,扣2分;报告了但工作负责人未同意即松液压丝杠,扣1分。 2)未检查新绝缘子受力情况,扣3分;检查了未报告,扣2分;报告了但工作负责人未同意即拆除液压丝杠,扣1分。 3)捆扎工具时,未正确使用绳结,扣1分。 4)高处坠物,扣2分/次。 5)工器具相互碰撞扣,扣1分/次;工器具与带电体或塔身相互碰撞,扣1分/次;绝缘绳缠绕,扣2分			

续表

序号	项目名称	质量要求	分值	扣分标准	扣分原因	扣分	得分
11	退出强电场	1)等电位电工检查作业部位无遗留物后,拆除并整理绝缘传递绳。 2)转移安全带主带,系挂在绝缘子串适当位置,越过均压环回到绝缘子串上,将电位转移棒钩紧均压环适当位置,沿绝缘子串向横担侧移动到均压环外三片绝缘子时,停止移动;一只手抓紧绝缘子,另一只手握紧电位转移棒,利用电位转移棒快速脱离等电位。 3)按照"跨二短三"作业方式沿绝缘子串到达横担	5	1)未向工作负责人申请即进行电位转移扣2分;申请了但未得同意即开始扣1分。 2)申请电位转移位置不合适扣1分。 3)等电位电工退出强电场动作不正确,反复放电扣2分。 4)未有效控制后备保护绳扣1分			
12	返回地面	塔上电工检查塔上无遗留物后,向工作负责人汇报,得到工作负责人同意后携带绝缘传递绳下塔	5	1)下塔过程未使用防坠装置扣2分。 2)塔上移位失去安全带保护的扣2分。 3)下塔抓塔钉,每处扣1分。 4)塔上有遗留物的,扣2分			
13	工作结束	1)工作负责人组织全体工作成员整理工器具和材料,将工器具清洁后放入专用的箱(袋)中;清理现场,做到"工完料尽场地清"。 2)召开班后会,工作负责人进行工作总结和点评工作。点评本次工作的施工质量;点评全体工作成员的安全措施落实情况。 3)工作负责人向值班调控人员汇报工作结束,申请恢复线路重合闸,终结工作票	10	1)工器具未清理扣2分。 2)工器具有遗漏扣2分。 3)未开班后会不得2分。 4)未拆除围栏扣2分。 5)未向调度汇报不得2分			
	合计		100				

模块 5　带电更换 1000kV 交流输电线路耐张玻璃绝缘子串任意单片绝缘子培训及考核标准

一、培训标准

（一）培训要求

模块名称	带电更换1000kV交流输电线路耐张玻璃绝缘子串任意单片绝缘子	培训类别	操作类
培训方式	实操培训	培训学时	21学时
培训目标	1.掌握沿耐张绝缘子串进、出1000kV强电场时采用"跨二短三"作业方式的电学意义。 2.能完成沿耐张绝缘子串进入1000kV中间电位作业点。 3.能独立完成带电更换1000kV交流输电线路耐张玻璃绝缘子串任意单片玻璃绝缘子的操作（中间电位作业法）		
培训场地	特高压交流实训线路		
培训内容	采用"跨二短三"作业方式沿耐张绝缘子串进入强电场，采用中间电位作业法带电更换1000kV交流输电线路耐张玻璃绝缘子串任意单片玻璃绝缘子		
适用范围	特高压交流输电线路检修人员		

（二）引用的规程规范

（1）《1000kV架空输电线路设计规范》（GB50665—2011）。

（2）《1000kV交流输电线路检修规范》（DL/T 209—2008）。

（3）《1000kV交流输电线路运行规范》（DL/T 307—2010）。

（4）《1000kV交流输电线路带电作业技术导则》（DL/T 392—2015）。

（5）《交流线路带电作业安全距离计算方法》（GB/T 19185—2008）。

（6）《带电作业用绝缘配合导则》（DL/T 867—2004）。

（7）《架空输电线路带电安装导则及作业工具设备》（DL/T 1007—2006）。

（8）《国家电网公司带电作业工作管理规定（试行）》（国家电网生〔2007〕751号）。

（9）《国家电网公司电力安全工作规程（线路部分）》（Q/GDW1799.2—2013）。

（10）《电工术语架空线路》（GB/T 2900.51—1998）。

（11）《电工术语带电作业》（GB/T 2900.55—2002）。

（12）《带电作业工具设备术语》（GB/T 14286—2002）。

（13）《带电作业用工具、装置和设备使用的一般要求》（DL/T 877—2004）。

（14）《带电作业工具、装置和设备预防性试验规程）》（DL/T 976—2005）。

（15）《带电作业用绝缘滑车》（GB/T 13034—2008）。

（16）《带电作业用绝缘绳索》（GB 13035—2008）。

（17）《1000kV交流带电作业用屏蔽服装》（GB/T 25726—2010）。

（三）培训教学设计

本设计以完成"带电更换1000kV交流输电线路耐张玻璃绝缘子串任意单片绝缘子"为工作任务，按工作任务完成的标准化作业流程来设计各个培训阶段，每个阶段包括了具体的培训目标、培训内容、培训学时、培训方法（培训资源）、培训环境和考核评价等内容，如表2-5-1所示。

表2-5-1 带电更换1000kV交流输电线路耐张玻璃绝缘子串任意单片玻璃绝缘子

培训流程	培训目标	培训内容	培训学时	培训方法与资源	培训环境	考核评价
1.理论教学	1.初步掌握沿绝缘子串进出1000kV强电场基本方法。2.熟悉输电线路耐张单片玻璃绝缘子更换方法	1.沿绝缘子进出强电场"跨二短三"作业方式的电学意义。2.输电线路耐张单片玻璃绝缘子更换方法和质量标准	2	培训方法：讲授法。培训资源：PPT、相关规程规范	多媒体教室	考勤、课堂提问和作业
2.准备工作	能完成作业前准备工作	1.作业现场查勘。2.编制培训标准化作业卡。3.填写培训操作工作票。4.完成本操作的工器具及材料准备	1	培训方法：1.现场查勘和工器具及材料清理采用现场实操方法。2.编写作业卡和填写工作票采用讲授方法。培训资源：1.1000kV实训线路。2.特高压工器具库房。3.空白工作票	1.特高压输电实训线路。2.多媒体教室	

续表

培训流程	培训目标	培训内容	培训学时	培训方法与资源	培训环境	考核评价
3.作业现场准备	能完成作业现场准备工作	1.作业现场复勘。 2.工作申请。 3.作业现场布置。 4.班前会。 5.工器具及材料检查	1	培训方法:演示与角色扮演法。 资源:1000kV实训线路	1000kV实训线路	
4.培训师演示	通过现场观摩,使学员初步领会本任务操作流程	1.中间电位电工沿耐张绝缘子串进、出强电场。 2.中间电位电工组装工器具。 3.中间电位电工完成单片玻璃绝缘子的更换工作	2	培训方法:演示法。 资源:1000kV实训线路	1000kV实训线路	
5.学员分组训练	1.能完成进、出1000kV强电场操作。 2.能完成单片玻璃绝缘子的更换工作	1.学员分组(6人一组)训练进、出1000kV强电场和更换绝缘子技能操作。 2.培训师对学员操作进行指导和安全监护	14	培训方法:角色扮演法。 资源:1000kV实训线路	1000kV实训线路	采用技能考核评分细则对学员操作评分
6.工作终结	1.使学员进一步辨析操作过程不足之处,便于后期提升。 2.培训学员安全文明生产的工作作风	1.作业现场清理。 2.向调度汇报工作终结。 3.班后会,对本次工作任务进行点评总结	1	培训方法:讲授和归纳法	1000kV实训线路	

(四)作业流程

1. 工作任务

带电更换1000kV交流输电线路耐张玻璃绝缘子串任意单片玻璃绝缘子。

2. 天气及作业现场要求

(1)带电更换1000kV交流输电线路耐张玻璃绝缘子串任意单片玻璃绝缘子应在良好的天气进行。

如遇雷电(听见雷声、看见闪电)、雪、雹、雨、雾等,禁止进行带电作业。风力大于5级,不宜进行带电作业;相对湿度大于80%的天气,若需进行带电作业,应采用具有防潮性能的绝缘工具。恶劣天气下必须开展带电抢修时,应组织有关人员充分讨论并编制必要的安全措施,经本单位批准后方可进行。

(2)作业人员精神状态良好,熟悉工作中保证安全的组织措施和技术措施,掌握高处应急救援及触电急救的方法;应持有在有效期内的带电作业资质证书。

(3)工作负责人应事先组织相关人员完成现场勘察,根据勘察结果确定本次作业方法和所需工器具,以及应采取的必要措施,并办理带电作业工作票。

(4)作业现场应合理设置围栏,并妥当布置警示标示牌,禁止非工作人员入内。

(5)本项目须停用线路重合闸装置。

(6)作业方式:中间电位作业。

(7)工作中安全距离及有效绝缘长度如表2-5-2所示。

表2-5-2 带电更换1000kV交流输电线路耐张玻璃绝缘子串任意单片玻璃绝缘子的安全距离

海拔高度/m	中间电位作业人员与带电体之间的最小安全距离/m		绝缘工器具的最小有效绝缘长度/m	最小组合间隙/m	
	中相	边相		中相	边相
$H \leq 1000$	6.8	6.0	6.8	6.9	6.7
$1000 < H \leq 2000$	7.4	6.6	7.2	7.6	7.3

注:表中数值不包括人体占位间隙,作业中需考虑人体占位间隙不得小于0.5m。

(8)中间电位作业人员沿耐张绝缘子串进入1000kV强电场时,人体短接绝缘子片数不得多于4片。耐张绝缘子串中扣除人体短接和不良绝缘子片数后,良好绝缘子最少片数应满足表2-5-3的规定。

表2-5-3 最小组合间隙和良好绝缘子的最小片数

海拔高度	单片玻璃绝缘子结构高度(mm)	良好绝缘子串的总长度最小值(m)	良好绝缘子的最少片数
$H \leq 1000$	170	7.2	43
	195		37
	205		36
$1000 < H \leq 2000$	170	8.0	47
	195		41
	205		39

注:表中数值不包括人体占位间隙,作业中需考虑人体占位间隙不得小于0.5m。

3. 准备工作

3.1 危险点及其预控措施

（1）危险点——触电伤害

预控措施：

①工作前，工作负责人应与值班调控人员联系，停用线路重合闸，并履行许可手续。

②塔上作业人员登塔前，必须仔细核对线路双重命名、杆塔编号、相别，确认无误后方可上塔。

③工作中，如遇线路突然停电，作业人员应视其仍然带电。工作负责人应尽快与值班调控人员联系，值班调控人员未与工作负责人取得联系前不准强送电。

④绝缘工具及绝缘绳索不得损坏、受潮、变形、失灵，不准使用非绝缘绳索（如棉纱绳、白棕绳、钢丝绳）。

⑤地面电工操作绝缘工具时应戴清洁、干燥的手套，进入作业现场应将使用的带电作业工具放置在防潮的帆布或绝缘垫上，防止绝缘工具在使用中脏污和受潮。

⑥中间电位电工应穿着阻燃内衣，衣服外面应穿戴合格全套屏蔽服（包括帽、衣裤、手套、袜和鞋），且各部分应连接良好。

⑦中间电位作业人员沿绝缘子串移动时，手与脚的位置必须保持对应一致，且人体和工具短接的绝缘子片数应符合表2-5-3规定。

⑧采用"跨二短三"作业方式（也称自由作业法）进入强电场。

⑨用绝缘绳索传递大件金属物品时，地面电工应将金属物品接地后再接触。

⑩带电作业过程中，工作负责人（监护人）应对作业人员进行不间断监护，随时纠正其不规范或违章动作。重点关注高处作业人员，使其保持足够的安全距离及组合间隙（符合表2-5-2的规定），禁止同时接触两个非连通的带电体或带电体与接地体。

（2）危险点——高处坠落

预控措施：

①高处作业人员登高前，必须具备符合本项作业要求的身体状况、精神状态和技能素质。

②高处作业人员应使用双保险安全带。上、下塔时，应手抓主材、脚踩脚钉，匀速行进。

③中间电位作业人员作业前应认真检查液压丝杠、闭式卡等，确保承力工具合格；沿绝缘子串移动时，手与脚的位置必须保持对应一致，安全带应系挂在手扶的绝缘子串上，并同步移动；更换绝缘子时，承力工具安装应可靠，荷载转移前、后应做冲击试验判定其可靠性，并及时向工作负责人汇报，得到工作负责人许可后方可实施。

④监护人员应随时纠正其不规范或违章动作，重点关注高处作业人员在转位的过程中不得失去安全带或绝缘后备保护绳的保护，严禁低挂高用。

(3)危险点——高处坠物伤人

预控措施：

①高处作业人员的个人工具及零星材料应装入工具袋，严禁在高处浮置物件、口中含物。

②地面电工必须正确佩戴安全帽，正确使用绳结，与作业点垂直下方距离不得小于坠落半径。

③作业现场设置围栏并挂好警示标示牌。监护人员应随时注意，禁止非工作人员及车辆进入作业区域。

3.2 工器具及材料选择

带电更换1000kV交流输电线路耐张玻璃绝缘子串任意单片玻璃绝缘子所需工器具及材料见表2-5-4。工器具出库前，应认真核对工器具的使用电压等级和试验周期，并检查确认外观良好、连接牢固、转动灵活，且符合本次工作任务的要求；工器具出库后，应存放在工具袋或工具箱内进行运输，防止脏污、受潮；金属工具和绝缘工器具应分开装运，防止因混装运输导致工器具变形、损伤等现象发生。

表2-5-4 带电更换1000kV交流输电线路耐张玻璃绝缘子串任意单片玻璃绝缘子所需工器具及材料表

序号	名称	规格型号	单位	数量	备注
1	屏蔽服	Ⅰ型	套	2	个人防护用具
2	导电鞋	尺码视穿着人员而定	双	2	个人防护用具
3	阻燃内衣	纯桑蚕丝	套	2	个人防护用具
4	双保险安全带	背带式	根	2	个人防护用具
5	安全帽		顶	7	个人防护用具
6	护目镜		副	2	个人防护用具
7	绝缘传递绳	$\phi14mm$，长度与起吊高度匹配	根	1	绝缘工具
8	绝缘后备保护绳	$\phi16mm$	根	3	绝缘工具
9	绝缘绳套	$\phi14mm$	根	2	绝缘工具
10	绝缘滑车	1T	个	1	绝缘工具
11	液压丝杠	8T	根	2	金属工具
12	闭式卡	Tc4	套	1	金属工具
13	拔销器		把	1	其他工具
14	绝缘电阻测试仪	2500V，电极宽2cm、极间宽2cm	套	1	其他工具
15	万用表		套	1	其他工具
16	风速、温湿度测试仪		只	1	其他工具
17	安全围网		套	若干	其他工具

续表

序号	名称	规格型号	单位	数量	备注
18	警示标示牌	"在此工作""从此进出""车辆慢行""车辆绕行"	套	1	其他工具
19	红马甲	"工作负责人""专责监护人"	件	1	其他工具
20	防潮苫布	3m×3m	块	2	其他工具
21	个人工具	扳手、老虎钳	套	1	其他工具
22	防坠器	与杆塔防坠落装置型号对应	只	2	其他工具
23	毛巾	棉质	条	1	其他工具
24	绝缘子	与被更换绝缘子同型号	片	1	材料

3.3 作业人员分工

本任务作业人员分工如表2-5-5所示。

表2-5-5 带电更换1000kV交流输电线路耐张玻璃绝缘子串任意单片玻璃绝缘子人员分工表

序号	工作岗位	数量(人)	工作职责
1	工作负责人	1	负责本次工作任务的人员分工、工作票的宣读、办理线路停用重合闸、办理工作许可手续、召开工作班前会、工作中突发情况的处理、工作质量的监督、工作后的总结
2	专责监护人	1	负责作业现场的安全把控
3	中间电位电工	1	负责工器具安装及绝缘子更换工作
4	地面电工	3	负责本次作业过程的地面辅助工作

4. 工作程序

本任务工作流程如表2-5-6所示。

表2-5-6 带电更换1000kV交流输电线路耐张玻璃绝缘子串任意单片玻璃绝缘子工作流程表

序号	作业内容	作业步骤及标准	安全措施及注意事项	责任人
1	现场复勘	工作负责人负责完成以下工作: (1)现场核对线路名称、杆塔编号,双重编号无误;基础及杆塔完好无异常;交叉跨越距离符合安全要求;确认缺陷情况及导地线规格型号等。 (2)检测风速、湿度等现场气象条件符合作业要求。 (3)检查地形环境符合作业要求。 (4)检查工作票所列安全措施与现场实际情况相符,必要时予以补充	(1)正确穿戴安全帽、工作服、工作鞋、劳保手套。 (2)不得在危及作业人员安全的气象条件下作业。 (3)严禁非工作人员、车辆进入作业现场	

续表

序号	作业内容	作业步骤及标准	安全措施及注意事项	责任人
2	工作许可	(1)工作负责人负责联系值班调控人员,按工作票内容申请停用线路重合闸。 (2)经值班调控人员许可后,方可开始带电作业工作	不得未经值班调控人员许可即开始工作	
3	现场布置	正确装设安全围栏并悬挂标示牌: (1)安全围栏范围应充分考虑高处坠物,以及对道路交通的影响。 (2)安全围栏出入口设置合理。 (3)妥当布置"从此进出""在此工作""车辆慢行"或"车辆绕行"等标示	对道路交通安全影响不可控时,应及时联系交通管理部门强化现场交通安全管控	
4	召开班前会	(1)全体工作成员列队。 (2)工作负责人宣读工作票,明确工作任务及人员分工;讲解工作中的安全措施和技术措施;查(问)全体工作成员精神状态;告知工作中存在的危险点及采取的预控措施。 (3)全体工作成员在工作票上签名确认	(1)工作票填写、签发和许可手续规范,签名完整。 (2)全体工作成员精神状态良好。 (3)全体工作成员明确任务分工、安全措施和技术措施	
5	检查工器具	(1)在防潮苫布上,将工器具按作业要求准备齐备,并分类定置摆放整齐。检查工器具外观和试验合格证,无遗漏。 (2)使用绝缘电阻测试仪检测绝缘工具及绝缘绳索的表面绝缘电阻值,方法正确,不得低于700MΩ。 (3)将新绝缘子擦拭干净,外观检查完好,不得有锈蚀、裂纹及破损。使用绝缘电阻测试仪测试其绝缘电阻值,方法正确,不得低于500 MΩ。 (4)使用万用表检测全套屏蔽服内阻,方法正确,不得大于20Ω。 (5)检查人员向工作负责人汇报各项检查结果符合作业要求	(1)防潮苫布数量足够,设置位置合理,保持清洁、干燥。 (2)金属、绝缘工器具在使用前,应仔细检查其是否无损伤、受潮、变形、失灵现象,合格证在有效期内。 (3)绝缘工具及绝缘绳索检测合格	
6	登塔	(1)中间电位电工再次核对线路双重名称及相别,检查并确认脚钉齐全、牢固;系好安全带、加挂防坠器;对安全带、防坠器做冲击试验,方法正确;工作负责人检查并确认中间电位电工穿戴的双保险安全带各部件的连接情况良好,包括肩带、胸带、腰带、腿带、后背保护绳、扣和环。 (2)背上工具包,携带绝缘传递绳(含绝缘滑车),方法正确。 (3)将安全带主带和后备保护绳斜跨肩上。 (4)清洁鞋底,经工作负责人许可后登塔。 (5)脚踩脚钉、手抓主材、匀步登塔至横担适当位置,系好安全带,脱离防坠器	(1)安全带、防坠器冲击试验合格。 (2)防止安全带、绝缘传递绳钩挂塔材。 (3)人体与导线保持的最小安全距离应符合表2-5-2的规定。 (4)禁止手抓脚钉。 (5)正确使用防坠器。 (6)转位时,不得失去安全带的保护	

91

续表

序号	作业内容	作业步骤及标准	安全措施及注意事项	责任人
7	进入强电场	(1)中间电位电工携带绝缘传递绳,转位至作业相耐张绝缘子串挂点处,将安全带主带系挂在绝缘子串连接金具上,将安全带后备保护绳系留在横担适当位置。 (2)中间电位电工再次检查并确认屏蔽服各部分连接良好、绝缘子串连接良好及故障绝缘子位置,经工作负责人许可后,双手抓扶一串,双脚踩另一串,采用"跨二短三"作业方式,沿绝缘子串平稳移动到作业点;手与脚的位置必须保持对应一致,安全带主带系挂在手扶的绝缘子串上,并同步移动。 (3)中间电位电工到达作业点后,在绝缘子串适当位置用绝缘绳套固定绝缘滑车,穿入绝缘传递绳;安装牢固可靠,便于工作	(1)防止安全带、绝缘传递绳钩挂塔材。 (2)转位时,不得失去安全带的保护。 (3)人体与带电体之间的安全距离、人体与接地体和带电体间的组合间隙不得小于表2-5-2的规定	
8	安装工具并转移导线张力	(1)地面电工使用绝缘传递绳将闭式卡、液压丝杠等分别传给中间电位电工。起吊过程平稳、无磕碰、无缠绕,正确使用绳结。 (2)中间电位电工将闭式卡前卡安装在需要更换绝缘子后两片绝缘子的卡槽内,后卡安装在需要更换绝缘子前一片绝缘子的钢帽上,并连接好液压丝杠。承力工具各部分安装牢固可靠。 (3)检查并确认承力工具各部分安装情况良好,经工作负责人许可后,操作液压丝杠使其逐渐受力,使需更换的绝缘子松弛。两根液压丝杠的受力应均匀	(1)人体与接地体和带电体间的组合间隙不得小于表2-5-2的规定。 (2)防止高处坠物。 (3)扣除劣质绝缘子、人体操作和工具短接的绝缘子后,良好绝缘子片数应符合表2-5-3的规定	
9	更换绝缘子	(1)中间电位电工做冲击试验,检查并确认承力工具受力正常;经工作负责人许可后,用绝缘传递绳系好旧绝缘子,取出旧绝缘子两端锁紧销,继续操作并收紧液压丝杠,直至拆除旧绝缘子。两根液压丝杠的受力应均匀,操作手柄不得敲击绝缘子。 (2)地面电工用绝缘传递绳的另一端系好新绝缘子,采用旧下、新上的方法,将新绝缘子传给中间电位电工。起吊过程平稳、无磕碰、无缠绕,正确使用绳结。 (3)中间电位电工安装新绝缘子,复位其两端锁紧销,并确认其安装到位	(1)人体与接地体和带电体间的组合间隙不得小于表2-5-2的规定。 (2)防止高处坠物	

续表

序号	作业内容	作业步骤及标准	安全措施及注意事项	责任人
10	拆除工具	(1)中间电位电工检查新绝缘子连接可靠，经工作负责人许可后，操作并松出液压丝杠，使更换的绝缘子逐渐受力。 (2)荷载转移完毕后，中间电位电工做冲击试验；检查并确认新绝缘子受力情况良好，经工作负责人许可后，拆除系在绝缘子上的绝缘传递绳，并将其系牢于承力工具适当位置，拆除液压丝杠、闭式卡等承力工具，在地面电工配合下传递至地面。传递过程平稳、无磕碰、无缠绕，正确使用绳结	(1)人体与接地体和带电体间的组合间隙不得小于表2-5-2的规定。 (2)防止高处坠物	
11	退出强电场	(1)中间电位电工检查作业部位无遗留物后，拆除绝缘传递绳。 (2)携带绝缘传递绳，按照"跨二短三"的作业方式沿绝缘子串回到横担上	(1)人体与接地体和带电体间的组合间隙不得小于表2-5-2的规定。 (2)转位时，不得失去安全带的保护	
12	撤离杆塔	中间电位电工检查塔上无遗留物，经工作负责人许可后，挂好防坠器，解开并整理好安全带，正确携带绝缘传递绳，脚踩脚钉、手抓主材、匀步下塔至地面	(1)转位时不得失去安全带保护。 (2)防止手滑脱、脚踏空、禁止手抓脚钉。 (3)正确使用防坠器。 (4)防止绝缘传递绳、安全带钩挂塔材或脚钉	
13	工作结束	(1)工作负责人组织全体工作成员整理工器具和材料，将工器具清洁后放入专用的箱(袋)中；清理现场，做到"工完料尽场地清"。 (2)召开班后会，工作负责人进行工作总结和点评工作。点评本次工作的施工质量；点评全体工作成员的安全措施落实情况。 (3)工作负责人向值班调控人员汇报工作结束，并申请恢复线路重合闸，终结工作票		

二、考核标准

特高压交流输电线路运检技能考核评分细则

考 生 填写栏	编号：	姓名：	所在岗位：	单位：	日 期：	年 月 日			
考评员 填写栏	成绩：	考评员：	考评组长：	开始时间：	结束时间：	操作时长：			
考核 模块	带电更换1000kV交流输电线路耐张 玻璃绝缘子串任意单片绝缘子			考核 对象	特高压交流输电线路检 修人员	考核 方式	操作	考核 时限	90min
任务 描述	带电更换1000kV交流输电线路耐张塔玻璃绝缘子串任意单片玻璃绝缘子								
工作规 范及要 求	1. 带电作业工作应在良好天气下进行。如遇雷、雨、雪、雾天气不得进行带电作业。风力大于5级时，不宜进行带电作业。湿度大于80%时，若需进行带电作业，应采用具有防潮性能的绝缘工具。 2. 本项作业需6人，其中工作负责人1名，专责监护人1名，中间电位电工1人，地面电工3名。 3. 工作负责(监护)人职责：负责本次工作任务的人员分工、工作票的宣读、办理线路停用重合闸、办理工作许可手续、召开工作班前会、负责作业过程中的安全监督、工作中突发情况的处理、工作质量的监督、工作后的总结。 4. 在带电作业中，遇雷、雨、大风或其他任何情况威胁到工作人员的安全时，工作负责人或监护人可根据情况临时停止工作。 给定条件： 1. 工作票已办理，安全措施已经完备(重合闸已停用)，工作开始、工作终结时应口头提出申请(调度或考评员)。 2. 安全、正确地使用仪器对绝缘工具进行检测。 3. 必须按工作程序进行操作，工序错误扣除应做项目分值，出现重大人身、器材和操作安全隐患，考评员可下令终止操作(考核)								
考核情 景准备	1. 塔形：1000kV交流耐张塔。 2. 所需作业工器具：双保险安全带2根，Ⅰ型屏蔽服2套，防潮布2张，万用表1块，绝缘电阻检测仪1个，风速仪、温湿度二合一1台，液压丝杠2根，绝缘绳2根，闭式卡1套，滑车1个，护目镜2个，拔销器1把，手动工具1套。 3. 作业现场做好监护工作，作业现场安全措施(围栏等)已全部落实；禁止非作业人员进入现场，工作人员进入作业现场必须戴安全帽。 4. 考生自备工作服，安全帽，线手套								
备注	1. 各项目得分均扣完为止，出现重大人身、器材和操作安全隐患，考评员可下令终止操作。 2. 设备、作业环境、安全带、安全帽、工器具、屏蔽服等不符合作业条件考评员可下令终止操作								

续表

序号	项目名称	质量要求	分值	扣分标准	扣分原因	扣分	得分
1	现场复勘	1)工作负责人到作业现场核对线路名称、杆塔编号、现场工作条件、缺陷部位等无误。 2)检测风速、湿度等现场气象条件符合作业要求。 3)检查工作票填写完整,无涂改,检查是否所列安全措施与现场实际情况相符,必要时予以补充	5	1)未核对线路名称、杆塔编号、现场工作条件、缺陷部位等,扣1分/项。 2)未检测风速、湿度等现场气象条件,扣1分/项。 3)工作票填写出现涂改,扣0.5分/处;工作票编号有误,扣1分;工作票填写不完整,扣1.5分			
2	工作许可	1)工作负责人联系值班调控人员(裁判),按工作票内容申请停用线路重合闸。 2)汇报内容规范、完整	2	1)未联系调度部门(裁判)停用重合闸,扣2分。 2)汇报专业用语不规范或不完整,扣1分			
3	现场布置	正确装设安全围栏并悬挂标示牌: 1)安全围栏范围应充分考虑高处坠物,以及对道路交通的影响。 2)安全围栏出入口设置合理。 3)妥当布置"从此进出""在此工作""从此上下"等标示	3	1)作业现场未装设围栏,扣1分。 2)未设立警示牌,扣1分。 3)未悬挂登塔作业标志,扣1分			
4	召开班前会	1)全体工作成员全体人员正确佩戴安全帽、工作服。 2)工作负责人穿红色背心,宣读工作票,明确工作任务及人员分工;讲解工作中的安全措施和技术措施;查(问)全体工作成员精神状态;告知工作中存在的危险点及采取的预控措施。 3)全体工作成员在工作票上签名确认	3	1)工作人员着装不整齐,扣0.5分/人。 2)未进行分工,扣3分;分工不明确,扣1分。 3)现场工作负责人未穿安全监护背心,扣1分。 4)工作票上工作班成员未签字或签字不全,扣1分			

续表

序号	项目名称	质量要求	分值	扣分标准	扣分原因	扣分	得分
5	工器具检查	1)在防潮苫布上,将工器具按作业要求准备齐备,并分类定置摆放整齐。检查工器具外观和试验合格证,无遗漏。 2)使用绝缘电阻测试仪检测绝缘工具及绝缘绳索的表面绝缘电阻值,方法正确,不得低于700MΩ。 3)将新绝缘子擦拭干净,外观检查完好,不得有锈蚀、裂纹及破损。使用绝缘电阻测试仪测试其绝缘电阻值,方法正确,不得低于500 MΩ。 4)使用万用表检测全套屏蔽服内阻,方法正确,不得大于20Ω。 5)检查人员向工作负责人汇报各项检查结果符合作业要求	7	1)未使用防潮苫布并定置摆放工器具,扣1分。 2)未检查工器具外观及试验合格证,扣0.5分/项。 3)未正确使用检测仪器对工器具进行检测,扣1分/项。 4)汇报检测结果不规范,扣1分;不完整,扣0.5分/项			
6	登塔	1)中间电位电工再次核对线路双重名称及相别,检查并确认脚钉齐全、牢固;系好安全带、加挂防坠器;对双保险安全带、防坠器做冲击试验,方法正确;并进行汇报。工作负责人检查并确认中间电位电工穿戴的双保险安全带各部件的连接情况良好,包括肩带、胸带、腰带、腿带、后背保护绳、扣和环。 2)背上工具包,携带绝缘传递绳(含绝缘滑车),方法正确。 3)登塔过程中系好防坠落保护装置,脚踩脚钉、手抓主材、匀步登塔至合适位置,系好安全带,脱离防坠器	5	1)中间电位电工未核对线路双重名称、杆号、相别、塔材情况,扣1分/项;核对完未汇报,扣1分。 2)双保险安全带及防坠器未进行冲击试验,扣2分/项。 3)现场工作负责人未对中间电位电工进行安全防护装备进行检查,扣1分。 4)手抓脚钉,扣0.5分/次。 5)滑车传递绳悬挂位置不合理,扣1分。 6)转位时失去安全带保护,扣5分			

序号	项目名称	质量要求	分值	扣分标准	扣分原因	扣分	得分
7	进入强电场	1)中间电位电工携带绝缘传递绳,转位至作业相耐张绝缘子串挂点处,将安全带主带系挂在绝缘子串连接金具上,将安全带后备保护绳系留在横担适当位置。 2)中间电位电工再次检查并确认屏蔽服各部分连接良好、绝缘子串连接良好及故障绝缘子位置,经工作负责人许可后,双手抓扶一串,双脚踩另一串,采用"跨二短三"作业方式,沿绝缘子串平稳移动到作业点;手与脚的位置必须保持对应一致,安全带主带系挂在手扶的绝缘子串上,并同步移动。 3)中间电位电工到达作业点后,在绝缘子串适当位置用绝缘绳套固定绝缘滑车,穿入绝缘传递绳;安装牢固可靠,便于工作	5	1)安全带后背保护绳系留位置不合理、使用不规范,扣2分。 2)中间电位电工未检查屏蔽服连接情况、绝缘子串连接情况及故障绝缘子位置,扣1分/项。 3)未得到工作负责人许可就进入强电场,扣5分。 4)中间电位电工进入强电场动作不正确、反复放电,扣2分/次。 5)绝缘传递绳安装位置不合理,扣1分。 6)高处坠物,扣2分/次。 7)转位时失去安全带保护,扣5分			
8	安装工具	1)地面电工使用绝缘传递绳将闭式卡、液压丝杠等工具分别传至中间电位电工,起吊过程平稳、无磕碰、无缠绕,正确使用绳结。 2)中间电位电工将闭式卡前卡安装在需要更换绝缘子后两片绝缘子的卡槽内,后卡安装在需要更换绝缘子前一片绝缘子的钢帽上,并连接好液压丝杠。承力工具各部分安装牢固可靠。 3)检查并确认承力工具各部分安装情况良好,经工作负责人许可后,操作液压丝杠使其逐渐受力,使需更换的绝缘子松弛。两根液压丝杠的受力应均匀	15	1)起吊过程不平稳,出现磕碰、缠绕,扣1分/次。 2)高处坠物,扣2分/次。 3)卡具安装不正确、固定不到位,扣2分。 4)未检查承力工具安装情况,扣3分;检查了未报告,扣1分;报告了但工作负责人未同意即开始收紧丝杠,扣1分。 5)作业过程短接绝缘子片数超过4片,扣3分/次。 6)安装卡具出现绝缘子碰撞破损,扣15分。 7)未均衡收紧丝杠,扣2分			

续表

序号	项目名称	质量要求	分值	扣分标准	扣分原因	扣分	得分
9	更换绝缘子	1)中间电位电工做冲击试验,检查并确认承力工具受力正常,经工作负责人许可后,用绝缘传递绳系好旧绝缘子,取出旧绝缘子两端锁紧销,继续操作并收紧液压丝杠,直至拆除旧绝缘子。两根液压丝杠的受力应均匀,操作手柄不得敲击绝缘子。 2)地面电工用绝缘传递绳的另一端系好新绝缘子,采用旧下、新上的方法,将新绝缘子传给中间电位电工。起吊过程平稳、无磕碰、无缠绕,正确使用绳结。 3)安装新绝缘子,并复位其两端锁紧销	20	1)未检查承力工具受力情况,扣3分;检查了未报告,扣2分;报告了但工作负责人未同意即取出旧绝缘子两端锁紧销,扣1分。 2)未均衡收紧丝杆,扣2分。 3)操作手柄敲击绝缘子,扣1分/次。 4)绳结错误,扣1分。 5)高处坠物,扣2分/次。 6)新旧绝缘子相互碰撞,扣1分。 7)传递绝缘子与塔身相互碰撞,扣1分/次。 8)绝缘绳缠绕,扣2分			
10	拆除工具	1)中间电位电工检查新绝缘子连接可靠,经工作负责人许可后,操作并松出液压丝杠,使更换的绝缘子逐渐受力。 2)荷载转移完毕后,中间电位电工做冲击试验,检查并确认新绝缘子受力情况良好,经工作负责人许可后,拆除系在绝缘子上的绝缘传递绳,并将其系牢于承力工具适当位置,拆除液压丝杠、闭式卡等承力工具,在地面电工配合下传递至地面。传递过程平稳、无磕碰、无缠绕,正确使用绳结	15	1)未检查新绝缘子连接情况,扣3分;检查了未报告,扣2分;报告了但工作负责人未同意即松液压丝杠,扣1分。 2)未检查新绝缘子受力情况,扣3分;检查了未报告,扣2分;报告了但工作负责人未同意即拆除液压丝杠,扣1分。 3)捆扎工具时,未正确使用绳结,扣1分。 4)高处坠物,扣2分/次。 5)工器具相互碰撞扣1分/次;工器具与带电体或塔身相互碰撞,扣1分/次;绝缘绳缠绕,扣2分			
11	退出强电场	1)中间电位电工检查作业部位无遗留物后,带好绝缘传递绳,作退出电位准备。 2)中间电位电工按照"跨二短三"作业方式退出等电位	5	1)未向工作负责人申请即退出强电场,扣2分;申请了但未得同意即开始退出强电场,扣1分。 2)中间电位电工退出强电场动作不正确,反复放电,扣2分/次。 3)未有效控制后备保护绳,扣1分			

续表

序号	项目名称	质量要求	分值	扣分标准	扣分原因	扣分	得分
12	返回地面	塔上电工检查塔上无遗留物后,经工作负责人许可后携带绝缘传递绳下塔	5	1)下塔过程未使用防坠器,扣5分。 2)塔上移位失去安全带保护,扣5分。 3)下塔手抓脚钉,扣1分/次。 4)塔上有遗留物,扣2分			
13	工作结束	1)工作负责人组织全体工作成员整理工器具和材料,将工器具清洁后放入专用的箱(袋)中;清理现场,做到"工完料尽场地清"。 2)召开班后会,工作负责人进行工作总结和点评工作。点评本次工作的施工质量;点评全体工作成员的安全措施落实情况。 3)工作负责人向值班调控人员汇报工作结束,申请恢复线路重合闸,终结工作票	10	1)工器具未清理,扣2分。 2)工器具有遗漏,扣2分。 3)未开班后会,扣10分。 4)未拆除围栏,扣2分。 5)未向调度汇报,扣2分			
	合计		100				

模块6 带电更换1000kV交流输电线路耐张玻璃绝缘子串任意段绝缘子培训及考核标准

一、培训标准

（一）培训要求

模块名称	带电更换1000kV交流输电线路耐张玻璃绝缘子串任意段绝缘子	培训类别	操作类
培训方式	实操培训	培训学时	21学时
培训目标	1.掌握带电更换1000kV流输电线路耐张玻璃绝缘子串任意段绝缘子的工作原理。 2.掌握1000kV交流输电线带电作业工器具的布置方法、采用"跨二短三"沿绝缘子串进、出强电场方法、更换任意段绝缘子(注：任意段绝缘子为连续7片及以下绝缘子串内单片或多片绝缘子)的方法。 3.能完成带电更换1000kV交流输电线路耐张玻璃绝缘子串任意段绝缘子工作		
培训场地	特高压交流实训线路		
培训内容	采用中间电位作业法带电更换1000kV交流输电线路耐张玻璃绝缘子串任意段绝缘子，作业相为中相		
适用范围	特高压交流输电线路检修人员		

（二）引用规程规范

（1）《电工术语》（GB/T 2900.55—2002）。

（2）《带电作业用绝缘滑车》（GB/T 13034—2008）。

（3）《带电作业用绝缘绳索》（GB/T 13035—2008）。

（4）《交流线路带电作业安全距离计算方法》（GB/T 18037—2000）。

（5）《1000kV交流带电作业用屏蔽服装》（GB/T 25726—2010）。

（6）《1000kV架空输电线路设计规范》（GB 50665—2011）。

（7）《1000kV交流输电线路检修规范》（DL/T 209—2008）。

（8）《1000kV交流输电线路运行规范》（DL/T 307—2010）。

（9）《1000kV交流输电线路带电作业技术导则》（DL/T 392—2015）。

（10）《带电作业工器具、装置和设备使用的一般要求》（DL/T 877—2004）。

（11）《国家电网公司电力安全工作规程（线路部分）》（Q/GDW 1799.2—2013）。

（三）培训教学设计

本设计以完成"带电更换1000kV交流输电线路耐张玻璃绝缘子串任意段绝缘子"为工作任务，按工作任务完成的标准化作业流程来设计各个培训阶段，每个阶段包括了具体的培训目标、培训内容、培训学时、培训方法（培训资源）、培训环境和考核评价等内容，如表2-6-1所示。

表2-6-1　带电更换1000kV交流输电线路耐张玻璃绝缘子串任意段绝缘子培训内容设计

培训流程	培训目标	培训内容	培训学时	培训方法与资源	培训环境	考核评价
1.理论教学	1.初步掌握沿绝缘子串进出1000kV强电场基本方法；2.熟悉更换任意段绝缘子工器具布置方法、更换多片绝缘子的方法。	1.采用"跨二短三"法沿绝缘子串进出强电场作业方式的理论依据和原理。2.1000kV交流输电线路更换任意段绝缘子的工器具布置方法和多片绝缘子的更换方法	2	培训方法：讲授法。培训资源：PPT、相关规程规范、视频演示	多媒体教室	考勤、课堂提问和作业
2.准备工作	能完成作业前准备工作	1.作业现场查勘。2.编制培训标准化作业卡。3.填写培训操作工作票。4.完成本操作的工器具及材料准备	1	培训方法：1.现场查勘和工器具及材料清理采用现场实操方法。2.编写作业卡和填写工作票采用讲授方法。培训资源：1. 1000kV实训线路。2.特高压工器具库房。3.空白工作票	1.特高压输电实训线路；2.多媒体教室	
3.作业现场准备	能完成作业现场准备工作	1.作业现场复勘。2.工作许可。3.作业现场布置。4.班前会。5.工器具及材料检查	1	培训方法：演示与角色扮演法。资源：1000kV实训线路	1000kV实训线路	

续表

培训流程	培训目标	培训内容	培训学时	培训方法与资源	培训环境	考核评价
4.培训师演示	通过现场观摩,使学员初步领会本任务操作流程	1.零值绝缘子的检测。 2.等电位电工沿耐张绝缘子串进、出强电场。 3.等电位电工(或地电位电工)与中间电位电工配合完成工器具安装。 4.中间电位电工完成任意段绝缘子的更换	2	培训方法:演示法。 资源:1000kV实训线路	1000kV实训线路	
5.学员分组训练	1.能完成带电作业工器具的布置操作。 2.能完成1000kV输电线路更换任意段绝缘子作业	1.学员分组(11人一组)训练工器具安装和绝缘子更换技能操作。 2.培训师对学员操作进行指导和安全监护	14	培训方法:角色扮演法。 资源:1000kV实训线路	1000kV实训线路	评分
6.工作终结	1.使学员进一步辨析操作过程不足之处,便于后期提升。 2.培养学员安全文明生产的工作作风	1.作业现场清理。 2.向调度汇报工作。 3.班后会,对本次工作任务进行点评总结	1	培训方法:讲授和归纳法	1000kV实训线路	

（四）作业流程

1. 工作任务

采用"跨二短三"的作业方式沿耐张绝缘子串进入强电场到达作业点，采用中间电位作业法带电更换1000kV交流输电线路耐张玻璃绝缘子串任意段绝缘子工作。

2. 天气及作业现场要求

（1）带电更换1000kV交流输电线路耐张玻璃绝缘子串任意段绝缘子应在良好的天气进行。如遇雷电（听见雷声、看见闪电）、雪、雹、雨、雾等，禁止进行带电作业。风力大于5级，或空气相对湿度大于80%时，不宜进行带电作业；恶劣天气下必须开展带电抢修时，应组织有关人员充分讨论并编制必要的安全措施，经本单位主管生产领导批准后方可进行。

（2）作业人员精神状态良好，熟悉工作中保证安全的组织措施和技术措施；应持有在有效期内的带电作业资质证书。

（3）工作负责人应事先组织相关人员完成现场勘察，根据勘察结果确定本次作业方法和所需工器具，以及应采取的必要措施，并办理带电作业工作票。

（4）作业现场应合理设置围栏，并妥当布置警示标示牌，禁止非工作人员入内。

（5）本项目需停用线路重合闸装置。

（6）工作中安全距离及有效绝缘长度如表2-6-2所示。

表2-6-2　带电更换1000kV交流输电线路耐张玻璃绝缘子串任意段绝缘子的安全距离

电压等级/m	人身与带电体安全距离/m	最小有效绝缘长度/m		最小组合间隙/m	转移电位时人体裸露部分与带电体的最小距离/m
		绝缘操作杆	绝缘承力工具、绝缘绳索		
1000kV	6.8	6.8	6.8	6.9	0.5

（7）在1000kV输电线路上作业，应保证作业相良好绝缘子片数不少于37片。

3. 准备工作

3.1　危险点及其预控措施

（1）危险点——触电伤害

预控措施：

①工作前，工作负责人应与值班调控人员联系，停用工作线路重合闸，并履行许可手续。

②塔上地电位作业人员登塔前，必须仔细核对线路名称、铁塔编号、相别，确认无误后方可上塔。

③工作中，如遇线路突然停电，作业人员应视其仍然带电。工作负责人应尽快与调控人员联系，值班调控人员未与工作负责人取得联系前不准强送电。

④绝缘工具及绝缘绳索不得损坏、受潮、变形、失灵，不准使用非绝缘绳索（如棉纱绳、白棕绳、钢丝绳）。

⑤地面电工操作绝缘工具时应戴清洁、干燥的手套，进入作业现场应将使用的带电作业工具放置在防潮苫布上，防止绝缘工具在使用中脏污和受潮。

⑥等电位作业人员应穿着阻燃内衣，衣服外面应穿戴全套合格的屏蔽服（包括屏蔽面罩、帽、衣裤、手套、袜和鞋），且各部分应连接良好。

⑦等电位作业人员在电位转移前，应得到工作负责人的许可，人体裸露部分与带电体的最小距离不小于0.5m；电位转移时，动作应迅速，严禁用头部充放电；与地电位作业人员传递工具和材料时，使用绝缘工具或绝缘绳索的有效长度不准小于表2-6-2的规定。

⑧用绝缘绳索传递大件金属物品时，地电位作业人员应将金属物品接地后再接触。

⑨工作负责人应对作业人员进行不间断监护，随时纠正其不规范动作和行为。重点监护高处作业人员，使其保持足够的安全距离（符合表2-6-2的规定）。

（2）危险点——高处坠落

预控措施：

①高处作业人员登塔前，必须具备符合本项作业要求的身体状况、精神状态和技能素质。

②塔上作业人员必须正确使用安全带，安全带应系在牢固部件上并位置合理，便于作业，登塔和塔上位移时，不得失去安全保护。

③塔上作业人员沿绝缘子串移动时，手与脚的位置必须保持对应一致，安全带应系在手扶的绝缘子串上，手脚同步移动。

④监护人员应随时纠正其不规范动作和行为，重点监护作业人员在转位的过程中不得失去安全带或绝缘后备保护绳的保护，严禁低挂高用。

（3）危险点——高处坠物伤人。

预控措施：

①高处作业人员的个人工具及零星材料应装入工具袋，严禁在高处浮置物件、口中含物。

②地面作业人员必须正确佩戴安全帽，正确使用绳结，与作业点正下方距离不得小于坠落半径。

③作业现场设置围栏并挂好警示标示牌。监护人员应随时注意，禁止非工作人员及车辆进入作业区域。

（4）危险点——卡具安装不到位。

预控措施：

①绝缘子在更换前认真详细检查端部卡、液压丝杠、闭式卡等，确保受力部件能正常良好工作，在绝缘子退出后，端部卡、液压丝杠、闭式卡能承载导线荷重。

②利用卡具更换绝缘子时，卡具安装应可靠，卡具受力转移导线荷载前应冲击判定其可靠性。经检查无误后方可更换绝缘子，进行绝缘子更换时，应向工作负责人汇报，得到工作负责人许可后方可进行。

3.2　工器具及材料选择

带电更换1000kV交流输电线路耐张玻璃绝缘子串任意段绝缘子所需工器具及材料见表2-6-3。工器具出库前，应认真核对工器具的使用电压等级和试验周期，并检查确认外观良好、连接牢固、转动灵活，且符合本次工作任务的要求；工器具出库后，应存放在工具袋或工具箱内进行运输，防止脏污、受潮；金属工具和绝缘工器具应分开装运，防止因混装运输导致工器具变形、损伤等现象发生。

表 2-6-3 带电更换 1000kV 交流输电线路耐张玻璃绝缘子串任意段绝缘子所需工器具及材料表

序号	名称	规格型号	单位	数量	备注
1	屏蔽服	Ⅰ型	套	4	
2	导电鞋	尺码视穿着人员而定	双	2	
3	阻燃内衣	纯桑蚕丝	套	2	
4	双保险安全带	背带式	根	2	
5	安全帽		顶	7	
6	绝缘传递绳	ϕ12mm，长度与起吊高度匹配	副	1	
7	绝缘传递绳	ϕ14mm，长度与起吊高度匹配	根	2	
8	绝缘后备保护绳	ϕ16mm	根	3	
9	绝缘绳套	ϕ14mm	根	3	
10	绝缘滑车	1T	个	3	
11	绝缘拉棒	8T	套	1	
12	滑车组	3-3	套	1	
13	机械丝杠	4T	根	2	
14	闭式卡	Tc4	套	1	
15	端部卡	Tc4	套	1	
16	电位转移棒		根	2	
17	拔销器		把	1	
18	绝缘电阻测试仪	2500V，电极宽2cm、极间宽2cm	套	1	
19	万用表		套	1	
20	风速、温湿度测试仪	HT-8321	只	1	
21	安全围网		套	若干	
22	警示标示牌	"在此工作""从此进出""车辆慢行""车辆绕行"	套	1	
23	红马甲	"工作负责人""专责监护人"	件	1	
24	防潮苫布	3m×3m	块	2	
25	个人工具	扳手、老虎钳	套	1	
26	防坠器	与铁塔防坠落装置型号对应	只	2	
27	毛巾	棉质	条	1	
28	绝缘子	U550BP/240T	片	若干	

3.3 作业人员分工

本任务作业人员分工如表 2-6-4 所示。

表2-6-4 带电补修1000kV交流输电线路导线人员分工表

序号	工作岗位	数量(人)	工作职责
1	工作负责人	1	负责本次工作任务的人员分工、工作票的宣读、办理线路停用重合闸、办理工作许可手续、召开工作班前会、工作中突发情况的处理、工作质量的监督、工作后的总结
2	专责监护人	1	负责作业现场的安全监护
3	中间电位电工	2	负责工器具安装及绝缘子更换工作
4	等电位电工	2	负责工器具安装工作
5	地面电工	6	负责传递工具、材料配合塔上电工安装工器具及绝缘子更换

4. 工作程序

本任务工作流程如表2-6-5所示。

表2-6-5 带电更换1000kV交流输电线路耐张玻璃绝缘子串任意段绝缘子工作流程表

序号	作业内容	作业步骤及标准	安全措施及注意事项	责任人
1	现场复勘	工作负责人负责完成以下工作： (1)现场核对线路名称、铁塔编号，双重编号无误；基础及铁塔完好无异常；交叉跨越距离符合安全要求；确认缺陷情况及导地线规格型号等。 (2)检测风速、湿度等现场气象条件符合作业要求。 (3)检查地形环境符合作业要求。 (4)检查工作票所列安全措施与现场实际情况相符，必要时予以补充	(1)正确穿戴安全帽、工作服、工作鞋、劳保手套。 (2)不得在危及作业人员安全的气象条件下作业。 (3)严禁非工作人员、车辆进入作业现场	
2	工作许可	(1)工作负责人负责联系值班调控人员，按工作票内容申请停用线路重合闸。 (2)经值班调控人员许可后，方可开始带电作业工作	不得未经值班调控人员许可即开始工作	
3	现场布置	正确装设安全围栏并悬挂标示牌： (1)安全围栏范围应充分考虑高处坠物，以及对道路交通的影响。 (2)安全围栏出入口设置合理。 (3)妥当布置"从此进出""在此工作""车辆慢行"或"车辆绕行"等标示	对道路交通安全影响不可控时，应及时联系交通管理部门强化现场交通安全管控	

第二部分 技能模块培训及考核标准

续表

序号	作业内容	作业步骤及标准	安全措施及注意事项	责任人
4	召开班前会	(1)全体工作成员列队。 (2)工作负责人宣读工作票,明确工作任务及人员分工;讲解工作中的安全措施和技术措施;查(问)全体工作成员精神状态;告知工作中存在的危险点及采取的预控措施。 (3)全体工作成员在工作票上签名确认	(1)工作票填写、签发和许可手续规范,签名完整。 (2)全体工作成员精神状态良好。 (3)全体工作成员明确任务分工、安全措施和技术措施	
5	检查工器具	(1)在防潮苫布上,将工器具按作业要求准备齐备,并分类定置摆放整齐。检查工器具外观和试验合格证,无遗漏。 (2)使用绝缘电阻测试仪检测绝缘工具及绝缘绳索的表面绝缘电阻值,方法正确,不得低于700MΩ。 (3)检测并清扫新绝缘子。 (4)使用万用表检测全套屏蔽服内阻,方法正确,不得大于20Ω。 (5)检查人员向工作负责人汇报各项检查结果符合作业要求	(1)防潮苫布数量足够,设置位置合理,保持清洁、干燥。 (2)金属、绝缘工器具在使用前,应仔细检查其是否无损伤、受潮、变形、失灵现象,合格证在有效期内。 (3)绝缘工具及绝缘绳索检测合格。 (4)新绝缘子应进行外观检查,并用绝缘电阻表在干燥、清洁的条件下检测,其阻值低于500 MΩ不得使用。	
6	登塔	(1)核对线路双重名称,检查铁塔基础。 (2)塔上电工携带绝缘传递绳登塔至横担耐张绝缘子串挂点处,系好安全带,并系好保护绳	(1)核对线路双重名称无误后,铁塔基础牢固可靠后方可塔上作业。 (2)登塔过程中应使用防坠器;铁塔上移位时,不准失去安全保护,作业人员必须攀抓牢固构件。 (3)中间电位电工必须穿全套合格的屏蔽服,且全套屏蔽服必须连接可靠	
7	进入强电场	(1)塔上电工将安全带转移到绝缘子连接金具上,并带好绝缘滑车和绝缘传递绳。 (2)等电位电工检查屏蔽服各部分连接良好后报经工作负责人同意,双手抓扶一串,双脚踩另一串,采用"跨二短三"方法沿绝缘子串进入绝缘子串,且移动到作业点。 (3)中间电位电工采用同样的方法到达作业点后,打好安全带并将绝缘传递绳安放在合适部位作起吊准备	(1)中间电位电工进入绝缘子串前必须得到工作负责人的许可。 (2)进入作业点位后安全带应系在不被更换绝缘子串侧并且位置合理,便于作业。 (3)中间电位电工进入绝缘子串必须系好保护绳,并调整绝缘传递绳。 (4)中间电位电工与接地体和带电体两部分间隙所组成的组合间隙不得小于中相6.9m(边相6.7m)。	

续表

序号	作业内容	作业步骤及标准	安全措施及注意事项	责任人
8	安装工具并转移导线张力	(1)地面电工将滑车组、闭式卡等分别传至中间电位电工,将液压丝杠和端部卡传递给等电位电工,中间电位电工和等电位电工分别安装好工具后,地面电工再将绝缘拉棒通过传递绳索同时传递给中间电位电工和等电位电工安装好。(2)中间电位电工在需更换绝缘子串前后合适位置安装好闭式卡。(3)塔上电工检查承力工具各部分安装情况无误后,分别向工作负责人汇报,得到工作负责人同意后,等电位电工收紧丝杠,待丝杠适当受力后,对受力工具进行冲击检查,经工作负责人同意后再收紧液压紧线系统,使需更换的绝缘子松弛	(1)上、下作业电工要密切配合,听从工作负责人的指挥。(2)上、下传递工具绑扎绳扣应正确可靠,等电位电工注意不得高空落物。(3)两液压丝杠受力应均匀。(4)扣除劣质绝缘子、人体操作和工具短接的绝缘子,其良好绝缘子(结构高度为195mm)不得少于37片	
9	更换绝缘子	(1)检查承力工具受力正常后,向工作负责人汇报,得到工作负责人同意后,中间电位电工取出需要更换绝缘子的上、下锁紧销,等电位电工继续收紧液压紧线系统,直至取出旧绝缘子。(2)中间电位电工用绝缘传递绳系好旧绝缘子,同时地面电工也用绝缘传递绳的另一侧系好绝缘子。采用旧下、新上的方法将新绝缘子传给地位电工。(3)中间电位电工换上新绝缘子,并复位上、下锁紧销	(1)绝缘子更换前,必须详细检查闭式卡、液压丝杠等部件的受力是否正常,检查确认无误后报经工作负责人同意后方可更换绝缘子。(2)对新绝缘子应使用5000V及以上绝缘电阻表进行测量,其绝缘电阻不小于500MΩ。并进行外观检查,如有锈蚀、破损、裂纹等不得使用。(3)操作时应均匀用力,尽量小幅度。(4)新、旧绝缘子上、下传递时不得碰撞。(5)卡具受力并完全转移导线荷载后,应进行冲击判定其可靠性	
10	拆除工具	(1)检查新绝缘子连接可靠,得到工作负责人同意后,等电位电工松出液压紧线系统。(2)中间电位电工再次检查新绝缘子受力情况无误后,得到工作负责人同意后,拆除更换工具并传递至地面	(1)新绝缘子更换完毕后,必须确认安装可靠,连接无误,锁紧销全部复位。(2)工具在传递过程中不得碰撞,系绳结应正确可靠	
11	退出电位	(1)塔上电工检查作业部位无遗留物后,带好绝缘传递绳,作退出电位准备。(2)塔上电工按照"跨二短三"的方法退出绝缘子串	(1)中间电位电工退出绝缘子串前必须得到工作负责人的许可。(2)中间电位电工在退出电位过程中组合间隙不得小于中相6.9m(边相6.7m)	

续表

序号	作业内容	作业步骤及标准	安全措施及注意事项	责任人
12	返回地面	塔上电工检查塔上无遗留物后,向工作负责人汇报,得到工作负责人同意后携带传递绳下塔	下塔过程中应使用防坠器;塔上移位时不得失去安全保护	
13	工作结束	(1)工作负责人组织全体工作成员整理工器具和材料,将工器具清洁后放入专用的箱(袋)中;清理现场,做到"工完料尽场地清"。 (2)召开班后会,工作负责人进行工作总结和点评工作。点评本次工作的施工质量;点评全体工作成员的安全措施落实情况。 (3)工作负责人向值班调控人员汇报工作结束,申请恢复线路重合闸,终结工作票	不得约时恢复线路重合闸	

二、考核标准

特高压交流输电线路运检技能考核评分细则

考生填写栏	编号:	姓名:	所在岗位:		单位:		日期:	年 月 日		
考评员填写栏	成绩:	考评员:		考评组长:		开始时间:	结束时间:		操作时长:	
考核模块	带电更换1000kV交流输电线路耐张玻璃绝缘子串任意段绝缘子			考核对象	特高压交流输电线路检修人员		考核方式	操作	考核时限	90min
任务描述	带电更换1000kV交流输电线路耐张玻璃绝缘子串任意段绝缘子									
工作规范及要求	1. 带电作业工作应在良好天气下进行。如遇雷、雨、雪、雾天气不得进行带电作业。风力大于5级时,相对湿度大于80%的天气,不宜进行带电作业。 2. 本项作业需11人,其中工作负责人1名,专责监护人1名,中间电位电工2人,等电位电工1人,地面电工6名,采用中间电位法完成带电更换1000kV交流输电线路耐张玻璃绝缘子串任意段绝缘子工作。 3. 工作负责人职责:负责本次工作任务的人员分工、工作票的宣读、办理线路停用重合闸、办理工作许可手续、召开工作班前会、工作中突发情况的处理、工作质量的监督、工作后的总结。 4. 专责监护人:负责作业现场的安全监护。 5. 中间电位电工:负责工器具安装及绝缘子更换工作 6. 等电位电工职责:负责工器具安装工作。 7. 地面电工职责:负责传递工具、材料配合等塔上电工安装工具及绝缘子更换工作。 8. 在带电作业中,如遇雷、雨、大风或其他任何情况威胁到工作人员的安全时,工作负责人或监护人可根据情况,临时停止工作。									

续表

工作规范及要求	给定条件： 1. 培训基地：特高压交流1000kV实训线路耐张塔B相大号侧绝缘子串，绝缘子型号：U550BP/240T。 2. 工作票已办理，安全措施已经完备（重合闸已停用），工作开始、工作终结时应口头提出申请（调度或考评员）。 3. 安全、正确地使用仪器对绝缘工具进行检测。 4. 必须按工作程序进行操作，工序错误扣除应做项目分值，出现重大人身、器材和操作安全隐患，考评员可下令终止操作（考核）
考核情景准备	1. 线路：特高压交流10000kV实训线路耐张塔B相，工作内容：带电更换1000kV任意段绝缘子，绝缘子型号：U550BP/240T。 2. 所需作业工器具：安全带（含绝缘后备保护绳）4根，Ⅰ屏蔽服4套，防潮苫布2张，万用表1，绝缘电阻检测仪1个，风速仪、温湿度二合一1台，液压丝杠2根，绝缘绳2根，闭式卡1套，滑车1个，电位转移棒2把，拔销器1把，手动工具1套； 3. 作业现场做好监护工作，作业现场安全措施（围栏等）已全部落实；禁止非作业人员进入现场，工作人员进入作业现场必须戴安全帽。 4. 考生自备工作服，阻燃内衣，安全帽，线手套，安全带（含绝缘后备保护绳）
备注	1. 各项目得分均扣完为止，出现重大人身、器材和操作安全隐患，考评员可下令终止操作。 2. 设备、作业环境、安全带、安全帽、工器具、屏蔽服等不符合作业条件考评员可下令终止操作

序号	项目名称	质量要求	分值	扣分标准	扣分原因	扣分	得分
1	现场复勘	1)工作负责人到作业现场核对线路名称和铁塔编号、现场工作条件、缺陷部位等。 2)检测风速、湿度等现场气象条件符合作业要求。 3)检查工作票填写完整，无涂改，检查是否所列安全措施与现场实际情况相符，必要时予以补充	5	1)未进行核对双重称号扣1分。 2)未核实现场工作条件（气象）、缺陷部位扣1分。 3)工作票填写出现涂改，每项扣0.5分，工作票编号有误，扣1分。工作票填写不完整，扣1.5分			
2	工作许可	1)工作负责人联系值班调控人员，按工作票内容申请停用线路重合闸。 2)汇报内容规范、完整	2	1)未联系调度部门（裁判）停用重合闸扣2分。 2)汇报专业用语不规范或不完整的各扣0.5分			

续表

序号	项目名称	质量要求	分值	扣分标准	扣分原因	扣分	得分
3	现场布置	正确装设安全围栏并悬挂标示牌： 1)安全围栏范围应充分考虑高处坠物，以及对道路交通的影响。 2)安全围栏出入口设置合理。 3)妥当布置"从此进出""在此工作""从此上下"等标示	3	1)作业现场未装设围栏扣0.5分。 2)未设立警示牌扣0.5分。 3)未悬挂登塔作业标志扣0.5分			
4	召开班前会	1)全体工作成员全体人员正确佩戴安全帽、工作服。 2)工作负责人穿红色背心，宣读工作票，明确工作任务及人员分工；讲解工作中的安全措施和技术措施；查(问)全体工作成员精神状态；告知工作中存在的危险点及采取的预控措施。 3)全体工作成员在工作票上签名确认	3	1)工作人员着装不整齐扣0.5分，工作人员着装不整齐每人次扣0.5分。 2)未进行分工本项不得分，分工不明扣1分。 3)现场工作负责人未穿安全监护背心扣0.5分。 4)工作票上工作班成员未签字或签字不全的扣1分			
5	工器具检查	1)工作人员按要求将工器具放在防潮苫布上；防潮苫布应清洁、干燥。 2)工器具应按定置管理要求分类摆放；绝缘工器具不能与金属工具、材料混放；对工器具进行外观检查。 3)绝缘工具表面不应磨损、变形损坏，操作应灵活。绝缘工具应使用2500V及以上绝缘电阻表进行分段绝缘检测，阻值应不低于700MΩ，并用清洁干燥的毛巾将其擦拭干净。 4)塔上电工按要求正确穿戴全套合格的屏蔽服、导电鞋，且各部分连接应良好，屏蔽服内不得贴身穿着化纤类衣服，并系好安全带；工作负责人应认真检查是否穿戴正确。 5)登塔人员再次核对双重名称、塔号、相别并报告	7	1)未使用防潮布并定置摆放工器具扣1分。 2)未检查工器具试验合格标签及外观检查扣每项0.5分。 3)未正确使用检测仪器对工器具进行检测每项扣1分。 4)作业人员未正确穿戴屏蔽服且各部位连接良好每人次扣2分。 5)现场工作负责人未对登塔作业人员进安全防护装备进行检查扣1分。 6)登塔人员未核对线路双重名称、杆号、相别每人扣2分。 7)登塔人员未报告核对结果每人扣2分			

续表

序号	项目名称	质量要求	分值	扣分标准	扣分原因	扣分	得分
6	登塔	1）塔上地电位电工、等电位电工穿好全套合格的屏蔽服，将安全带做冲击试验后，系好安全带后携带绝缘传递绳相继登塔。 2）登塔过程中系好防坠落保护装置，登塔至合适位置，系好安全带，布置好绝缘传递绳，然后配合地面电工将绝缘传递绳分开作起吊准备。 3）登塔过程中应系好防坠落保护装置，匀速登塔，手抓主材，将安全带挂在肩上并与带电体保持6.8m以上安全距离，工作负责人加强作业监护	5	1）未系安全带或安全带及后备保护绳未进行冲击试验各扣2分。 2）手抓脚钉扣2分。 3）滑车传递绳悬挂位置不便工具取用扣1分。 4）传递时金属工具难以保证安全距离扣2分；工具绑扎不牢扣2分。 5）传递时高空落物扣2分。 6）传递过程工具与塔身磕碰扣2分。 7）传递工具绳索打结、缠绕扣1分。 8）工作负责人监护不到位扣2分。 9）塔上电工操作不正确扣2分			
7	检测绝缘子	1）地面电工将绝缘操作杆及绝缘子检测仪传至塔上。等电位电工协助塔上地电位电工对绝缘子进行检测。 2）检测绝缘子工作必须逐串逐片进行，接触必须可靠，由导线侧依次向横担侧进行。 3）当任意一串良好绝缘子少于37片时，立即停止检测，并停止本次带电作业工作	2	1）未进行绝缘子测零扣2分。 2）测零方法错误扣2分。 3）测零结果未汇报扣2分			
8	进入强电场	1）等电位电工进入绝缘子串前必须系好保护绳，并调整好绝缘传递绳和电位转移棒。 2）采用"跨二短三"作业方式沿绝缘子串进入强电场。 3）等电位电工沿绝缘子串进入强电场过程中组合间隙不得小于中相6.9m，进入强电场必须用电位转移棒进行电位转移	8	1）等电位电工电位转移过程中裸露部分距离不够扣2分。 2）等电位电工进入强电场动作不正确，反复放电每次扣2分。 3）转移电位动作不熟练扣1分，电位转移过程未使用电位转移棒的扣3分。 4）未得到工作负责人许可就进行电位转移的扣5分			

续表

序号	项目名称	质量要求	分值	扣分标准	扣分原因	扣分	得分
9	安装工具	1)塔上电工达到作业点位后布置传递绳索,地面电工将滑车组、闭式卡等分别传至中间电位电工,将液压丝杠和端部卡传递给等电位电工,中间电位电工和等电位电工分别安装好工具后,地面电工再将绝缘拉棒通过传递绳索同时传递给中间电位电工和等电位电工安装好。2)中间电位电工在需更换绝缘子串前后合适位置安装好闭式卡。塔上电工检查承力工具各部分安装情况无误后,分别向工作负责人汇报,得到工作负责人同意后,等电位电工预收紧丝杠,待丝杠适当受力后,对受力工具进行冲击检查,经工作负责人同意后再收紧液压紧线系统,使需更换的绝缘子松弛	17	1)卡具固定不到位扣2分;2)未检查绝缘子串安装情况即开始松丝杠扣分;检查了未报告扣1分;报告了但作负责人未同意即开始松丝杠扣1分;3)作业过程短接绝缘子片数超过4片扣3分/次;4)安装卡具出现绝缘子碰撞破损扣2分;5)卡具安装到其他更换错误绝缘子不得分			
10	更换绝缘子	1)塔上电工检查承力工具受力正常后,向工作负责人汇报,得到工作负责人同意后,中间电位电工取出需要更换绝缘子的上、下锁紧销,等电位电工继续收紧液压紧线系统,直至取出旧绝缘子。2)中间电位电工用绝缘传递绳系好旧绝缘子,同时地面电工也用绝缘传递绳的另一侧系好绝缘子。采用旧下、新上的方法将新绝缘子传给地电位电工。中间电位电工换上新绝缘子,并复位上、下锁紧销	17	1)未检查承力工具安装情况即开始收紧丝杠扣3分;检查了未报告扣2分;报告了但工作负责人未同意即开始收紧丝杠扣1分。2)预收紧丝杠后未检查承力工具受力情况扣3分;检查了未报告扣2分;报告了但工作负责人未同意即继续收紧丝杠扣1分。3)未均衡收紧丝杠扣2分。4)绳结错误扣1分。5)传递时高空落物扣1分/次。6)新旧绝缘子相互碰撞扣1分。7)传递绝缘子与塔身相互碰撞扣1分。8)绝缘绳缠绕扣0.5分			

续表

序号	项目名称	质量要求	分值	扣分标准	扣分原因	扣分	得分
11	拆除工具	1)中间电位电工检查新绝缘子连接可靠,得到工作负责人同意后,等电位电工松出液压紧线系统。 2)中间电位电工再次检查新绝缘子受力情况无误后,得到工作负责人同意后,拆除更换工具并传递至地面	6	1)捆扎工具时,未正确使用绳结扣1分。 2)传递时高处落物扣1分/次件。 3)工器具相互碰撞扣1分/次;工器具与带电4)体或塔身相互碰撞扣1分/次;绝缘绳缠绕扣0.5分/次。 5)地面人员踩踏防潮苫布扣1分/次;地面人员站在作业点正下方扣1分/人.次			
12	退出强电场	1)经检查作业点无遗留物后经工作负责人许可,等电位电工携带绝缘传递绳作退出电位准备。 2)等电位电工利用电位转移棒钩紧均压环,并进入距均压环的第3片绝缘子,一只手抓紧绝缘子,另一只手握电位转移棒,利用电位转移棒快速脱离电位。 3)等电位电工按照"跨二短三"作业方式退出强电场	10	1)未向工作负责人申请即进行电位转移扣2分;申请了但未得同意即开始扣1分。 2)等电位电工电位转移过程中裸露部分距离不够扣2分。 3)等电位电工退出强电场动作不正确,反复放电扣2分。 4)转移电位动作不熟练扣2分,电位转移未使用电位转移棒扣3分			
13	返回地面	塔上电工检查塔上无遗留物后,向工作负责人汇报,得到工作负责人同意后携带绝缘传递绳下塔	5	1)下塔过程未使用防坠装置扣2分。 2)塔上移位失去安全带保护的扣2分。 3)下塔抓塔钉,每处扣1分。 4)塔上有遗留物的,扣2分			
14	工作结束	1)工作负责人组织全体工作成员整理工器具和材料,将工器具清洁后放入专用的箱(袋)中;清理现场,做到"工完料尽场地清"。 2)召开班后会,工作负责人进行工作总结和点评工作。点评本次工作的施工质量;点评全体工作成员的安全措施落实情况。 3)工作负责人向值班调控人员汇报工作结束,申请恢复线路重合闸,终结工作票	10	1)工器具未清理扣2分。 2)工器具有遗漏扣2分。 3)未开班后会不得2分。 4)未拆除围栏扣2分。 5)未向调度汇报不得2分			
	合计		100				

模块 7　带电更换 1000kV 交流输电线路导线间隔棒培训及考核标准

一、培训标准

(一) 培训要求

模块名称	带电更换1000千伏交流输电线路导线间隔棒	培训类别	操作类
培训方式	实操培训	培训学时	21学时
培训目标	1.掌握沿耐张绝缘子串进、出1000kV强电场时采用"跨二短三"作业方式的电学意义。 2.能完成沿耐张绝缘子串进入1000kV等电位作业点。 3.能独立完成更换导线间隔棒的操作(等电位作业法)		
培训场地	特高压交流实训线路		
培训内容	采用"跨二短三"作业方式沿耐张绝缘子串进入强电场,采用等电位作业法带电更换1000kV交流输电线路八分裂导线间隔棒		
适用范围	特高压交流输电线路检修人员		

(二) 引用规程规范

(1)《1000kV架空输电线路设计规范》(GB 50665—2011)。

(2)《1000kV交流输电线路检修规范》(DL/T 209—2008)。

(3)《1000kV交流输电线路运行规范》(DL/T 307—2010)。

(4)《1000kV交流输电线路带电作业技术导则》(DL/T 392—2015)。

(5)《交流线路带电作业安全距离计算方法》(GB/T 19185—2008)。

(6)《国家电网公司带电作业工作管理规定(试行)》(国家电网生〔2007〕751号)。

(7)《国家电网公司电力安全工作规程(线路部分)》(Q/GDW1799.2—2013)。

(8)《电工术语 架空线路》(GB/T 2900.51—1998)。

(9)《电工术语 带电作业》(GB/T 2900.55—2002)。

(10)《带电作业工具设备术语》(GB/T 14286—2002)。

(11)《带电作业用绝缘滑车》(GB/T 13034—2008)。

(12)《带电作业用绝缘绳索》(GB 13035—2008)。

(13)《带电作业用工具、装置和设备使用的一般要求》(DL/T 877—2004)。

(14)《带电作业工具、装置和设备预防性试验规程)》(DL/T 976—2005)。

(15)《1000kV交流带电作业用屏蔽服装》(GB/T 25726—2010)。

(16)《1000kV交流输电线路带电作业技术导则》(DL/T 392—2010)。

(三)培训教学设计

本设计以完成"带电更换1000kV交流输电线路导线间隔棒"为工作任务,按工作任务完成的标准化作业流程来设计各个培训阶段,每个阶段包括了具体的培训目标、培训内容、培训学时、培训方法(培训资源)、培训环境和考核评价等内容,如表2-7-1所示。

表2-7-1 带电更换1000kV交流输电线路导线间隔棒培训内容设计

培训流程	培训目标	培训内容	培训学时	培训方法与资源	培训环境	考核评价
1.理论教学	1.初步掌握沿绝缘子串进出1000kV强电场基本方法。2.熟悉电位转移的方法。3.熟悉输电线路受损导线间隔棒更换方法。4.熟悉特高压交流线路带电作业的安全距离、危险点辨识及预控	1.沿绝缘子进出强电场"跨二短三"作业方式的电学意义。2.进、出特高压强电场时电位转移棒的使用方法。3.输电线路导线间隔棒更换方法和质量标准。4.特高压交流线路带电作业安全距离、危险点分析及预控措施	2	培训方法:讲授法。培训资源:PPT、相关规程、规范及技术导则	多媒体教室	考勤、课堂提问和作业
2.准备工作	能完成作业前准备工作	1.作业现场查勘。2.编制培训标准化作业卡。3.填写培训操作工作票。4.完成本操作的工器具及材料准备	1	培训方法:1.现场查勘和工器具及材料准备采用现场实操方法。2.编写作业卡和填写工作票采用讲授方法。培训资源:1.1000kV实训线路。2.特高压工器具库房。3.空白工作票	1.1000kV实训线路2.多媒体教室	

续表

培训流程	培训目标	培训内容	培训学时	培训方法与资源	培训环境	考核评价
3.作业现场准备	能完成作业现场准备工作	1.作业现场复勘。2.工作申请。3.作业现场布置。4.班前会。5.工器具及材料检查。6.间隔棒专用扳手使用方法	1	培训方法:演示与角色扮演法。资源:1000kV实训线路	1000kV实训线路	
4.培训师演示	通过现场观摩,使学员初步领会本任务操作流程	1.零值绝缘子的检测。2.等电位电工沿耐张绝缘子串进、出强电场及电位转移。3.等电位电工采用走线方式到达间隔棒更换位置。4.等电位电工用专用工具完成导线间隔棒更换	2	培训方法:演示法。资源:1000kV实训线路	1000kV实训线路	
5.学员分组训练	1.能完成沿绝缘子串进、出1000kV强电场及电位转移操作。2.能完成1000kV输电线路导线间隔棒更换作业	1.学员分组(6人一组)训练进、出1000kV强电场、电位转移和更换导线间隔棒技能操作。2.培训师对学员操作进行指导和安全监护	14	培训方法:角色扮演法。资源:1000kV实训线路	1000kV实训线路	采用技能考核评分细则对学员操作评分
6.工作终结	1.使学员进一步辨析操作过程不足之处,便于后期提升。2.培养学员安全文明生产的工作作风	1.作业现场清理。2.向调度汇报工作。3.班后会,对本次工作任务进行点评总结	1	培训方法:讲授和归纳法	1000kV实训线路	

(四)作业流程

1. 工作任务

采用"跨二短三"的作业方式沿耐张绝缘子串进入强电场、到达作业点,采用等电位作业法带电更换1000kV交流输电线路八分裂导线间隔棒。

(本作业任务适用于海拔1000m及以下地区1000kV交流单回输电线路耐张塔边相导线第一个间隔棒作业点位)

2. 天气及作业现场要求

(1)带电更换1000kV交流输电线路导线八分裂间隔棒应在良好的天气进行。

如遇雷电(听见雷声、看见闪电)、雪、雹、雨、雾等,禁止进行带电作业。风力大于5

级,或空气相对湿度大于80%时,不宜进行带电作业;恶劣天气下必须开展带电抢修时,应组织有关人员充分讨论并编制必要的安全措施,经本单位批准后方可进行。

(2)作业人员精神状态良好,熟悉工作中保证安全的组织措施和技术措施;应持有在有效期内的带电作业资质证书。

(3)工作负责人应事先组织相关人员完成现场勘察,根据勘察结果确定本次作业方法和所需工器具,以及应采取的必要措施,并办理带电作业工作票。

(4)作业现场应合理设置围栏,并妥当布置警示标示牌,禁止非工作人员入内。

(5)本项目需停用线路重合闸装置。

(6)工作中安全距离及有效绝缘长度如表2-7-2所示。

表2-7-2 带电更换1000kV交流输电线路导线间隔棒的安全距离

电压等级	人身与带电体安全距离/m	与邻相导线的最小距离/m	最小有效绝缘长度/m		最小组合间隙/m	转移电位时人体裸露部分与带电体的最小距离/m
			绝缘操作杆	绝缘承力工具、绝缘绳索		
1000kV	6.0 (6.8)	6.9 (7.2)	6.8	6.8	6.7 (6.9)	0.5

注:①海拔高度1000m及以下时,1000kV交流单回输电线路带电作业中相的安全距离采用括号内6.8m的数据、最小距离采用括号内7.2m的数据、组合间隙采用括号内6.9m的数据。

②因为1000kV特高压线路相间距离足够大(一般控制相地距离即可),不做重要安全因素考虑,所以《安规》没有给出"与邻相导线的最小距离"的数据,实际工作中可以参考750kV的数据。

(7)在1000kV输电线路上作业,应保证作业相良好绝缘子片数不少于37片。

3. 准备工作

3.1 危险点及其预控措施

(1)危险点——触电伤害

预控措施:

①工作前,工作负责人应与值班调控人员联系,停用线路重合闸,并履行许可手续。

②塔上等电位作业人员登塔前,必须仔细核对线路名称、杆塔编号、相别,确认无误后方可上塔。

③工作中,如遇线路突然停电,作业人员应视其仍然带电。工作负责人应尽快与调控人员联系,值班调控人员未与工作负责人取得联系前不准强送电。

④地面电工操作绝缘工具时应戴清洁、干燥的防汗手套,绝缘工具及绝缘绳索不得损坏、受潮、变形、失灵,不准使用非绝缘绳索(如棉纱绳、白棕绳、钢丝绳),现场所使用

的带电作业工具应放置在防潮苫布上，防止绝缘工具在使用中脏污和受潮。

⑤等电位作业人员应穿着阻燃内衣，衣服外面应穿戴全套1000kV带电作业用屏蔽服（包括连衣裤帽、面罩、手套、导电袜和导电鞋），且各部分应连接良好，全套屏蔽服衣裤最远端点之间的电阻值不得大于20Ω。

⑥等电位作业人员在电位转移前，应得到工作负责人的许可，人体裸露部分与带电体的最小距离不小于0.5m；电位转移时，应使用电位转移棒，动作应迅速，严禁用头部充放电；与地电位作业人员传递工具和材料时，使用绝缘工具或绝缘绳索的有效长度不准小于表2-7-2的规定。

⑦用绝缘绳索传递大件金属物品时，地电位作业人员应将金属物品接地后再接触。

⑧专责监护人应对作业人员进行不间断监护，随时纠正其不规范或违章动作。重点关注高处作业人员，使其保持足够的安全距离（符合表2-7-2的规定），禁止同时接触两个非连通的带电体或带电体与接地体。

（2）危险点——高处坠落

预控措施：

①高处作业人员登高前，必须具备符合本项作业要求的身体状况、精神状态和技能素质。

②高处作业人员登塔前对安全带和防坠器进行外观检查和冲击试验检查，确保其机械强度符合要求。

③高处作业人员应先检查脚钉是否齐全牢固、鞋底是否清洁，防坠装置是否牢固可靠并加挂防坠器；上下塔时，手抓主材、脚踩脚钉、匀步登（下）塔。

④监护人员应随时纠正其不规范或违章动作，重点关注高处作业人员在转位的过程中不得失去安全带或绝缘后备保护绳的保护，安全带系在牢固部件上，严禁低挂高用。

⑤等电位作业人员在绝缘子串上平行移动通常采取双手抓扶一串，双脚踩另一串的姿势匀速进入强电场，移动过程中，后备保护绳兜住二串绝缘子，避免大挥手、大迈步等动作发生。

⑥等电位作业人员沿导线走线，必须系好安全带，并使后备保护绳需将子导线全部兜住；走线过程中应控制重心，防止导线翻转。

（3）危险点——高处坠物伤人。

预控措施：

①高处作业人员的个人工具及零星材料应装入工具袋，严禁在高处浮置物件、口中含物。

②地面作业人员必须正确佩戴安全帽，正确使用绳结传递工器具及材料，与作业点垂直下方距离不得小于坠落半径。

③作业现场设置围栏并挂好警示标示牌。监护人员应时刻注意，禁止非工作人员及车辆进入作业区域。

3.2 工器具及材料选择

带电更换1000kV交流输电线路八分裂导线间隔棒所需工器具及材料见表2-7-3。工器具出库前,应认真核对工器具的使用电压等级和试验周期,并检查确认外观良好、连接牢固、转动灵活,且符合本次工作任务的要求;工器具出库后,应存放在工具袋或工具箱内进行运输,防止脏污、受潮;金属工具和绝缘工器具应分开装运,防止因混装运输导致工器具变形、损伤等现象发生。

表2-7-3 带电更换1000kV交流输电线路导线间隔棒所需工器具及材料表

序号	名称	规格型号	单位	数量	备注
1	绝缘传递绳	TJS-12,长度与起吊高度匹配	根	2	绝缘工具
2	绝缘后备保护绳	TJS-16,加缓冲器	根	2	绝缘工具
3	绝缘滑车	JH10-0.5	只	2	绝缘工具
4	绝缘绳套	TJS-14	根	2	绝缘工具
5	电位转移棒	0.4m	根	1	绝缘工具
6	绝缘千斤		根	4	绝缘工具
7	I型屏蔽服(连衣裤帽、面罩、手套和导电袜)	屏蔽效率>=60dB(屏蔽面罩屏蔽效率>=20dB)	套	2	个人防护用具
8	导电鞋	尺码视穿着人员而定	双	2	个人防护用具
9	阻燃内衣	纯桑蚕丝	套	2	个人防护用具
10	双保险安全带	全身背带式	副	2	个人防护用具
11	防坠器	与杆塔防坠器装置型号对应	只	2	个人防护用具
12	安全帽		顶	6	个人防护用具
13	间隔棒专用扳手	八分裂间隔棒用	个	1	专用工具
14	绝缘子检测仪		套	1	其他工具
15	绝缘电阻测试仪	5000V,电极宽2cm、极间宽2cm	套	1	其他工具
16	风速、温湿度测试仪	HT-8321	套	1	其他工具
17	万用表		只		其他工具
18	对讲机	视工作需要	套	2	其他工具
19	防潮苫布	2m×4m	块	2	其他工具
20	安全围栏		套	若干	其他工具
21	警示标示牌	"在此工作""从此进出""从此上下"	套	1	其他工具
22	红马甲	"工作负责人"	件	1	其他工具
23	清洁毛巾	棉质	条	1	其他工具

续表

序号	名称	规格型号	单位	数量	备注
24	鞋套		双	若干	其他工具
25	工作手套		双	若干	其他工具
26	个人工具	工具袋、平口钳、记号笔	套	2	其他工具
27	八分裂间隔棒	与被更换间隔棒同型号	只	1	材料

注：绝缘工器具的电气及机械强度应满足《电力安全工作规程（线路部分）》要求，试验合格并在有效期内。

3.3 作业人员分工

本任务作业人员分工如表2-7-4所示。

表2-7-4 带电更换1000kV交流输电线路导线间隔棒人员分工表

序号	工作岗位	数量（人）	工作职责
1	工作负责人	1	负责本次工作任务的人员分工、工作票的宣读、办理线路停用重合闸、办理工作许可手续、召开工作班前会、工作中突发情况的处理、工作质量的监督、工作后的总结
2	专责监护人	1	负责作业过程中的安全监督及把控
3	等电位电工	1	负责进入等电位更换八分裂导线间隔棒工作
4	塔上地电位电工	1	负责检测零值或低值绝缘子、协助等电位进出强电场
5	地面电工	2	负责执行现场安全措施、布置作业现场、检查工器具、传递工具及材料，配合等电位电工进出等电位

4. 工作程序

本任务工作流程如表2-7-5所示。

表2-7-5 带电更换1000kV交流输电线路导线间隔棒工作流程表

序号	作业内容	作业步骤及标准	安全措施及注意事项	责任人
1	现场复勘	工作负责人负责完成以下工作： (1)现场核对线路名称、杆塔编号，相别无误；基础及杆塔完好无异常；交叉跨越距离符合安全要求；确认缺陷情况及导地线规格型号等。 (2)检测风速、湿度等现场气象条件符合作业要求。 (3)检查地形环境符合作业要求。 (4)检查工作票所列安全措施与现场实际情况相符，必要时予以补充	(1)正确穿戴安全帽、工作服、工作鞋、劳保手套。 (2)不得在危及作业人员安全的气象条件下作业。 (3)严禁非工作人员、车辆进入作业现场	

续表

序号	作业内容	作业步骤及标准	安全措施及注意事项	责任人
2	工作许可	(1)工作负责人负责联系值班调控人员,按工作票内容申请停用线路重合闸。 (2)经值班调控人员许可后,方可开始带电作业工作	不得未经值班调控人员许可即开始工作	
3	现场布置	正确装设安全围栏并悬挂标示牌: (1)安全围栏范围应充分考虑高处坠物,以及对道路交通的影响。 (2)安全围栏出入口设置合理。 (3)妥当布置齐备"从此进出""在此工作""从此上下"等标示	对道路交通安全影响不可控时,应及时联系交通管理部门强化现场交通安全管控	
4	召开班前会	(1)全体工作成员列队。 (2)工作负责人宣读工作票,明确工作任务及人员分工;讲解工作中的安全措施和技术措施;查(问)全体工作成员精神状态;告知工作中存在的危险点及采取的预控措施。 (3)全体工作成员在工作票上签名确认	(1)工作票填写、签发和许可手续规范,签名完整。 (2)全体工作成员精神状态良好。 (3)全体工作成员明确任务分工、安全措施和技术措施	
5	检查工器具	(1)塔上地电位电工和等电位电工正确地穿好屏蔽服并检测合格,由负责人监督检查。 (2)正确佩戴个人安全用具(大小合适,锁扣自如),由负责人监督检查。 (3)测量风速风向、湿度,检查绝缘工具的绝缘性能,并做好记录	(1)金属、绝缘工具使用前,应仔细检查其是否损坏、变形、失灵。绝缘工具应使用2500V及以上绝缘电阻测试仪进行分段绝缘检测,阻值应不低于700MΩ,并用清洁干燥的毛巾将其擦拭干净。 (2)用万用表测量屏蔽服衣裤最远端点之间的电阻值不得大于20Ω。工作负责人认真检查作业电工屏蔽服的连接情况。 (3)检查工具组装情况并确认连接可靠。 (4)现场所使用的带电作业工具应放置在防潮苫布上	

续表

序号	作业内容	作业步骤及标准	安全措施及注意事项	责任人
6	登塔	(1)核对线路名称、杆塔编号无误后,塔上地电位电工和等电位电工冲击检查安全带、防坠器受力情况。 (2)塔上地电位电工携带绝缘传递绳登塔,等电位电工随后登塔,两人至横担作业点,选择合适位置系好安全带,塔上地电位电工将绝缘滑车和绝缘传递绳安装在横担合适位置。然后配合地面电工将绝缘传递绳分开作起吊准备	(1)核对线路名称和杆塔编号无误后,方可登塔作业。 (2)登塔过程中应使用塔上安装的防坠装置;杆塔上移动及转位时,不准失去安全保护,作业人员必须攀抓牢固构件。 (3)作业电工必须穿全套合格的屏蔽服,且全套屏蔽服必须连接可靠。在横担进入等电位前,等电位电工再次检查确认屏蔽服各部位连接可靠后方能进行下一步操作	
7	检测绝缘子	(1)地面电工将绝缘操作杆及绝缘子检测仪传至塔上。等电位电工协助塔上地电位电工对绝缘子进行检测。 (2)检测顺序由导线侧向横担侧进行,并作好记录。 (3)检测完毕后,地面电工配合塔上电工将绝缘操作杆传至地面,并将电位转移棒传至塔上	(1)如果是玻璃绝缘子则确认是否良好,不用检测;如果是瓷质绝缘子则如下(2)、(3)。 (2)检测绝缘子工作必须逐串逐片进行,接触必须可靠。 (3)当任意一串良好绝缘子少于37片时,立即停止检测,并停止本次带电作业工作。 (4)上下传递工具时,绑扎绳扣应正确可靠,防止高处坠物	
8	进入强电场	(1)等电位电工将安全带转移到绝缘子连接金具上,并携带电位转移棒、绝缘滑车和绝缘传递绳。 (2)等电位电工检查屏蔽服各部分连接良好后报经工作负责人同意,双手抓扶一串,双脚踩另一串,采用"跨二短三"作业方式沿绝缘子串进入强电场。 (3)当作业人员平行移动至距导线侧均压环3片绝缘子时,应停止移动,利用电位转移棒进行电位转移	(1)等电位电工进入电位前必须得到工作负责人的许可。 (2)等电位电工进入绝缘子串时应交替使用安全带和后备保护绳(用后备保护绳兜住两串绝缘子、手抓扶其中一串,脚踩另一串),不得失去安全带的保护;并调整好绝缘传递绳和电位转移棒。 (3)等电位电工在进入电位过程中手和脚应协调配合,速度均匀,避免大挥手、大迈步等动作发生;与接地体和带电体两部分间隙所组成的组合间隙边相大于6.7m(中相大于6.9m)。 (4)与相邻导线的最小距离大于6.9(中相7.2)m。 (5)等电位电工进行电位转移前应检查电位转移棒与屏蔽服的电气连接是否可靠,人体裸露部分与带电体的最小距离大于0.5m,并得到工作负责人的许可;电位转移时不得失去安全带的保护,进入强电场瞬间动作准确、平稳、迅速	

续表

序号	作业内容	作业步骤及标准	安全措施及注意事项	责任人
9	更换导线八分裂间隔棒	(1)等电位电工进入等电位后,将安全带系在上子导线上,并装好走线绝缘保护绳(需将子导线全部兜住)。 (2)等电位电工携带绝缘传递绳沿导线走线至更换间隔棒作业点,先将绝缘绳套安装在子导线上合适位置,其次连接绝缘滑车和绝缘传递绳,再将绝缘滑车钩挂在绝缘绳套内。 (3)等电位电工对导线上旧间隔棒安装点使用记号笔4点对称画印进行标记。 (4)等电位电工在旧间隔棒旁合适位置采用二二对应的方式安装4根绝缘千斤,将子导线可靠固定。 (5)等电位电工先利用绝缘传递绳采用活结的方式绑牢旧间隔棒,然后利用间隔棒专用扳手将旧间隔棒拆除,与地面电工配合利用绝缘传递绳将其放至地面。 (6)地面电工起吊新间隔棒至等电位电工处,等电位电工对应画印标记,正确安装新间隔棒,安装完毕后,应保持间隔棒的平面与子导线垂直。 (7)等电位电工依次拆除绝缘千斤、绝缘滑车、绝缘传递绳及绳套。 (8)等电位作业过程中不得掉落工器具和材料	(1)等电位电工不得失去安全带的保护。 (2)等电位电工与地面电工要密切配合,听从工作负责人的指挥。 (3)与相邻导线的最小距离大于6.9(中相7.2)m。 (4)导线间隔棒在上下传递过程中,不得磕碰,绝缘传递绳索不得相互缠绕。 (5)上下传递工具时,绑扎绳结应正确可靠,防止高空坠物。 (6)传递工器具及材料过程中,地面电工禁止站立在等电位电工工作点位正下方	
10	退出强电场	(1)经检查间隔棒安装牢固,作业点无遗留物后经工作负责人许可,等电位电工携带绝缘传递绳沿导线返回均压环处,作退出电位准备。 (2)等电位电工利用电位转移棒钩紧均压环,并进入距均压环的第3片绝缘子,一只手抓紧绝缘子,另一只手握电位转移棒,利用电位转移棒快速脱离等电位。 (3)等电位电工按照"跨二短三"作业方式退出强电场	(1)等电位电工退出电位前必须得到工作负责人的许可。 (2)等电位电工返回绝缘子串时应交替使用安全带和后备保护绳,电位转移时不得失去安全带的保护,退出强电场瞬间动作准确、平稳、迅速。 (3)等电位电工在脱离电位过程中手和脚应协调配合,速度均匀,避免大挥手、大迈步等动作发生;与接地体和带电体两部分间隙所组成的组合间隙边相大于6.7m(中相大于6.9m);人体裸露部分与带电体的最小距离大于0.5m。 (4)等电位电工沿绝缘子串移动时,用后备保护绳兜住两串绝缘子、手要抓牢,脚要踏实。 (5)与相邻导线的最小距离大于6.9(中相7.2)m。 (6)等电位电工返回横担时不得同时失去安全带或后备保护绳的保护	

续表

序号	作业内容	作业步骤及标准	安全措施及注意事项	责任人
11	返回地面	塔上电工检查塔上无遗留物后,向工作负责人汇报,得到工作负责人同意后携带绝缘传递绳下塔	下塔过程中应使用塔上安装的防坠装置,杆塔上移动及转位时,不得失去安全保护,作业人员必须攀抓牢固构件	
12	工作结束	(1)工作负责人组织全体工作成员整理工器具和材料,将工器具清洁后放入专用的箱(袋)中;清理现场,做到"工完料尽场地清"。 (2)召开班后会,工作负责人进行工作总结和点评工作。点评本次工作的施工质量;点评全体工作成员的安全措施落实情况。 (3)工作负责人向值班调控人员汇报工作结束,申请恢复线路重合闸,终结工作票	严禁约时恢复线路重合闸	

二、考核标准

特高压交流输电线路运检技能考核评分细则

考生填写栏	编号:	姓名:	所在岗位:	单位:	日 期: 年 月 日			
考评员填写栏	成绩:	考评员:	考评组长:	开始时间:	结束时间:	操作时长:		
考核模块	带电更换1000kV交流输电线路导线间隔棒		考核对象	特高压交流输电线路检修人员	考核方式	操作	考核时限	90min
任务描述	沿耐张绝缘子串进入强电场对1000kV交流输电线路受损导线八分裂间隔棒进行带电更换(等电位作业法)							
工作规范及要求	1. 带电作业工作应在良好天气下进行。如遇雷、雨、雪、雾天气不得进行带电作业。风力大于5级、湿度大于80%时,一般不宜进行带电作业。 2. 本项作业需工作负责人1名,专责监护人1人、塔上地电工1人,等电位电工1人,地面辅助电工2人,采用沿绝缘子串进入强电场对1000kV交流输电线路受损导线八分裂间隔棒进行带电更换。 3. 工作负责人职责:负责本次工作任务的人员分工、工作票的宣读、办理线路停用重合闸、办理工作许可手续、召开工作班前会、工作中突发情况的处理、工作质量的监督、工作后的总结。 4. 专责监护人:负责作业过程中的安全监督及把控。 5. 等电位电工职责:负责沿绝缘子串进入强电场对导线八分裂间隔棒进行更换。 6. 塔上地电工职责:负责检测零值或低值绝缘子、协助等电位进出强电场。 7. 地面电工职责:负责执行现场安全措施、布置作业现场、检查工器具、传递工具及材料,配合等电位电工进出等电位。 8. 在带电作业中,如遇雷、雨、大风或其他任何情况威胁到工作人员的安全时,工作负责人或监护人可根据情况,临时停止工作。							

续表

工作规范及要求	给定条件： 1. 培训基地：特高压交流1000kV实训线路耐张塔大号侧A相导线第1个八分裂间隔棒，导线型号：8×JL/G1A-630/45。 2. 工作票已办理，安全措施已经完备（重合闸已停用），工作开始、工作终结时应口头提出申请（调度或考评员）。 3. 作业现场装设安全围栏，悬挂"在此工作""从此进出"等标示牌，安全措施已完备。 4. 安全、正确地使用仪器仪表对绝缘工具进行检测。 5. 上下塔过程中应使用防坠落装置，防止高处坠落。 6. 必须按标准化作业程序进行操作，工序错误扣除应做项目分值，出现重大人身、器材和操作安全隐患，考评员可下令终止操作（考核），本模块考核成绩记为"不合格"
考核情景准备	1. 线路：特高压交流1000kV实训线路耐张塔大号侧A相导线，工作内容：带电更换1000kV受损导线间隔棒，导线型号：8×JL/G1A-630/45。 2. 所需作业工器具：绝缘传递绳2根（TJS-12），绝缘后备保护绳2根（TJS-16，加缓冲器），绝缘滑车2只（JH10-0.5），绝缘绳套2根（TJS-14），绝缘千斤4根，电位转移棒1根（0.4m），I型屏蔽服2套（连衣裤帽、面罩、手套和导电袜），导电鞋2双，防坠器2只，间隔棒专用扳手1个，绝缘子检测仪1套，绝缘电阻测试仪1套（5000V型），风速、温湿度测试仪1套（HT-8321），万用表1只，防潮苫布2块（2m×4m），红马甲1件（工作负责人），清洁毛巾2条，个人工具2套（工具袋、平口钳、记号笔），同型号八分裂间隔棒1只。 3. 作业现场做好监护工作，作业现场安全措施（围栏等）已全部落实；禁止非作业人员进入现场，工作人员进入作业现场必须戴安全帽。 4. 考生自备工作服，阻燃纯棉内衣，安全帽，线手套，安全带（含二保绳）
备注	1. 本模块总分为100分，各项得分均以对应分值扣完即止，在规定时间内不能完成任务应立即终止考试，本模块成绩按已完成项实际得分统计，未完成项不得分。 2. 考核过程中因设备、作业环境、安全措施、安全防护、安全距离等不符合作业要求，或人为误操作，出现可能危及作业安全的任意情况，考评员应下令终止操作。 3. 考试前统一组织参考人员进行现场查勘，并提前办理工作票

序号	项目名称	质量要求	分值	扣分标准	扣分原因	扣分	得分
1	现场复勘	1）工作负责人到作业现场核对线路名称和杆塔编号、现场工作条件、缺陷部位等。 2）检测风速、湿度等现场气象条件符合作业要求。 3）检查工作票填写完整，无涂改，检查是否所列安全措施与现场实际情况相符，必要时予以补充	5	1）无工作票，本项不得分。 2）未核对双重称号，扣1分。 3）未核实现场工作条件（气象）、缺陷部位，每项扣1分。 4）工作票填写出现涂改、不整洁，每处扣0.5分，工作票编号有误，扣1分。工作票填写漏项，每项扣1分			

续表

序号	项目名称	质量要求	分值	扣分标准	扣分原因	扣分	得分
2	工作许可	1)工作负责人联系值班调控人员,按工作票内容申请停用线路重合闸。 2)汇报内容规范、完整,声音清楚宏亮。 3)及时履行相关许可手续	2	1)未取得调度部门(考评员)工作许可擅自开工,本项不得分。 2)汇报专业用语不规范、不完整或声音清楚宏亮各扣0.5分。 3)未申请停用重合闸,扣1分。 4)未复诵许可内容,扣1分。 5)复诵内容漏项,每项扣0.5分(许可人姓名、许可时间、工作任务、重合闸状态)。 6)未及时完善工作票,扣1分			
3	现场布置	正确装设安全围栏并悬挂标示牌: 1)安全围栏范围应充分考虑高处坠物,以及对道路交通的影响。 2)安全围栏出入口设置合理。 3)妥当布置"从此进出""在此工作""从此上下"等标示	3	1)未装设作业现场安全围栏,扣2分。 2)作业现场安全围栏设置不合理,每处扣1分。 3)未设悬挂标示牌,扣1.5分。 4)悬挂标示牌不齐,每块0.5分。 5)非作业人员进入围栏区,每人扣0.5分			
4	召开班前会	1)全体工作成员正确佩戴安全帽、工作服。 2)工作负责人穿(戴)安全红马甲,宣读工作票,明确工作任务及人员分工;讲解工作中的安全措施和技术措施;查(问)全体工作成员精神状态;告知工作中存在的危险点及采取的预控措施。 3)全体工作成员在工作票上签名确认	3	1)工作成员安全帽佩戴不正确,每人扣0.5分,着装不整齐,每人次扣0.5分。 2)工作负责人和专责监护人未穿(戴)安全红马甲,每人扣0.5分。 3)未明确工作任务及分工,本项不得分。 4)人员分工不明确,扣1分。 5)安全措施、预控措施交代不全,扣1分。 6)未告知工作中存在的危险点,扣1分。 7)未确认工作班成员精神状态,扣1分。 8)工作班成员未签字或签字不全,扣1分			

续表

序号	项目名称	质量要求	分值	扣分标准	扣分原因	扣分	得分
5	工器具检查	1)工作班成员在合适位置正确设置防潮苫布,防潮苫布应清洁、干燥,严禁踩踏苫布。2)工器具应按定置管理要求分类、整齐摆放于防潮苫布上;绝缘工器具不能与金属工具、材料混放;对工器具及仪器仪表进行外观检查。3)各种工具均试验合格,并在试验有效时间内。绝缘工具表面不应磨损、变形损坏,操作应灵活。绝缘工具应使用2500V及以上绝缘电阻表进行分段绝缘检测,阻值应不低于700MΩ,并用清洁干燥的毛巾将其擦拭干净。4)塔上地电位和等电位人员按要求正确穿戴全套合格的屏蔽服、导电鞋,且各部分连接应良好,屏蔽服内不得贴身穿着化纤类衣服,并系好安全带;工作负责人应认真检查确认是否穿戴正确、各部连接良好。5)全套屏蔽服应使用万用表进行测试,其最远两点之间阻值不大于20Ω,单件不大于15Ω。6)登塔电工对安全带及后备保护绳、防坠器进行外观检查,并经冲击试验合格。	7	1)防潮苫布设置位置不合适,扣1分。2)踩踏防潮苫布,每次扣0.5分。3)工器具未分类定置摆放,扣1分。4)未检查工器具及仪器仪表试验合格标签和外观检查,每件扣1分。5)工器具及仪器仪表检查漏项,每件扣0.5分。6)仪器仪表、工器具检查方法不正确,每件扣0.5分。7)未对硬质绝缘工具进行清洁、擦拭,每件扣0.5分。8)未戴清洁干燥的棉线手套持、拿绝缘工具,每次扣0.5分。9)未正确使用检测仪器对工器具及全套屏蔽服进行检测,每项扣1分。10)登塔电工未正确穿戴屏蔽服或连接部分未检查,每人扣2分。11)安全带及后备保护绳、防坠器未外观检查和冲击试验(或方式不正确),每项扣1分。12)工作负责人未检查或漏查登塔电工安全防护装备,每项扣1分			
6	登塔	1)登塔人员再次核对双重名称、杆号、相别并向工作负责人报告及申请登塔。2)塔上地电位电工、等电位电工携带绝缘传递绳相继登塔。3)登塔过程中必须使用防坠落装置,手抓主材、匀步登塔,安全带及后备保护绳挂在肩上并与带电体保持6m以上安全距离,工作负责人加强作业监护。	5	1)登塔电工未核对线路双重名称、塔号、相别,每项扣1分。2)登塔电工未报告核对结果,未申请登塔,每项扣1分。3)未使用防坠落装置登塔,考评员应下令终止操作(考核)。4)手抓脚钉登塔,每次扣0.5分。5)登塔踩滑、踏空,每次扣1分。			

续表

序号	项目名称	质量要求	分值	扣分标准	扣分原因	扣分	得分
6	登塔	4)登塔至合适位置,正确使用安全带,布置好绝缘传递绳,然后塔上地电位电工配合地面电工将绝缘传递绳分开作起吊准备。 5)工作负责人认真监护并提醒整个登塔过程		6)安全带主带和后备保护绳未挂肩上,每人扣1分。 7)安全带及后备保护绳发生缠绕、勾住,每次扣1分。 8)安全带及后备保护绳低挂高用,每次扣1分。 9)安全带及后备保护绳系在同一构件上,每次扣1分。 10)高处作业失去安全带的保护,本项不得分。 12)安装滑车未使用绝缘绳套,扣1分。 13)滑车传递绳悬挂位置不便于工具取用,扣1分。 14)工作负责人监护及提醒不到位,扣1分。 15)安全距离不够,考评员应下令终止操作(考核)			
7	检测绝缘子	1)地面电工将绝缘操作杆及绝缘子检测仪传至塔上。地电位电工对火花间隙检测仪进行调试,间隙距离0.4mm。 2)等电位电工协助塔上地电位电工对绝缘子逐串逐片进行检测,接触必须可靠,由导线侧向横担侧进行。 3)遇有零值或低值绝缘子应复测3次。 4)当任意一串良好绝缘子少于37片时,立即停止检测,并停止本次带电作业工作。 5)塔上电工应与带电体保持不小于6m的安全距离,绝缘绳索的有效绝缘长度不小于6.8m,绝缘检测杆的最小绝缘长度不小于6.8m。 6)检测完毕后,地面电工配合塔上电工将绝缘操作杆传至地面,并将电位转移棒传至塔上。 7)工作负责人认真监护并提醒本操作过程	10	1)传递绳绑扎工器具绳结使用错误,扣0.5分。 2)传递工器具发生碰撞,每件扣0.5分。 3)传递绳索缠绕,每次扣0.5分。 4)传递时高处坠物,每次扣1分。 5)未进行绝缘子测零,本项不得分。 6)塔上电工操作方法错误,扣2分。 7)检测绝缘子串顺序错误,扣2分。 8)零值绝缘子未进行复测,扣2分。 9)作业安全距离不够,考评员应下令终止操作。 10)测零结果未汇报,扣2分。 11)未传递电位转移棒,扣2分。 12)工作负责人监护及提醒不到位,扣2分			

续表

序号	项目名称	质量要求	分值	扣分标准	扣分原因	扣分	得分
8	进入强电场	1)进入电位前检查屏蔽服各部分连接良好后,报经工作负责人同意。 2)等电位电工将安全带转移到绝缘子连接金具上,并携带电位转移棒、绝缘滑车和绝缘传递绳。 3)等电位电工进入绝缘子串前必须系好保护绳(用后备保护绳兜住两串绝缘子、双手抓扶其中一串,脚踩另一串)。 4)采用"跨二短三"作业方式沿绝缘子串平行移动至距导线侧均压环3片绝缘子时,应停止移动,得到工作负责人许可后,利用电位转移棒进行电位转移。 5)等电位电工在进入电位过程中与接地体和带电体两部分间隙所组成的组合间隙边相不得小于6.7m(中相不得小于6.9m),进入强电场必须用电位转移棒进行电位转移,人体裸露部分与带电体的最小距离不得小0.5m。 6)进入强电场过程不得失去安全带的保护	10	1)等电位电工未检查屏蔽服各连接部分,扣2分。 2)等电位电工转移到绝缘子连接金具上失去安全带的保护,扣2分。 3)等电位电工与接地体和带电体两部分间隙所组成的组合间隙不够,扣2分。 4)等电位电工进入强电场动作不正确、不熟练,每次扣2分。 5)等电位电工电位转移过程中裸露部分距离不够,导致反复放电,扣2分。 6)转移电位动作不熟练,扣1分。 7)电位转移过程未使用电位转移棒,扣5分。 8)未得到工作负责人许可就进行电位转移,扣3分。 9)等电位电工进入强电场过程失去安全带的保护,扣2分。 10)工作负责人监护及提醒不到位,扣2分			
9	更换导线八分裂间隔棒	1)等电位电工进入等电位后,将围杆带系在上子导线上,并装好走线绝缘保护绳(需将子导线全部兜住)。 2)等电位电工携带绝缘传递绳走线至更换间隔棒作业点,将绝缘绳套正确装在子导线合适位置,并挂好绝缘滑车和绝缘传递绳。 3)等电位电工使用记号笔对旧间隔棒固定点对称画印,做好标记。 4)等电位电工在旧间隔棒旁合适位置采用二二对应的方式安装4根绝缘千斤,将子导线可靠固定。	30	1)绝缘保护绳未将子导线全部兜住,扣3分。 2)走线过程围杆带未系在上子,扣2分。 3)走线动作不熟练,扣2分。 4)绝缘滑车直接钩挂在导线上,扣5分。 5)绝缘绳套安装方式错误或位置不合适,每项扣1分。 6)绝缘滑车钩挂绝缘绳套内后未闭锁,扣1分。 7)未画印及标记,扣3分。 8)未安装4根绝缘千斤进行子导线固定,每根扣2分。 9)未系绳结就开始拆除间隔棒,扣5分。			

续表

序号	项目名称	质量要求	分值	扣分标准	扣分原因	扣分	得分
9	更换导线八分裂间隔棒	5)等电位电工首先采用活结的方式将绝缘传递绳绑牢于旧间隔棒上,然后利用间隔棒专用扳手拆除旧间隔棒,与地面电工配合将其放至地面防潮苫布上。 6)地面电工起吊新间隔棒传递给等电位电工。 7)等电位电工对应画印及标记,正确使用间隔棒专用扳手安装新间隔棒,安装质量应保持间隔棒的平面与子导线垂直。 8)等电位电工依次拆除绝缘千斤、绝缘滑车及绝缘传递绳、绳套。 9)等电位作业过程中不得掉落工器具和材料	30	10)拆除工器具时动作慌乱,扣2分。 11)发生高处坠物,每件扣3分。 12)发生高处抛掷旧间隔棒,本模块不得分。 13)地面电工站立在等电位电工工作点位正下方,每人次扣2分。 14)旧间隔棒未落放在防潮苫布上,扣1分。 15)上下传递工具时绑扎绳结方式错误,扣1分。 16)未安装完新间隔棒就解开绳结,扣5分。 17)安装新间隔棒偏离原间隔棒位置,扣3分。 18)安装完后,新间隔棒的平面与子导线不垂直,扣3分。 19)未拆除绝缘千斤、绝缘滑车及绝缘传递绳、绳套,每件扣1分。 20)未按顺序拆除绝缘千斤、绝缘滑车及绝缘传递绳、绳套,扣2分。 21)未完成新旧间隔棒的的更换工作,本模块不得分。 22)与相邻导线安全距离不够,考评员应下令终止操作			
10	退出强电场	1)经检查间隔棒安装牢固,作业点无遗留物后经工作负责人许可,等电位电工携带绝缘传递绳沿导线返回均压环处,作退出电位准备。 2)等电位电工利用电位转移棒钩紧均压环,并进入距均压环的第3片绝缘子,一只手抓紧绝缘子,另一只手握电位转移棒,利用电位转移棒快速脱离等电位。 3)退出强电场过程不得失去安全带的保护。	10	1)作业点有遗留物,每件扣2分。 2)未向工作负责人申请即进行电位转移,扣3分。 3)未得到工作负责人许可就进行电位转移,扣2分。 4)电位转移过程未使用电位转移棒,扣5分。 5)转移电位动作不熟练,扣1分。 6)等电位电工退出强电场过程失去安全带的保护,扣2分。			

续表

序号	项目名称	质量要求	分值	扣分标准	扣分原因	扣分	得分
10	退出强电场	4)等电位电工按照"跨二短三"作业方式沿绝缘子串退出强电场,转移到横担上。 5)等电位电工在退出电位过程中与接地体和带电体两部分间隙所组成的组合间隙边相不得小于6.7m(中相不得小于6.9m),退出强电场必须用电位转移棒进行电位转移,人体裸露部分与带电体的最小距离不得小0.5m	10	7)等电位电工电位转移过程中裸露部分距离不够,导致反复放电,扣2分。 8)等电位电工退出强电场动作不正确、不熟练,每次扣2分。 9)等电位电工与接地体和带电体两部分间隙所组成的组合间隙不够,扣2分。 10)等电位电工转移到横担上失去安全带的保护,扣2分。 11)工作负责人监护及提醒不到位,扣2分			
11	返回地面	1)塔上电工检查塔上无遗留物后,向工作负责人汇报,得到工作负责人同意后,携带绝缘传递绳相继下塔。 2)下塔过程中必须使用防坠落装置,手抓主材、匀步下塔,安全带及后备保护绳挂在肩上并与带电体保持6m以上安全距离,工作负责人加强作业监护	5	1)塔上有遗留物,每件扣2分。 2)登塔电工未报告遗留物检查结果,未申请下塔,每项扣1分。 3)未使用防坠落装置下塔,考评员应下令终止操作(考核)。 4)手抓脚钉下塔,每次扣0.5分。 5)下塔踩滑、踏空,每次扣1分。 6)安全带主带和后备保护绳未挂肩上,每人扣1分			
12	工作结束	1)工作负责人组织全体工作成员整理工器具和材料,将工器具清洁后放入专用的箱(袋)中;清理现场,做到"工完料尽场地清"。 2)召开班后会,工作负责人进行工作总结和点评。点评本次工作的施工质量;点评全体工作成员的安全措施落实情况。 3)工作负责人向值班调控人员汇报工作结束,申请恢复线路重合闸,终结工作票	10	1)对绝缘工器具未进行清洁、擦拭,每件扣0.5分。 2)工器具未归类整理摆放,每件扣1分。 3)工器具乱丢乱扔或踩踏防潮苫布,每次扣0.5分。 4)未拆除围栏及标示牌或有遗留物,每件扣1分。 5)未开班后会,扣2分。 6)集合站队不整齐、注意力不集中,扣1分。 7)工作班成员参加班后会人员不齐,缺一人扣1分。 8)点评不到位,扣1分。 9)未向调度部门(考评员)汇报工作结束,申请恢复线路重合闸,扣1分。 10)汇报专业用语不规范、不完整或声音不宏亮,各扣0.5分			

续表

序号	项目名称	质量要求	分值	扣分标准	扣分原因	扣分	得分
12	工作结束		10	11)未复诵许可内容,扣1分。 12)复诵内容漏项,每项扣0.5分(单位名称、负责人姓名、时间、线路名称、工作完成情况、设备已恢复正常、人员已撤离、可恢复重合闸)。 13)未及时完善工作票终结手续或填写错误,每项扣1分			
	合计		100				

模块8 带电更换1000kV交流输电线路导线防振锤培训及考核标准

一、培训标准

(一) 培训要求

模块名称	带电更换1000kV交流输电线路导线防振锤	培训类别	操作类
培训方式	实操培训	培训学时	21学时
培训目标	1.掌握沿耐张绝缘子串进、出1000kV强电场时采用"跨二短三"作业方式的电学意义。 2.能完成沿耐张绝缘子串进入1000kV等电位作业点。 3.能独立完成更换导线防振锤的操作(等电位作业法)		
培训场地	特高压交流实训线路		
培训内容	采用"跨二短三"作业方式沿耐张绝缘子串进入强电场,采用等电位作业法带电更换1000kV交流输电线路导线防振锤		
适用范围	特高压交流输电线路检修人员		

(二) 引用规程规范

(1)《1000kV架空输电线路设计规范》(GB 50665—2011)。

(2)《1000kV交流输电线路检修规范》(DL/T 209—2008)。

(3)《1000kV交流输电线路运行规范》(DL/T 307—2010)。

(4)《1000kV交流输电线路带电作业技术导则》(DL/T 392—2015)。

(5)《交流线路带电作业安全距离计算方法》(GB/T 19185—2008)。

(6)《带电作业用绝缘配合导则》(DL/T 867—2004)。

(7)《架空输电线路带电安装导则及作业工具设备》(DL/T 1007—2006)。

(8)《国家电网公司带电作业工作管理规定(试行)》(国家电网生〔2007〕751号)。

(9)《国家电网公司电力安全工作规程(线路部分)》(Q/GDW1799.2—2013)。

(10)《电工术语 架空线路》(GB/T 2900.51—1998)。

(11)《电工术语 带电作业》(GB/T 2900.55—2002)。

(12)《带电作业工具设备术语》(GB/T 14286—2002)。

(13)《带电作业用工具、装置和设备使用的一般要求》(DL/T 877—2004)。

(14)《带电作业工具、装置和设备预防性试验规程)》(DL/T 976—2005)。

(15)《带电作业用绝缘滑车》(GB/T 13034—2008)。

(16)《带电作业用绝缘绳索》(GB 13035—2008)。

(17)《1000kV交流带电作业用屏蔽服装》(GB/T 25726—2010)。

(三)培训教学设计

本设计以完成"带电更换1000kV交流输电线路导线防振锤"为工作任务,按工作任务完成的标准化作业流程来设计各个培训阶段,每个阶段包括了具体的培训目标、培训内容、培训学时、培训方法(培训资源)、培训环境和考核评价等内容,如表2-8-1所示。

表2-8-1 带电更换1000kV交流输电线路导线防振锤培训内容设计

培训流程	培训目标	培训内容	培训学时	培训方法与资源	培训环境	考核评价
1.理论教学	1.初步掌握沿绝缘子串进出1000kV强电场基本方法。2.熟悉电位转移的方法。3.熟悉输电线路受损导线防振锤更换方法。4.熟悉特高压交流线路带电作业的安全距离、危险点辨识及预控	1.沿绝缘子进出强电场"跨二短三"作业方式的电学意义。2.进、出特高压强电场时电位转移棒的使用方法。3.输电线路导线防振锤更换方法和质量标准。4.特高压交流线路带电作业安全距离、危险点分析及预控措施	2	培训方法:讲授法。培训资源:PPT、相关规程、规范及技术导则	多媒体教室	考勤、课堂提问和作业
2.准备工作	能完成作业前准备工作	1.作业现场查勘。2.编制培训标准化作业卡。3.填写培训操作工作票。4.完成本操作的工器具及材料准备	1	培训方法:1.现场查勘和工器具及材料准备采用现场实操方法。2.编写作业卡和填写工作票采用讲授方法。培训资源:1.1000kV实训线路。2.特高压工器具库房。3.空白工作票	1.1000kV实训线路。2.多媒体教室	

续表

培训流程	培训目标	培训内容	培训学时	培训方法与资源	培训环境	考核评价
3.作业现场准备	能完成作业现场准备工作	1.作业现场复勘。 2.工作申请。 3.作业现场布置。 4.班前会。 5.工器具及材料检查。 6.防振锤专用扳手使用方法	1	培训方法： 演示与角色扮演法。 资源： 1000kV 实训线路	1000kV 实训线路	
4.培训师演示	通过现场观摩，使学员初步领会本任务操作流程	1.零值绝缘子的检测。 2.等电位电工沿耐张绝缘子串进、出强电场及电位转移。 3.等电位电工采用走线方式到达防振锤更换位置。 4.等电位电工用专用工具完成导线防振锤更换	2	培训方法： 演示法。 资源： 1000kV 实训线路	1000kV 实训线路	
5.学员分组训练	1.能完成沿绝缘子串进、出1000kV强电场及电位转移操作。 2.能完成 1000kV 输电线路导线防振锤更换作业	1.学员分组（6 人一组）训练进、出1000kV强电场、电位转移和更换导线防振锤技能操作。 2.培训师对学员操作进行指导和安全监护	14	培训方法： 角色扮演法。 资源： 1000kV 实训线路	1000kV 实训线路	采用技能考核评分细则对学员操作评分
6.工作终结	1.使学员进一步辨析操作过程不足之处，便于后期提升。 2.培养学员安全文明生产的工作作风	1.作业现场清理。 2.向调度汇报工作。 3.班后会，对本次工作任务进行点评总结	1	培训方法： 讲授和归纳法	1000kV 实训线路	

（四）作业流程

1. 工作任务

采用"跨二短三"的作业方式沿耐张绝缘子串进入强电场、到达作业点，采用等电位作业法带电更换1000kV交流输电线路导线防振锤。

（本作业任务适用于海拔1000m及以下地区1000kV交流单回输电线路耐张塔边相导线第一组防振锤作业点位）

2. 天气及作业现场要求

（1）带电更换1000kV交流输电线路导线防振锤应在良好的天气进行。

如遇雷电（听见雷声、看见闪电）、雪、雹、雨、雾等，不应进行带电作业。风力大于5级时，不宜进行带电作业；相对湿度大于80%的天气，如需进行带电作业应采用具有防潮性能的绝缘工具；恶劣天气下必须开展带电抢修时，应组织有关人员充分讨论并编制必要的安全措施，经本单位批准后方可进行。

（2）作业人员精神状态良好，熟悉工作中保证安全的组织措施和技术措施；应持有在有效期内的带电作业资质证书。

（3）工作负责人应事先组织相关人员完成现场勘察，根据勘察结果确定本次作业方法和所需工器具，以及应采取的必要措施，并办理带电作业工作票。

（4）作业现场应合理设置围栏，并妥当布置警示标示牌，禁止非工作人员入内。

（5）本项目需停用线路重合闸装置。

（6）工作中安全距离及有效绝缘长度如表2-8-2所示。

表2-8-2 带电更换1000kV交流输电线路导线防振锤的安全距离

电压等级/m	人身与带电体安全距离/m	与邻相导线的最小距离/m	最小有效绝缘长度/m		最小组合间隙/m	转移电位时人体裸露部分与带电体的最小距离/m
			绝缘操作杆	绝缘承力工具、绝缘绳索		
1000kV	6.0 (6.8)	6.9 (7.2)	6.8	6.8	6.7 (6.9)	0.5

注：①海拔高度1000m及以下时，1000kV交流单回输电线路带电作业中相的安全距离采用括号内6.8m的数据、最小距离采用括号内7.2m的数据、组合间隙采用括号内6.9m的数据。
②因为1000kV特高压线路相间距离足够大（一般控制相地距离即可），不做重要安全因素考虑，所以《安规》没有给出"与邻相导线的最小距离"的数据，实际工作中可以参考750kV的数据。

（7）在1000kV输电线路上作业，应保证作业相良好绝缘子片数不少于37片。

3. 准备工作

3.1 危险点及其预控措施

（1）危险点——触电伤害

预控措施：

①工作前，工作负责人应与值班调控人员联系，停用线路重合闸，并履行许可手续。

②塔上等电位作业人员登塔前，必须仔细核对线路名称、杆塔编号、相别，确认无误后方可上塔。

③工作中，如遇线路突然停电，作业人员应视其仍然带电。工作负责人应尽快与调控人员联系，值班调控人员未与工作负责人取得联系前不准强送电。

④地面电工操作绝缘工具时应戴清洁、干燥的防汗手套，绝缘工具及绝缘绳索不得损坏、受潮、变形、失灵，不准使用非绝缘绳索（如棉纱绳、白棕绳、钢丝绳），现场所使用

的带电作业工具应放置在防潮苫布上，防止绝缘工具在使用中脏污和受潮。

⑤等电位作业人员应穿着阻燃内衣，衣服外面应穿戴全套1000kV带电作业用屏蔽服（包括连衣裤帽、面罩、手套、导电袜和导电鞋），且各部分应连接良好，全套屏蔽服衣裤最远端点之间的电阻值不得大于20Ω。

⑥等电位作业人员在电位转移前，应得到工作负责人的许可，人体裸露部分与带电体的最小距离不小于0.5m；电位转移时，应使用电位转移棒，动作应迅速，严禁用头部充放电；与地电位作业人员传递工具和材料时，使用绝缘工具或绝缘绳索的有效长度不得小于表2-8-2的规定。

⑦用绝缘绳索传递大件金属物品时，地电位作业人员应将金属物品接地后再与绝缘绳接触。

⑧专责监护人应对作业人员进行不间断监护，随时纠正其不规范或违章动作。重点关注高处作业人员，使其保持足够的安全距离（符合表2-8-2的规定），禁止同时接触两个非连通的带电体或带电体与接地体。

（2）危险点——高处坠落

预控措施：

①高处作业人员登高前，必须具备符合本项作业要求的身体状况、精神状态和技能素质。

②高处作业人员登塔前对安全带和防坠器进行外观检查和冲击试验检查，确保其机械强度符合要求。

③高处作业人员应先检查脚钉是否齐全牢固、鞋底是否清洁，防坠装置是否牢固可靠并加挂防坠器；上下塔时，手抓主材、脚踩脚钉、匀步登（下）塔。

④监护人员应随时纠正其不规范或违章动作，重点关注高处作业人员在转位的过程中不得失去安全带或绝缘后备保护绳的保护，安全带系在牢固部件上，严禁低挂高用。

⑤等电位作业人员在绝缘子串上平行移动通常采取双手抓扶一串，双脚踩另一串的姿势匀速进入强电场，移动过程中，后备保护绳兜住二串绝缘子，避免大挥手、大迈步等动作发生。

⑥等电位作业人员沿导线走线，必须系好安全带，并使后备保护绳需将子导线全部兜住；走线过程中应控制重心，防止导线翻转。

（3）危险点——高处坠物伤人。

预控措施：

①高处作业人员的个人工具及零星材料应装入工具袋，严禁在高处浮置物件、口中含物。

②地面作业人员必须正确佩戴安全帽，正确使用绳结传递工器具及材料，与作业点垂直下方距离不得小于坠落半径。

③作业现场设置围栏并挂好警示标示牌。监护人员应时刻注意，禁止非工作人员及车辆进入作业区域。

3.2 工器具及材料选择

带电更换1000kV交流输电线路导线防振锤所需工器具及材料见表2-8-3。工器具出库前，应认真核对工器具的使用电压等级和试验周期，并检查确认外观良好、连接牢固、转动灵活，且符合本次工作任务的要求；工器具出库后，应存放在工具袋或工具箱内进行运输，防止脏污、受潮；金属工具和绝缘工器具应分开装运，防止因混装运输导致工器具变形、损伤等现象发生。

表2-8-3 带电更换1000kV交流输电线路导线防振锤所需工器具及材料表

序号	名称	规格型号	单位	数量	备注
1	绝缘传递绳	TJS-12,长度与起吊高度匹配	根	2	绝缘工具
2	绝缘后备保护绳	TJS-16,加缓冲器	根	2	绝缘工具
3	绝缘滑车	JH10-0.5	只	2	绝缘工具
4	绝缘绳套	TJS-14	根	2	绝缘工具
5	电位转移棒	0.4m	根	1	绝缘工具
6	I型屏蔽服（连衣裤帽、面罩、手套和导电袜）	屏蔽效率≥60dB（屏蔽面罩屏蔽效率≥20dB）	套	2	个人防护用具
7	导电鞋	尺码视穿着人员而定	双	2	个人防护用具
8	阻燃内衣	纯桑蚕丝	套	2	个人防护用具
9	双保险安全带	全身背带式	副	2	个人防护用具
10	防坠器	与杆塔防坠器装置型号对应	只	2	个人防护用具
11	安全帽		顶	6	个人防护用具
12	防振锤专用扳手		个	1	专用工具
13	绝缘子检测仪		套	1	其他工具
14	绝缘电阻测试仪	5000V,电极宽2cm、极间宽2cm	套	1	其他工具
15	风速、温湿度测试仪	HT-8321	套	1	其他工具
16	万用表		只		其他工具
17	对讲机	视工作需要	套	2	其他工具
18	防潮苫布	2m×4m	块	2	其他工具
19	安全围栏		套	若干	其他工具
20	警示标示牌	"在此工作""从此进出""从此上下"	套	1	其他工具
21	红马甲	"工作负责人"	件	1	其他工具
22	清洁毛巾	棉质	条	1	其他工具
23	鞋套		双	若干	其他工具

续表

序号	名称	规格型号	单位	数量	备注
24	工作手套		双	若干	其他工具
25	个人工具	工具袋、平口钳、记号笔	套	2	其他工具
26	防振锤	与被更换防振锤同型号	只	1	材料
27	铝包带		米	适量	材料

注：绝缘工器具的电气及机械强度应满足《电力安全工作规程（线路部分）》要求，试验合格并在有效期内。

3.3 作业人员分工

本任务作业人员分工如表2-8-4所示。

表2-8-4　带电更换1000kV交流输电线路导线防振锤人员分工表

序号	工作岗位	数量（人）	工作职责
1	工作负责人	1	负责本次工作任务的人员分工、工作票的宣读、办理线路停用重合闸、办理工作许可手续、召开工作班前会、工作中突发情况的处理、工作质量的监督、工作后的总结
2	专责监护人	1	负责作业过程中的安全监督及把控
3	等电位电工	1	负责进入等电位更换导线防振锤工作
4	塔上地电位电工	1	负责检测零值或低值绝缘子、协助等电位进出强电场
5	地面电工	2	负责执行现场安全措施、布置作业现场、检查工器具、传递工具及材料，配合等电位电工进出等电位

4. 工作程序

本任务工作流程如表2-8-5所示。

表2-8-5　带电更换1000kV交流输电线路导线防振锤工作流程表

序号	作业内容	作业步骤及标准	安全措施及注意事项	责任人
1	现场复勘	工作负责人负责完成以下工作： (1)现场核对线路名称、杆塔编号、相别无误；基础及杆塔完好无异常；交叉跨越距离符合安全要求；确认缺陷情况及导地线规格型号等。 (2)检测风速、湿度等现场气象条件符合作业要求。 (3)检查地形环境符合作业要求。 (4)检查工作票所列安全措施与现场实际情况相符，必要时予以补充	(1)正确穿戴安全帽、工作服、工作鞋、劳保手套。 (2)不得在危及作业人员安全的气象条件下作业。 (3)严禁非工作人员、车辆进入作业现场	

续表

序号	作业内容	作业步骤及标准	安全措施及注意事项	责任人
2	工作许可	(1)工作负责人负责联系值班调控人员,按工作票内容申请停用线路重合闸。 (2)经值班调控人员许可后,方可开始带电作业工作	不得未经值班调控人员许可即开始工作	
3	现场布置	正确装设安全围栏并悬挂标示牌: (1)安全围栏范围应充分考虑高处坠物,以及对道路交通的影响。 (2)安全围栏出入口设置合理。 (3)妥当布置齐备"从此进出""在此工作""从此上下"等标示	对道路交通安全影响不可控时,应及时联系交通管理部门强化现场交通安全管控	
4	召开班前会	(1)全体工作成员列队。 (2)工作负责人宣读工作票,明确工作任务及人员分工;讲解工作中的安全措施和技术措施;查(问)全体工作成员精神状态;告知工作中存在的危险点及采取的预控措施。 (3)全体工作成员在工作票上签字确认	(1)工作票填写、签发和许可手续规范,签字完整。 (2)全体工作成员精神状态良好。 (3)全体工作成员明确任务分工、安全措施和技术措施	
5	检查工器具	(1)塔上地电位电工和等电位电工正确地穿戴好屏蔽服并检测合格,由负责人监督检查。 (2)正确佩戴个人安全用具(大小合适,锁扣自如),由负责人监督检查。 (3)测量风速风向、湿度,检查绝缘工具的绝缘性能,并做好记录	(1)金属、绝缘工具使用前,应仔细检查其是否损坏、变形、失灵。绝缘工具应使用2500V及以上绝缘电阻测试仪进行分段绝缘检测,阻值应不低于700MΩ,并用清洁干燥的毛巾将其擦拭干净。 (2)用万用表测量屏蔽服衣裤最远端点之间的电阻值不得大于20Ω。工作负责人认真检查作业电工屏蔽服的连接情况。 (3)检查工具组装情况并确认连接可靠。 (4)现场所使用的带电作业工具应放置在防潮苫布上	
6	登塔	(1)核对线路名称、杆塔编号无误后,塔上地电位电工和等电位电工冲击检查安全带、防坠器受力情况。 (2)塔上地电位电工携带绝缘传递绳登塔、等电位电工随后登塔,两人至横担作业点,选择合适位置系好安全带,塔上地电位电工将绝缘滑车和绝缘传递绳安装在横担合适位置。然后配合地面电工将绝缘传递绳分开作起吊准备	(1)核对线路名称和杆塔编号无误后,方可登塔作业。 (2)登塔过程中应使用塔上安装的防坠装置;杆塔上移动及转位时,不准失去安全保护,作业人员必须攀抓牢固构件。 (3)作业电工必须穿全套合格的屏蔽服,且全套屏蔽服必须连接可靠。在横担进入等电位前,等电位电工再次检查确认屏蔽服各部位连接可靠后方能进行下一步操作	

续表

序号	作业内容	作业步骤及标准	安全措施及注意事项	责任人
7	检测绝缘子	(1)地面电工将绝缘操作杆及绝缘子检测仪传至塔上。等电位电工协助塔上地电位电工对绝缘子进行检测。 (2)检测顺序由导线侧向横担侧进行,并作好记录。 (3)检测完毕后,地面电工配合塔上电工将绝缘操作杆传至地面,并将电位转移棒传至塔上	(1)如果是玻璃绝缘子则确认是否良好,不用检测;如果是瓷质绝缘子则如下(2)、(3)。 (2)检测绝缘子工作必须逐串逐片进行,接触必须可靠。 (3)当任意一串良好绝缘子少于37片时,立即停止检测,并停止本次带电作业工作。 (4)上下传递工具时,绑扎绳扣应正确可靠,防止高处坠物	
8	进入强电场	(1)等电位电工将安全带转移到绝缘子连接金具上,并携带电位转移棒、绝缘滑车和绝缘传递绳。 (2)等电位电工检查屏蔽服各部分连接良好后报经工作负责人同意,双手抓扶一串,双脚踩另一串,采用"跨二短三"作业方式沿绝缘子串进入等电位。 (3)当作业人员平行移动至距导线侧均压环3片绝缘子时,应停止移动,利用电位转移棒进行电位转移	(1)等电位电工进入电位前必须得到工作负责人的许可。 (2)等电位电工进入绝缘子串时应交替使用安全带和后备保护绳(用后备保护绳兜住两串绝缘子、手抓扶其中一串,脚踩另一串),不得失去安全带的保护;并调整好绝缘传递绳和电位转移棒。 (3)等电位电工在进入强电场过程中手和脚应协调配合,速度均匀,避免大挥手、大迈步等危险动作;与接地体和带电体两部分间隙所组成的组合间隙边相大于6.7m(中相大于6.9m)。 (4)与相邻导线的最小距离大于6.9m(中相7.2m)。 (5)等电位电工进行电位转移前应检查电位转移棒与屏蔽服的电气连接是否可靠,人体裸露部分与带电体的最小距离应大于0.5m,并得到工作负责人的许可;电位转移时不得失去安全带的保护,进入强电场瞬间动作准确、平稳、迅速	
9	更换子导线防振锤	(1)等电位电工进入等电位后,将安全带系在上子导线上,并装好走线绝缘保护绳(需将子导线全部兜住)。 (2)等电位电工携带绝缘传递绳沿导线走线至更换防振锤作业点,先将绝绳套安装在子导线上合适位置,其次连接绝缘滑车和绝缘传递绳,再将绝缘滑车钩挂在绝缘绳套内。 (3)等电位电工对导线上旧防振锤安装点使用记号笔两端画印进行标记。	(1)等电位电工不得失去安全带的保护。 (2)等电位电工与地面电工要密切配合,听从工作负责人的指挥。 (3)与相邻导线的最小距离大于6.9m(中相7.2m)。 (4)导线防振锤在上下传递过程中,不得磕碰,两侧绝缘传递绳不得相互缠绕。	

续表

序号	作业内容	作业步骤及标准	安全措施及注意事项	责任人
9	更换子导线防振锤	(4)等电位电工先利用绝缘传递绳采用活结的方式绑牢旧防振锤,然后利用防振锤专用扳手将旧防振锤拆除,与地面电工配合利用绝缘传递绳将其放至地面。 (5)等电位电工拆除铝包带,放于工具包中;对应画印标记缠绕新铝包带,缠绕应平整、紧密,其绕向应与外层导线绞制方向一致。 (6)地面电工起吊新防振锤至等电位电工处,等电位电工正确安装新防振锤。 (7)对安装质量进行检查,防振锤方向竖直向下,锤球与主线平行;安装位移不应超过±30mm;铝包带两端断头应回压到防振锤夹内,两端应露出10mm,螺栓处弹簧垫圈应紧平,螺栓穿向与其他子导线防振锤一致。 (8)等电位电工依次拆除绝缘滑车、绝缘传递绳及绳套。 (9)等电位作业过程中不得掉落工器具和材料	(5)上下传递工具时,绑扎绳结应正确可靠,防止高处坠物。 (6)传递工器具及材料过程中,地面电工禁止站立在等电位电工工作点位正下方	
10	退出强电场	(1)经检查防振锤安装牢固,作业点无遗留物后经工作负责人许可,等电位电工携带绝缘传递绳沿导线返回均压环处,作退出电位准备。 (2)等电位电工利用电位转移棒钩紧均压环,并进入距均压环的第3片绝缘子,一只手抓紧绝缘子,另一只手握电位转移棒,利用电位转移棒快速脱离等电位。 (3)等电位电工按照"跨二短三"作业方式退出强电场	(1)等电位电工退出电位前必须得到工作负责人的许可。 (2)等电位电工返回绝缘子串时应交替使用安全带和后备保护绳,电位转移时不得失去安全带的保护,退出强电场瞬间动作准确、平稳、迅速。 (3)等电位电工在脱离电位过程中手和脚应协调配合,速度均匀,避免大挥手、大迈步等动作发生;与接地体和带电体两部分间隙所组成的组合间隙边相大于6.7m(中相大于6.9m);人体裸露部分与带电体的最小距离大于0.5m。 (4)等电位电工沿绝缘子串移动时,用后备保护绳兜住两串绝缘子、手要抓牢,脚要踏实。 (5)与相邻导线的最小距离大于6.9m(中相7.2m)。 (6)等电位电工返回横担时不得同时失去安全带或后备保护绳的保护	

续表

序号	作业内容	作业步骤及标准	安全措施及注意事项	责任人
11	返回地面	塔上电工检查塔上无遗留物后,向工作负责人汇报,得到工作负责人同意后携带绝缘传递绳下塔	下塔过程中应使用塔上安装的防坠装置,杆塔上移动及转位时,不得失去安全保护,作业人员必须攀抓牢固构件	
12	工作结束	(1)工作负责人组织全体工作成员整理工器具和材料,将工器具清洁后放入专用的箱(袋)中;清理现场,做到"工完料尽场地清"。 (2)召开班后会,工作负责人进行工作总结和点评工作。点评本次工作的施工质量;点评全体工作成员的安全措施落实情况。 (3)工作负责人向值班调控人员汇报工作结束,申请恢复线路重合闸,终结工作票	严禁约时恢复线路重合闸	

二、考核标准

特高压交流输电线路运检技能考核评分细则

考生填写栏	编号:	姓 名:		所在岗位:		单 位:		日 期:		年 月 日
考评员填写栏	成绩:	考评员:		考评组长:		开始时间:		结束时间:		操作时长:
考核模块	带电更换1000kV交流输电线路导线防振锤		考核对象	特高压交流输电线路检修人员		考核方式	操作	考核时限		90min
任务描述	沿耐张绝缘子串进入强电场对1000kV交流输电线路受损导线防振锤进行带电更换(等电位作业法)									
工作规范及要求	1. 带电作业工作应在良好天气下进行。如遇雷、雨、雪、雾天气不得进行带电作业。风力大于5级、湿度大于80%时,一般不宜进行带电作业。 2. 本项作业需工作负责人1名,专责监护人1人,塔上地电工1人,等电位电工1人,地面辅助电工2人,采用沿绝缘子串进入强电场对1000kV交流输电线路受损导线防振锤进行带电更换。 3. 工作负责人职责:负责本次工作任务的人员分工、工作票的宣读、办理线路停用重合闸、办理工作许可手续、召开工作班前会、工作中突发情况的处理、工作质量的监督、工作后的总结。 4. 专责监护人:负责作业过程中的安全监督及把控。 5. 等电位电工职责:负责沿绝缘子串进入强电场对导线防振锤进行更换。 6. 塔上地电工职责:负责检测零值或低值绝缘子、协助等电位进出强电场。 7. 地面电工职责:负责执行现场安全措施、布置作业现场、检查工器具、传递工具及材料,配合等电位电工进出等电位。 8. 在带电作业中,如遇雷、雨、大风或其他任何情况威胁到工作人员的安全时,工作负责人或监护人可根据情况,临时停止工作。									

续表

给定条件：	1. 培训基地：特高压交流1000kV实训线路耐张塔大号侧A相子导线防振锤，防振锤型号：FRYJ-4/6。 2. 工作票已办理，安全措施已经完备（重合闸已停用），工作开始、工作终结时应口头提出申请（调度或考评员）。 3. 作业现场装设安全围栏，悬挂"在此工作""从此进出"等标示牌，安全措施已完备。 4. 安全、正确地使用仪器仪表对绝缘工具进行检测。 5. 上下塔过程中应使用防坠落装置，防止高处坠落。 6. 必须按标准化作业程序进行操作，工序错误扣除应做项目分值，出现重大人身、器材和操作安全隐患，考评员可下令终止操作（考核），本模块考核成绩记为"不合格"
考核情景准备	1. 线路：特高压交流1000kV实训线路耐张塔大号侧A相导线，工作内容：带电更换1000kV受损导线防振锤，防振锤型号：FRYJ-4/6。 2. 所需作业工器具：绝缘传递绳2根(TJS-12)，绝缘后备保护绳2根(TJS-16、加缓冲器)，绝缘滑车2只(JH10-0.5)，绝缘绳套2根(TJS-14)，电位转移棒1根(0.4m)，I型屏蔽服2套(连衣裤帽、面罩、手套和导电袜)，导电鞋2双，防坠器2只，防振锤专用扳手1个，绝缘子检测仪1套，绝缘电阻测试仪1套(5000V型)，风速、温湿度测试仪1套(HT-8321)，万用表1只，防潮苫布2块(2m×4m)，红马甲1件(工作负责人)，清洁毛巾2条，个人工具2套(工具袋、平口钳、记号笔)，同型号防振锤1只，铝包带若干。 3. 作业现场做好监护工作，作业现场安全措施（围栏等）已全部落实；禁止非作业人员进入现场，工作人员进入作业现场必须戴安全帽。 4. 考生自备工作服，阻燃纯棉内衣，安全帽，线手套，安全带（含二保绳）
备注	1. 本模块总分为100分，各项得分均以对应分值扣完即止，在规定时间内不能完成任务应立即终止考试，本模块成绩按已完成项实际得分统计，未完成项不得分。 2. 考核过程中因设备、作业环境、安全措施、安全防护、安全距离等不符合作业要求，或人为误操作，出现可能危及作业安全的任意情况，考评员应下令终止操作。 3. 考试前统一组织参考人员进行现场查勘，并提前办理工作票

序号	项目名称	质量要求	分值	扣分标准	扣分原因	扣分	得分
1	现场复勘	1）工作负责人到作业现场核对线路名称和杆塔编号、现场工作条件、缺陷部位等。 2）检测风速、湿度等现场气象条件符合作业要求。 3）检查工作票填写完整，无涂改，检查是否所列安全措施与现场实际情况相符，必要时予以补充	5	1）无工作票，本项不得分。 2）未核对双重称号，扣1分。 3）未核实现场工作条件（气象）、缺陷部位，每项扣1分。 4）工作票填写出现涂改、不整洁，每处扣0.5分，工作票编号有误，扣1分。工作票填写漏项，每项扣1分			
2	工作许可	1）工作负责人联系值班调控人员，按工作票内容申请停用线路重合闸。 2）汇报内容规范、完整，声音清楚宏亮。 3）及时履行相关许可手续	2	1）未取得调度部门（考评员）工作许可擅自开工，本项不得分。 2）汇报专业用语不规范、不完整或声音清楚宏亮各扣0.5分。 3）未申请停用重合闸，扣1分。 4）未复诵许可内容，扣1分。 5）复诵内容漏项，每项扣0.5分（许可人姓名、许可时间、工作任务、重合闸状态）。 6）未及时完善工作票，扣1分			

续表

序号	项目名称	质量要求	分值	扣分标准	扣分原因	扣分	得分
3	现场布置	正确装设安全围栏并悬挂标示牌： 1）安全围栏范围应充分考虑高处坠物，以及对道路交通的影响。 2）安全围栏出入口设置合理。 3）妥当布置"从此进出""在此工作""从此上下"等标示	3	1）未装设作业现场安全围栏，扣2分。 2）作业现场安全围栏设置不合理，每处扣1分。 3）未设悬挂标示牌，扣1.5分。 4）悬挂标示牌不齐，每块0.5分。 5）非作业人员进入围栏区，每人扣0.5分。			
4	召开班前会	1）全体工作成员正确佩戴安全帽、工作服。 2）工作负责人穿安全红马甲，宣读工作票，明确工作任务及人员分工；讲解工作中的安全措施和技术措施；查（问）全体工作成员精神状态；告知工作中存在的危险点及采取的预控措施。 3）全体工作成员在工作票上签字确认	3	1）工作成员安全帽佩戴不正确，每人扣0.5分，着装不整齐，每人次扣0.5分。 2）工作负责人和专责监护人未穿安全红马甲，每人扣0.5分。 3）未明确工作任务及分工，本项不得分。 4）人员分工不明确，扣1分。 5）安全措施、预控措施交代不全，扣1分。 6）未告知工作中存在的危险点，扣1分。 7）未确认工作班成员精神状态，扣1分。 8）工作班成员未签字或签字不全，扣1分			
5	工器具检查	1）工作班成员在合适位置正确设置防潮苫布，防潮苫布应清洁、干燥，严禁踩踏苫布。 2）工器具应按定置管理要求分类、整齐摆放于防潮苫布上；绝缘工器具不能与金属工具、材料混放；对工器具及仪器仪表进行外观检查。 3）各种工具均试验合格，并在试验有效时间内。绝缘工具表面不应磨损、变形损坏，操作应灵活。绝缘工具应使用2500V及以上绝缘电阻表进行分段绝缘检测，阻值应不低于700MΩ，并用清洁干燥的毛巾将其擦拭干净。	7	1）防潮苫布设置位置不合适，扣1分。 2）踩踏防潮苫布，每次扣0.5分。 3）工器具未分类定置摆放，扣1分。 4）未检查工器具及仪器仪表试验合格标签和外观检查，每件扣1分。 5）工器具及仪器仪表检查漏项，每件扣0.5分。 6）仪器仪表、工器具检查方法不正确，每件扣0.5分。 7）未对硬质绝缘工具进行清洁、擦拭，每件扣0.5分。 8）未戴清洁干燥的棉线手套持、拿绝缘工具，每次扣0.5分。			

续表

序号	项目名称	质量要求	分值	扣分标准	扣分原因	扣分	得分
5	工器具检查	4)塔上地电位和等电位人员按要求正确穿戴全套合格的屏蔽服、导电鞋,且各部分连接应良好,屏蔽服内不得贴身穿着化纤类衣服,并系好安全带;工作负责人应认真检查确认是否穿戴正确、各部连接良好。 5)全套屏蔽服应使用万用表进行测试,其最远两点之间阻值不大于20Ω,单件不大于15Ω。 6)登塔电工对安全带及后备保护绳、防坠器进行外观检查,并经冲击试验合格。		9)未正确使用检测仪器对工器具及全套屏蔽服进行检测,每项扣1分。 10)登塔电工未正确穿屏蔽服或连接部分未检查,每人扣2分。 11)安全带及后备保护绳、防坠器未外观检查和冲击试验(或方式不正确),每项扣1分。 12)工作负责人未检查或漏查登塔电工安全防护装备,每项扣1分			
6	登塔	1)登塔人员再次核对双重名称、杆号、相别并向工作负责人报告及申请登塔。 2)塔上地电位电工、等电位电工携带绝缘传递绳相继登塔。 3)登塔过程中必须使用防坠落装置,手抓主材、匀步登塔,安全带及后备保护绳挂在肩上并与带电体保持6m以上安全距离,工作负责人加强作业监护。 4)登塔至合适位置,正确使用安全带,布置好绝缘传递绳,然后塔上地电位电工配合地面电工将绝缘传递绳分开作起吊准备。 5)工作负责人认真监护并提醒整个登塔过程	5	1)登塔电工未核对线路双重名称、塔号、相别,每项扣1分。 2)登塔电工未报告核对结果,未申请登塔,每项扣1分。 3)未使用防坠落装置登塔,考评员应下令终止操作(考核)。 4)手抓脚钉登塔,每次扣0.5分。 5)登塔踩滑、踏空,每次扣1分。 6)安全带主带和后备保护绳未挂肩上,每人扣1分。 7)安全带及后备保护绳发生缠绕、勾住,每次扣1分。 8)安全带及后备保护绳低挂高用,每次扣1分。 9)安全带及后备保护绳系在同一构件上,每次扣1分。 10)高处作业失去安全带的保护,本项不得分。 12)安装滑车未使用绝缘绳套,扣1分。 13)滑车传递绳悬挂位置不便于工具取用,扣1分。 14)工作负责人监护及提醒不到位,扣1分。 15)安全距离不够,考评员应下令终止操作(考核)			

续表

序号	项目名称	质量要求	分值	扣分标准	扣分原因	扣分	得分
7	检测绝缘子	1)地面电工将绝缘操作杆及绝缘子检测仪传至塔上。地电位电工对火花间隙检测仪进行调试,间隙距离0.4mm。 2)等电位电工协助塔上地电位电工对绝缘子逐串逐片进行检测,接触必须可靠,由导线侧向横担侧进行。 3)遇有零值或低值绝缘子应复测3次。 4)当任意一串良好绝缘子少于37片时,立即停止检测,并停止本次带电作业工作。 5)塔上电工应与带电体保持不小于6m的安全距离,绝缘绳索的有效绝缘长度不小于6.8m,绝缘检测杆的最小绝缘长度不小于6.8m。 6)检测完毕后,地面电工配合塔上电工将绝缘操作杆传至地面,并将电位转移棒传至塔上。 7)工作负责人认真监护并提醒本操作过程	10	1)传递工器具时绳结使用错误,扣0.5分。 2)传递工器具发生碰撞,每件扣0.5分。 3)传递绳索缠绕,每次扣0.5分。 4)传递时高处坠物,每次扣1分。 5)未进行绝缘子测零,本项不得分。 6)塔上电工操作方法错误,扣2分。 7)检测绝缘子串顺序错误,扣2分。 8)零值绝缘子未进行复测,扣2分。 9)作业安全距离不够,考评员应下令终止操作。 10)测零结果未汇报,扣2分。 11)未传递电位转移棒,扣2分。 12)工作负责人监护及提醒不到位,扣2分			
8	进入强电场	1)进入电位前检查屏蔽服各部分连接良好后,报经工作负责人同意。 2)等电位电工将安全带转移到绝缘子连接金具上,并携带电位转移棒、绝缘滑车和绝缘传递绳。 3)等电位电工进入绝缘子串前必须系好保护绳(用后备保护绳兜住两串绝缘子、双手抓扶其中一串,脚踩另一串)。 4)采用"跨二短三"作业方式沿绝缘子串平行移动至距导线侧均压环3片绝缘子时,应停止移动,得到工作负责人许可后,利用电位转移棒进行电位转移。 5)等电位电工在进入电位过程中与接地体和带电体两部分间隙所组成的组合间隙边相不得小于6.7m(中相不得小于6.9m),进入强电场必须用电位转移棒进行电位转移,人体裸露部分与带电体的最小距离不得小0.5m。 6)进入强电场过程不得失去安全带的保护	10	1)等电位电工未检查屏蔽服各连接部分,扣2分。 2)等电位电工转移到绝缘子连接金具上失去安全带的保护,扣2分。 3)等电位电工与接地体和带电体两部分间隙所组成的组合间隙不够,扣2分。 4)等电位电工进入强电场动作不正确、不熟练,每次扣2分。 5)等电位电工电位转移过程中裸露部分距离不够,导致反复放电,扣2分。 6)转移电位动作不熟练,扣1分。 7)电位转移过程未使用电位转移棒,扣5分。 8)未得到工作负责人许可就进行电位转移,扣3分。 9)等电位电工进入强电场过程失去安全带的保护,扣2分。 10)工作负责人监护及提醒不到位,扣2分			

续表

序号	项目名称	质量要求	分值	扣分标准	扣分原因	扣分	得分
9	更换子导线防振锤	1)等电位电工进入等电位后,将围杆带系在上子导线上,并装好走线绝缘保护绳(需将子导线全部兜住)。 2)等电位电工携带绝缘传递绳走线至更换防振锤作业点,将绝缘绳套正确装在子导线合适位置,并挂好绝缘滑车和绝缘传递绳。 3)等电位电工使用记号笔对旧防振锤安装位置两端画印,做好标记。 4)等电位电工先利用绝缘传递绳采用活结的方式绑牢旧防振锤,然后利用防振锤专用扳手将旧防振锤拆除,与地面电工配合利用绝缘传递绳将其放至地面防潮苫布上。 5)等电位电工拆除铝包带,放于工具包中;对应画印标记缠绕新铝包带,缠绕应平整、紧密,其绞制方向应与外层导线绞制方向一致。 6)地面电工起吊新防振锤至等电位电工处。 7)等电位电工正确安装新防振锤,并确保防振锤竖直向下,锤球与主线平行,安装位移不应超过±30mm;铝包带两端断头应回压到防振锤夹内,两端应露出10mm,螺栓处弹簧垫圈应紧平,螺栓穿向与其他子导线防振锤一致。 8)等电位电工依次拆除绝缘滑车、绝缘传递绳及绳套。 9)等电位作业过程中不得掉落工器具和材料	30	1)绝缘保护绳未将子导线全部兜住,扣3分。 2)走线过程围杆带未系在上子,扣1分。 3)走线动作不熟练,扣2分。 4)绝缘滑车直接钩挂在导线上,扣5分。 5)绝缘绳套安装方式错误或位置不合适,每项扣1分。 6)绝缘滑车钩挂绝缘绳套内后未闭锁,扣1分。 7)未画印及标记,扣3分。 8)未系绳结就开始拆除防振锤,扣5分。 9)拆除工器具时动作慌乱,扣2分。 10)发生高处坠物,每件扣3分。 11)发生高处抛掷旧防振锤,本模块不得分。 12)地面电工站立在等电位电工工作点位正下方,每人次扣2分。 13)旧防振锤未落放在防潮苫布上,扣1分。 14)上下传递工具时绑扎绳结方式错误,扣1分。 15)未安装完新防振锤就解开绳结,扣5分。 16)安装新防振锤偏离原位置超过允许值,扣3分。 17)安装完后,防振锤没有竖直向下或锤球与主线不平行,各扣3分。 18)安装完后,铝包带两端断头应未压到防振锤夹内,扣2分。 19)安装完后,铝包两端未露出10mm,扣2分。 20)防振锤螺栓穿向错误,每处扣1分。 21)未拆除绝缘滑车、绝缘传递绳及绳套,每件扣1分。 22)未按顺序拆绝缘滑车、绝缘传递绳及绳套,扣2分。 23)未完成新旧防振锤的更换工作,本模块不得分。 24)与相邻导线安全距离不够,考评员应下令终止操作			

续表

序号	项目名称	质量要求	分值	扣分标准	扣分原因	扣分	得分
10	退出强电场	1)经检查防振锤安装牢固，作业点无遗留物后经工作负责人许可，等电位电工携带绝缘传递绳沿导线返回均压环处，作退出电位准备。 2)等电位电工利用电位转移棒钩紧均压环，并进入距均压环的第3片绝缘子，一只手抓紧绝缘子，另一只手握电位转移棒，利用电位转移棒快速脱离等电位。 3)退出强电场过程不得失去安全带的保护。 4)等电位电工按照"跨二短三"作业方式沿绝缘子串退出强电场，转移到横担上。 5)等电位电工在退出电位过程中与接地体和带电体两部分间隙所组成的组合间隙边相不得小于6.7m(中相不得小于6.9m)，退出强电场必须用电位转移棒进行电位转移，人体裸露部分与带电体的最小距离不得小0.5m	10	1)作业点有遗留物，每件扣2分。 2)未向工作负责人申请即进行电位转移，扣3分。 3)未得到工作负责人许可就进行电位转移，扣2分。 4)电位转移过程未使用电位转移棒，扣5分。 5)转移电位动作不熟练，扣1分。 6)等电位电工退出强电场过程失去安全带的保护，扣2分。 7)等电位电工电位转移过程中裸露部分距离不够，导致反复放电，扣2分。 8)等电位电工退出强电场动作不正确、不熟练，每次扣2分。 9)等电位电工与接地体和带电体两部分间隙所组成的组合间隙不够，扣2分。 10)等电位电工转移到横担上失去安全带的保护，扣2分。 11)工作负责人监护及提醒不到位，扣2分			
11	返回地面	1)塔上电工检查塔上无遗留物后，向工作负责人汇报，得到工作负责人同意后，携带绝缘传递绳相继下塔。 2)下塔过程中必须使用防坠落装置，手抓主材、匀步下塔，安全带及后备保护绳挂在肩上并与带电体保持6m以上安全距离，工作负责人加强作业监护。	5	1)塔上有遗留物，每件扣2分。 2)登塔电工未报告遗留物检查结果，未申请下塔，每项扣1分。 3)未使用防坠落装置下塔，考评员应下令终止操作(考核)。 4)手抓脚钉下塔，每次扣0.5分。 5)下塔踩滑、踏空，每次扣1分。 6)安全带主带和后备保护绳未挂肩上，每人扣1分			

续表

序号	项目名称	质量要求	分值	扣分标准	扣分原因	扣分	得分
12	工作结束	1)工作负责人组织全体工作成员整理工器具和材料,将工器具清洁后放入专用的箱(袋)中;清理现场,做到"工完料尽场地清"。 2)召开班后会,工作负责人进行工作总结和点评。点评本次工作的施工质量;点评全体工作成员的安全措施落实情况。 3)工作负责人向值班调控人员汇报工作结束,申请恢复线路重合闸,终结工作票	10	1)对绝缘工器具未进行清洁、擦拭,每件扣0.5分。 2)工器具未归类整理摆放,每件扣1分。 3)工器具乱丢乱扔或踩踏防潮苫布,每次扣0.5分。 4)未拆除围栏及标示牌或有遗留物,每件扣1分。 5)未开班后会,扣2分。 6)集合站队不整齐、注意力不集中,扣1分。 7)工作班成员参加班后会人员不齐,缺一人扣1分。 8)点评不到位,扣1分。 9)未向调度部门(考评员)汇报工作结束,申请恢复线路重合闸,扣1分。 10)汇报专业用语不规范、不完整或声音不洪亮,各扣0.5分。 11)未复诵许可内容,扣1分。 12)复诵内容漏项,每项扣0.5分(单位名称、负责人姓名、时间、线路名称、工作完成情况、设备已恢复正常、人员已撤离、可恢复重合闸)。 13)未及时完善工作票终结手续或填写错误,每项扣1分			
	合计		100				

模块9 带电修补1000kV交流输电线路导线培训及考核标准

一、培训标准

(一) 培训要求

模块名称	带电修补1000kV交流输电线路导线	培训类别	操作类
培训方式	实操培训	培训学时	14学时
培训目标	1.掌握沿耐张绝缘子串进、出1000kV强电场时采用"跨二短三"作业方式的电学意义。 2.能完成沿耐张绝缘子串进入1000kV等电位作业点。 3.能独立完成用预绞式护线条修补导线的操作(等电位作业法)		
培训场地	特高压交流实训线路		
培训内容	采用"跨二短三"作业方式沿耐张绝缘子串进入强电场,采用等电位作业法带电修补1000kV交流输电线路导线		
适用范围	特高压交流输电线路检修人员		

(二) 引用规程规范

(1)《1000kV架空输电线路设计规范》(GB 50665—2011)。

(2)《1000kV交流输电线路检修规范》(DL/T 209—2008)。

(3)《1000kV交流输电线路运行规范》(DL/T 307—2010)。

(4)《1000kV交流输电线路带电作业技术导则》(DL/T 392—2015)。

(5)《交流线路带电作业安全距离计算方法》(GB/T 19185—2008)。

(6)《带电作业用绝缘配合导则》(DL/T 867—2004)。

(7)《架空输电线路带电安装导则及作业工具设备》(DL/T 1007—2006)。

(8)《国家电网公司带电作业工作管理规定(试行)》(国家电网生〔2007〕751号)。

(9)《国家电网公司电力安全工作规程(线路部分)》(Q/GDW1799.2—2013)。

(10)《电工术语 架空线路》(GB/T 2900.51—1998)。

(11)《电工术语 带电作业》(GB/T 2900.55—2002)。

(12)《带电作业工具设备术语》(GB/T 14286—2002)。

（13）《带电作业用工具、装置和设备使用的一般要求》（DL/T 877—2004）。

（14）《带电作业工具、装置和设备预防性试验规程）》（DL/T 976—2005）。

（15）《带电作业用绝缘滑车》（GB/T 13034—2008）。

（16）《带电作业用绝缘绳索》（GB/T 13035—2008）。

（17）《1000kV交流带电作业用屏蔽服装》（GB/T 25726—2010）。

（三）培训教学设计

本设计以完成"带电修补1000kV交流输电线路导线"为工作任务，按工作任务完成的标准化作业流程来设计各个培训阶段，每个阶段包括了具体的培训目标、培训内容、培训学时、培训方法（培训资源）、培训环境和考核评价等内容，如表2-9-1所示。

表2-9-1　带电修补1000kV交流输电线路导线培训内容设计

培训流程	培训目标	培训内容	培训学时	培训方法与资源	培训环境	考核评价
1.理论教学	1.初步掌握沿绝缘子串进出1000kV强电场基本方法； 2.熟悉电位转移的方法； 3.熟悉输电线路受损害导线修补方法	1.沿绝缘子进出强电场"跨二短三"作业方式的电学意义。 2.进、出特高压强电场时电位转移棒的使用方法。 3.输电线路导线修补方法和质量标准	2	培训方法：讲授法。 培训资源：PPT、相关规程规范	多媒体教室	考勤、课堂提问和作业
2.准备工作	能完成作业前准备工作	1.作业现场查勘。 2.编制培训标准化作业卡。 3.填写培训操作工作票。 4.完成本操作的工器具及材料准备	1	培训方法： 1.现场查勘和工器具及材料清理采用现场实操方法； 2.编写作业卡和填写工作票采用讲授方法。 培训资源： 1.1000kV实训线路。 2.特高压工器具库房。 3.空白工作票	1.特高压输电实训线路； 2.多媒体教室	

续表

培训流程	培训目标	培训内容	培训学时	培训方法与资源	培训环境	考核评价
3.作业现场准备	能完成作业现场准备工作	1.作业现场复勘。 2.工作申请。 3.作业现场布置。 4.班前会。 5.工器具及材料检查	1	培训方法：演示与角色扮演法。资源：1000kV实训线路	1000kV实训线路	
4.培训师演示	通过现场观摩，使学员初步领会本任务操作流程	1.零值绝缘子的检测。 2.等电位电工沿耐张绝缘子串进、出强电场。 3.等电位电工采用走线方式到达导线修补位置。 4.等电位电工用预绞丝完成导线修补	2	培训方法：演示法。资源：1000kV实训线路	1000kV实训线路	
5.学员分组训练	1.能完成进、出1000kV强电场操作。 2.能完成1000kV输电线路导线修补作业	1.学员分组（6人一组）训练进、出1000kV强电场和修补导线技能操作。 2.培训师对学员操作进行指导和安全监护	7	培训方法：角色扮演法。资源：1000kV实训线路	1000kV实训线路	采用技能考核评分细则对学员操作评分
6.工作终结	1.使学员进一步辨析操作过程不足之处，便于后期提升。 2.培训学员安全文明生产的工作作风	1.作业现场清理。 2.向调度汇报工作。 3.班后会，对本次工作任务进行点评总结	1	培训方法：讲授和归纳法	1000kV实训线路	

（四）作业流程

1. 工作任务

采用"跨二短三"的作业方式沿耐张绝缘子串进入强电场、到达作业点，采用等电位作业法带电修补1000kV交流输电线路导线。

2. 天气及作业现场要求

（1）带电修补1000kV交流输电线路导线应在良好的天气进行。

如遇雷电（听见雷声、看见闪电）、雪、雹、雨、雾等，禁止进行带电作业。风力大于 5 级，或空气相对湿度大于 80%时，不宜进行带电作业；恶劣天气下必须开展带电抢修时，应组织有关人员充分讨论并编制必要的安全措施，经本单位批准后方可进行。

（2）作业人员精神状态良好，熟悉工作中保证安全的组织措施和技术措施；应持有在有效期内的带电作业资质证书。

（3）工作负责人应事先组织相关人员完成现场勘察，根据勘察结果确定本次作业方法和所需工器具，以及应采取的必要措施，并办理带电作业工作票。

（4）作业现场应合理设置围栏，并妥当布置警示标示牌，禁止非工作人员入内。

（5）本项目需停用线路重合闸装置。

（6）工作中安全距离及有效绝缘长度如表2-9-2所示。

表2-9-2　带电补修1000kV交流输电线路导线的安全距离

电压等级	人身与带电体安全距离/m	最小有效绝缘长度/m		最小组合间隙/m	转移电位时人体裸露部分与带电体的最小距离/m
		绝缘操作杆	绝缘承力工具、绝缘绳索		
1000kV	6.8	6.8	6.8	6.9	0.5

（7）在1000kV输电线路上作业，应保证作业相良好绝缘子片数不少于37片。

3. 准备工作

3.1 危险点及其预控措施

（1）危险点——触电伤害

预控措施：

①工作前，工作负责人应与值班调控人员联系，停用线路重合闸，并履行许可手续。

②塔上地电位作业人员登塔前，必须仔细核对线路名称、杆塔编号、相别，确认无误后方可上塔。

③工作中，如遇线路突然停电，作业人员应视其仍然带电。工作负责人应尽快与调控人员联系，值班调控人员未与工作负责人取得联系前不准强送电。

④绝缘工具及绝缘绳索不得损坏、受潮、变形、失灵，不准使用非绝缘绳索（如棉纱绳、白棕绳、钢丝绳）。

⑤等电位作业人员应穿着阻燃内衣，衣服外面应穿戴全套屏蔽服（包括帽、衣裤、手套、袜和鞋），且各部分应连接良好。

⑥等电位作业人员在电位转移前，应得到工作负责人的许可，人体裸露部分与带电体的最小距离不小于0.5m；电位转移时，动作应迅速，严禁用头部充放电；与地电位作业人员传递工具和材料时，使用绝缘工具或绝缘绳索的有效长度不准小于表2-9-2的规定。

⑦用绝缘绳索传递大件金属物品时，地电位作业人员应将金属物品接地后再接触。

⑧专责监护人应对作业人员进行不间断监护，随时纠正其不规范或违章动作。重点关注高处作业人员，使其保持足够的安全距离（符合表2-9-2的规定），禁止同时接触两个非连通的带电体或带电体与接地体。

（2）危险点——高处坠落

预控措施：

①高处作业人员登高前，必须具备符合本项作业要求的身体状况、精神状态和技能素质。

②监护人员应随时纠正其不规范或违章动作，重点关注作业人员在转位的过程中不得失去安全带或绝缘后备保护绳的保护，严禁低挂高用。

（3）危险点——高处坠物伤人。

预控措施：

①高处作业人员的个人工具及零星材料应装入工具袋，严禁在高处浮置物件、口中含物。

②地面作业人员必须正确佩戴安全帽，正确使用绳结，与作业点垂直下方距离不得小于坠落半径。

③作业现场设置围栏并挂好警示标示牌。监护人员应随时注意，禁止非工作人员及车辆进入作业区域。

3.2 工器具及材料选择

带电补修1000kV交流输电线路导线所需工器具及材料见表2-9-3。工器具出库前，应认真核对工器具的使用电压等级和试验周期，并检查确认外观良好、连接牢固、转动灵活，且符合本次工作任务的要求；工器具出库后，应存放在工具袋或工具箱内进行运输，防止脏污、受潮；金属工具和绝缘工器具应分开装运，防止因混装运输导致工器具变形、损伤等现象发生。

表2-9-3 带电补修1000kV交流输电线路导线所需工器具及材料表

序号	名称	规格型号	单位	数量	备注
1	绝缘传递绳	TJS-12	根	2	
2	绝缘保护绳	TJS-16	根	2	
3	绝缘子检测仪		套	1	
4	绝缘滑车	JH10-1	个	2	
5	安全帽		顶	6	
6	电位转移棒		根	1	
7	绝缘电阻表	5000V	块	1	
8	风速风向仪		块	1	

续表

序号	名称	规格型号	单位	数量	备注
9	温湿度仪		块	1	
10	万用表		块	1	
11	防潮帆布	2m×4m	块	2	
12	绝缘绳套		根	4	
13	屏蔽服	屏蔽效率≥60dB（屏蔽面罩屏蔽效率≥20dB）	套	2	
14	防坠器	与杆塔防坠器装置型号对应	只	2	
15	安全带		副	2	
16	安全围栏		套	若干	
17	警示标示牌	"在此工作""从此进出""从此上下"	套	1	
18	红马甲	"工作负责人"	件	1	
19	预绞式护线条		套	1	
20	导电膏		盒	1	
21	砂纸		张	1	
22	清洁毛巾		条	1	
23	对讲机		台	4	
24	操作杆		根	1	

3.3 作业人员分工

本任务作业人员分工如表2-9-4所示。

表2-9-4 带电补修1000kV交流输电线路导线人员分工表

序号	工作岗位	数量(人)	工作职责
1	工作负责人	1	负责本次工作任务的人员分工、工作票的宣读、办理线路停用重合闸、办理工作许可手续、召开工作班前会、工作中突发情况的处理、工作质量的监督、工作后的总结
2	专责监护人	1	负责作业现场的安全把控
3	等电位电工	1	负责进入等电位补修导线工作
4	地电位电工	1	负责检测绝缘子、协助等电位进出强电场
5	地面电工	2	负责传递工具、材料配合等电位电工进出等电位

4. 工作程序

本任务工作流程如表2-9-4所示。

表2-9-4 带电补修1000kV交流输电线路导线工作流程表

序号	作业内容	作业步骤及标准	安全措施及注意事项	责任人
1	现场复勘	工作负责人负责完成以下工作: (1)现场核对线路名称、杆塔编号,相别无误;基础及杆塔完好无异常;交叉跨越距离符合安全要求;确认缺陷情况及导地线规格型号等。 (2)检测风速、湿度等现场气象条件符合作业要求。 (3)检查地形环境符合作业要求。 (4)检查工作票所列安全措施与现场实际情况相符,必要时予以补充	(1)正确穿戴安全帽、工作服、工作鞋、劳保手套。 (2)不得在危及作业人员安全的气象条件下作业。 (3)严禁非工作人员、车辆进入作业现场	
2	工作许可	(1)工作负责人负责联系值班调控人员,按工作票内容申请停用线路重合闸。 (2)经值班调控人员许可后,方可开始带电作业工作	不得未经值班调控人员许可即开始工作	
3	现场布置	正确装设安全围栏并悬挂标示牌: (1)安全围栏范围应充分考虑高处坠物,以及对道路交通的影响。 (2)安全围栏出入口设置合理。 (3)妥当布置"从此进出""在此工作""从此上下"等标示	对道路交通安全影响不可控时,应及时联系交通管理部门强化现场交通安全管控	
4	召开班前会	(1)全体工作成员列队。 (2)工作负责人宣读工作票,明确工作任务及人员分工;讲解工作中的安全措施和技术措施;查(问)全体工作成员精神状态;告知工作中存在的危险点及采取的预控措施。 (3)全体工作成员在工作票上签名确认	(1)工作票填写、签发和许可手续规范,签名完整。 (2)全体工作成员精神状态良好。 (3)全体工作成员明确任务分工、安全措施和技术措施	
5	检查工具	(1)地电位电工和等电位电工正确地穿好屏蔽服并检测合格,由负责人监督检查。 (2)正确佩戴个人安全用具(大小合适,锁扣自如),由负责人监督检查。 (3)测量风速风向、湿度,检查绝缘工具的绝缘性能,并做好记录	(1)金属、绝缘工具使用前,应仔细检查其是否损坏、变形、失灵。绝缘工具应使用2500V及以上绝缘电阻表进行分段绝缘检测,阻值应不低于700MΩ,并用清洁干燥的毛巾将其擦拭干净。 (2)用万用表测量屏蔽服衣裤最远端点之间的电阻值不得大于20Ω。工作负责人认真检查作业电工屏蔽服的连接情况。 (3)检查工具组装情况并确认连接可靠。 (4)现场所使用的带电作业工具应放置在防潮帆布上	

第二部分 技能模块培训及考核标准

续表

序号	作业内容	作业步骤及标准	安全措施及注意事项	责任人
6	登塔	(1)核对线路名称、杆塔编号无误后,地电位电工和等电位冲击检查安全带、防坠器手里情况。 (2)地电位电工携带绝缘传递绳登塔、等电位电工随后登塔,两人至横担作业点,选择合适位置系好安全带,地电位电工将绝缘滑车和绝缘传递绳安装在横担合适位置。然后配合地面电工将绝缘传递绳分开作起吊准备	(1)核对线路线路名称和杆塔编号无误后,方可登塔作业。 (2)登塔过程中应使用塔上安装的防坠装置;杆塔上移动及转位时,不准失去安全保护,作业人员必须攀抓牢固构件。 (3)作业电工必须穿全套合格的屏蔽服,且全套屏蔽服必须连接可靠。在横担进入等电位前,等电位电工要检查确认屏蔽服个部位连接可靠后方能进行下一步操作	
7	检测绝缘子	(1)地面电工将绝缘操作杆及绝缘子检测仪传至塔上。等电位电工协助地电位电工对绝缘子进行检测。 (2)检测工作由横担侧向带电侧进行,并作好记录。 (3)检测完毕后,地面电工配合塔上电工将绝缘操作杆传至地面,并将电位转移棒传至塔上	(1)如果是玻璃绝缘子则确认是否良好,不用检测;如果是瓷质绝缘子则如下(2)、(3)。 (2)检测绝缘子工作必须逐串逐片进行,接触必须可靠。 (3)当任意一串良绝缘子少于37片时,立即停止检测,并停止本次带电作业工作	
8	进入强电场	(1)等电位电工将安全带转移到绝缘子连接金具上,并携带电位转移棒、绝缘滑车和绝缘传递绳。 (2)等电位电工检查屏蔽服各部分连接良好后报经工作负责人同意,双手抓扶一串,双脚踩另一串,采用"跨二短三"作业方式沿绝缘子串进入等电位。 (3)当作业人员平行移动至距导线侧均压环三片绝缘子时,应停止移动,利用电位转移棒进行电位转移	(1)等电位电工进入电位前必须得到工作负责人的许可。 (2)等电位电工进入绝缘子串前必须系好保护绳(用后备保护绳兜住脚踩绝缘子串),并调整好绝缘传递绳和电位转移棒。 (3)等电位电工在进入电位过程中与接地体和带电体两部分间隙所组成的组合间隙不得小于中相6.9m(边相6.7m)	
9	损伤导线表面处理	(1)等电位电工进入等电位后,将安全带系在上子导线上,并装好走线绝缘保护绳(需将子导线全部兜住)。 (2)等电位电工携带绝缘传递绳走线至作业点,将绝缘滑车和绝缘传递绳安装在子导线上。 (3)等电位电工检查导线损伤情况,并对损伤点进行处理,用0号砂纸将损伤部位毛刺打磨平整。 (4)等电位电工用抹布将打磨后的导线表面处理干净,并将带电膏均匀涂抹在导线受伤处	(1)等电位电工对导线损伤点进行打磨处理时,用力不得过大,不得使损伤程度扩大。 (2)导线打磨后,要将表面充分清洁干净。 (3)导电膏均匀涂抹在导线表面	

续表

序号	作业内容	作业步骤及标准	安全措施及注意事项	责任人
10	导线修补	(1)地面电工利用传递绳将预绞丝传给等电位电工。 (2)等电位电工利用预绞丝对导线损伤部位进行补强	(1)预绞式护线条的规格型号应与导线匹配。 (2)预绞式护线条的中心应位于损伤最严重处。 (3)预绞式护线条的长度应将损伤部位全部覆盖,且护线条端部距损伤部位边缘的单位长度不得小于100mm。 (4)预绞式护线条绑扎紧密接触,不得抛股、漏股、散股	
11	退出强电场	(1)经检查受损带线已补强良好,作业点无遗留物后经工作负责人许可,等电位电工携带绝缘传递绳走线返回均压环处,作退出电位准备。 (2)等电位电工利用电位转移棒钩紧均压环,并进入距均压环的第3片绝缘子,一只手抓紧绝缘子,另一只手握电位转移棒,利用电位转移棒快速脱离电位。 (3)等电位电工按照"跨二短三"作业方式的方法退出等电位	(1)等电位电工退出电位前必须得到工作负责人的许可。 (2)等电位电工在退出电位过程中与接地体和带电体两部分间隙所组成的组合间隙不得小于中相6.9m(边相6.7m)。 (3)沿绝缘子串移动时,手要抓牢,脚要踏实	
12	返回地面	塔上电工检查塔上无遗留物后,向工作负责人汇报,得到工作负责人同意后携带绝缘传递绳下塔	下塔过程中应使用塔上安装的防坠装置,杆塔上移动及转位时,不准失去安全保护,作业人员必须攀抓牢固构件	
13	工作结束	(1)工作负责人组织全体工作成员整理工器具和材料,将工器具清洁后放入专用的箱(袋)中;清理现场,做到"工完料尽场地清"。 (2)召开班后会,工作负责人进行工作总结和点评工作。点评本次工作的施工质量;点评全体工作成员的安全措施落实情况。 (3)工作负责人向值班调控人员汇报工作结束,申请恢复线路重合闸,终结工作票	不得约时恢复线路重合闸	

第二部分
技能模块培训及考核标准

二、考核标准

特高压交流输电线路运检技能考核评分细则

考 生 填写栏	编号：	姓 名：		所在岗位：	单 位：		日 期：	年 月 日		
考评员 填写栏	成绩：	考评员：		考评组长：	开始时间：		结束时间：	操作时长：		
考核 模块	带电修补1000kV交流输电线路导线		考核 对象	特高压交流输电线路检修人员			考核 方式	操作	考核 时限	60min
任务 描述	沿耐张绝缘子串进入强电场对1000kV受损导线进行带电修补									
工作规范及要求	1. 带电作业工作应在良好天气下进行。如遇雷、雨、雪、雾天气不得进行带电作业。风力大于5级、湿度大于80%时，一般不宜进行带电作业。 2. 本项作业需工作负责人1名，专责监护人1人，地电位电工1人，等电位电工1人，地面辅助电工2人，采用沿绝缘子串进入强电场对受损子导线进行带电修补。 3. 工作负责人职责：负责本次工作任务的人员分工、工作票的宣读、办理线路停用重合闸、办理工作许可手续、召开工作班前会、工作中突发情况的处理、工作质量的监督、工作后的总结。 4. 专责监护人：负责作业现场的安全把控。 5. 等电位电工职责：负责沿绝缘子串进入强电场对受损导线进行修补。 6. 地电位电工职责：负责检测绝缘子、协助等电位进出强电场。 7. 地面电工职责：负责传递工具、材料配合等电位电工进出等电位。 8. 在带电作业中，如遇雷、雨、大风或其他任何情况威胁到工作人员的安全时，工作负责人或监护人可根据情况，临时停止工作。 给定条件： 1. 培训线路：特高压1000kV交流实训线路A相八分裂导线某子导线，导线型号：8×JL/G1A-630/45。 2. 工作票已办理，安全措施已经完备(重合闸已停用)，工作开始、工作终结时应口头提出申请(调度或考评员)。 3. 安全、正确地使用仪器对绝缘工具进行检测。 4. 必须按工作程序进行操作，工序错误扣除应做项目分值，出现重大人身、器材和操作安全隐患，考评员可下令终止操作(考核)。									
考核情景准备	1. 线路：特高压1000kV交流实训线路A相子导线，工作内容：带电修补1000kV受损导线，导线型号：8×JL/G1A-630/45。 2. 所需作业工器具：绝缘传递绳1根(TJS-12)，绝缘保护绳(TJS-16)，绝缘滑车1个(JH10-1)，绝缘检测仪，电位转移棒(1根)，绝缘电阻表(5000V型)，屏蔽服(屏蔽效率≥60Db)2套，万用表1块，苫布1块，绝缘测试仪1台，温湿度表、风速仪各1台，纯棉毛巾2条，预绞丝1组，0号砂纸1张，木榔头1把、操作杆(1根)。 3. 作业现场做好监护工作，作业现场安全措施(围栏等)已全部落实；禁止非作业人员进入现场，工作人员进入作业现场必须戴安全帽。 4. 考生自备工作服，阻燃纯棉内衣，安全帽，线手套，安全带(含二保绳)									
备注	1. 各项目得分均扣完为止，出现重大人身、器材和操作安全隐患，考评员可下令终止操作。 2. 设备、作业环境、安全带、安全帽、工器具、屏蔽服等不符合作业条件考评员可下令终止操作									

续表

序号	项目名称	质量要求	分值	扣分标准	扣分原因	扣分	得分
1	现场复勘	1)工作负责人到作业现场核对线路名称和杆塔编号、现场工作条件、缺陷部位等。 2)检测风速、湿度等现场气象条件符合作业要求。 3)检查工作票填写完整,无涂改,检查是否所列安全措施与现场实际情况相符,必要时予以补充	5	1)未核对线路名称、杆塔编号、现场工作条件、缺陷部位等,扣1分/项。 2)未检测风速、湿度等现场气象条件,扣1分/项。 3)工作票填写出现涂改,扣0.5分/处;工作票编号有误,扣1分;工作票填写不完整,扣1.5分			
2	工作许可	1)工作负责人联系值班调控人员,按工作票内容申请停用线路重合闸。 2)汇报内容规范、完整	2	1)未联系调度部门(裁判)停用重合闸,扣2分。 2)汇报专业用语不规范或不完整,扣1分			
3	现场布置	正确装设安全围栏并悬挂标示牌: 1)安全围栏范围应充分考虑高处坠物,以及对道路交通的影响。 2)安全围栏出入口设置合理。 3)妥当布置"从此进出""在此工作""从此上下"等标示	3	1)作业现场未装设围栏,扣1分。 2)未设立警示牌,扣1分。 3)未悬挂登塔作业标志,扣1分			
4	召开班前会	1)全体工作成员全体人员正确佩戴安全帽、工作服。 2)工作负责人穿红色背心,宣读工作票,明确工作任务及人员分工;讲解工作中的安全措施和技术措施;查(问)全体工作成员精神状态;告知工作中存在的危险点及采取的预控措施。 3)全体工作成员在工作票上签名确认	3	1)工作人员着装不整齐,扣0.5分/人。 2)未进行分工,扣3分;分工不明确,扣1分。 3)现场工作负责人未穿安全监护背心,扣1分。 4)工作票上工作班成员未签字或签字不全,扣1分			

续表

序号	项目名称	质量要求	分值	扣分标准	扣分原因	扣分	得分
5	工器具检查	1)工作人员按要求将工器具放在防潮苫布上；防潮苫布应清洁、干燥。 2)工器具应按定置管理要求分类摆放；绝缘工器具不能与金属工具、材料混放；对工器具进行外观检查。 3)绝缘工具表面不应磨损、变形损坏，操作应灵活。绝缘工具应使用2500V及以上绝缘电阻表进行分段绝缘检测，阻值应不低于700MΩ，并用清洁干燥的毛巾将其擦拭干净。 4)塔上地电位和登电位人员按要求正确穿全套合格的屏蔽服、导电鞋，且各部分连接应良好，屏蔽服内不得贴身穿着化纤类衣服，并系好安全带；工作负责人应认真检查是否穿戴正确。 5)登塔人员再次核对双重名称、杆号、相别并报告	7	1)未使用防潮苫布并定置摆放工器具，扣1分。 2)未检查工器具外观及试验合格证，扣0.5分/项。 3)未正确使用检测仪器对工器具进行检测，扣1分/项。 4)作业人员未正确穿戴屏蔽服且各部位连接良好，扣2分/人次。 5)现场工作负责人未对登塔作业人员进安全防护装备进行检查，扣1分。 6)登塔人员未核对线路双重名称、杆号、相别，扣2分/人。 7)汇报检测结果不规范，扣1分；不完整，扣0.5分/项。			
6	登塔	1)地电位电工、等电位电工穿好全套合格的屏蔽服，将安全带做冲击试验后，系好安全带后携带绝缘传递绳相继登塔。 2)登塔过程中系好防坠落保护装置，登塔至合适位置，系好安全带，布置好绝缘传递绳，然后配合地面电工将绝缘传递绳分开作起吊准备。 3)登塔过程中应系好防坠落保护装置，匀速登塔，手抓主材，将安全带挂在肩上并与带电体保持6.8m以上安全距离，工作负责人加强作业监护。	5	1)未系安全带或安全带及后备保护绳或未进行冲击试验，扣2分/项。 2)手抓脚钉，扣0.5分/次。 3)现场工作负责人未对地电位、中间电位电工进行安全防护装备进行检查，扣1分/项。 4)传递时高处落物，扣1分/次。 5)传递过程工具与塔身磕碰，扣1分/项。 6)塔上电工转位时失去安全带保护，扣5分			
7	检测绝缘子	1)地面电工将绝缘操作杆及绝缘子检测仪传至塔上。等电位电工协助地电位电工对绝缘子进行检测。 2)检测绝缘子工作必须逐串逐片进行，接触必须可靠，由横担侧向带电侧进行。 3)当任意一串良好绝缘子少于37片时，立即停止检测，并停止本次带电作业工作	2	1)未进行绝缘子测零，扣2分。 2)测零方法错误，扣2分。 3)测零结果未汇报，扣2分			

续表

序号	项目名称	质量要求	分值	扣分标准	扣分原因	扣分	得分
8	进入强电场	1)等电位电工进入绝缘子串前必须系好保护绳(用后备保护绳兜住脚踩绝缘子串),并调整好绝缘传递绳和电位转移棒。2)等电位电工检查屏蔽服各部分连接良好后报经工作负责人同意,双手抓扶一串,双脚踩另一串,采用"跨二短三"作业方式沿绝缘子串进入强电场。3)等电位电工在进入电位过程中与接地体和带电体两部分间隙所组成的组合间隙不得小于中相6.9m(边相6.7m),进入强电场必须用电位转移棒进行电位转移	8	1)等电位电工电位转移过程中裸露部分距离不够,扣2分。2)等电位电工进入强电场动作不正确,反复放电,扣2分/次。3)转移电位动作不熟练,扣1分/次;电位转移过程未使用电位转移棒,扣4分/次。4)未得到工作负责人许可就进行电位转移,扣4分/次。			
9	损伤导线表面处理	1)等电位电工携带绝缘传递绳走线至作业点,将绝缘滑车和绝缘传递绳安装在子导线上,并将子导线用绝缘绳索全部箍住。2)等电位电工检查导线损伤情况,并对损伤点进行处理,用0号砂纸将损伤部位毛刺打磨平整。3)等电位电工用抹布将打磨后的导线表面处理干净,并将带电膏均匀涂抹在导线受伤处表面	20	1)未将子导线用绝缘绳索全部箍住,扣4分。2)未向工作负责人汇报导线损伤情况,扣2分。3)对受损导线未处理平整,扣2分。4)不清除表面氧化物或未涂抹导电膏,扣2分/项			
10	导线修补	1)地面电工利用传递绳将预绞丝传给等电位电工。等电位电工利用预绞丝对导线损伤部位进行补强。2)预绞式护线条的规格型号应与导线匹配,护线条的中心应位于损伤最严重处。3)预绞式护线条的长度应将损伤部位全部覆盖,且护线条端部距损伤部位边缘的单位长度不得小于100mm。4)预绞式护线条绑扎紧密接触,不得抛股、漏股、散股	30	1)补修中心误差超过5mm,扣3分。2)预绞丝安装出现缝隙,扣2分/处。3)若因操作不当致使预绞丝变形,扣10分。4)端头不平齐的,扣1分/根。5)预绞式护线条绑扎不紧密,出现抛股、漏股、散股,扣0.5分/处			

续表

序号	项目名称	质量要求	分值	扣分标准	扣分原因	扣分	得分
11	退出强电场	1）经检查受损带线已补强良好，作业点无遗留物后经工作负责人许可，等电位电工携带绝缘传递绳走线返回均压环处，作退出电位准备。2）等电位电工利用电位转移棒钩紧均压环，并进入距均压环的第3片绝缘子，一只手抓紧绝缘子，另一只手握电位转移棒，利用电位转移棒快速脱离电位。3）等电位电工按照"跨二短三"作业方式的方法退出等电位	10	1）未向工作负责人申请即进行电位转移，扣2分；申请了但未得同意即开始，扣1分。2）等电位电工电位转移过程中裸露部分与带电体距离不够，扣2分/次。3）等电位电工退出强电场动作不正确，反复放电，扣2分/次。4）转移电位动作不熟练，扣2分；电位转移未使用电位转移棒，扣3分			
12	返回地面	塔上电工检查塔上无遗留物后，向工作负责人汇报，得到工作负责人同意后携带绝缘传递绳下塔	5	1）下塔过程未使用防坠器，扣5分。2）塔上移位失去安全带保护，扣5分。3）下塔手抓脚钉，扣1分/次。4）塔上有遗留物，扣2分			
13	工作结束	1）工作负责人组织全体工作成员整理工器具和材料，将工器具清洁后放入专用的箱（袋）中；清理现场，做到"工完料尽场地清"。2）召开班后会，工作负责人进行工作总结和点评工作。点评本次工作的施工质量；点评全体工作成员的安全措施落实情况。3）工作负责人向值班调控人员汇报工作结束，申请恢复线路重合闸，终结工作票	10	1）工器具未清理，扣2分。2）工器具有遗漏，扣2分。3）未开班后会，扣10分。4）未拆除围栏，扣2分。5）未向调度汇报，扣2分			
	合计		100				

模块10　带电处理1000kV交流输电线路导线引流板发热缺陷培训及考核标准

一、培训标准

（一）培训要求

模块名称	带电处理1000kV交流输电线路导线引流板发热缺陷	培训类别	操作类
培训方式	实操培训	培训学时	14学时
培训目标	1.掌握采用"跨二短三"的作业方式沿耐张绝缘子串进、出1000kV强电场。 2.能完成沿耐张绝缘子串进入1000kV等电位作业点。 3.能独立完成采用等电位作业法带电处理1000kV交流输电线路导线引流板发热缺陷（等电位作业法）		
培训场地	特高压交流实训线路		
培训内容	采用"跨二短三"作业方式沿耐张绝缘子串进入强电场，采用等电位作业法带电处理1000kV交流输电线路导线引流板发热缺陷		
适用范围	特高压交流输电线路带电检修人员		

（二）引用规程规范

（1）《1000kV架空输电线路设计规范》（GB 50665—2011）。

（2）《1000kV交流输电线路检修规范》（DL/T 209—2008）。

（3）《1000kV交流输电线路运行规范》（DL/T 307—2010）。

（4）《1000kV交流输电线路带电作业技术导则》（DL/T 392—2015）。

（5）《交流线路带电作业安全距离计算方法》（GB/T 19185—2008）。

（6）《带电作业用绝缘配合导则》（DL/T 867—2004）。

（7）《架空输电线路带电安装导则及作业工具设备》（DL/T 1007—2006）。

（8）《国家电网公司带电作业工作管理规定（试行）》（国家电网生〔2007〕751号）。

（9）《国家电网公司电力安全工作规程（线路部分）》（Q/GDW1799.2—2013）。

（10）《电工术语 架空线路》（GB/T 2900.51—1998）。

(11)《电工术语 带电作业》(GB/T 2900.55—2002)。

(12)《带电作业工具设备术语》(GB/T 14286—2002)。

(13)《带电作业用工具、装置和设备使用的一般要求》(DL/T 877—2004)。

(14)《带电作业工具、装置和设备预防性试验规程)》(DL/T 976—2005)。

(15)《带电作业用绝缘滑车》(GB/T 13034—2008)。

(16)《带电作业用绝缘绳索》(GB/T 13035—2008)。

(17)《1000kV交流带电作业用屏蔽服装》(GB/T 25726—2010)。

(三)培训教学设计

本设计以完成"带电处理1000kV交流输电线路导线引流板发热缺陷"为工作任务,按工作任务完成的标准化作业流程来设计各个培训阶段,每个阶段包括了具体的培训目标、培训内容、培训学时、培训方法(培训资源)、培训环境和考核评价等内容,如表2-10-1所示。

表2-10-1 带电处理1000kV交流输电线路导线引流板发热缺陷培训内容设计

培训流程	培训目标	培训内容	培训学时	培训方法与资源	培训环境	考核评价
1.理论教学	1.初步掌握沿绝缘子串进出1000kV交流强电场基本方法。 2.熟悉电位转移的方法。 3.熟悉输电线路导线引流板发热缺陷的处理方法	1.采用"跨二短三"的方式沿绝缘子串进、出强电场。 2.进、出特高压强电场时电位转移棒的使用方法。 3.输电线路导线引流板发热缺陷的处理方法和质量标准	2	培训方法:讲授法。 培训资源:PPT、相关规程规范	多媒体教室	考勤、课堂提问和作业
2.准备工作	能完成作业前准备工作	1.作业现场查勘。 2.编制培训标准化作业卡。 3.填写培训操作工作票。 4.完成本操作的工器具及材料准备。 5.值班调控人员联系申请停用工作线路重合闸装置	1	培训方法: 1.现场查勘和工器具及材料清理采用现场实操方法。 2.编写作业卡和填写工作票采用讲授方法。 培训资源: 1.1000kV实训线路。 2.特高压工器具库房。 3.空白工作票	1.特高压输电实训线路; 2.多媒体教室	

续表

培训流程	培训目标	培训内容	培训学时	培训方法与资源	培训环境	考核评价
3.作业现场准备	能完成作业现场准备工作	1.重合闸已装置停用,得到调度许可。 2.作业现场复勘。 3.作业现场布置。 4.班前会。 5.工器具及材料检查	1	培训方法: 演示与角色扮演法。 资源: 1000kV实训线路	1000kV实训线路	
4.培训师演示	通过现场观摩,使学员初步领会本任务操作流程	1.零值绝缘子检测。 2.等电位电工沿耐张绝缘子串进、出强电场,到达引流线连接处。 3.等电位电工用套筒扳手对导线引流线螺栓进行紧固	2	培训方法: 演示法。 资源: 1000kV实训线路	1000kV实训线路	
5.学员分组训练	1.能完成进、出1000kV强电场操作。 2.能完成1000kV输电线路导线引流线发热带电处理	1.学员分组(6人一组)训练进、出1000kV强电场和导线引流线发热处理技能操作。 2.培训师对学员操作进行指导和安全监护	7	培训方法: 角色扮演法。 资源: 1000kV实训线路	1000kV实训线路	采用技能考核评分细则对学员操作评分
6.工作终结	1.使学员进一步辨析操作过程不足之处,便于后期提升。 2.培养学员安全文明生产的工作作风	1.作业现场清理。 2.向调度汇报工作结束并申请恢复重合闸装置。 3.班后会,对本次工作任务进行点评总结	1	培训方法: 讲授和归纳法	1000kV实训线路	

(四)作业流程

1. 工作任务

采用"跨二短三"法沿绝缘子串进入强电场到达作业点,采用等电位作业法带电处理1000kV交流输电线路导线引流板发热缺陷。

2. 天气及作业现场要求

(1) 带电处理1000kV交流输电线路导线引流板发热缺陷应在良好的天气进行。

如遇雷电(听见雷声、看见闪电)、雪、雹、雨、雾等,禁止进行带电作业。风力大于5级,或空气相对湿度大于80%时,不宜进行带电作业;恶劣天气下必须开展带电抢修时,应组织有关人员充分讨论并编制必要的安全措施,经本单位批准后方可进行。

（2）作业人员精神状态良好，熟悉工作中保证安全的组织措施和技术措施；应持有在有效期内的特高压交流带电作业资质证书。

（3）工作负责人应事先组织相关人员完成现场勘察，根据勘察结果确定本次作业方法和所需工器具，以及应采取的必要措施，并办理带电作业工作票。

（4）作业现场应合理设置围栏，并妥当布置警示标示牌，禁止非工作人员入内。

（5）本项目需停用线路重合闸装置。

（6）工作中安全距离及有效绝缘长度如表2-10-2所示。

（7）在1000kV输电线路上作业，应保证作业相良好绝缘子片数不少于37片。

表2-10-2　带电补修1000kV交流输电线路导线的安全距离

海拔高度/m	最小安全距离/m		最小组合间隙/m		绝缘工具最小有效绝缘长度/m	转移电位时人体裸露部分与带电体的最小距离/m
	中相	边相	中相	边相		
$H \leqslant 500$	6.5	5.8	6.7	6.4	6.8	0.5
$500 < H \leqslant 1000$	6.8	6.0	6.9	6.7	6.8	0.5
$1000 < H \leqslant 1500$	7.0	6.3	7.2	7.0	7.2	0.5
$1500 < H \leqslant 2000$	7.4	6.6	7.6	7.3	7.2	0.5

3. 准备工作

3.1　危险点及其预控措施

（1）危险点——触电伤害

预控措施：

①工作前，工作负责人应与值班调控人员联系，停用线路重合闸，并履行许可手续。

②塔上地电位作业人员登塔前，必须仔细核对线路名称，确认无误后方可上塔。

③工作中，如遇线路突然停电，作业人员应视其仍然带电。工作负责人应尽快与调控人员联系，值班调控人员未与工作负责人取得联系前不准强送电。

④绝缘工具及绝缘绳索不得损坏、受潮、变形、失灵，不准使用非绝缘绳索（如棉纱绳、白棕绳、钢丝绳）。

⑤等电位作业人员应穿着阻燃内衣，衣服外面应穿戴全套合格的屏蔽服（包括帽、衣裤、手套、袜、面罩和鞋），且各部分应连接良好。

⑥等电位作业人员作业时保证最小安全距离中相6.8m，边相6.0m；最小组合间隙中相6.9m，边相6.7m；绝缘工具最小有效绝缘长度6.8m；在电位转移前，应得到工作负责人的许可，人体裸露部分与带电体的最小距离不小于0.5m；电位转移时，动作应迅速，严禁用头部

充放电；与地电位作业人员传递工具和材料时，使用绝缘工具或绝缘绳索的有效长度不准小于表2-10-2的规定。

⑦监护人、工作负责人应对作业人员进行不间断监护，随时纠正其不规范动作和行为。重点监护高处作业人员，使其保持足够的安全距离（符合表2-10-2的规定），禁止同时接触两个非连通的带电体或带电体与接地体。

（2）危险点——高处坠落。

①高处作业人员登高前，必须具备符合本项作业要求的身体状况、精神状态和技能素质。

②等电位作业人员进入强电场过程中应手脚一致，速度均匀，动作平稳；移动全过程应使用人体后备保护绳。

③监护人员应随时纠正其不规范动作和行为，重点监护作业人员在转位的过程中不得失去安全带或绝缘后备保护绳的保护，严禁低挂高用。

（3）危险点——高处坠物伤人。

预控措施：

①高处作业人员的个人工具及零星材料应装入工具袋，严禁在高处浮置物件、口中含物。

②地面作业人员必须正确佩戴安全帽，正确使用绳结，与作业点正下方距离不得小于坠落半径。

③作业现场设置围栏并挂好警示标示牌。监护人员应随时注意，禁止非工作人员及车辆进入作业区域。

3.2 工器具及材料选择

带电处理1000kV交流输电线路导线引流板发热缺陷所需工器具及材料见表2-10-3。工器具出库前，应认真核对工器具的使用电压等级和试验周期，并检查确认外观良好、连接牢固、转动灵活，且符合本次工作任务的要求；工器具出库后，应存放在工具袋或工具箱内进行运输，防止脏污、受潮；金属工具和绝缘工器具应分开装运，防止因混装运输导致工器具变形、损伤等现象发生。

表2-10-3 带电处理1000kV交流输电线路导线引流板发热缺陷所需工器具及材料表

序号	名称	规格型号	单位	数量	备注
1	绝缘传递绳	$\phi 12$	根	2	
2	绝缘后备保护绳	$\phi 16$	根	2	
3	绝缘子检测仪		套	1	
4	绝缘滑车	1T	个	2	
5	套筒扳手	与引流板螺栓型号一致	套	1	
6	电位转移棒		根	1	

续表

序号	名称	规格型号	单位	数量	备注
7	绝缘电阻测试仪	5000V	台	1	
8	风速风向仪		台	1	
9	温湿度仪		台	1	
10	万用表		台	1	
11	防潮帆布	2m×4m	张	1	
12	屏蔽服	屏蔽效率≥60dB（屏蔽面罩屏蔽效率≥20dB）	套	2	
13	防坠器	与铁塔防坠器装置型号对应	只	2	
14	安全围网		套	若干	
15	警示标示牌	"在此工作""从此进出""从此上下"	套	1	
16	红马甲	"工作负责人""专责监护人"	件	2	
17	螺栓	与引流板螺栓型号一致	个	若干	备用
18					
19	背负式安全带	带后备保护绳	套	2	
20	清洁毛巾		条	1	
21	绝缘绳套	$\phi12$	根	2	
22	安全帽		顶	6	
23	检测杆		根	1	
24	工具包		个	2	
25	对讲机		个	4	

3.3 作业人员分工

本任务作业人员分工如表2-10-4所示。

表2-10-4 带电处理1000kV交流输电线路导线引流板发热缺陷人员分工表

序号	工作岗位	数量(人)	工作职责
1	工作负责人	1	负责组织作业现场的各项工作
2	专责监护人	1	负责作业现场的安全监护
3	等电位电工	1	负责进入等电位处理导线引流板发热缺陷
4	地电位电工	1	负责传递工具、材料配合等电位电工进、出等电位
4	地面电工	2	负责传递工具、材料配合等电位电工进、出等电位

4. 工作程序

本任务工作流程如表2-10-5所示。

表2-10-5　带电处理1000kV交流输电线路导线引流板发热缺陷工作流程表

序号	作业内容	作业步骤及标准	安全措施及注意事项	责任人
1	工作许可	（1）工作负责人负责联系值班调控人员，按工作票内容申请停用线路重合闸。（2）经值班调控人员许可后，方可开始带电作业工作。	不得未经值班调控人员许可即开始工作	
2	现场复勘	工作负责人负责完成以下工作：（1）现场核对线路名称无误；基础及铁塔完好无异常；交叉跨越距离符合安全要求；确认缺陷情况及导地线规格型号等。（2）检查地形环境符合作业要求。（3）检查工作票所列安全措施与现场实际情况相符，必要时予以补充	（1）正确穿戴安全帽、工作服、工作鞋、劳保手套。（2）不得在危及作业人员安全的气象条件下作业。（3）严禁非工作人员、车辆进入作业现场	
3	现场布置	正确装设安全围栏并悬挂标示牌：（1）安全围栏范围应充分考虑高处坠物，以及对道路交通的影响。（2）安全围栏出入口设置合理。（3）妥当布置"从此进出""在此工作""从此上下"等标示	对道路交通安全影响不可控时，应及时联系交通管理部门强化现场交通安全管控	
4	召开班前会	（1）全体工作成员列队。（2）工作负责人宣读工作票，明确工作任务及人员分工；讲解工作中的安全措施和技术措施；查（问）全体工作成员精神状态；告知工作中存在的危险点及采取的预控措施。（3）全体工作成员在工作票上签名确认	（1）工作票填写、签发和许可手续规范，签名完整。（2）全体工作成员精神状态良好。（3）全体工作成员明确任务分工、安全措施和技术措施	
5	检查工具	（1）等电位电工及地电位电工正确地穿戴好屏蔽服并检测合格，由负责人监督检查。（2）正确佩戴个人安全用具（大小合适，锁扣自如），由负责人监督检查。（3）测量风速、湿度，检查绝缘工具的绝缘性能，并做好记录	（1）金属、绝缘工具使用前，应仔细检查其是否损坏、变形、失灵。绝缘工具应使用5000V绝缘电阻表进行分段绝缘检测，阻值应不低于700MΩ，并用清洁干燥的毛巾将其擦拭干净。（2）用万用表测量屏蔽服衣裤最远端点之间的电阻值不得大于20Ω。工作负责人认真检查作业电工屏蔽服的连接情况。（3）检查工具组装情况并确认连接可靠。（4）现场所使用的带电作业工器具应放置在防潮帆布上	

续表

序号	作业内容	作业步骤及标准	安全措施及注意事项	责任人
6	登塔	(1)核对线路名称无误后,塔上电工对安全带、防坠器做冲击试验检查。 (2)塔上电工携带绝缘传递绳登塔至横担作业点,选择合适位置系好安全带,将绝缘滑车和绝缘传递绳安装在横担合适位置。然后配合地面电工将绝缘传递绳分开作起吊准备	(1)核对线路名称无误后,方可登塔作业。 (2)登塔过程中应使用塔上安装的防坠装置;塔上移动及转位时,不准失去安全保护,作业人员必须攀抓牢固构件。 (3)作业电工必须穿全套合格的屏蔽服,且全套屏蔽服必须连接可靠。在进入等电位前,等电位电工要检查确认屏蔽服个部位连接可靠后方能进行下一步操作	
7	检测绝缘子	(1)地面电工将绝缘操作杆及绝缘子检测仪传至塔上。等电位电工对绝缘子进行检测。 (2)地电位电工配合等电位电工手持绝缘操作杆从导线侧依次向横担侧检测,遇有低、零值绝缘子时应复测2-3次,并作好记录; (3)地电位电工将绝缘操作杆及绝缘子检测仪传至塔下	(1)检测绝缘子工作必须逐片进行,接触必须可靠。 (2)扣除人体短接后,当良好绝缘子少于37片时,立即停止作业。 (3)作业电工应清楚如何判断(无声为零值绝缘子,声音较小为低值绝缘子)并区分好电晕声和放电声	
8	进入强电场	(1)地面电工将电位转移棒传至塔上。 (2)等电位电工携带电位转移棒移动到绝缘子连接金具上,并带好绝缘滑车和绝缘传递绳。 (3)等电位电工检查屏蔽服各部分连接良好后报经工作负责人同意,双手抓扶一串,双脚踩另一串,采用"跨二短三"方法沿绝缘子串进入强电场。 (4)当等电位电工平行移动至距导线侧均压环三片绝缘子时,应停止移动,利用电位转移棒进行电位转移	(1)等电位电工进入电位前必须得到工作负责人的许可。 (2)等电位电工进入绝缘子串前必须系好后备保护绳,并调整好绝缘传递绳。 (3)等电位电工在进入电位过程中与接地体和带电体两部分间隙所组成的组合间隙不得小于中相6.9m(边相6.7m)	
9	处理导线引流板发热缺陷	(1)等电位电工进入等电位后,将安全带系在子导线上,地面电工利用传递绳将套筒扳手传给等电位电工。 (2)等电位电工利用套筒扳手对导线引流板连接螺栓进行紧固。 (3)等电位电工利用传递绳将套筒扳手传递至塔下	(1)等电位电工注意避免动作幅度过大,避免将肢体伸向绝缘子。 (2)等电位电工注意避免被烫伤	

续表

序号	作业内容	作业步骤及标准	安全措施及注意事项	责任人
10	退出电位	(1)经检查作业点无遗留物后经工作负责人许可,等电位电工带好绝缘传递绳,作退出电位准备。 (2)等电位电工利用电位转移棒钩紧均压环,并进入距均压环的第3片绝缘子,一只手抓紧绝缘子,另一只手握电位转移棒,利用电位转移棒快速脱离电位。 (3)等电位电工按照"跨二短三"的方法退出等电位	(1)等电位电工退出电位前必须得到工作负责人的许可。 (2)等电位电工在退出电位过程中与接地体和带电体两部分间隙所组成的组合间隙不得小于中相6.9m(边相6.7m)。 (3)沿绝缘子串移动时,手要抓牢,脚要踏实	
11	返回地面	塔上电工利用传递绳将塔上工具传递至塔下。检查塔上无遗留物后,向工作负责人汇报,得到工作负责人同意后携带绝缘传递绳下塔	下塔过程中应使用塔上安装的防坠装置,塔上移动及转位时,不准失去安全保护,作业人员必须攀抓牢固构件	
12	工作结束	(1)工作负责人组织全体工作成员整理工器具和材料,将工器具清洁后放入专用的箱(袋)中;清理现场,做到"工完料尽场地清"。 (2)召开班后会,工作负责人进行工作总结和点评工作。点评本次工作的施工质量;点评全体工作成员的安全措施落实情况。 (3)工作负责人向值班调控人员汇报工作结束,申请恢复线路重合闸,终结工作票	不得约时恢复线路重合闸	

二、考核标准

特高压交流输电线路运检技能考核评分细则

考生填写栏	编号:		姓名:		所在岗位:		单位:		日期:		年 月 日	
考评员填写栏	成绩:		考评员:		考评组长:		开始时间:		结束时间:		操作时长:	
考核模块	带电处理1000kV交流输电线路导线引流板发热缺陷		考核对象		特高压交流输电线路检修人员			考核方式	操作	考核时限		60min
任务描述	采用"跨二短三"沿绝缘子串进强电场带电处理1000kV交流输电线路导线引流板发热缺陷											

续表

工作规范及要求	1. 带电作业工作应在良好天气下进行。如遇雷、雨、雪、雾天气不得进行带电作业。风力大于5级、湿度大于80%时，一般不宜进行带电作业。 2. 本项作业需6人，其中工作负责人1名，专责监护人1名，等电位电工1名，地电位电工1名，地面电工2名，采用耐张串沿绝缘子串进出强电场法带电处理1000kV交流输电线路导线引流板发热缺陷。 3. 工作负责(监护)人职责：负责本次工作任务的人员分工、工作票的宣读、办理线路停用重合闸、办理工作许可手续、召开工作班前会、负责作业过程中的安全监督、工作中突发情况的处理、工作质量的监督、工作后的总结。 4. 等电位电工职责：负责本次作业过程中的主要作业，根据作业的位置安装、拆除作业工器具，进入等电位处理导线引流板发热缺陷。 5. 地电位电工职责：负责传递工具、材料配合等电位电工进出等电位。 6. 地面电工职责：负责传递工具、材料。 7. 在带电作业中，如遇雷、雨、大风或其他任何情况威胁到工作人员的安全时，工作负责人或监护人可根据情况，临时停止工作。 给定条件： 1. 1000kV实训线路耐张塔一边相导线。 2. 工作票已办理，安全措施已经完备(重合闸已停用)，工作开始、工作终结时应口头提出申请(调度或考评员)。 3. 安全、正确地使用仪器对绝缘工具进行检测。 4. 必须按工作程序进行操作，工序错误扣除应做项目分值，出现重大人身、器材和操作安全隐患，考评员可下令终止操作(考核)
考核情景准备	1. 塔形：1000kV实训线路耐张塔一边相导线，工作内容：带电处理1000kV交流输电线路导线引流板发热缺陷。 2. 所需作业工器具：φ12绝缘传递绳2根、φ16绝缘保护绳2根、绝缘子检测仪1套、1T绝缘滑车2个、各不同型号套筒扳手1套、电位转移棒1根、5000V绝缘电阻测试仪1台、风速风向仪1台、温湿度仪1台、万用表1台、2m×4m防潮帆布1张、II型屏蔽服2套、防坠器2只、安全围网1套、警示标示牌1套、红马甲2件、各不同型号螺栓若干、各不同型号销子若干、双保险安全带2条、清洁毛巾1条、φ12绝缘绳套2根、安全帽6顶、操作杆1根、工具包2个、对讲机4个。 3. 作业现场做好监护工作，作业现场安全措施(围栏等)已全部落实；禁止非作业人员进入现场，工作人员进入作业现场必须戴安全帽。 4. 考生自备工作服，阻燃纯棉内衣，安全帽，线手套，安全带(含二保绳)。
备注	1. 各项目得分均扣完为止，出现重大人身、器材和操作安全隐患，考评员可下令终止操作 2. 设备、作业环境、安全带、安全帽、工器具、屏蔽服等不符合作业条件考评员可下令终止操作

序号	项目名称	质量要求	分值	扣分标准	扣分原因	扣分	得分
1	现场复勘	1)工作负责人到作业现场核对线路名称和杆塔编号、现场工作条件、缺陷部位等。 2)检测风速、湿度等现场气象条件符合作业要求。 3)检查工作票填写完整，无涂改，检查是否所列安全措施与现场实际情况相符，必要时予以补充	5	1)未核对线路名称、杆塔编号、现场工作条件、缺陷部位等，扣1分/项。 2)未检测风速、湿度等现场气象条件，扣1分/项。 3)工作票填写出现涂改，扣0.5分/处；工作票编号有误，扣1分；工作票填写不完整，扣1.5分			

续表

序号	项目名称	质量要求	分值	扣分标准	扣分原因	扣分	得分
2	工作许可	1）工作负责人联系值班调控人员，按工作票内容申请停用线路重合闸。 2）汇报内容规范、完整。	2	1）未联系调度部门（裁判）停用重合闸，扣2分。 2）汇报专业用语不规范或不完整，扣1分。			
3	现场布置	正确装设安全围栏并悬挂标示牌： 1）安全围栏范围应充分考虑高处坠物，以及对道路交通的影响。 2）安全围栏出入口设置合理。 3）妥当布置"从此进出""在此工作""从此上下"等标示	3	1）作业现场未装设围栏，扣1分。 2）未设立警示牌，扣1分。 3）未悬挂登塔作业标志，扣1分。			
4	召开班前会	1）全体工作成员全体人员正确佩戴安全帽、工作服。 2）工作负责人穿红色背心，宣读工作票，明确工作任务及人员分工；讲解工作中的安全措施和技术措施；查（问）全体工作成员精神状态；告知工作中存在的危险点及采取的预控措施。 3）全体工作成员在工作票上签名确认	3	1）工作人员着装不整齐，扣0.5分/人。 2）未进行分工，扣3分；分工不明确，扣1分。 3）现场工作负责人未穿安全监护背心，扣1分。 4）工作票上工作班成员未签字或签字不全，扣1分。			
5	工器具检查	1）工作人员按要求将工器具放在防防潮苫布上；防潮苫布应清洁、干燥。 2）工器具应按定置管理要求分类摆放；绝缘工器具不能与金属工具、材料混放；对工器具进行外观检查。 3）绝缘工具表面不应磨损、变形损坏，操作应灵活。绝缘工具应使用5000V绝缘电阻表进行分段绝缘检测，阻值应不低于700MΩ，并用清洁干燥的毛巾将其擦拭干净。 4）塔上地电位和登电位人员按要求正确穿全套合格的屏蔽服、导电鞋，且各部分连接应良好，屏蔽服内不得贴身穿着化纤类衣服，并系好安全带；工作负责人应认真检查是否穿戴正确。 5）登塔人员再次核对双重名称、杆号、相别并报告	7	1）未使用防潮苫布并定置摆放工器具，扣1分。 2）未检查工器具外观及试验合格证，扣0.5分/项。 3）未正确使用检测仪器对工器具进行检测，扣1分/项。 4）作业人员未正确穿屏蔽服且各部位连接良好，扣2分/人次。 5）现场工作负责人未对登塔作业人员进安全防护装备进行检查，扣1分。 6）登塔人员未核对线路双重名称、杆号、相别，扣2分/人。 7）汇报检测结果不规范，扣1分；不完整，扣0.5分/项			

续表

序号	项目名称	质量要求	分值	扣分标准	扣分原因	扣分	得分
6	登塔	1)地电位电工、等电位电工穿好全套合格的屏蔽服,将安全带做冲击试验后,系好安全带后携带绝缘传递绳相继登塔。2)登塔过程中系好防坠落保护装置,登塔至合适位置,系好安全带,布置好绝缘传递绳,然后配合地面电工将绝缘传递绳分开作起吊准备。3)登塔过程中应系好防坠落保护装置,匀速登塔,手抓主材,将安全带挂在肩上并与带电体保持6.8m以上安全距离,工作负责人加强作业监护。	5	1)未系安全带或安全带及后备保护绳或未进行冲击试验,扣2分/项。2)手抓脚钉,扣0.5分/次。3)现场工作负责人未对地电位、中间电位电工进行安全防护装备进行检查,扣1分/项。4)传递时高处落物,扣1分/次。5)传递过程工具与塔身磕碰,扣1分/项。6)塔上电工转位时失去安全带保护,扣5分			
7	检测绝缘子	1)地面电工将绝缘操作杆及绝缘子检测仪传至塔上。等电位电工协助地电位电工对绝缘子进行检测。2)检测绝缘子工作必须逐串逐片进行,接触必须可靠,由横担侧向带电侧进行。3)当任意一串良好绝缘子少于37片时,立即停止检测,并停止本次带电作业工作	2	1)未进行绝缘子测零,扣2分。2)测零方法错误,扣2分。3)测零结果未汇报,扣2分			
8	进入强电场	1)等电位电工进入绝缘子串前必须系好保护绳(用后备保护绳兜住脚踩绝缘子串),并调整好绝缘传递绳和电位转移棒。2)等电位电工检查屏蔽服各部分连接良好后报经工作负责人同意,双手抓扶一串,双脚踩另一串,采用"跨二短三"作业方式沿绝缘子串进入强电场。3)等电位电工在进入电位过程中与接地体和带电体两部分间隙所组成的组合间隙不得小于中相6.9m(边相6.7m),进入强电场必须用电位转移棒进行电位转移	15	1)等电位电工电位转移过程中裸露部分距离不够,扣2分。2)等电位电工进入强电场动作不正确,反复放电,扣2分/次。3)转移电位动作不熟练,扣1分/次;电位转移过程未使用电位转移棒,扣4分/次。4)未得到工作负责人许可就进行电位转移,扣4分/次			

序号	项目名称	质量要求	分值	扣分标准	扣分原因	扣分	得分
9	损伤导线表面处理	1)等电位电工进入等电位后,将安全带系在子导线上,地面电工利用传递绳将套筒扳手传给等电位电工。 2)等电位电工利用套筒扳手对导线引流板连接螺栓进行紧固。 3)等电位电工利用传递绳将套筒扳手传递至塔下	30	1)等电位电工注意避免动作幅度过大,避免将肢体伸向绝缘子,否则扣3分/次。 2)传递工具和材料应使用绝缘工具和绝缘绳索,否则扣2分/次。 3)作业过程中不得掉落工器具和材料,否则扣3分/次。 4)作业转位时,不得失去安全带保护,否则扣3分/次。 5)转移电位时,人体裸露部分与带电体的距离不小于0.5米,否则扣2分/次。 6)不得遗留工器具和材料,否则扣3分/次			
10	退出强电场	1)经检查发热点消失,作业点无遗留物后经工作负责人许可,等电位电工携带绝缘传递绳走线返回均压环处,作退出电位准备。 2)等电位电工利用电位转移棒钩紧均压环,并进入距均压环的第3片绝缘子,一只手抓紧绝缘子,另一只手握电位转移棒,利用电位转移棒快速脱离电位。 3)等电位电工按照"跨二短三"作业方式的方法退出等电位	13	1)未向工作负责人申请即进行电位转移,扣2分;申请了但未得同意即开始,扣1分。 2)等电位电工电位转移过程中裸露部分与带电体距离不够,扣2分/次。 3)等电位电工退出强电场动作不正确,反复放电,扣2分/次。 4)转移电位动作不熟练,扣2分;电位转移未使用电位转移棒,扣3分			
11	返回地面	塔上电工检查塔上无遗留物后,向工作负责人汇报,得到工作负责人同意后携带绝缘传递绳下塔	5	1)下塔过程未使用防坠器,扣5分。 2)塔上移位失去安全带保护,扣5分。 3)下塔手抓脚钉,扣1分/次。 4)塔上有遗留物,扣2分			

续表

序号	项目名称	质量要求	分值	扣分标准	扣分原因	扣分	得分
12	工作结束	1)工作负责人组织全体工作成员整理工器具和材料,将工器具清洁后放入专用的箱(袋)中;清理现场,做到"工完料尽场地清"。 2)召开班后会,工作负责人进行工作总结和点评工作。点评本次工作的施工质量;点评全体工作成员的安全措施落实情况。 3)工作负责人向值班调控人员汇报工作结束,申请恢复线路重合闸,终结工作票	10	1)工器具未清理,扣2分。 2)工器具有遗漏,扣2分。 3)未开班后会,扣10分。 4)未拆除围栏,扣2分。 5)未向调度汇报,扣2分。			
	合计		100				

模块 11 带电更换 1000kV 交流输电线路直线塔地线联接金具培训及考核标准

一、培训标准

(一) 培训要求

模块名称	带电更换1000kV交流输电线路直线塔地线联接金具	培训类别	操作类
培训方式	实操培训	培训学时	14学时
培训目标	1.掌握带电更换1000kV交流输电线路直线塔地线联接金具的操作流程。 2.能完成转移1000kV交流输电线路直线塔地线联接金具荷载的操作。 3.能独立完成更换1000kV交流输电线路直线塔地线联接金具的操作（地电位作业法）		
培训场地	特高压交流实训线路		
培训内容	采用地电位作业法带电更换1000kV交流输电线路直线塔地线联接金具		
适用范围	特高压交流输电线路检修人员		

(二) 引用规程规范

(1)《电工术语》(GB/T 2900.55—2002)

(2)《带电作业用绝缘滑车》(GB/T 13034—2008)。

(3)《带电作业用绝缘绳索》(GB/T 13035—2008)。

(4)《交流线路带电作业安全距离计算方法》(GB/T 18037—2000)。

(5)《1000kV交流带电作业用屏蔽服装》(GB/T 25726—2010)。

(6)《1000kV架空输电线路设计规范》(GB 50665—2011)。

(7)《1000kV交流输电线路检修规范》(DL/T 209—2008)。

(8)《1000kV交流输电线路运行规范》(DL/T 307—2010)。

(9)《1000kV交流输电线路带电作业技术导则》(DL/T 392—2015)。

(10)《带电作业工器具、装置和设备使用的一般要求》(DL/T 877—2004)。

(11)《国家电网公司电力安全工作规程（线路部分）》(Q/GDW 1799.2—2013)。

(三)培训教学设计

本设计以完成"带电更换1000kV交流输电线路直线塔地线联接金具"为工作任务,按工作任务完成的标准化作业流程来设计各个培训阶段,每个阶段包括了具体的培训目标、培训内容、培训学时、培训方法(培训资源)、培训环境和考核评价等内容,如表2-11-1所示。

表2-11-1 带电更换1000kV交流输电线路直线塔地线联接金具培训内容设计

培训流程	培训目标	培训内容	培训学时	培训方法与资源	培训环境	考核评价
1.理论教学	1.掌握地电位作业法的基本原理。 2.熟悉转移1000kV交流输电线路直线塔地线联接金具荷载的方法。 3.熟悉更换1000kV交流输电线路直线塔地线联接金具的方法	1.地电位作业法的基本原理和等效电路图。 2.转移荷载时地线提线器的使用方法。 3.地线联接金具更换的方法和质量标准	2	培训方法:讲授法。 培训资源:PPT、相关规程规范	多媒体教室	考勤、课堂提问和作业
2.准备工作	能完成作业前准备工作	1.作业现场查勘。 2.编制培训标准化作业卡。 3.填写培训操作工作票。 4.完成本操作的工器具及材料准备。 5.工作申请	1	培训方法: 1.现场查勘和工器具及材料清理采用现场实操方法。 2.编写作业卡和写工作票采用讲授方法。 培训资源: 1.1000kV实训线路; 2.特高压工器具库房; 3.空白工作票	1.特高压输电实训线路; 2.多媒体教室	
3.作业现场准备	能完成作业现场准备工作	1.作业现场布置。 2.班前会。 3.工器具及材料检查	1	培训方法:演示与角色扮演法。 资源:1000kV实训线路	1000kV实训线路	
4.培训师演示	通过现场观摩,使学员初步领会本任务操作流程	1.登塔,安装绝缘滑车,作起吊准备。 2.地电位电工将作业点两端用接地线可靠接地。 3.安装提线工具并转移地线荷载。 4.拆除原地线金具。 5.安装新地线金具。 6.拆除工具返回地面	2	培训方法:演示法。 资源:1000kV实训线路	1000kV实训线路	

续表

培训流程	培训目标	培训内容	培训学时	培训方法与资源	培训环境	考核评价
5.学员分组训练	1.能完成转移1000kV交流输电线路直线塔地线联接金具荷载的操作。2.能完成更换地线直线金具的操作	1.学员分组（5人一组）训练更换1000kV输电线路直线塔地线联接金具技能操作。2.培训师对学员操作进行指导和安全监护	7	培训方法：角色扮演法。资源：1000kV实训线路	1000kV实训线路	采用技能考核评分细则对学员操作评分
6.工作终结	1.使学员进一步辨析操作过程不足之处，便于后期提升。2.培养学员安全文明生产的工作作风	1.作业现场清理。2.向调度汇报工作。3.班后会，对本次工作任务进行点评总结	1	培训方法：讲授和归纳法	1000kV实训线路	

（四）作业流程

1. 工作任务

地电位电工携带绝缘传递绳登塔至架空地线作业点，采用地电位作业法更换1000kV交流输电线路直线塔地线联接金具。

2. 天气及作业现场要求

（1）带电更换1000kV交流输电线路直线塔地线联接金具应在良好的天气下进行。

如遇雷电（听见雷声、看见闪电）、雪、雹、雨、雾等，不应进行带电作业。风力大于5级时，不宜进行带电作业；相对湿度大于80%的天气，如需进行带电作业应采用具有防潮性能的绝缘工具。恶劣天气下必须开展带电抢修时，应组织有关人员充分讨论并编制必要的安全措施，经本单位主管生产领导批准后方可进行。

（2）作业人员精神状态良好，熟悉工作中保证安全的组织措施和技术措施；应持有在有效期内的特高压交流带电作业资质证书。

（3）工作负责人应事先组织有特高压交流带电作业经验的人员到现场进行勘察，根据勘察结果确定本次作业方法和所需工器具，以及应采取的必要措施，并办理带电作业工作票，编写标准化作业指导书。

（4）作业现场应合理设置围栏，并妥当布置警示标示牌，禁止非工作人员入内。

（5）本项目不需向调度申请停用线路重合闸装置。在带电作业过程中如设备突然停电，作业电工应视设备仍然带电。

（6）地电位电工与带电体的安全距离不得小于6.8（中相）/6.0（边相）m，绝缘工具有效绝缘长度不得小于6.8m。

3. 准备工作

3.1 危险点及其预控措施

（1）危险点——触电伤害

预控措施：

①工作前，工作负责人应与值班调控人员联系，履行工作许可手续。

②地电位作业人员登塔前，必须仔细核对线路双重名称、杆塔编号、地线位置，确认无误后方可登塔。

③工作中，如遇线路突然停电，作业人员应视其仍然带电。工作负责人应尽快与调控人员联系，值班调控人员未与工作负责人取得联系前不准强送电。

④绝缘工具及绝缘绳索不得损坏、受潮、变形、失灵，不准使用非绝缘绳索（如棉纱绳、白棕绳、钢丝绳）。

⑤地电位电工应穿着阻燃内衣，衣服外面应穿戴全套屏蔽服（包括帽、衣裤、手套、袜、鞋和面罩），且各部分应连接良好。

⑥地电位电工在接触架空地线前，应将作业点架空地线两端可靠接地。

⑦地电位电工在杆塔上移动及转位时与带电体安全距离不得小于6.8（中相）/6.0（边相）m；与地面作业人员传递工具和材料时，使用绝缘工具或绝缘绳索的有效长度不得小于6.8m。

⑧用绝缘绳索传递大件金属物品时，地面电工应将金属物品接地后再接触。

（2）危险点——高处坠落

预控措施：

①地电位电工登塔前，必须具备符合本项作业要求的身体状况、精神状态和技能素质。

②地电位电工登塔至作业点后，安全带、保护绳应分别系在牢固构件上。

③监护人员应随时纠正其不规范动作和行为，重点监护作业人员在转位的过程中不得失去安全带或绝缘后备保护绳的保护，严禁低挂高用。

（3）危险点——高处坠物伤人。

预控措施：

①地电位电工的个人工具及零星材料应装入工具袋，严禁在高处浮置物件、口中含物。

②地面电工必须正确佩戴安全帽，正确使用绳结，与作业点正下方距离不得小于坠落半径。

③作业现场设置围栏并挂好警示标示牌。监护人员应随时注意，禁止非工作人员及车辆进入作业区域。

3.2 工器具及材料选择

带电更换1000kV交流输电线路直线塔地线联接金具所需工器具及材料见表2-11-2。工器具出库前,应认真核对工器具的使用电压等级和试验周期,并检查确认外观良好、连接牢固、转动灵活,且符合本次工作任务的要求;工器具出库后,应存放在工具袋或工具箱内进行运输,防止脏污、受潮;金属工具和绝缘工器具应分开装运,防止因混装运输导致工器具变形、损伤等现象发生。

表2-11-2 带电更换1000kV交流输电线路直线塔地线联接金具所需工器具及材料表

序号	名称	规格型号	单位	数量	备注
1	屏蔽服	屏蔽效率≥60dB	套	1	个人防护用具
2	导电鞋	尺码视穿着人员而定	双	1	个人防护用具
3	阻燃内衣	纯桑蚕丝	套	1	个人防护用具
4	双保险安全带	背负式	根	1	个人防护用具
5	安全帽		顶	5	个人防护用具
6	地线提线器		套	1	金属工具
7	架空地线专用接地棒		套	2	金属工具
8	绝缘传递绳	ϕ12mm,长度与起吊高度匹配	根	1	绝缘工具
9	绝缘后备保护绳	ϕ16mm,长度与作业现场需要匹配	根	1	绝缘工具
10	绝缘绳套	ϕ14 mm	根	1	绝缘工具
11	绝缘滑车	0.5T	个	1	绝缘工具
12	便携式通话系统		套	1	其他工具
13	绝缘电阻测试仪	5000V,电极宽2cm、极间宽2cm	套	1	其他工具
14	地线后备保护装置		套	1	其他工具
15	万用表		套	1	其他工具
16	风速、温湿度测试仪	HT-8321	只	1	其他工具
17	安全围栏		套	若干	其他工具
18	警示标示牌	"在此工作""从此进出""车辆慢行""车辆绕行"	套	1	其他工具
19	红马甲	"工作负责人"	件	1	其他工具
20	防潮苫布	2m×4m	块	1	其他工具
21	个人工具	平口钳、专用扳手、记号笔	套	1	其他工具
22	毛巾	棉质	条	2	其他工具
23	直角挂板	ZH-10	件	1	材料
24	挂点金具	GD-12S	件	1	材料
25	U型挂板	U-10	件	1	材料
26	挂板	ZS-10	件	1	材料
27	悬垂线夹	XGU-2F	套	1	材料

3.3 作业人员分工

本任务作业人员分工如表2-11-3所示。

表2-11-3 带电更换1000kV交流输电线路直线塔地线联接金具人员分工表

序号	工作岗位	数量(人)	工作职责
1	工作负责人（监护人）	1	负责本次工作任务的人员分工、工作前的现场查勘、作业方案的制订、工作票的填写、现场复勘、办理工作许可手续、召开工作班前会、落实现场安全措施、负责作业过程中的安全监督、工作中突发情况的处理、工作质量的监督、工作后的总结
2	专责监护人	1	工作前，对被监护人员交待监护范围内的安全措施、告知危险点和安全注意事项；监督被监护人员遵守本规程和执行现场安全措施，及时纠正被监护人员的不安全动作和行为
3	地电位电工	1	负责地电位更换架空地线金具
4	地面电工	2	负责传递作业时所用材料及工器具

4. 工作程序

本任务工作流程如表2-11-4所示。

表2-11-4 带电更换1000kV交流输电线路直线塔地线联接金具工作流程表

序号	作业内容	作业步骤及标准	安全措施及注意事项	责任人
1	现场复勘	工作负责人负责完成以下工作： (1)现场核对线路名称、杆塔编号无误；基础及杆塔完好无异常；交叉跨越距离符合安全要求；确认缺陷情况及导地线规格型号等。 (2)检查地形环境符合作业要求。 (3)检查工作票所列安全措施与现场实际情况相符，必要时予以补充	(1)正确穿戴安全帽、工作服、工作鞋、劳保手套。 (2)不得在危及作业人员安全的气象条件下作业。 (3)严禁非工作人员、车辆进入作业现场	
2	工作许可	工作负责人负责联系值班调控人员，经值班调控人员许可后，方可开始带电作业工作	不得未经值班调控人员许可即开始工作	
3	现场布置	正确装设安全围栏并悬挂标示牌： (1)安全围栏范围应充分考虑高处坠物，以及对道路交通的影响。 (2)安全围栏出入口设置合理。 (3)妥当布置"从此进出""在此工作""从此上下"等标示	对道路交通安全影响不可控时，应及时联系交通管理部门强化现场交通安全管控	

续表

序号	作业内容	作业步骤及标准	安全措施及注意事项	责任人
4	召开班前会	(1)全体工作成员列队。 (2)工作负责人宣读工作票,明确工作任务及人员分工;讲解工作中的安全措施和技术措施;查(问)全体工作成员精神状态;告知工作中存在的危险点及采取的预控措施。 (3)全体工作成员在工作票上签名确认	(1)工作票填写、签发和许可手续规范,签名完整。 (2)全体工作成员精神状态良好。 (3)全体工作成员明确任务分工、安全措施和技术措施	
5	检查工器具	(1)在防潮苫布上,将工器具按作业要求准备齐备,并分类定置摆放整齐。检查工器具外观和试验合格证,无遗漏。 (2)使用绝缘电阻测试仪检测绝缘工具表面绝缘电阻值,方法正确,阻值不得低于700MΩ。 (3)使用万用表检测全套屏蔽服内阻,方法正确,阻值不得大于20Ω。 (4)测量风速,湿度,检查绝缘工具的绝缘性能,并做好记录。 (5)检查人员向工作负责人汇报各项检查结果符合作业要求。 (6)地电位电工正确穿戴好检测合格的屏蔽服,由负责人监督检查。 (7)正确佩戴个人安全用具(大小合适,锁扣自如),由负责人监督检查	(1)防潮苫布设置位置合理,保持清洁、干燥。 (2)工器具外观检查合格,无损伤、受潮、变形、失灵现象,合格证在有效期内。 (3)绝缘工具及绝缘绳索检测合格。 (4)检查工具组装连接情况	
6	登塔	(1)核对线路双重名称无误后,地电位电工对安全带、防坠器做冲击试验检查。 (2)地电位电工携带绝缘传递绳登塔至地线支架作业点,选择合适位置系好安全带、后备保护绳,将绝缘滑车和绝缘传递绳安装在地线支架合适位置。 (3)地面电工起吊地线接地棒,地电位电工将工作点两端地线可靠接地	(1)核对线路双重名称无误后,方可登塔作业。 (2)登塔作业过程中应手抓主材;塔上移动及转位时,不准失去安全带保护,作业人员必须攀抓牢固构件。 (3)地电位电工必须穿全套合格的屏蔽服,且全套屏蔽服必须连接可靠后放能进行下一步操作。 (4)接地过程中应先接地端,后接地线端,保证接地线的绝缘棒绝缘良好	
7	安装工具并转移地线荷载	(1)地面电工将地线后备保护装置和地线提线器传给地电位电工,地电位电工正确安装提线工具。 (2)工具安装完毕检查无误后,地电位电工收紧地线提线器丝杆,将地线金具上的垂直荷载转移到地线提线器	(1)上、下作业电工要密切配合,地面电工要听从地电位电工的指挥。 (2)地电位电工对带电体的最小安全距离不得小于6.8m。绝缘绳的有效绝缘长度不得小于6.8m。 (3)杆塔上下传递工器具应绑扎牢固,地电位电工在作业过程中禁止高处落物。 (4)地线提线器受力应均匀	

续表

序号	作业内容	作业步骤及标准	安全措施及注意事项	责任人
8	拆除原地线金具	对地线提线器做冲击试验检查,无误后,拆除原地线金具	地线金具更换前,对地线提线器做冲击试验检查无误,并经工作负责人同意后方可拆除原地线金具	
9	安装地线金具	地面电工起吊新地线金具至地线支架附近,地电位电工安装新地线金具	(1)作业时注意不得撞击地线提线器。(2)工具在传递过程中不得碰撞,系绳扣方法正确	
10	拆除工具返回地面	(1)松出地线提线器,将地线提升器上的荷载全部转移到新地线金具,并对地线和金具做冲击试验检查,无误并得到工作负责人许可后,拆除地线提线器和地线后备保护装置传至地面。(2)拆除接地线。(3)地电位电工对塔上进行全面检查无误,得到工作负责人许可后携带绝缘传递绳下塔	(1)拆除接地线过程中应先拆地线端,后拆接地端。(2)下塔时,手要抓牢,防止踏空踩滑	
11	工作结束	(1)清理现场及工具,现场无遗留物,工作负责人全面检查工作完成情况,清点人数。(2)召开班后会,工作负责人进行工作总结和点评工作。点评本次工作的施工质量;点评全体工作成员的安全措施落实情况。(3)工作负责人向值班调控人员汇报工作结束,终结工作票,集齐作业人员离开作业现场		

二、考核标准

特高压交流输电线路运检技能考核评分细则

考生填写栏	编号:	姓名:	所在岗位:	单位:	日期:	年 月 日
考评员填写栏	成绩:	考评员:	考评组长:	开始时间:	结束时间:	操作时长:

考核模块	带电更换1000kV交流输电线路直线塔地线联接金具	考核对象	特高压交流输电线路检修人员	考核方式	操作	考核时限	60min
任务描述	带电更换1000kV交流输电线路直线塔地线联接金具						

续表

工作规范及要求	1. 带电作业工作应在良好天气下进行。如遇雷、雨、雪、雾天气不得进行带电作业。风力大于5级，湿度大于80%，一般不宜进行带电作业。 2. 本项作业需5人，其中工作负责人1名，专责监护人1名，地电位电工1名，地面电工2名。 3. 地电位电工与带电体的安全距离不得小于6.8(中相)/6.0(边相)m，绝缘工具有效长度不得小于6.8m。 4. 工作负责(监护)人职责：负责本次工作任务的人员分工、工作票的宣读、办理工作许可手续、召开工作班前会、负责作业过程中的安全监督、工作中突发情况的处理、工作质量的监督、工作后的总结。 5. 在带电作业中，遇雷、雨、大风或其他任何情况威胁到工作人员的安全时，工作负责人或监护人可根据情况，临时停止工作 给定条件： 1. 培训线路：1000kV交流实训线路直线塔，地线型号：JLB20A-150。 2. 工作票已办理，安全措施已经完备，工作开始、工作终结时应口头提出申请(调度或考评员)。 3. 安全、正确地使用仪器对绝缘工具进行检测。 4. 必须按工作程序进行操作，工序错误扣除应做项目分值，出现重大人身、器材和操作安全隐患，考评员可下令终止操作(考核)
考核情景准备	1. 线路：1000kV交流实训线路直线塔。工作内容：带电更换1000kV交流输电线路直线塔地线联接金具。地线型号：JLB20A-150。 2. 所需作业工器具：屏蔽服1套、导电鞋1双、地线提线器1套、架空地线专用接地棒2套、绝缘传递绳1根、绝缘后备保护绳1根、地线后备保护装置1套、绝缘绳套1根、绝缘滑车1个、便携式通话系1套、绝缘电阻测试仪1套、万用表1套、风速、温湿度测试仪1只、安全围栏若干套、警示标示牌1套、红马甲1件、防潮苫布1块、个人工具1套、毛巾2条。 3. 作业现场做好监护工作，作业现场安全措施(围栏等)已全部落实；禁止非作业人员进入现场，工作人员进入作业现场必须戴好安全帽。 4. 考生自备工作服，阻燃纯棉内衣，安全帽，线手套，安全带(含后备保护绳)
备注	1. 各项目得分均扣完为止，出现重大人身、器材和操作安全隐患，考评员可下令终止操作。 2. 设备、作业环境、安全带、安全帽、工器具、屏蔽服等不符合作业条件考评员可下令终止操作

序号	项目名称	质量要求	分值	扣分标准	扣分原因	扣分	得分
1	现场勘察、资料调查	线路名称和杆塔号，现场工作条件、安全距离、导线型号	2	1)未进行现场勘察不得分。 2)现场查勘不详漏一项扣0.5分			
2	填写带电作业工作票	1)填写带电作业工作票，内容正确、完整、票面整洁。 2)标准化作业指导书内容正确、完整、整洁	3	1)无工作票不得分。 2)工作票填写错误一项扣0.5分。 3)书面不整洁扣1分			
3	工作许可	与值班调控人员联系得到工作许可	2	1)申请许可工作不全面每漏一项扣0.5分			

续表

序号	项目名称	质量要求	分值	扣分标准	扣分原因	扣分	得分
4	召开班前会	1)全体人员正确佩戴安全帽、工作服,列队宣读工作票、交代本次工作的安全措施和技术措施、危险点及预控措施,查(问)工作班组成员精神状态,工作任务明确后签字确认。 2)人员分工准确、全面,工作负责人检查参加工作的人员是否清楚、明白	4	1)未进行分工不得分。 2)未按规程要求着装扣1分。 3)分工不明扣1分。 4)未佩戴袖标(穿红马甲)扣1分。 5)工作票现场签字确认,缺签扣0.5分/人			
5	布置工作现场	1)在作业工器具摆放位置四周设置好安全护栏和作业标志,安全围栏的范围应考虑作业中高处落物的影响以及道路交通。 2)围栏的出入口应设置合理。 3)警示标示应包括"从此进出""在此工作"等,道路两侧应有"车辆慢行"或"车辆绕行"标示或路障	2	未布置现场不得分			
6	定置摆放工器具并检查	1)按要求将工器具放在防潮苫布上。 2)防潮苫布应清洁、干燥。 3)工器具应按定置管理要求分类摆放。 4)绝缘工器具不能与金属工具、材料混放。 5)对工器具进行外观检查;检查人员应戴清洁、干燥的线手套;绝缘工具表面不应磨损、变形损坏,操作应灵活。 6)安全带应在试验合格期内,并对外观进行检查和作冲击试验;使用绝缘电阻检测仪分段检测绝缘工具的表面绝缘电阻值;测量电极应符合规程要求(极宽2cm、极间距2cm)。 7)正确使用绝缘电阻检测仪,绝缘电阻值不得低于700 MΩ,全套合格屏蔽服内阻不大于20Ω。 8)绝缘工器具检查完毕,向工作负责人汇报检查结果	6	1)未定置摆放扣1分。 2)未检查试验合格标签扣0.5分。 3)检测工具使用方法不正确扣1分。 4)未检测绝缘工具扣2分。 5)作业人员裸手持、拿绝缘工具扣1分。 6)未用毛巾清洁硬质绝缘工具扣0.5分。 7)未摇测屏蔽服电阻扣1分。 8)安全带和防坠器未做冲击检查扣1分。 9)测试结果未汇报扣0.5分。 10)检测漏项扣1分/项			

续表

序号	项目名称	质量要求	分值	扣分标准	扣分原因	扣分	得分
7	登塔前准备	1)按要求正确穿戴全套合格的屏蔽服、导电鞋,各部分连接应良好,屏蔽服内穿阻燃内衣,并系好安全带。2)工作负责人应认真检查是否穿戴正确	4	1)未穿屏蔽服不得分。2)未连接好屏蔽服扣1分。3)未检查扣1分。4)屏蔽服穿戴不规范扣1分			
8	核对线路名称、杆号、相别	核对线路名称、塔号、地线位置并报告	2	1)未核对线路名称、塔号、地线位置不得分。2)核对漏项扣1分。3)未报告核对结果扣1分			
9	登塔	1)地电位电工携带绝缘传递绳登塔至地线支架作业点,选择合适位置系好安全带、后备保护绳,将绝缘滑车绝缘传递绳安装在地线支架合适位置。2)地面电工起吊地线接地棒,地电位电工将工作点两端地线可靠接地	10	1)登塔电工未向工作负责人申请即开始登塔扣2分;申请了但未得同意即开始扣1分。2)登塔作业过程中手未抓主材扣2分。3)安全带未挂肩上扣1分,绝缘绳、安全带缠绕、勾住扣2分/次。4)上塔手抓脚钉、手脚打滑、踏空、扣1分/次。5)未系好安全带扣2分。6)安全带扣环未扣牢扣1分。7)接地过程中应先接接地端,后接地线端,不符合要求者扣2分。8)塔上移位时,不准失去安全保护,不符合要求者扣1分			
10	安装工具并转移地线荷载	1)地面电工将地线后备保护装置和地线提线器传给地电位电工,地电位电工正确地安装好全部工具。2)工具安装完毕检查无误后,地电位电工收紧地线提线器丝杆,将地线金具上的垂直荷载转移到地线提线器	15	1)地面电工和地电位电工配合不协调扣1分,未向工作负责人报告扣1分。2)工具安装完毕未检查扣2分。3)地电位电工未收紧地线提线器丝杆扣2分。4)地电位电工在作业过程中发生高处落物扣2分。5)未安装地线后备保护装置扣4分			

第二部分 技能模块培训及考核标准

续表

序号	项目名称	质量要求	分值	扣分标准	扣分原因	扣分	得分
11	拆除原地线金具	对地线提线器做冲击试验检查,无误后,拆除原地线金具	15	1)未对地线提线器做冲击试验,扣2分。 2)未向工作负责人汇报即拆除地线金具扣2分。 3)向工作负责人汇报了但未经同意即拆除地线金具扣1分			
12	安装新地线金具	地面电工起吊新地线金具至地线支架附近,地电位电工安装新地线金具	15	1)操作时不能撞击地线提线器,未按要求扣2分。 2)工具在传递过程中发生碰撞,扣2分。 3)系绳扣不正确,扣1分。 4)系绳扣不牢固,扣1分			
13	拆除工具,返回地面	1)松出地线提线器,将地线提升器上的荷载全部转移到新地线金具,并对地线和金具做冲击试验检查,无误后得到工作负责人许可后,拆除地线提线器和地线后备保护装置传至地面。 2)拆除接地线。 3)地电位电工对塔上进行全面检查无误,得到工作负责人许可后携带绝缘传递绳下塔	5	1)未对地线和金具做冲击试验,扣1分。 2)未向工作负责人汇报即拆除地线提线器和地线后备保护扣2分。 3)向工作负责人汇报了但未经同意即拆除地线提线器和地线后备保护扣1分。 4)拆除接地线过程中应先拆地线端,后拆接地端,未按要求拆除扣2分。 5)未向工作负责人申请即开始下塔扣2分。 6)下塔手脚打滑、踏空、手抓脚钉扣0.5分/次。 7)绝缘绳、安全带缠绕、挂住现象扣0.5分/次。 8)塔上有遗留工具扣1分/件			
14	清理工具和现场	1)清理所有作业工器具并归类放好。 2)经检查无遗漏问题(含材料、工器具等)	6	1)未清理扣不得分。 2)有遗漏扣2分			
15	班后会	1)工作负责人作工作总结汇报。 2)撤除围栏、工作人员离开现场	6	1)未开班后会不得分。 2)未拆除围栏扣2分			
16	汇报工作结束	工作负责人向调度汇报工作结束	3	未向调度汇报不得分			
	合计		100				

模块12 停电更换1000kV交流输电线路直线塔Ⅰ型复合绝缘子培训及考核标准

一、培训标准

（一）培训要求

模块名称	停电更换1000kV交流输电线路直线塔Ⅰ型复合绝缘子	培训类别	操作类
培训方式	实操培训	培训学时	21学时
培训目标	1.掌握各类工器具、机具的使用方案和受力结构，以及更换复合绝缘子技术要点。 2.能熟练掌握停电更换1000kV交流输电线路直线Ⅰ型复合绝缘子的操作流程、技术方法和施工作业危险点。 3.作为主要作业人员，能熟练完成1000kV交流输电线路直线Ⅰ型复合绝缘子的更换。		
培训场地	特高压交流实训线路		
培训内容	正确使用各类受力工器具的操作方法，正确安装各类工器具，采用停电作业法更换1000kV交流输电线路直线Ⅰ型复合绝缘子。		
适用范围	特高压交流输电线路检修人员		

（二）引用规程规范

（1）《架空送电线路运行规程》（DL/T 741—2010）。

（2）《110～500kV架空送电线路设计技术规程》（DL/T 5092—1999）。

（3）《国家电网公司电力安全工作规程（线路部分）》（Q/GDW1799.2—2013）。

（4）《1000kV架空输电线路设计规范》（GB 50665—2011）。

（5）《1000kV交流输电线路检修规范》（DL/T 209—2008）。

（6）《1000kV交流输电线路运行规范》（DL/T 307—2010）。

（7）《110（66）kV～500kV架空输电线路检修规范》（国家电网公司）。

（8）《架空输电线路状态检修导则》（DLT 1248—2013）。

（9）《输变电设备状态检修管理规定》（国家电网公司）。

（10）《输变电设备状态检修试验规程》（国家电网公司）。

（三）培训教学设计

本设计以完成"停电更换1000kV交流输电线路直线塔Ⅰ型复合绝缘子"为工作任务，按工作任务完成的标准化作业流程来设计各个培训阶段，每个阶段包括了具体的培训目标、培训内容、培训学时、培训方法（培训资源）、培训环境和考核评价等内容，如表2-12-1所示。

表2-12-1　停电更换1000kV交流输电线路直线Ⅰ型复合绝缘子培训内容设计

培训流程	培训目标	培训内容	培训学时	培训方法与资源	培训环境	考核评价
1.理论教学	1.掌握各类工器具、机具的使用方案和受力结构，以及更换复合绝缘子技术要点。 2.能熟练掌握停电更换1000 kV交流输电线路直线Ⅰ型复合绝缘子的操作流程、技术方法和施工作业危险点	1.正确使用各类受力工器具的操作方法。 2.正确安装各类工器具。 3.采用停电作业法更换1000kV交流输电线路直线Ⅰ型复合绝缘子。	2	培训方法：讲授法。 培训资源：PPT、相关规程规范	多媒体教室	考勤、课堂提问和作业
2.准备工作	能完成作业前准备工作	1.作业现场查勘。 2.编制培训标准化作业卡。 3.填写培训操作工作票。 4.完成本操作的工器具及材料准备	1	培训方法： 1.现场查勘和工器具及材料清理采用现场实操方法。 2.编写作业卡和填写工作票采用讲授方法。 培训资源： 1. 1000kV实训线路。 2.特高压工器具库房。 3.空白工作票	1.特高压输电实训线路； 2.多媒体教室	
3.作业现场准备	能完成作业现场准备工作	1.作业现场复勘。 2.工作申请。 3.作业现场布置。 4.班前会。 5.工器具及材料检查	1	培训方法： 演示与角色扮演法。 资源： 1000kV实训线路	1000kV实训线路	

续表

培训流程	培训目标	培训内容	培训学时	培训方法与资源	培训环境	考核评价
4.培训师演示	通过现场观摩，使学员初步领会本任务操作流程	1.各类工器具使用方法讲解。2.演示更换复合绝缘子的塔上工器具连接方式。3.高空作业人员配合演示更换复合复合绝缘子的操作流程。4.利用地面人员配合更换1000kV交流输电线路直线Ⅰ型复合绝缘子。	2	培训方法：演示法。资源：1000kV实训线路	1000kV实训线路	
5.学员分组训练	1.能掌握各类受力工器具的使用方法和注意事项。2.掌握更换复合整串绝缘子的全部操作流程。3.能完成1000kV交流输电线路直线Ⅰ型复合绝缘子的更换	1.学员分组（高空3人、地面配合5人）训练工器具、机具的操作方法和更换Ⅰ型复合绝缘子的现场实际操作。2.培训师对学员操作进行指导和安全监护	14	培训方法：角色扮演法。资源：1000kV实训线路	1000kV实训线路	采用技能考核评分细则对学员操作评分
6.工作终结	1.使学员进一步辨析操作过程不足之处，便于后期提升。2.培训学员安全文明生产的工作作风	1.作业现场清理。2.向调度汇报工作。3.班后会，对本次工作任务进行点评总结	1	培训方法：讲授和归纳法	1000kV实训线路	

（四）作业流程

1. 工作任务

完成停电更换1000kV交流输电线路直线塔Ⅰ型复合绝缘子。

2. 天气及作业现场要求

（1）停电更换1000kV交流输电线路直线塔Ⅰ型复合绝缘子应在良好的天气进行。

在5级及以上的大风以及暴雨、雷电、冰雹、大雾、沙尘暴等恶劣天气下，应停止露天高处作业。特殊情况下，确须在恶劣天气进行抢修时，应组织有关人员充分讨论必要的安全措施，经本单位批准后方可进行。

（2）作业人员精神状态良好，工作班成员认真学习工作票和安全技术措施，所有人员做到"四清楚"(作业任务精楚、危险点清楚、作业程序清楚、安全措施清楚)。

(3) 作业前停送电联系人必须与调度联系履行工作许可手续，严禁约时停送电。工作负责人必须在得到许可人的许可工作命令后，方可在需检修的线路上验电、挂设接地线和进行检修工作。

(4) 停电后，工作负责人应认真做好记录。

(5) 登杆前应检查塔上是否有蜂窝，发现蜂窝严禁登塔。

(6) 塔上作业人员必须使用双保险安全带，并佩戴护目镜。

3. 准备工作

3.1 危险点及其预控措施

(1) 危险点——误登带电线路。

预控措施：

①登杆塔作业前，工作负责人、工作班成员应共同认真核查双重名称和识别标记（色标、判别标志等）与停电线路名称相符。

②登杆塔前应检查铁塔根部、基础等，必须牢固可靠。

③登杆塔前应检查登高工器具和设施，如安全带、脚钉、塔材等必须完整牢靠。

④不涉及挂设接地线的中间作业人员，应认真核实线路相序、色标、名称、编号与停电线路相符，确认线路名称无刷错、刷反等情况后，方可登杆。

(2) 危险点——登塔时、塔上作业时违反安规进行操作，可能引起高空坠落。

预控措施：

①攀爬过程中，为防止登杆人员串落，登杆作业人员间距不得小于1.6m。

②攀爬铁塔前应将脚底泥土清除干净，检查工具包完整，攀爬过程中不得掉落物件伤人。

③作业人员攀登杆塔时应戴好安全帽，穿软底鞋，动作不能过大，匀步攀登。

④攀爬过程中，安全带应收拾妥当，长尾绳放置在工具包内，主带应挂在肩上，防止攀爬过程中安全带勾挂脚钉和塔材，致使作业人员高空坠落。

⑤杆塔上移位时，不得失去安全带保护，做到踩稳抓牢。

⑥到达作业点位置，系好安全带（绳），应牢固可靠，不得低挂高用。

⑦未验电前，人体、绳索等与导线的安全距离必须不小于9.5m，工作中应设专人监护。

(3) 危险点——高处坠物伤人。

预控措施：

①地面人员不得站在作业点垂直下方。塔上人员应防止落物伤人，使用的工具、材料应用绳索传递。

②在高处作业应使用工具袋，较大的工具应固定在牢固的构件上，不准随便乱放。

③使用绞磨起吊过程中，应设专人指挥，统一配合，绝缘子串刚离地后应进行冲击检查。

（4）危险点——防止感应电伤人。

①为防感应电伤人，塔上作业人员应穿全套屏蔽服。

②如需接触架空地线，在架空地线接触前应进行可靠接地。

（5）危险点——现场作业安全监护。

①自作业开始至作业终结，安全监护人必须始终在现场对作业人员进行不间断的安全监护。

②工作负责人，监护人必须穿安全监护背心。

（6）危险点——交通安全。

出车时应注意车辆行驶安全，谨慎驾驶车辆，禁止违法行车。

3.2 工器具及材料选择

停电更换1000kV耐张整串绝缘子所需工器具及材料见表2-12-2。工器具出库前，应认真核对工器具的使用电压等级和试验周期，并检查确认外观良好、连接牢固、转动灵活，且符合本次工作任务的要求；工器具出库后，应防止脏污、受潮；金属工具和绝缘工器具应分开装运，防止因混装运输导致工器具变形、损伤等现象发生。

表2-12-2 停电更换1000kV交流输电线路耐张整串绝缘子所需工器具及材料表

序号	名称	规格型号	单位	数量	备注
1	接地线	1000kV专用	组	2	绝缘工具
2	绝缘手套	10kV	副	2	绝缘工具
3	验电器	1000kV专用	支	2	其他工具
4	全身式安全带	含带缓冲包长20M的后保绳	套	3	个人防护用具
5	个人保安线	（直径不小于16mm²）	根	1	其他工具
6	钢丝套	Φ22	根	4	其他工具
7	八勾卡	适用于八分裂900mm²及以上导线	套	2	金属工具
8	卸扣	10T	个	4	金属工具
9	软梯	长度不小于15M	副	1	其他工具
10	手扳葫芦	9T	副	2	金属工具
11	个人手动工具		套	3	绝缘工具
12	对讲机		台	5	其他工具
13	吊绳滑车	1T	个	1	其他工具
14	传递绳	φ16	套	1	其他工具
15	拔销器		把	2	其他工具
16	安全背心		件	2	其他工具
17	安全围栏		卷	4	其他工具

续表

序号	名称	规格型号	单位	数量	备注
18	防潮苫布		张	1	其他工具
19	垫木		块	若干	其他工具
20	复合绝缘子	FXBW-1000/420	串	1	材料

3.3 作业人员分工

本任务作业人员分工如表2-12-3所示。

表2-12-3　停电更换1000kV 输电线路Ⅰ型复合绝缘子人员分工表

序号	工作岗位	数量(人)	工作职责
1	工作负责人	1	负责本次工作任务的人员分工、工作前的现场查勘、作业方案的制订、工作票的填写、现场复勘、办理工作许可手续、召开工作班前会、落实现场安全措施、负责作业过程中的安全监督、工作中突发情况的处理、工作质量的监督、工作后的总结
2	安全监护人	2	负责本次工作过程中的安全监护工作
3	高空作业人员	3	负责本次停电更换1000kV Ⅰ串绝缘子操作
4	地面辅助人员	5	负责本次作业过程的地面辅助工作

4. 工作程序

本任务工作流程如表2-12-4所示。

表2-12-4　停电更换1000kV输电线路Ⅰ型复合绝缘子工作流程表

序号	作业内容	作业标准	安全注意事项	责任人
1	工作许可	作业前停送电联系人必须与调度联系履行工作许可手续。	(1)不得未经工作许可人许可即开始工作 (2)严禁约时停送电	
2	现场布置	正确装设安全围栏并悬挂标示牌： (1)安全围栏范围应充分考虑高处坠物，以及对道路交通的影响。 (2)安全围栏出入口设置合理。 (3)妥当布置"从此进出""在此工作""车辆慢行"或"车辆绕行"等标示	对道路交通安全影响不可控时，应及时联系交通管理部门强化现场交通安全管控	
3	召开班前会	(1)全体工作成员列队。 (2)工作负责人宣读工作票，明确工作任务及人员分工；讲解工作中的安全措施和技术措施；查(问)全体工作成员精神状态；告知工作中存在的危险点及采取的预控措施。 (3)全体工作成员在工作票上签名确认	(1)工作票填写、签发和许可手续规范，签名完整。 (2)全体工作成员精神状态良好。 (3)全体工作成员明确任务分工、安全措施和技术措施	

续表

序号	作业内容	作业标准	安全注意事项	责任人
4	检查工器具	(1)在防潮苫布上,将工器具按作业要求准备齐备,并分类定置摆放整齐。检查工器具外观和试验合格证,无遗漏。 (2)检查人员向工作负责人汇报各项检查结果符合作业要求	(1)防潮苫布数量足够,设置位置合理,保持清洁、干燥。 (2)工器具外观检查合格,无损伤、受潮、变形、失灵现象,合格证在有效期内。	
5	登杆塔	(1)登杆塔作业前,必须先核对线路名称及编号。对同塔多回线路,工作负责人、工作班成员应共同认真核查双重名称和识别标记(色标、判别标志等)。 (2)登杆塔前应检查铁塔根部、基础等,必须牢固可靠。 (3)攀爬过程,为防止登杆人员串落,登杆作业人员间距不得小于1.6m,安全带收拾妥当,后保绳放置在工具包内,主带应挂在肩上,防止攀爬过程中安全带勾攀脚钉和塔材,致使作业人员高空坠落。 (4)登杆塔至横担处时,监护人和作业人员应再次核对停电线路的识别标记和双重名称,确实无误后方可进入作业点位。	(1)作业人员攀登杆塔时应戴好安全帽,穿软底鞋,动作不能过大,匀步攀登。 (2)攀爬过程中,安全带应收拾妥当,长尾绳放置在工具包内,主带应挂在肩上,防止攀爬过程中安全带勾挂脚钉和塔材,致使作业人员高空坠落。 (3)杆塔上移位时,不得失去安全带保护,做到踩稳抓牢。 (4)到达作业点位置,系好安全带(绳),应牢固可靠,不得低挂高用。 (5)未验电前,人体、传递绳等与导线的安全距离必须不小于9.5m,工作中应设专人监护。	
6	验电、装设接地线	(1)登杆就位后,将安全带系在牢固可靠构件或电杆上,必须检查扣环是否正确就位。验电杆(笔)等工器具必须使用绳索传递。 (2)检查接地线完好,按程序(先接接地端,后导线端)装设好接地线。 (3)装设接地线时,必须使用绝缘绳或绝缘手柄进行操作,禁止直接用手操作接地线的金属部分的方式装设设接地线。确认接地线的夹头与导线连接紧密可靠。	(1)验电器在领用时和使用前应检查是否正常。 (2)禁止以缠绕导线的方式装设接地线。	

续表

序号	作业内容	作业标准	安全注意事项	责任人
7	更换I串复合绝缘子	(1)高空作业人员到达作业点位后,架设软梯,利用软梯进入导线,将两套八勾卡与9T手板葫芦固定在垂直于线路方向的绝缘子串两侧,并保留足够的人员操作空间。 (2)确认固定后,高空人员同时收紧手板葫芦,待手板葫芦完全承受复合绝缘子的拉力后,应再次检查横担有无变形,连接部位有无异常,确认一切正常后,才能进行下一步操作。 (3)使用绳索将复合绝缘子绑扎牢固,取下复合绝缘子与导线端连接金具的连接部分后,铁塔上的高空人员配合取下复合绝缘子横担端连接金具的连接部分。 (4)将旧复合绝缘子缓慢放至地面,地面人员取下旧复合绝缘子,更换完好复合绝缘子,将复合绝缘子绑扎牢固后,方可传递至塔上作业人员处。将复合绝缘子拉至作业点位后,塔上作业人员应先连接横担端连接金具的连接部分,并确保销子到位后方可继续进行连接复合绝缘子导线端的连接工作。 (5)装好复合绝缘子后,应检查球头、碗头是否安装到位、销子是否安装到位,并进行冲击试验,确认安装正确,连接可靠后方可同时缓慢放松2把手板葫芦	(1)使用手扳葫芦、八勾卡更换绝缘子串过程中,在手扳葫芦开始承受导线荷载后,必须检查手扳葫芦、钢丝绳套(吊装带)、卸扣的连接和受力情况,并做冲击试验,确认完全可靠后方可继续收紧手扳葫芦。 (2)在使用手扳葫芦、软梯过程中,应防止与复合绝缘子碰撞,避免损伤复合绝缘子。 (3)在使用传递绳吊放绝缘子串时,地面工作人员和塔上的工作人员必须密切配合,防止起吊绳缠绕,绝缘子串碰撞损伤导线。吊放绝缘子串时,地面人员不得站在垂直下方。	
8	拆除工器具	拆除八勾卡、手扳葫芦、钢丝套等工器具		
9	工作结束	(1)工作负责人组织全体工作成员整理工器具和材料,清理现场,做到"工完料尽场地清"。 (2)召开班后会,工作负责人进行工作总结和点评工作。点评本次工作的施工质量;点评全体工作成员的安全措施落实情况。 (3)工作负责人向工作许可人汇报工作结束,恢复停电线路送电,终结工作票		

二、考核标准

特高压交流输电线路运检技能考核评分细则

考生填写栏	编号： 姓名： 所在岗位： 单位： 日 期： 年 月 日						
考评员填写栏	成绩： 考评员： 考评组长： 开始时间： 结束时间： 操作时长：						
考核模块	停电更换1000千伏交流输电线路直线塔Ⅰ型复合绝缘子	考核对象	特高压交流输电线路检修人员	考核方式	操作	考核时限	150min
任务描述	停电更换1000kV输电线路C相Ⅰ型复合绝缘子						
工作规范及要求	1. 给定条件：1000kV交流实训线路C相Ⅰ串复合绝缘子老化，需要更换。线路已经停电、验电、挂接地线，所使用复合绝缘子已经过测试，工作票已办理，安全措施已经完备。 2. 整个过程主要操作流程由工作负责人1名，专责监护人1人，塔上电工3人配合完成，地面辅助工5人，协助参考人员完成工器具、材料的上、下传递工作以及其他非技术性工作。 3. 操作前参考人员应作必要的安全检查。 4. 工作开始应口头提出申请，工作结束时应口头汇报						
考核情景准备	1. 工器具：八勾卡2套、软梯1副、个人保安线1根、Φ22钢丝套4根、10T卸扣4个、9T手扳葫芦2把、传递绳1根、1T滑车1个、绳套1根、对讲机5个、双保险安全带3套、防潮苫布1张。 2. 材料：同型号复合绝缘子一串。 3. 在培训线路上操作						
备注	1. 个人工器具由参考人员自备。 2. 各项目得分均扣完为止						

序号	项目名称	质量要求	分值	扣分标准	扣分原因	扣分	得分
1	工具材料准备						
1.1	个人工具检查	活动扳手、平口钳、拔销钳、工具包符合质量要求	2	错漏1项扣1分			
1.2	受力工具检查	八勾卡、手扳葫芦、软梯检查，在试验合格期内	3	错漏1项扣1分			
1.3	安全工器具检查	双保险安全带、个人保安线、绝缘手套符合质量要求，并在试验合格周期内	3	错漏1项扣1分			
1.4	材料检查	核对复合绝缘子串型号，外观检查符合要求	2	错漏1项扣1分			

第二部分 技能模块培训及考核标准

续表

序号	项目名称	质量要求	分值	扣分标准	扣分原因	扣分	得分
2	场地布置						
2.1	场地围栏	场地围栏布置	2	未布置扣2分			
3	登塔及横担上的操作						
3.1	登塔	(1)检查杆塔基础无异常。 (2)正确携带传递绳(吊绳头折双、打死结、斜挎肩上)。 (3)沿脚钉侧主材正确登塔	6	(1)未检查1项扣1分。 (2)未携带传递绳扣2分,吊绳携带方式不规范扣1分。 (3)手抓脚钉1次扣1分。 (4)未沿脚钉侧主材登塔扣2分			
3.2	进入横担上工作点	由塔身到横担上工作点不得失去安全带保护。	3	未正确使用安全带扣3分			
3.3	安装滑车、个人保安线	传递滑车安装位置正确,方便操作,并加挂个人保安线	3	(1)滑车安装不规范扣1分。 (2)个人保安线漏挂扣2分			
4	绝缘子串上操作						
4.1	进入工作点	(1)将双保险安全带的后保绳系在横担适当位置。 (2)安装好软梯并沿软梯进入作业点,将围杆带系在导线上。 (3)检查复合绝缘子串锁紧销	5	(1)未使用双保险安全带扣5分。 (2)未正确使用软梯扣5分。 (3)未正确使用双保险安全带一次扣3分。 (4)未检查扣3分			
4.2	安装工器具	(1)正确安装钢丝绳套和9T手板葫芦。 (2)正确安装八勾卡。 (3)钢丝绳套、卸扣型号选用正确、安装可靠。 (4)安装的手板葫芦、个人保安线、软梯不影响人员操作	9	(1)未对塔材采取保护1处扣1分。 (2)手板葫芦、八勾卡安装不对1处扣1分。 (3)手板葫芦受力后有碰撞、缠绕,1处扣1分。 (4)钢丝绳套、卸扣型号选用不正确一处扣1分。 (5)手板葫芦漏1套扣4分			
4.3	收紧复合绝缘子串	(1)同时收紧2把9T手板葫芦,确保手板葫芦受力均衡 (2)9T手板葫芦受力后,检查滑车、卸扣、钢丝绳等连接部位,确认连接可靠。 (3)确认受力无误后继续收紧手板葫芦,直至复合绝缘子松动,作冲击试验。 (4)使用手板葫芦过程中,应防止手板葫芦损伤复合绝缘子	7	(1)碰响绝缘子一次扣1分。 (2)未冲击试验扣3分。 (3)使用手板葫芦不正确扣2分。 (4)碰伤绝缘子1处扣2分			

续表

序号	项目名称	质量要求	分值	扣分标准	扣分原因	扣分	得分
4.4	更换复合I串绝缘子	(1)将传递绳绑扎在复合绝缘子的合适位置,高空人员先取脱复合绝缘子导线端。 (2)地面辅助人员收紧传递绳,配合高空作业人员取脱复合绝缘子横担端,并传送至地面。 (3)将新复合绝缘子通过传递绳传递到杆塔上,高空作业人员先连接横担端,并将销子安装到位。 (4)高空作业人员安装复合绝缘子导线端,安装金具R销子,并检查是否安装到位。 (5)确认连接无误后,松出手扳葫芦,使复合绝缘子受力,并做冲击试验	15	(1)未冲击试验扣2分。 (2)传递绳索缠绕扣2分。 (3)传递物有撞击现象一次扣2分。 (4)掉落物件扣5分。 (5)未装锁紧销1个扣1分。 (6)拆装导线端和横担端顺序错误扣4分			
4.5	撤除工器具	(1)检查锁紧销、球头是否齐全到位,碗口朝向正确。 (2)拆除手扳葫芦、八勾卡及钢丝绳套,传递至地面	7	(1)绳索缠绕扣2分。 (2)传递物有撞击现象一次扣(2分)。 (3)锁紧销位置不正确扣1分。 (4)碗口朝向不正确扣1分			
4.6	清理杆塔上工器具	确认无遗留物。	4	有遗留物扣4分			
4.7	从导线到塔身	(1)沿软梯到进入铁塔横担 (2)攀爬软梯过程中不得失去安全带保护	8	(1)失去安全带保护扣4分。 (2)未正确使用软梯扣4分			
5	下塔	(1)必须沿脚钉侧主材正确下塔。 (2)正确携带传递绳(吊绳头折双、打死结,斜挎肩上)	8	(1)未携带传递绳扣4分,传递绳携带方式不规范扣2分。 (2)手抓脚钉1次扣1分。 (3)未沿脚钉侧主材登塔扣4分			
6	其他要求						
6.1	塔上作业	(1)严禁高处坠物。 (2)在操作过程中应双手协调配合操作。 (3)严禁浮置物品。 (4)严禁口中含物	5	(1)高处坠物一件扣5分。 (2)动作不协调扣2分。 (3)浮置物品扣2分。 (4)口中含物扣2分			
6.2	着装	工作服、工作胶鞋、安全帽、劳保手套穿戴正确。	2	漏一项扣2分			
6.3	清理现场	完工后清理作业现场,符合文明生产要求	2	未清理作业现场扣2分			
6.4	完成时间	在规定时间内按要求完成	3	超过时间10min扣1分,达到180min即终止操作,只记完成部分得分			
	合计		100				

模块13 停电更换1000kV交流输电线路直线塔单V型复合绝缘子培训及考核标准

一、培训标准

(一) 培训要求

模块名称	停电更换1000kV交流输电线路直线塔单V型复合绝缘子	培训类别	操作类
培训方式	实操培训	培训学时	21学时
培训目标	1.掌握各类工器具、机具的使用方案和受力结构,以及更换整串绝缘子技术要点。 2.能熟练掌握停电更换1000kV交流输电线路直线塔单V型复合绝缘子的操作流程、技术方法和施工作业危险点。 3.作为主要作业人员,能熟练完成1000kV交流输电线路直线塔单V型复合绝缘子的更换		
培训场地	特高压交流实训线路		
培训内容	正确使用各类受力工器具的操作方法正确安装各类工器具,采用停电作业法更换1000kV交流输电线路线塔单V型复合绝缘子		
适用范围	特高压交流输电线路检修人员		

(二) 引用规程规范

(1)《架空送电线路运行规程》(DL/T 741—2010)。

(2)《110~500kV架空送电线路设计技术规程》(DL/T 5092—1999)。

(3)《国家电网公司电力安全工作规程(线路部分)》(Q/GDW1799.2—2013)。

(4)《1000kV架空输电线路设计规范》(GB 50665—2011)。

(5)《1000kV交流输电线路检修规范》(DL/T 209—2008)。

(6)《1000kV交流输电线路运行规范》(DL/T 307—2010)。

(7)《110(66)kV~500kV架空输电线路检修规范》(国家电网公司)。

(8)《架空输电线路状态检修导则》(DLT 1248—2013)。

(9)《输变电设备状态检修管理规定》(国家电网公司)。

(10)《输变电设备状态检修试验规程》(国家电网公司)。

（三）培训教学设计

本设计以完成"停电更换 1000kV 交流输电线路直线塔单 V 型复合绝缘子"为工作任务，按工作任务完成的标准化作业流程来设计各个培训阶段，每个阶段包括了具体的培训目标、培训内容、培训学时、培训方法（培训资源）、培训环境和考核评价等内容，如表 2-13-1 所示。

表 2-13-1　1000kV 交流输电线路直线 V 型复合绝缘子培训内容设计

培训流程	培训目标	培训内容	培训学时	培训方法与资源	培训环境	考核评价
1. 理论教学	1.掌握各类工器具、机具的使用方案和受力结构，以及更换复合绝缘子技术要点。2.能熟练掌握停电更换 1000kV 交流输电线路直线塔单 V 型复合绝缘子的操作流程、技术方法和施工作业危险点	1.正确使用各类受力工器具，熟悉绞磨等机具的操作方法。2.正确安装各类工器具。3.采用停电作业法更换 1000kV 交流输电线路直线塔单 V 型复合绝缘子	2	培训方法：讲授法。培训资源：PPT、相关规程规范	多媒体教室	考勤、课堂提问和作业
2. 准备工作	能完成作业前准备工作	1.作业现场查勘。2.编制培训标准化作业卡。3.填写培训操作工作票。4.完成本操作的工器具及材料准备	1	培训方法：1.现场查勘和工器具及材料清理采用现场实操方法。2.编写作业卡和填写工作票采用讲授方法。培训资源：1. 1000kV 实训线路。2.特高压工器具库房。3.空白工作票	1.特高压输电实训线路；2.多媒体教室	
3. 作业现场准备	能完成作业现场准备工作	1.作业现场复勘。2.工作申请。3.作业现场布置。4.班前会。5.工器具及材料检查	1	培训方法：演示与角色扮演法。资源：1000kV 实训线路	1000kV 实训线路	

续表

培训流程	培训目标	培训内容	培训学时	培训方法与资源	培训环境	考核评价
4.培训师演示	通过现场观摩,使学员初步领会本任务操作流程	1.各类工器具使用方法讲解。 2.演示更换复合绝缘子的塔上工器具连接方式。 3.高空作业人员配合演示更换复合绝缘子的操作流程。 4.利用地面人员配合更换1000kV交流输电线路直线塔单V型复合绝缘子。	2	培训方法:演示法。 资源:1000kV实训线路	1000kV实训线路	
5.学员分组训练	1.能掌握各类受力工器具的使用方法和注意事项。 2.掌握更换复合绝缘子的全部操作流程。 3.能完成1000kV交流输电线路直线塔单V型复合绝缘子的更换	1.学员分组(高空3人、地面配合5人)训练工器具、机具的操作方法和更换V型复合绝缘子的现场实际操作。 2.培训师对学员操作进行指导和安全监护	14	培训方法:角色扮演法。 资源:1000kV实训线路	1000kV实训线路	采用技能考核评分细则对学员操作评分
6.工作终结	1.使学员进一步辨析操作过程不足之处,便于后期提升。 2.培训学员安全文明生产的工作作风	1.作业现场清理。 2.向调度汇报工作。 3.班后会,对本次工作任务进行点评总结	1	培训方法:讲授和归纳法	1000kV实训线路	

(四)作业流程

1. 工作任务

完成停电更换1000kV交流输电线路直线塔单V型复合绝缘子。

2. 天气及作业现场要求

(1)停电更换1000kV V型复合绝缘子应在良好的天气进行。

在5级及以上的大风以及暴雨、雷电、冰雹、大雾、沙尘暴等恶劣天气下,应停止露天高处作业。特殊情况下,确需在恶劣天气进行抢修时,应组织有关人员充分讨论必要的安全措施,经本单位批准后方可进行。

（2）作业人员精神状态良好，工作班成员认真学习工作票和安全技术措施，所有人员做到"四清楚"（作业任务精楚、危险点清楚、作业程序清楚、安全措施清楚）。

（3）作业前停送电联系人必须与调度联系履行工作许可手续，严禁约时停送电。工作负责人必须在得到许可人的许可工作命令后，方可在需检修的线路上验电、挂设接地线和进行检修工作。

（4）停电后，工作负责人应认真做好记录。

（5）登杆前应检查塔上是否有蜂窝，发现蜂窝严禁登塔。

（6）塔上作业人员必须使用双保险安全带，并佩戴护目镜。

3. 准备工作

3.1 危险点及其预控措施

（1）危险点——误登带电线路。

预控措施：

①登杆塔作业前，工作负责人、工作班成员应共同认真核查双重名称和识别标记（色标、判别标志等）与停电线路名称相符。

②登杆塔前应检查铁塔根部、基础等，必须牢固可靠。

③登杆塔前应检查登高工器具和设施，如安全带、脚钉、塔材等必须完整牢靠。

④不涉及挂设接地线的中间作业人员，应认真核实线路相序、色标、名称、编号与停电线路相符，确认线路名称无刷错、刷反等情况后，方可登杆。

（2）危险点——登塔时、塔上作业时违反安规进行操作，可能引起高空坠落。

预控措施：

①攀爬过程中，为防止登杆人员串落，登杆作业人员间距不得小于1.6m。

②攀爬铁塔前应将脚底泥土清除干净，检查工具包完整，攀爬过程中不得掉落物件伤人。

③作业人员攀登杆塔时应戴好安全帽，穿软底鞋，动作不能过大，匀步攀登。

④攀爬过程中，安全带应收拾妥当，长尾绳放置在工具包内，主带应挂在肩上，防止攀爬过程中安全带勾挂脚钉和塔材，致使作业人员高空坠落。

⑤杆塔上移位时，不得失去安全带保护，做到踩稳抓牢。

⑥到达作业点位置，系好安全带（绳），应牢固可靠，不得低挂高用。

⑦未验电前，人体、绳索等与导线的安全距离必须不小于9.5m，工作中应设专人监护。

（3）危险点——高处坠物伤人。

预控措施：

①地面人员不得站在作业点垂直下方。塔上人员应防止落物伤人，使用的工具、材料应用绳索传递。

②在高处作业应使用工具袋，较大的工具应固定在牢固的构件上，不准随便乱放。

③使用绞磨起吊过程中，应设专人指挥，统一配合，复合绝缘子串刚离地后应进行冲击检查。

（4）危险点——防止感应电伤人。

①为防感应电伤人，塔上作业人员应穿全套屏蔽服。

②如需接触架空地线，在架空地线接触前应进行可靠接地。

（5）危险点——现场作业安全监护。

①自作业开始至作业终结，安全监护人必须始终在现场对作业人员进行不间断的安全监护。

②工作负责人，监护人必须穿安全监护背心。

（6）危险点——交通安全。

出车时应注意车辆行驶安全，谨慎驾驶车辆，禁止违法行车。

3.2 工器具及材料选择

停电更换1000kV输电线路直线塔单V型复合绝缘子所需工器具及材料见表2-13-2。工器具出库前，应认真核对工器具的使用电压等级和试验周期，并检查确认外观良好、连接牢固、转动灵活，且符合本次工作任务的要求；工器具出库后，应防止脏污、受潮；金属工具和绝缘工器具应分开装运，防止因混装运输导致工器具变形、损伤等现象发生。

表2-13-2　停电更换1000kV交流输电线路单V型复合绝缘子所需工器具及材料表

序号	名称	规格型号	单位	数量	备注
1	接地线	1000kV	组	2	绝缘工具
2	绝缘手套	10kV	副	2	绝缘工具
3	验电器	1000kV专用	支	2	其他工具
4	全身式安全带	含带缓冲包长20M的后保绳	套	3	个人防护用具
5	个人保安线	（直径不小于16mm²）	根	1	其他工具
6	钢丝套	$\phi 22$	根	4	其他工具
7	八勾卡	适用于八分裂900mm²及以上导线	套	2	金属工具
8	卸扣	10T	个	4	金属工具
9	软梯	长度不小于15M	副	1	其他工具
10	手扳葫芦	9T	副	2	金属工具
11	个人手动工具		套	3	绝缘工具
12	对讲机		台	5	其他工具
13	吊绳滑车	1T	个	1	其他工具
14	传递绳	$\phi 16$	套	1	其他工具

续表

序号	名称	规格型号	单位	数量	备注
15	拔销器		把	2	其他工具
16	安全背心		件	2	其他工具
17	安全围栏		卷	4	其他工具
18	垫木		块	若干	其他工具
19	防潮苫布		张	1	其他工具
20	复合绝缘子	FXBW-1000/420	串	1	材料

3.3 作业人员分工

本任务作业人员分工如表2-13-3所示。

表2-13-3 停电更换1000kV XX线V型复合绝缘子人员分工表

序号	工作岗位	数量(人)	工作职责
1	工作负责人	1	负责本次工作任务的人员分工、工作前的现场查勘、作业方案的制订、工作票的填写、现场复勘、办理工作许可手续、召开工作班前会、落实现场安全措施、负责作业过程中的安全监督、工作中突发情况的处理、工作质量的监督、工作后的总结
2	安全监护人	2	负责本次工作过程中的安全监护工作
3	高空作业人员	3	负责本次停电更换1000kV V型复合绝缘子操作
4	地面辅助人员	5	负责本次作业过程的地面辅助工作

4. 工作程序

本任务工作流程如表2-13-4所示。

表2-13-4 停电更换1000kV交流输电线路耐张整串绝缘子工作流程表

序号	作业内容	作业标准	安全注意事项	责任人
1	工作许可	作业前停送电联系人必须与调度联系履行工作许可手续。	(1)不得未经工作许可人许可即开始工作 (2)严禁约时停送电	
2	现场布置	正确装设安全围栏并悬挂标示牌: (1)安全围栏范围应充分考虑高处坠物,以及对道路交通的影响。 (2)安全围栏出入口设置合理。 (3)妥当布置"从此进出""在此工作""车辆慢行"或"车辆绕行"等标示	对道路交通安全影响不可控时,应及时联系交通管理部门强化现场交通安全管控	

续表

序号	作业内容	作业标准	安全注意事项	责任人
3	召开班前会	(1)全体工作成员列队。 (2)工作负责人宣读工作票,明确工作任务及人员分工;讲解工作中的安全措施和技术措施;查(问)全体工作成员精神状态;告知工作中存在的危险点及采取的预控措施。 (3)全体工作成员在工作票上签名确认	(1)工作票填写、签发和许可手续规范,签名完整。 (2)全体工作成员精神状态良好。 (3)全体工作成员明确任务分工、安全措施和技术措施	
4	检查工器具	(1)在防潮苫布上,将工器具按作业要求准备齐备,并分类定置摆放整齐。检查工器具外观和试验合格证,无遗漏。 (2)检查人员向工作负责人汇报各项检查结果符合作业要求	(1)防潮苫布数量足够,设置位置合理,保持清洁、干燥。 (2)工器具外观检查合格,无损伤、受潮、变形、失灵现象,合格证在有效期内	
5	登杆塔	(1)登杆塔作业前,必须先核对线路名称及编号。对同塔多回线路,工作负责人、工作班成员应共同认真核查双重名称和识别标记(色标、判别标志等)。 (2)登杆塔前应检查铁塔根部、基础等,必须牢固可靠。 (3)攀爬过程,为防止登杆人员串落,登杆作业人员间距不得小于1.6m,安全带收拾妥当,后保绳放置在工具包内,主带应挂在肩上,防止攀爬过程中安全带勾攀脚钉和塔材,致使作业人员高空坠落。 (4)登杆塔至横担处时,监护人和作业人员应再次核对停电线路的识别标记和双重名称,确实无误后方可进入作业点位	(1)作业人员攀登杆塔时应戴好安全帽,穿软底鞋,动作不能过大,匀步攀登。 (2)攀爬过程中,安全带应收拾妥当,长尾绳放置在工具包内,主带应挂在肩上,防止攀爬过程中安全带勾挂脚钉和塔材,致使作业人员高空坠落。 (3)杆塔上移位时,不得失去安全带保护,做到踩稳抓牢。 (4)到达作业点位置,系好安全带(绳),应牢固可靠,不得低挂高用。 (5)未验电前,人体、无头绳等与导线的安全距离必须不小于9.5m,工作中应设专人监护	
6	验电、装设接地线	杆就位后,将安全带系在牢固可靠构件或电杆上,必须检查扣环是否正确就位。验电器等工器具必须使用传递绳传递。 (2)检查接地线完好,按程序(先接接地端,后导线端)装设好接地线。 (3)装设接地线时,必须使用绝缘绳或绝缘手柄进行操作,禁止直接用手操作接地线的金属部分的方式装设接地线。确认接地线的夹头与导线连接紧密可靠	(1)验电器在领用时和使用前应检查是否正常。 (2)禁止以缠绕导线的方式装设接地线	

续表

序号	作业内容	作业标准	安全注意事项	责任人
7	更换V型复合绝缘子	(1)高空作业人员到达作业点位后,在塔上合适位置架设软梯,1名高空人员利用软梯进入导线,将两套八勾卡与9T手扳葫芦固定在垂直于线路方向的复合绝缘子金具两侧后,架设牢固,并保留足够的人员操作空间。 (2)确认所有连接部位已固定后,2名高空人员同时收紧2把9T手扳葫芦,待9T手扳葫芦完全承受绝缘子串的拉力后,应再次检查横担有无变形,连接部位有无异常,确认一切正常后,使用传递绳将复合绝缘子绑扎牢固后,再取下复合绝缘子与导线端的连接金具的连接部分,铁塔端高空人员配合取下复合绝缘子铁塔端连接金具的连接部分。 (3)待复合绝缘子的连接部分均已取下后,方可将复合绝缘子缓慢下放,地面人员取下复合绝缘子,更换完好复合绝缘子,并将复合绝缘子绑扎牢固到位后,方可传至塔上作业人员处。 (4)将复合绝缘子拉至作业点位后,塔上作业人员应先连接铁塔端连接金具的连接部分,并确保销子到位后方可继续进行连接复合绝缘子导线端的连接工作。 (5)安装好复合绝缘子后,应检查连接金具是否安装到位、销子是否安装到位,并进行冲击试验,确认安装正确,连接可靠后方可缓慢放松2把9T手扳葫芦	(1)使用手扳葫芦、八勾卡更换绝缘子串过程中,在手扳葫芦开始承受导线荷载后,必须检查手扳葫芦、钢丝绳套(吊装带)、卸扣的连接和受力情况,并做冲击试验,确认完全可靠后方可继续收紧手扳葫芦。 (2)在使用手扳葫芦、软梯过程中,应防止与复合绝缘子碰撞,避免损伤复合绝缘子。 (3)在使用传递绳吊放绝缘子串时,地面工作人员和塔上的工作人员必须密切配合,防止传递绳缠绕,绝缘子串碰撞损伤导线。吊放绝缘子串时,地面人员不得站在垂直下方。	
8	拆除工器具	拆除八勾卡、手扳葫芦、钢丝套等工器具		
9	工作结束	(1)工作负责人组织全体工作成员整理工器具和材料,清理现场,做到"工完料尽场地清"。 (2)召开班后会,工作负责人进行工作总结和点评工作。点评本次工作的施工质量;点评全体工作成员的安全措施落实情况。 (3)工作负责人向工作许可人汇报工作结束,恢复停电线路送电,终结工作票		

二、考核标准

特高压交流输电线路运检技能考核评分细则

考生填写栏	编号：	姓名：	所在岗位：	单位：	日期：	年 月 日			
考评员填写栏	成绩：	考评员：	考评组长：	开始时间：	结束时间：	操作时长：			
考核模块	停电更换1000kV交流输电线路直线塔单V型复合绝缘子		考核对象	特高压交流输电线路检修人员	考核方式	操作	考核时限	150min	
任务描述	停电更换1000kV交流输电线路C相V型复合绝缘子								
工作规范及要求	1.给定条件：1000kV交流实训线路C相V型复合绝缘子老化，需要更换。线路已经停电、验电、挂接地线，所使用绝缘子已经过测试，工作票已办理，安全措施已经完备。 2.整个过程主要操作流程由工作负责人1名、专责监护人1人、塔上电工3人配合完成，地面辅助工5人，协助参考人员完成工器具、材料的上、下传递工作以及其他非技术性工作。 3.操作前参考人员应作必要的安全检查。 4.工作开始应口头提出申请，工作结束时应口头汇报								
考核情景准备	1.工器具：八勾卡2套、软梯1副、个人保安线1根、Φ22钢丝套4根、10T卸扣4个、9T手扳葫芦2把、1.5T手扳葫芦1把、传递绳1根、1T滑车1个、绳套1根、对讲机5个、双保险安全带3套、防潮苦布1张。 2.材料：同型号复合绝缘子一串。 3.在培训线路上操作								
备注	1.个人工器具由参考人员自备。 2.各项目得分均扣完为止								

序号	项目名称	质量要求	分值	扣分标准	扣分原因	扣分	得分
1	工具材料准备						
1.1	个人工具检查	活动扳手、平口钳、拔销钳、工具包符合质量要	2	错漏1项扣1分			
1.2	受力工具检查	八勾卡、手扳葫芦、软梯检查，在试验合格期内	3	错漏1项扣1分			
1.3	安全工器具检查	双保险安全带、个人保安线、绝缘手套符合质量要求，并在试验合格周期内	3	错漏1项扣1分			
1.4	材料检查	核对复合绝缘子串型号，外观检查符合要求	2	错漏1项扣1分			
2	场地布置						
2.1	场地围栏	场地围栏布置	2	未布置扣2分			
3	登塔及横担上的操作						

续表

序号	项目名称	质量要求	分值	扣分标准	扣分原因	扣分	得分
3.1	登塔	1)检查杆塔基础无异常。 2)正确携带传递绳(吊绳头折双、打死结、斜挎肩上)。 3)沿脚钉侧主材正确登塔	6	1)未检查1项扣1分。 2)未携带传递绳扣2分,传递绳携带方式不规范扣1分。 3)手抓脚钉1次扣1分。 4)未沿脚钉侧主材登塔扣2分			
3.2	进入横担上工作点	由塔身到横担上工作点不得失去安全带保护	3	未正确使用安全带扣3分			
3.3	安装滑车、个人保安线	传递滑车安装位置正确,方便操作,并加挂个人保安线	3	滑车安装不规范扣1分;个人保安线漏挂扣2分			
4	绝缘子串上操作						
4.1	进入工作点	1)将双保险安全带的安全绳系在横担适当位置。 2)沿绝缘子串进入作业点,将围杆带系在绝缘子串上。 3)检查绝缘子串锁紧销,连接金具	6	1)未使用双保险安全带扣4分。 2)未正确使用双保险安全带一次扣2分。 3)未检查扣2分			
4.2	安装工器具	1)正确安装钢丝绳套和9T手板葫芦。 2)正确安装八勾卡。 3)钢丝绳套、卸扣型号选用正确、安装可靠 4)安装的手扳葫芦、个人保安线、软梯不影响人员操作	9	1)未对塔材采取保护1处扣1分。 2)手扳葫芦、八勾卡安装不对1处扣1分。 3)手扳葫芦受力后有碰撞、缠绕,1处扣1分。 4)钢丝绳套、卸扣型号选用不正确一处扣1分。 5)手扳葫芦漏1套扣4分			
4.3	收紧复合绝缘子串	1)同时收紧2把9T手扳葫芦,确保手扳葫芦受力均衡 2)手扳葫芦串受力后,检查滑车、卸扣、钢丝绳等连接部位,确认连接可靠,并对绝缘子串做冲击 3)确认受力无误后继续收紧手扳葫芦,直至复合绝缘子串松弛。	6	1)碰响绝缘子一次扣1分。 2)未冲击试验扣3分。 3)使用手扳葫芦不正确扣2分			
4.4	更换复合I串绝缘子	1)将传递绳绑扎在复合绝缘子的合适位置,高空人员先取脱复合绝缘子导线端。 2)地面辅助人员收紧绝缘子配合高空作业人员取下绝缘子铁塔端,并传送至地面。 3)将新复合绝缘子通过传递绳传递到杆塔上,塔上作业先连接铁塔端,并将销子安装到位。	16	1)未冲击试验扣2分。 2)传递绳索缠绕扣2分。 3)传递物有撞击现象一次扣2分。 4)掉落物件扣5分。 5)未装锁紧销1个扣1分。 6)拆装导线端和铁塔端顺序错误扣4分			

续表

序号	项目名称	质量要求	分值	扣分标准	扣分原因	扣分	得分
4.4	更换复合I串绝缘子	4)继续安装复合绝缘子导线端,安装金具R销子,并检查是否安装到位。 5)确认连接无误后,松出手扳葫芦,使复合绝缘子受力,并做冲击试验	16				
4.5	撤除工器具	1)检查锁紧销、球头是否齐全到位,碗口朝向正确,清洁绝缘子串表面污垢。 2)拆除手扳葫芦、八勾卡及钢丝绳套,传递至地面	7	1)绳索缠绕扣2分。 2)传递物有撞击现象一次扣2分。 3)锁紧销位置不正确扣1分。 4)碗口朝向不正确扣1分。 5)未清洁绝缘子串扣1分			
4.6	清理杆塔上工器具	确认无遗留物	4	有遗留物扣4分			
4.7	从导线到塔身	1)沿软梯到进入铁塔横担。 2)攀爬软梯过程中不得失去安全带保护	8	1)失去安全带保护扣4分 2)未正确使用软梯扣4分			
5	下塔	1)必须沿脚钉侧主材正确下塔。 2)正确携带传递绳(吊绳头折双、打死结,斜挎肩上)	8	1)未携带传递绳扣4分,传递绳携带方式不规范扣2分 2)手抓脚钉1次扣1分 3)未沿脚钉侧主材登塔扣4分			
6	其他要求						
6.1	塔上作业	1)严禁高处坠物。 2)在操作过程中应双手协调配合操作。 3)严禁浮置物品。 4)严禁口中含物	5	1)高处坠物一件扣5分。 2)动作不协调扣2分。 3)浮置物品扣2分。 4)口中含物扣2分			
6.2	着装	工作服、工作胶鞋、安全帽、劳保手套穿戴正确	2	漏一项扣2分			
6.3	清理现场	完工后清理作业现场,符合文明生产要求	2	未清理作业现场扣2分			
6.4	完成时间	在规定时间内按要求完成	3	超过时间10min扣1分,达到480min即终止操作,只记完成部分得分			
	合计		100				

模块14 停电更换1000kV交流输电线路耐张整串绝缘子培训及考核标准

一、培训标准

(一) 培训要求

模块名称	停电更换1000kV交流输电线路耐张整串绝缘子	培训类别	操作类
培训方式	实操培训	培训学时	28学时
培训目标	1.掌握各类工器具、机具的使用方案和受力结构,以及更换整串绝缘子技术要点。 2.能熟练掌握停电更换1000kV交流输电线路耐张整串绝缘子的操作流程、技术方法和施工作业危险点。 3.作为高空主要作业人员,能熟练完成1000kV交流输电线路耐张整串绝缘子的更换。		
培训场地	特高压交流实训线路		
培训内容	正确使用各类受力工器具,熟练绞磨等机具的操作方法,采用"滑轮组"作业方式正确安装各类工器具,采用停电作业法更换1000kV交流输电线路耐张整串绝缘子。		
适用范围	特高压交流输电线路检修人员		

(二) 引用规程规范

(1)《架空送电线路运行规程》(DL/T 741—2010)。

(2)《110~500kV架空送电线路设计技术规程》(DL/T 5092—1999)。

(3)《国家电网公司电力安全工作规程(线路部分)》(Q/GDW1799.2—2013)。

(4)《1000kV架空输电线路设计规范》(GB 50665—2011)。

(5)《1000kV交流输电线路检修规范》(DL/T 209—2008)。

(6)《1000kV交流输电线路运行规范》(DL/T 307—2010)。

(7)《110(66)kV~500kV架空输电线路检修规范》(国家电网公司)。

(8)《架空输电线路状态检修导则》(DLT 1248—2013)。

(9)《输变电设备状态检修管理规定》(国家电网公司)。

(10)《输变电设备状态检修试验规程》(国家电网公司)。

(三) 培训教学设计

本设计以完成"停电更换1000kV交流输电线路耐张整串绝缘子"为工作任务,按工作任务完成的标准化作业流程来设计各个培训阶段,每个阶段包括了具体的培训目标、培训内容、培训学时、培训方法(培训资源)、培训环境和考核评价等内容,如表2-14-1所示。

表2-14-1　停电更换1000kV交流输电线路耐张整串绝缘子培训内容设计

培训流程	培训目标	培训内容	培训学时	培训方法与资源	培训环境	考核评价
1.理论教学	1.掌握各类工器具、机具的使用方案和受力结构,以及更换整串绝缘子技术要点。 2.能熟练掌握停电更换1000kV交流输电线路耐张整串绝缘子的操作流程、技术方法和施工作业危险点	1.正确使用各类受力工器具,熟悉绞磨等机具的操作方法。 2.采用"滑轮组"作业方式正确安装各类工器具。 3.采用停电作业法更换1000kV交流输电线路耐张整串绝缘子	2	培训方法:讲授法。 培训资源:PPT、相关规程规范	多媒体教室	考勤、课堂提问和作业
2.准备工作	能完成作业前准备工作	1.作业现场查勘。 2.编制培训标准化作业卡。 3.填写培训操作工作票。 4.完成本操作的工器具及材料准备	1	培训方法: 1.现场查勘和工器具及材料清理采用现场实操方法。 2.编写作业卡和填写工作票采用讲授方法。 培训资源: 1.1000kV实训线路。 2.特高压工器具库房。 3.空白工作票	1.特高压输电实训线路; 2.多媒体教室	
3.作业现场准备	能完成作业现场准备工作	1.作业现场复勘。 2.工作申请。 3.作业现场布置。 4.班前会。 5.工器具及材料检查	3	培训方法:演示与角色扮演法。 资源:1000kV实训线路	1000kV实训线路	

续表

培训流程	培训目标	培训内容	培训学时	培训方法与资源	培训环境	考核评价
4.培训师演示	通过现场观摩，使学员初步领会本任务操作流程	1.各类工器具使用方法讲解。2.演示更换整串绝缘子的塔上工器具连接方式。3.高空作业人员配合演示更换整串绝缘子的操作流程。4.利用绞磨配合更换1000kV线路耐张整串绝缘子	7	培训方法：演示法。资源：1000kV实训线路	1000kV实训线路	
5.学员分组训练	1.能掌握各类受力工器具、绞磨机具的使用方法和注意事项。2.掌握更换整串绝缘子的全部操作流程。3.能完成1000kV输电线路耐张整串绝缘子的更换	1.学员分组（高空6人、地面配合9人一组）训练工器具、机具的操作方法和更换整串绝缘子的现场实际操作。2.培训师对学员操作进行指导和安全监护	14	培训方法：角色扮演法。资源：1000kV实训线路	1000kV实训线路	采用技能考核评分细则对学员操作评分
6.工作终结	1.使学员进一步辨析操作过程不足之处，便于后期提升。2.培训学员安全文明生产的工作作风	1.作业现场清理。2.向调度汇报工作。3.班后会，对本次工作任务进行点评总结	1	培训方法：讲授和归纳法	1000kV实训线路	

（四）作业流程

1. 工作任务

完成停电更换1000kV交流输电线路耐张整串绝缘子。

2. 天气及作业现场要求

（1）停电更换1000kV耐张绝缘子应在良好的天气进行。

在5级及以上的大风以及暴雨、雷电、冰雹、大雾、沙尘暴等恶劣天气下，应停止露天高处作业。特殊情况下，确需在恶劣天气进行抢修时，应组织有关人员充分讨论必要的安全措施，经本单位批准后方可进行。

（2）作业人员精神状态良好，工作班成员认真学习工作票和安全技术措施，所有人员做到"四清楚"（作业任务精楚、危险点清楚、作业程序清楚、安全措施清楚）。

（3）作业前停送电联系人必须与调度联系履行工作许可手续，严禁约时停送电。工作负责人必须在得到许可人的许可工作命令后，方可在需检修的线路上验电、挂设接地线和进行检修工作。

（4）停电后，工作负责人应认真做好记录。

（5）登杆前应检查塔上是否有蜂窝，发现蜂窝严禁登塔。

（6）塔上作业人员必须使用双保险安全带，并佩戴护目镜。

3. 准备工作

3.1 危险点及其预控措施

（1）危险点——误登带电线路。

预控措施：

①登杆塔作业前，工作负责人、工作班成员应共同认真核查双重名称和识别标记（色标、判别标志等）与停电线路名称相符。

②登杆塔前应检查铁塔根部、基础等，必须牢固可靠。

③登杆塔前应检查登高工器具和设施，如安全带、脚钉、塔材等必须完整牢靠。

④不涉及挂设接地线的中间作业人员，应认真核实线路相序、色标、名称、编号与停电线路相符，确认线路名称无刷错、刷反等情况后，方可登杆。

（2）危险点——登塔时、塔上作业时违反安规进行操作，可能引起高空坠落。

预控措施：

①攀爬过程中，为防止登杆人员串落，登杆作业人员间距不得小于1.6m。

②攀爬铁塔前应将脚底泥土清除干净，检查工具包完整，攀爬过程中不得掉落物件伤人。

③作业人员攀登杆塔时应戴好安全帽，穿软底鞋，动作不能过大，匀步攀登。

④攀爬过程中，安全带应收拾妥当，长尾绳放置在工具包内，主带应挂在肩上，防止攀爬过程中安全带勾挂脚钉和塔材，致使作业人员高空坠落。

⑤杆塔上移位时，不得失去安全带保护，做到踩稳抓牢。

⑥到达作业点位置，系好安全带（绳），应牢固可靠，不得低挂高用。

⑦未验电前，人体、无头绳等与导线的安全距离必须不小于9.5m，工作中应设专人监护。

（3）危险点——高处坠物伤人。

预控措施：

①地面人员不得站在作业点垂直下方。塔上人员应防止落物伤人，使用的工具、材料应

用绳索传递。

②在高处作业应使用工具袋，较大的工具应固定在牢固的构件上，不准随便乱放。

③使用绞磨起吊过程中，应设专人指挥，统一配合，绝缘子串刚离地后应进行冲击检查。

（4）危险点——防止感应电伤人。

①为防感应电伤人，塔上作业人员应穿全套屏蔽服。

②如需接触架空地线，在架空地线接触前应进行可靠接地。

（5）危险点——现场作业安全监护。

①自作业开始至作业终结，安全监护人必须始终在现场对作业人员进行不间断的安全监护。

②工作负责人，监护人必须穿安全监护背心。

（6）危险点——交通安全。

出车时应注意车辆行驶安全，谨慎驾驶车辆，禁止违法行车。

3.2 工器具及材料选择

停电更换1000kV耐张整串绝缘子所需工器具及材料见表2-14-2。工器具出库前，应认真核对工器具的使用电压等级和试验周期，并检查确认外观良好、连接牢固、转动灵活，且符合本次工作任务的要求；工器具出库后，应防止脏污、受潮；金属工具和绝缘工器具应分开装运，防止因混装运输导致工器具变形、损伤等现象发生。

表2-14-2 停电更换1000kV交流输电线路耐张整串绝缘子所需工器具及材料表

序号	名称	规格型号	单位	数量	备注
1	接地线	1000kV专用	组	2	绝缘工具
2	绝缘手套	10kV	副	2	绝缘工具
3	验电器	1000kV专用	支	2	其他工具
4	全身式安全带	含带缓冲包长24m的后保绳	套	6	个人防护用具
5	个人保安线	（直径不小于16mm²）	根	1	其他工具
6	钢丝套	Φ24	根	20	其他工具
7	铁滑车	15T	个	12	金属工具
8	卸扣	18T	个	12	金属工具
9	铁滑车	5T	个	8	金属工具
10	钢丝绳	Φ24	m	150	其他工具
11	磨绳	Φ17.5	m	600	其他工具
12	绞磨	5T	台	2	机动工具

续表

序号	名称	规格型号	单位	数量	备注
13	手扳葫芦	9T	副	2	金属工具
14	个人手动工具		套	6	其他工具
15	对讲机		台	8	其他工具
16	吊绳滑车	1T	个	2	金属工具
17	传递绳	$\phi16$	套	2	其他工具
18	拔销器		把	3	金属工具
19	安全背心		件	3	其他工具
20	护目镜		副	6	其他工具
21	安全围栏		卷	5	其他工具
22	垫木		块	若干	其他工具
23	防潮苫布		张	1	其他工具
24	瓷质绝缘子	U550BP/240T	片	56	材料

3.3 作业人员分工

本任务作业人员分工如表2-14-3所示。

表2-14-3 停电更换1000kV交流输电线路耐张整串绝缘子人员分工表

序号	工作岗位	数量(人)	工作职责
1	工作负责人	1	负责本次工作任务的人员分工、工作前的现场查勘、作业方案的制订、工作票的填写、现场复勘、办理工作许可手续、召开工作班前会、落实现场安全措施、负责作业过程中的安全监督、工作中突发情况的处理、工作质量的监督、工作后的总结
2	安全监护人	2	负责本次工作过程中的安全监护工作
3	高空作业人员	6	负责本次停电更换1000kV耐张整串绝缘子操作
4	地面辅助人员	5	负责本次作业过程的地面辅助工作
5	绞磨操作人员	2	负责本次作业过程的绞磨操作工作
6	信号指挥人员	2	负责2台绞磨启停的指挥工作

4. 工作程序

本任务工作流程如表2-14-4所示。

表2-14-4 停电更换1000kV交流输电线路耐张整串绝缘子工作流程表

序号	作业内容	作业标准	安全注意事项	责任人
1	工作许可	作业前停送电联系人必须与调度联系履行工作许可手续	(1)不得未经工作许可人许可即开始工作。 (2)严禁约时停送电	
2	现场布置	正确装设安全围栏并悬挂标示牌： (1)安全围栏范围应充分考虑高处坠物,以及对道路交通的影响。 (2)安全围栏出入口设置合理。 (3)妥当布置"从此进出""在此工作""车辆慢行"或"车辆绕行"等标示	对道路交通安全影响不可控时,应及时联系交通管理部门强化现场交通安全管控	
3	召开班前会	(1)全体工作成员列队。 (2)工作负责人宣读工作票,明确工作任务及人员分工;讲解工作中的安全措施和技术措施;查(问)全体工作成员精神状态;告知工作中存在的危险点及采取的预控措施。 (3)全体工作成员在工作票上签名确认	(1)工作票填写、签发和许可手续规范,签名完整。 (2)全体工作成员精神状态良好。 (3)全体工作成员明确任务分工、安全措施和技术措施	
4	检查工器具	(1)在防潮苫布上,将工器具按作业要求准备齐备,并分类定置摆放整齐。检查工器具外观和试验合格证,无遗漏。 (2)检查人员向工作负责人汇报各项检查结果符合作业要求	(1)防潮苫布数量足够,设置位置合理,保持清洁、干燥。 (2)工器具外观检查合格,无损伤、受潮、变形、失灵现象,合格证在有效期内。	
5	登杆塔	(1)登杆塔作业前,必须先核对线路名称及编号。对同塔多回线路,工作负责人、工作班成员应共同认真核查双重名称和识别标记(色标、判别标志等)。 (2)登杆塔前应检查铁塔根部、基础等,必须牢固可靠。 (3)攀爬过程,为防止登杆人员串落,登杆作业人员间距不得小于1.6m,安全带收拾妥当,长尾绳放置在工具包内,主带应挂在肩上,防止攀爬过程中安全带勾攀脚钉和塔材,致使作业人员高空坠落。 (4)登杆塔至横担处时,监护人和作业人员应再次核对停电线路的识别标记和双重名称,确实无误后方可进入停电线路侧的横担	(1)作业人员攀登杆塔时应戴好安全帽,穿软底鞋,动作不能过大,匀步攀登。 (2)攀爬过程中,安全带应收拾妥当,长尾绳放置在工具包内,主带应挂在肩上,防止攀爬过程中安全带勾挂脚钉和塔材,致使作业人员高空坠落。 (3)杆塔上移时,不得失去安全带保护,做到踩稳抓牢。 (4)到达作业点位置,系好安全带(绳),应牢固可靠,不得低挂高用。 (5)未验电前,人体、无头绳等与导线的安全距离必须不小于9.5m,工作中应设专人监护	

续表

序号	作业内容	作业标准	安全注意事项	责任人
6	验电、装设接地线	(1)杆就位后,将安全带系在牢固可靠构件或电杆上,必须检查扣环是否正确就位。验电杆(笔)等工器具必须使用绳索传递。 (2)查接地线完好,按程序(先接接地端,后接导线端)装设好接地线。 (3)设接地线时,必须使用绝缘绳或绝缘手柄进行操作,禁止直接用手操作接地线的金属部分的方式装设设接地线。确认接地线的夹头与导线连接紧密可靠	(1)验电杆在领用时和使用前应检查是否正常。 (2)禁止以缠绕导线的方式装设接地线	
7	更换耐张整串绝缘子	(1)高空作业人员到达作业点位后,3名高空人员在横担端的2串耐张绝缘子之间的牢固塔材上安装3个15T的铁滑车,3名高空人员在对应的带电侧联板上安装3个15T的铁滑车,将钢丝套与手扳葫芦勾卡固定于横担侧的3个铁滑车的一边,再将手板葫芦链条连接的钢丝绳依次穿入两边的6个滑车中,构成一套滑轮组,并保留足够的人员操作空间,确认手扳葫芦与滑轮组连接部位已连接牢固后,再收紧9T手扳葫芦,当9T手板葫芦已完全承受绝缘子串的拉力后,应再次检查横担有无变形,连接部位有无异常,确认一切正常。 (2)3名高空人员将1号绞磨上的磨绳穿过设置于横担上5T滑轮中,再将其固定于绝缘子串的横担端的第三片处,固定稳固后,再将2号绞磨上的磨绳依次过设置于横担端与带电端连接金具上的2个5T滑轮后,再将2号绞磨的磨绳固定于带电端的第3片绝缘子处,待所有连接部位都已连接稳固后,方可开动绞磨。 (3)地面绞磨操作人员开动2号绞磨,应先将绝缘子串带电端拉紧后,待高空人员取下带电端绝缘子串的连接部分后,再缓慢将带电端的绝缘子串向下松出,待绝缘子串垂直于地面后,再开动1号绞磨,将绝缘子串拉紧提起一端后,高空人员取下横担端绝缘子串的连接部分,然后两端同时缓慢松出,松至地面,地面人员配合取下绝缘子。待地面人员取下绝缘子,更换完好绝缘子,先开动1号绞磨,再开动2号,安装顺序与拆除顺序相反,先将横担端安装完毕后,再紧固带电端。 (4)绝缘子到位后,塔上作业人员应先连接铁塔端连接金具,并确保销子到位后方可继续进行金具与铁塔、导线的连接工作。 (5)安装好绝缘子串后,应检查连接金具是否安装到位、销子是否安装到位,并进行冲击试验,确认安装正确,连接可靠后方可缓慢松出9T手扳葫芦	(1)使用手扳葫芦卡更换绝缘子串过程中,在手扳葫芦开始承受导线荷载时,必须检查手扳葫芦、钢丝绳套(吊装带)、卸扣的连接和受力情况,并做冲击试验,确认完全可靠后方可继续收紧手扳葫芦。 (2)在使用手扳葫芦时,应防止与绝缘子碰撞,避免损伤绝缘子。 (3)在使用传递绳吊放整串绝缘子串时,地面工作人员和塔上的工作人员必须密切配合,防止起吊绳缠绕,绝缘子串碰撞损伤导线。吊放绝缘子串时,地面人员不得站在垂直下方。 (4)绞磨起吊整串绝缘子时,应有专人指挥,2台绞磨应密切配合,绝缘子串一起地,应再次检查绝缘子串是否绑扎牢固,并做冲击试验,确认无误后方可继续起吊	

续表

序号	作业内容	作业标准	安全注意事项	责任人
8	拆除工器具	依序拆除磨绳、手扳葫芦、钢丝套、滑车等工器具		
9	工作结束	(1)工作负责人组织全体工作成员整理工器具和材料,清理现场,做到"工完料尽场地清"。 (2)召开班后会,工作负责人进行工作总结和点评工作。点评本次工作的施工质量;点评全体工作成员的安全措施落实情况。 (3)工作负责人向工作许可人汇报工作结束,恢复停电线路送电,终结工作票		

二、考核标准

特高压交流输电线路运检技能考核评分细则

考生填写栏	编号:		姓名:		所在岗位:		单位:		日期:		年　月　日	
考评员填写栏	成绩:		考评员:		考评组长:		开始时间:		结束时间:		操作时长:	
考核模块	停电更换1000kV交流输电线路耐张整串绝缘子			考核对象	特高压交流输电线路检修人员			考核方式	操作	考核时限	360min	
任务描述	停电更换1000kV交流输电线路C相右串整串瓷质绝缘子											
工作规范及要求	1. 给定条件:1000kV交流实训线路C相右串整串瓷质绝缘子老化,需要更换。线路已经停电、验电、挂接地线,所使用绝缘子已经过测试,工作票已办理,安全措施已经完备。 2. 整个过程主要操作流程由工作负责人1名、专责监护人1人、塔上电工6人配合完成,地面辅助工7人,绞磨操作人员2人,协助参考人员完成工器具、材料的上、下传递工作,以及其他非技术性工作。 3. 操作前参考人员应作必要的安全检查。 4. 更换整串绝缘子时采用的工具应满足受力要求。 5. 工作开始应口头提出申请,工作结束时应口头汇报											
考核情景准备	1. 工器具:ϕ24钢丝套150m、15T铁滑车12个、5T铁滑车8个、18T卸扣12个、ϕ24钢丝绳150m、ϕ16磨绳一捆、9T手扳葫芦2把、5T绞磨2台、吊绳滑车2套、绳套2根、对讲机若干、双保险安全带6套。 2. 材料:同型号瓷质绝缘子一串。 3. 在培训线路上操作											
备注	1. 个人工器具由参考人员自备。 2. 各项目得分均扣完为止											

续表

序号	项目名称	质量要求	分值	扣分标准	扣分原因	扣分	得分
1	工具材料准备						
1.1	个人工具检查	活动扳手、平口钳、拔销钳、工具包符合质量要求	2	错漏1项扣1分			
1.2	机具检查	绞磨试机,检查档位是否正常;手扳葫芦检查	2	错漏1项扣1分			
1.3	钢丝绳、滑车检查	对钢丝绳、滑车进行检查,确认连接可靠,受力满足要求	2	错漏1项扣1分			
1.4	安全带	双保险安全带符合质量要求,在试验周期内	1	未检查扣1分			
1.5	专用工具检查	个人保安线、绝缘手套外观检查符合要求,在试验周期内	1	错漏1项扣0.5分			
1.6	材料检查	清洁绝缘子,核对绝缘子串数量,外观检查符合要求	2	错漏1项扣1分			
2	场地布置						
2.1	绞磨场地布置	绞磨位置布置、转角滑车位置布置	2	错漏1项扣1分			
2.2	场地围栏	场地围栏布置	1	未布置扣1分			
3	登塔及横担上的操作						
3.1	登塔	1)检查杆塔基础无异常。2)正确携带传递绳(吊绳头折双、打死结、斜挎肩上)。3)沿脚钉侧主材正确登塔	6	1)未检查1项扣1分。2)未携带传递绳扣2分,传递绳携带方式不规范扣1分。3)手抓脚钉1次扣1分。4)未沿脚钉侧主材登塔扣2分			
3.2	进入横担上工作点	由塔身到横担上工作点不得失去安全带保护。	3	未正确使用安全带扣3分。			
3.3	传递滑车安装	传递滑车安装位置正确,方便操作	1	安装不规范扣1分			
4	绝缘子串上操作						
4.1	进入工作点	1)将双保险安全带的安全绳系在横担适当位置。2)沿绝缘子串进入作业点,将围杆带系在绝缘子串上。3)检查绝缘子串锁紧销	5	1)未使用双保险安全带扣5分。2)未正确使用双保险安全带一次扣3分。3)未检查扣3分			

续表

序号	项目名称	质量要求	分值	扣分标准	扣分原因	扣分	得分
4.2	安装3-3滑车组	1)导线端和铁塔端3-3滑车组位置安装正确。2)钢丝绳穿向正确，不缠绕。3)钢丝绳套、卸扣型号选用正确、安装可靠	6	1.滑车组安装不对1处扣1分。2.钢丝绳穿向位置不对1处扣1分。3.钢丝绳受力后有碰撞、缠绕，1处扣1分。4.钢丝绳套、卸扣型号选用不正确一处扣1分			
4.3	安装手扳葫芦	1)将手扳葫芦传递至塔上，并正确安装。2)将手扳葫芦与3-3滑车组连接，使其略为受力	5	1)绳索缠绕扣1分2)位置不正确扣2分3)碰响绝缘子一次扣1分			
4.4	收紧绝缘子串	1)收紧手扳葫芦，手扳葫芦受力后检查滑车、卸扣、钢丝绳等连接部位，确认连接可靠，并对绝缘子串做冲击。2)确认受力无误后继续收紧手扳葫芦，直至绝缘子串松弛。	5	1)未冲击试验扣3分2)使用手扳葫芦不正确扣2分			
4.5	更换绝缘子串	1)在已松弛的绝缘子串上正确的安装6T辅助手扳葫芦。2)在绝缘子串上适当位置连接起吊磨绳。3)收紧9T手扳葫芦，受力后进行冲击试验，无异常，取出R销。4)取下绝缘子串两端连接金具，收紧起吊磨绳。5)磨绳受力后，检查磨绳连接部位，并冲击试验。6.拆除9T手扳葫芦，通过起吊磨绳将绝缘子串传送至地面。7.将新绝缘子串传递到杆塔上，安装9T手扳葫芦。8)收紧手扳葫芦，安装两端R销子，并检查是否安装到位。	18	1)手扳葫芦安装位置不对、使用操作不对，1次扣2分2)磨绳起吊滑车位置不正确扣2分3)未冲击试验扣3分4)传递绳索缠绕扣2分5)传递物有撞击现象一次扣2分6)掉落物件扣5分7)未装锁紧销1个扣2分。			
4.6	撤除工器具	1)检查锁紧销、球头是否齐全到位，碗口朝正确，清洁绝缘子串。2)松出3-3滑车组绝缘子串受力后，对绝缘子串做冲击试验。3)拆除手扳葫芦、滑车组及钢丝绳，传递至地面。	10	1)绳索缠绕扣2分。2)传递物有撞击现象一次扣2分。3)未转移围杆带到绝缘子串上扣5分。4)锁紧销位置不正确扣2分。5)绝缘子大口朝向不正确扣2分。6)未作冲击试验扣3分。7)未清洁绝缘子串扣1分			

续表

序号	项目名称	质量要求	分值	扣分标准	扣分原因	扣分	得分
4.7	清理杆塔上工器具	确认无遗留物。	4	有遗留物扣4分			
4.8	从绝缘子串到塔身	由绝缘子串到塔身上不得失去安全带保护。	5	失去安全带保护扣5分			
5	下塔	1)必须沿脚钉侧主材正确下塔。2)正确携带传递绳(吊绳头折双、打死结,斜挎肩上)。	8	1)未携带传递绳扣4分,吊绳携带方式不规范扣2分。2)手抓脚钉1次扣1分。3)未沿脚钉侧主材登塔扣4分			
6	其他要求						
6.1	塔上作业	1)严禁高处坠物。2)在操作过程中应双手协调配合操作。3)严禁浮置物品。4)严禁口中含物。	5	1)高处坠物一件扣5分。2)动作不协调扣2分。3)浮置物品扣2分。4)口中含物扣2分			
6.2	着装	工作服、工作胶鞋、安全帽、劳保手套穿戴正确	2	漏一项扣2分			
6.3	清理现场	完工后清理作业现场,符合文明生产要求	2	未清理作业现场扣2分			
6.4	完成时间	在规定时间内按要求完成	2	超过时间10min扣1分,达到480min即终止操作,只记完成部分得分			
	合计		100				

模块15　停电修补1000kV交流输电线路架空地线培训及考核标准

一、培训标准

（一）培训要求

模块名称	停电修补1000kV交流输电线路架空地线	培训类别	操作类
培训方式	实操培训	培训学时	14学时
培训目标	1. 能使用飞车沿1000kV交流输电线路架空地线到达指定作业位置； 2. 能独立完成用预绞丝补修条修补架空地线的操作		
培训场地	特高压交流实训线路		
培训内容	使用飞车沿1000kV输电线路架空地线到达指定作业位置，用预绞丝补修条修补1000kV交流输电线路架空地线		
适用范围	特高压交流输电线路检修人员		

（二）引用规程规范

（1）《电工术语 架空线路》（GB/T 2900.51—1998）。

（2）《1000kV架空输电线路设计规范》（GB 50665—2011）。

（3）《1000kV交流输电线路检修规范》（DL/T 209—2008）。

（4）《1000kV交流输电线路运行规范》（DL/T 307—2010）。

（5）《国家电网公司电力安全工作规程（线路部分）》（Q/GDW1799.2—2013）。

（三）培训教学设计

本设计以完成"停电修补1000kV交流输电线路架空地线"为工作任务，按工作任务完成的标准化作业流程来设计各个培训阶段，每个阶段包括了具体的培训目标、培训内容、培训学时、培训方法（培训资源）、培训环境和考核评价等内容，如表2-15-1所示。

表 2-15-1　停电修补 1000kV 交流输电线路架空地线培训内容设计

培训流程	培训目标	培训内容	培训学时	培训方法与资源	培训环境	考核评价
1.理论教学	1.初步掌握使用飞车沿 1000kV 交流输电线路架空地线到达指定作业位置基本方法。2.熟悉输电线路架空地线受损的修补方法	1.飞车的分类、结构及其使用注意事项。2.使用飞车沿 1000kV 交流输电线路架空地线到达指定作业位置方法。3.输电线路架空地线修补方法和质量标准	2	培训方法：讲授法。培训资源：PPT、相关规程规范	多媒体教室	考勤、课堂提问和作业
2.准备工作	能完成作业前准备工作	1.作业现场查勘。2.编制培训标准化作业卡。3.填写输电线路第一种工作票。4.完成本操作的工器具及材料准备	1	培训方法：1.现场查勘和工器具及材料清理采用现场实操方法。2.编写作业卡和填写工作票采用讲授方法。培训资源：1. 1000kV 实训线路；2.特高压工器具库房；3.空白工作票	1.特高压输电实训线路；2.多媒体教室	
3.作业现场准备	能完成作业现场准备工作	1.工作申请。2.作业现场复勘。3.作业现场布置。4.班前会。5.工器具及材料检查	1	培训方法：演示与角色扮演法。资源：1000kV 实训线路	1000kV 实训线路	
4.培训师演示	通过现场观摩，使学员初步领会本任务操作流程	1.装设接地线。2.安装作业飞车。3.使用飞车沿 1000kV 交流输电线路架空地线到达指定作业位置。4.用预绞丝补修条完成架空地线修补	2	培训方法：演示法。资源：1000kV 实训线路	1000kV 实训线路	
5.学员分组训练	1.能完成飞车的正确安装。2.能使用飞车沿 1000kV 交流输电线路架空地线到达指定作业位置。3.能完成 1000kV 输电线路架空地线修补作业	1.学员分组（6 人一组）训练飞车的使用和修补架空地线技能操作。2.培训师对学员操作进行指导和安全监护	7	培训方法：角色扮演法。资源：1000kV 实训线路	1000kV 实训线路	采用技能考核评分细则对学员操作评分

续表

培训流程	培训目标	培训内容	培训学时	培训方法与资源	培训环境	考核评价
6.工作终结	1.使学员进一步辨析操作过程不足之处，便于后期提升。 2.培养学员安全文明生产的工作意识	1.作业现场清理。 2.向调度汇报工作。 3.班后会，对本次工作任务进行点评总结	1	培训方法： 讲授和归纳法	1000kV 实训线路	

（四）作业流程

1. 工作任务

完成停电修补1000kV交流输电线路架空地线任务。

2. 天气及作业现场要求

（1）停电修补1000kV交流输电线路架空地线应在良好的天气进行。

如遇雷电（听见雷声、看见闪电）、雪、雹、雨、雾等，风力大于5级，不得进行作业；恶劣天气下必须开展作业时，应组织有关人员充分讨论并编制必要的安全措施，经本单位主管生产领导批准后方可进行。

（2）作业人员精神状态良好，工作班成员认真学习工作票和安全技术措施，所有人员做到"四清楚"（作业任务清楚、危险点清楚、作业程序清楚、安全措施清楚）。

（3）作业前工作负责人必须与调度联系履行工作许可手续，严禁约时停送电。工作负责人必须在得到许可人的许可工作命令后，方可在需检修的线路上验电、装设接地线和进行检修工作。

（4）停电后，工作负责人应认真做好记录。

（5）登塔前应检查塔上是否有蜂窝，发现蜂窝严禁登塔。

（6）塔上作业人员必须使用安全带。

3. 准备工作

3.1 危险点及其预控措施

（1）危险点——误登带电线路。

预控措施：

①登塔作业前，工作负责人、工作班成员应认真核对双重称号和识别标记（色标、判别标志等）与停电线路名称相符。

②登塔前应检查铁塔根部、基础等，必须牢固可靠。

③塔前应检查脚钉、塔材等必须完整牢靠。

(2）危险点——登塔和塔上作业时违反安规进行操作，可能引起高处坠落。

预控措施：

①登塔过程中，为防止登塔人员相互碰撞，登塔作业人员间距不得小于1.6m。

②登塔前应将脚底泥土清除干净，检查工具包完整，登塔过程中不得掉负重登杆。

③作业人员登塔时应戴好安全帽，穿软底工作鞋，动作不宜过大，匀步攀登。

④登塔过程中，安全带应收拾妥当，后背保护绳放置在工具包内，主带挂在肩上，防止登塔过程中安全带勾挂脚钉和塔材，致使作业人员高处坠落。

⑤塔上移位时，不得失去安全带保护，做到踩稳抓牢。

⑥到达作业点位置，系好安全带，将安全带、后备保护绳应分别系在牢固构件上，不得低挂高用。

⑦未验电前，人体、传递绳等与导线的安全距离不小于9.5m，工作中应设专人监护。

（3）危险点——高处坠物伤人。

预控措施：

①地面人员不得站在作业点正下方。塔上人员应防止坠物伤人，使用的工具、材料应用绳索传递。

②高处作业应使用工具袋，较大的工器具应固定在牢固的构件上，不准随便乱放。

（4）危险点——防止感应电伤人。

①接触架空地线前，应将地线接地端可靠接地。

②挂接地线时，作业人员戴绝缘手套，手握绝缘部位。

（5）危险点——现场作业安全监护。

①作业过程中，安全监护人对作业人员进行不间断监护。

②工作负责人，监护人必须有符合身份的明显标识。

（6）危险点——交通安全。

出车时应注意车辆行驶安全，谨慎驾驶车辆，禁止违法行车。

3.2 工器具及材料选择

停电修补1000kV交流输电线路架空地线所需工器具及材料见表2-15-2。工器具出库前，应认真核对工器具的使用电压等级和试验周期，并检查确认外观良好、连接牢固、转动灵活，且符合本次工作任务的要求；工器具出库后，应防止脏污、受潮；金属工具和绝缘工器具应分开装运，防止因混装运输导致工器具变形、损伤等现象发生。

表2-15-2 停电修补1000kV交流输电线路架空地线所需工器具及材料表

序号	名称	规格型号	单位	数量	备注
1	作业飞车		副	1	金属工具
2	安全带	含带缓冲包长20M的后保绳	根	2	个人防护用具
3	绝缘手套		双	1	绝缘工具
4	传递绳	$\phi 16$	根	1	其他工具
5	铁滑车	0.5T	个	1	金属工具
6	钢丝套	$\phi 8$	根	1	其他工具
7	个人工具		套	2	其他工具
8	砂纸		张	1	其他工具
9	钢卷尺		把	1	其他工具
10	记号笔		根	1	其他工具
11	木榔头		把	1	其他工具
12	对讲机		台	3	其他工具
13	安全背心		件	2	其他工具
14	安全围栏		卷	4	其他工具
15	防潮苫布		张	1	其他工具
16	抛挂式接地线	1000kV	组	3	其他工具
17	地线接地线	$25mm^2$	组	1	其他工具
18	验电器	1000kV	个	1	其他工具
19	防坠器	T形	个	1	其他工具
20	预绞丝	与地线型号对应	组	1	材料

3.3 作业人员分工

本任务作业人员分工如表2-15-3所示。

表2-15-3 停电修补1000kV交流输电线路架空地线人员分工表

序号	工作岗位	数量(人)	工作职责
1	工作负责人	1	负责本次工作任务的人员分工、工作前的现场查勘、作业方案的制订、工作票的填写、现场复勘、办理工作许可手续、召开工作班前会、落实现场安全措施、负责作业过程中的安全监督、工作中突发情况的处理、工作质量的监督、工作后的总结
2	安全监护人	1	工作前,向被监护人员交待监护范围内的安全措施、告知危险点和安全注意事项;监督被监护人员遵守本规程并严格执行现场安全措施,及时纠正被监护人员不安全动作和行为。
3	高处作业人员	2	负责本次停电修补1000kV交流输电线路架空地线操作
4	地面辅助人员	2	负责本次作业过程的地面辅助工作

4. 工作程序

本任务工作流程如表2-15-4所示。

表2-15-4 停电修补1000kV交流输电线路架空地线工作流程表

序号	作业内容	作业步骤及标准	安全措施及注意事项	责任人
1	工作许可	(1)作业前工作负责人必须与调度联系履行工作许可手续	(1)不得未经工作许可人许可即开始工作。 (2)严禁约时停送电	
2	现场复勘	工作负责人负责完成以下工作： (1)现场核对线路名称、铁塔编号无误；基础及塔身完好无异常；交叉跨越距离符合安全要求；确认缺陷情况及地线规格型号等； (2)检查地形环境符合作业要求； (3)检查工作票所列安全措施与现场实际情况相符，必要时予以补充	(1)正确穿戴安全帽、工作服、工作鞋、劳保手套。 (2)严禁非工作人员、车辆进入作业现场	
3	现场布置	正确设置安全围栏并悬挂标示牌： (1)安全围栏范围应充分考虑高处坠物，以及对道路交通的影响； (2)安全围栏出入口设置合理； (3)妥当布置"从此进出""在此工作""车辆慢行"或"车辆绕行"等标示	对道路交通安全影响不可控时，应及时联系交通管理部门强化现场交通安全管控	
4	召开班前会	(1)全体工作成员列队； (2)工作负责人宣读工作票，明确工作任务及人员分工；讲解工作中的安全措施和技术措施；查问全体工作成员精神状态；告知工作中存在的危险点及采取的预控措施； (3)全体工作成员在工作票上签名确认	(1)工作票填写、签发和许可手续规范，签名完整。 (2)全体工作成员精神状态良好。 (3)全体工作成员明确任务分工、安全措施和技术措施	
5	检查工器具	(1)在防潮垫布上，将工器具按作业要求准备齐备，并分类定置摆放整齐。检查工器具外观和试验合格证，无遗漏； (2)检查人员向工作负责人汇报各项检查结果符合作业要求	(1)防潮垫布设置位置合理，保持清洁、干燥。 (2)工器具外观检查合格，无损伤、受潮、变形、失灵现象，合格证在有效期内	

续表

序号	作业内容	作业步骤及标准	安全措施及注意事项	责任人
6	登塔	(1)登塔作业前,必须先核对线路双重名称及编号。对同塔多回线路,工作负责人、工作班成员应共同认真核查双重名称和识别标记(色标、判别标志等); (2)登塔前应检查塔身、基础等,必须牢固可靠; (3)登塔过程,为防止登塔人员相互碰撞,登塔作业人员相互间距不得小于1.6米,安全带应收拾妥当,后备保护绳放置在工具包内,主带应挂在肩上。 (4)登塔至横担处时,看清楚行走通道,与导线的安全距离不小于9.5m	(1)作业人员登塔时应戴好安全帽,穿软底工作鞋,匀步攀登。 (2)登塔过程中,安全带应收拾妥当,后背保护绳放置在工具包内,主带应挂在肩上,防止登塔过程中安全带勾挂脚钉和塔材,致使作业人员高处坠落。 (3)塔上移位时,不得失去安全带保护,做到踩稳抓牢。 (4)到达作业点位置,将安全带、后备保护绳应分别系在牢固构件上,不得低挂高用。 (5)未验电前,人体、传递绳等与导线的安全距离不小于9.5m,工作中应设专人监护	
7	验电、装设导线接地线	(1)登塔至指定位置后,将安全带系在牢固的构件上,检查扣环是否扣牢;滑车安装位置正确,方便操作;工器具必须使用绳索传递。 (2)使用验电器验电前,戴好绝缘手套自检其声光信号正常。使用伸缩式验电器时,应将其各段绝缘杆全部拉出到位,以保证绝缘杆的有效绝缘长度;验电时,作业人员应手持验电器绝缘手柄,保证人体与导线间的足够安全距离;先下层后验上层,先验近侧后验远侧,线路验电应逐相进行。 (3)验明确无电压后立即装设导线接地线,先用砂纸打磨接地端安装位置,先安装接地端,后装导线端。 (4)装设导线接地线时,必须使用绝缘手套采用抛挂方式进行操作,禁止直接用手操作接地线的金属部分的方式装设导线接地线。确认接地线的挂钩与导线连接紧密可靠	(1)验电器在领用时和使用前应检查是否正常。 (2)接地线与导线及塔材接触良好,禁止以缠绕导线的方式装设接地线。 (3)人体与导线保持足够安全距离	
8	装设地线接地线	(1)登塔至指定位置后,将安全带系在牢固的构件上,检查扣环是否扣牢。使用绳索传递传递工器具。 (2)检查接地线完好,先用砂纸打磨塔材接地端,并先安装接地端后装导线端。 (3)装设地线接地线时,必须使用绝缘手套进行操作,禁止直接用手操作接地线的金属部分的方式装设地线接地线。确认地线接地线的挂钩与地线连接紧密可靠	(1)禁止以缠绕导线的方式装设地线接地线。 (2)挂接地线时,作业人员戴绝缘手套,手握绝缘棒	

第二部分
技能模块培训及考核标准

续表

序号	作业内容	作业步骤及标准	安全措施及注意事项	责任人
9	安装作业飞车	(1)作业人员进入地线前应检查连接金具锈蚀情况。 (2)检查地线绝缘子是否完好,销子是否齐全。 (3)对地线做冲击试验。 (4)塔上辅助人员配合在地线上安装作业飞车	(1)作业人员进入地线前应检查连接金具锈蚀情况。 (2)安装作业飞车前,应对地线做冲击试验	
10	进入作业点	正确使用作业飞车,移动速度均匀,无危险动作;在地线上移动时,不得失去安全带的保护	(1)整个过程不能失去安全带保护。 (2)移动飞车速度不能过快	
11	修补断股地线	(1)到达作业点位后固定好作业飞车。 (2)作业人员对地线损伤处进行打磨处理。 (3)量出预绞丝长度,用钢卷尺在导线损伤处的一端量出1/2预绞丝长的位置画印。 (4)预绞丝中心应安装在导线损伤部位严重处,不得有缝隙。 (5)用木榔头轻敲预绞丝端头,端头应平整	(1)正确使用作业飞车,移动速度均匀,无危险动作;在地线上移动时,不得失去安全带的保护。 (2)断股地线修补前应对损伤点进行打磨处理。 (3)预绞丝应与断股地线绞制方向一致,缠绕应平滑、紧密,损伤部位应位于预绞丝的中间位置。 (4)缠绕预绞丝时,不得用力强扭、撬动,防止其变形,缠绕时两端应保持平整	
12	进入横担	(1)作业人员移动飞车沿地线返回横担。 (2)进入横担过程中不得失去安全带的保护。 传递绳将作业飞车传至地面	(1)整个过程不能失去安全带保护。 (2)移动飞车速度不能过快	
13	清理铁塔上工器具	(1)依次拆作业飞车、地线接地线、导线接地线、钢丝套、滑车等工器具。 (2)拆除地线上的接地线和导线上的接地线,先拆导(地)线端、后拆接地端。 确认无遗留物	(1)拆除地线上的接地线和导线上的接地线,作业人员戴绝缘手套,手握绝缘棒。 (2)人体与导线保持足够安全距离。 (3)防止高处坠物	
14	下塔	(1)沿脚钉侧主材正确下塔。 (2)正确携带传递绳(传递绳头折双、打死结,斜挎肩上)	防止高处坠落	

续表

序号	作业内容	作业步骤及标准	安全措施及注意事项	责任人
15	工作结束	(1)工作负责人组织全体工作成员整理工器具和材料,清理现场,做到"工完料尽场地清"。 (2)召开班后会,工作负责人进行工作总结和点评工作。点评本次工作的施工质量;点评全体工作成员的安全措施落实情况。 (3)工作负责人向工作许可人汇报工作结束,恢复停电线路送电,终结工作票	现场不能有遗留物	

二、考核标准

特高压交流输电线路运检技能考核评分细则

考生填写栏	编号: 姓 名: 所在岗位: 单 位: 日 期: 年 月 日				
考评员填写栏	成绩: 考评员: 考评组长: 开始时间: 结束时间: 操作时长:				
考核模块	停电修补1000kV交流输电线路架空地线	考核对象	特高压交流输电线路检修人员	考核方式 操作	考核时限 100min
任务描述	停电修补1000kV交流输电线路左架空地线				
工作规范及要求	1. 给定条件:1000kV交流实训线路左架空地线需要修补。线路已经停电、验电、装设接地线,所使用工器具已经过测试,工作票已办理,安全措施已经完备。 2. 操作前参考人员应作必要的安全检查。 3. 作业飞车应满足受力要求。 4. 整个过程主要操作流程由工作负责人1名、专责监护人1人、塔上电工1人、塔上辅助工1人、地面辅助工2人完成。工作负责人职责:负责本次工作任务的人员分工、工作前的现场查勘、作业方案的制订、工作票的填写、现场复勘、办理工作许可手续、召开工作班前会、落实现场安全措施、负责作业过程中的安全监督、工作中突发情况的处理、工作质量的监督、工作后的总结。安全监护人责任:工作前,对被监护人员交待监护范围内的安全措施、告知危险点和安全注意事项;监督被监护人员遵守本规程和执行现场安全措施,及时纠正被监护人员的不安全动作和行为。塔上作业人员:负责停电修补1000kV交流输电线路架空地线。塔上辅助工:负责传递工具、材料配合作业人员安装飞车。地面电工职责:协助参考人员完成工器具、材料的上、下传递工作,以及其他辅助工作。 给定条件: 1. 培训线路:特高压1000kV交流实训线路左架空地线,架空地线型号:JLB20A-170。 2. 必须按工作程序进行操作,工序错误扣除相应项目分值,出现重大人身、器材和操作安全隐患,考评员可下令终止考核				

续表

考核情景准备	1. 线路:特高压1000kV交流实训线路左架空地线,工作内容:停电修补1000kV交流输电线路左架空地线,架空地线型号:JLB20A-170。 2. 所需作业工器具:作业飞车1副、安全带2套、对讲机3个、导线接地线1根、地线接地线1根、绝缘手套1双、传递绳1根、0.5T滑车1个、钢丝套1根、砂纸1张、个人工具2套、木榔头1个。 3. 材料:与地线型号相对应的预绞丝一组
备注	1. 个人工器具由考生自备。 2. 各项目得分均扣完为止

序号	项目名称	质量要求	分值	扣分标准	扣分原因	扣分	得分
1	着装	工作服、工作鞋、安全帽、劳保手套穿戴正确	4	漏一项扣1分			
2	个人工具检查	活动扳手、平口钳、工具包符合质量要求	3	未检查1项扣1分			
3	安全带检查	安全带符合质量要求,在试验周期内	2	(1)未进行外观检查扣1分; (2)未检查出厂合格证和试验合格证扣1分			
4	作业工具检查	作业飞车检查符合要求,在试验周期内	2	(1)未进行外观检查扣1分; (2)未检查出厂合格证和试验合格证扣1分			
5	材料检查	确认预绞丝型号与地线相对应,外观检查符合要求无损伤	2	(1)错选预绞丝型号扣1分; (2)未进行检查扣1分			
6	登塔	(1)现场核对线路名称、铁塔编号无误;基础及塔身完好无异常;交叉跨越距离符合安全要求;确认缺陷情况及地线规格型号等; (2)正确携带传递绳(传递绳头折双、打死结、斜肩上); (3)沿脚钉侧主材正确登塔; (4)登塔至指定位置后,将安全带系在牢固的构件上,检查扣环是否扣牢	9	(1)未确认线路双重称号扣2分,未检查基础、塔身、交叉跨越,每项扣1分,未确认地线缺陷扣1分; (2)未携带传递绳扣2分,携带方式不规范扣1分; (3)手抓脚钉1次扣1分; (4)未沿脚钉侧主材登塔扣2分; (5)安全带使用不规范扣2分; (6)踏空、踩滑每次扣1分			

续表

序号	项目名称	质量要求	分值	扣分标准	扣分原因	扣分	得分
7	验电、挂设导线接地线	(1)登塔至指定位置后,将安全带系在牢固的构件上,检查扣环是否扣牢;滑车安装位置正确,方便操作;工器具必须使用绳索传递; (2)使用验电器前,再次检查其声光信号正常。使用伸缩式验电器时,应将其各段绝缘杆全部拉出到位,以保证绝缘杆的有效绝缘长度;验电时,作业人员应手持验电器绝缘手柄,保证人体与导线间的足够安全距离;先验下层后验上层,先验近侧后验远侧,线路验电应逐相进行; (3)验明确无电压后立即装设导线接地线,先用砂纸打磨接地端安装的塔材位置,先安装接地端,后装导线端;装设顺序为先中相,后两边相; (4)装设导线接地线时,必须使用绝缘手套进行操作,禁止直接用手操作接地线的金属部分的方式装设导线接地线。确认接地线的挂钩与导线连接紧密可靠	8	(1)滑车安装不规范扣2分; (2)安全带使用不规范,扣2分; (3)验电、安装接地线未戴绝缘手套扣2分; (4)验电器使用不规范扣2分,验电顺序错误扣2分; (5)接地线两端安装顺序错误扣2分,连接处不牢固扣1分,未对安装好的端部进行检查扣1分; (6)接地线缠绕,扣2分			
8	挂设地线接地线	(1)登塔至指定位置后,将安全带系在牢固的构件上,检查扣环是否扣牢。使用绳索传递传递工器具; (2)检查接地线完好,先用砂纸打磨塔材接地端,并先安装接地端后装导线端; (3)装设地线接地线时,必须使用绝缘手套进行操作,禁止直接用手操作接地线的金属部分的方式装设地线接地线。确认地线接地线的挂钩与地线连接紧密可靠	6	(1)安全带使用不规范,扣2分; (2)安装接地线未戴绝缘手套扣2分; (3)接地线两端安装顺序错误扣2分,连接处不牢固扣1分,未对安装好的端部进行检查扣1分; (4)接地线缠绕,扣2分			

续表

序号	项目名称	质量要求	分值	扣分标准	扣分原因	扣分	得分
9	安装作业飞车	(1)作业人员进入地线前应检查连接金具锈蚀情况； (2)检查地线绝缘子是否完好，销子是否齐全； (3)对地线做冲击试验； (4)塔上辅助人员配合安装作业飞车	9	(1)未检查连接金具锈蚀情况扣2分； (2)未检查地线绝缘子是否完好、销子是否齐全，每项扣1分； (3)未对地线做冲击试验扣2分； (4)作业飞车安装不规范扣2分			
10	进入作业点	正确使用作业飞车，移动速度均匀，无危险动作；在地线上移动时，不得失去安全带的保护	9	(1)失去安全带保护1次扣3分； (2)移动飞车速度过快扣2分； (3)移动过程中有危险动作1次扣3分			
11	修补断股地线	(1)到达作业点位后固定好作业飞车； (2)损伤地线修复、打磨处理平整； (3)量出预绞丝长度，用钢卷尺在导线损伤处的一端量出1/2预绞丝长的位置画印； (4)预绞丝中心应安装在导线损伤部位严重处，不得有缝隙； (5)用木榔头轻敲预绞丝端头，端头应平整	16	(1)未在合适位置固定作业飞车扣3分； (2)地线未打磨处理平整扣3分； (3)未正确画印扣2分； (4)预绞丝安装顺序不规范扣4分； (5)预绞丝末端若不平整，扣2分； (6)预绞丝安装有缝隙，每处扣1分			
12	进入横担	(1)作业人员移动飞车沿地线返回横担； (2)进入横担过程中不得失去安全带的保护； (3)用传递绳将作业飞车传至地面	8	(1)作业飞车移动速度控制不当扣2分； (2)进入横担过程中失去安全带的保护1次扣2分； (3)传递飞车时，传递不规范扣2分			
13	清理铁塔上工器具	(1)依次拆作业飞车、地线接地线、导线接地线、钢丝套、滑车等工器具； (2)拆除地线上的接地线和导线上的接地线，先拆导(地)线端、后拆接地端； (3)确认无遗留物	7	(1)拆除顺序错误扣3分； (2)有遗留物扣4分			
14	下塔	(1)携带传递绳沿脚钉匀步下塔； (2)正确携带传递绳(传递绳头折双、打死结，斜挎肩上)	8	(1)未携带传递绳扣4，传递绳携带方式不规范扣2分； (2)手抓脚钉1次扣1分； (3)未沿脚钉侧主材登塔扣4分			

续表

序号	项目名称	质量要求	分值	扣分标准	扣分原因	扣分	得分
15	摆放工器具	完工后清理作业现场,按要求摆放工器具整齐	2	未按要求摆放工器具扣2分			
16	文明施工	(1)严禁高处坠物； (2)严禁浮置物品； (3)严禁口中含物	5	(1)高处坠物扣5分； (2)浮置物品扣2分； (3)口中含物扣2分			
17	完成时间	在规定时间内按要求完成。		在规定时间内按要求完成,超过时间10分钟即终止操作,只记完成部分得分			
18	合计		100				

模块16　停电更换1000kV交流输电线路子导线培训及考核标准

一、培训标准

(一) 培训要求

模块名称	停电更换1000kV交流输电线路子导线	培训类别	操作类
培训方式	实操培训	培训学时	21学时
培训目标	1. 熟悉1000kV交流输电线路子导线的结构和安装方式 2. 能够在合适位置对1000kV交流输电线路导线验电，并挂设接地线 3. 能完成停电更换1000kV交流输电线路八分裂子导线工作		
培训场地	特高压交流实训线路		
培训内容	停电更换1000kV交流输电线路子导线		
适用范围	特高压交流输电线路检修人员		

(二) 引用规程规范

(1)《电工术语 架空线路》(GB/T 2900.51—1998)。

(2)《1000kV架空输电线路设计规范》(GB 50665—2011)。

(3)《1000kV交流输电线路检修规范》(DL/T 209—2008)。

(4)《1000kV交流输电线路运行规范》(DL/T 307—2010)。

(5)《国家电网公司电力安全工作规程（线路部分）》(Q/GDW1799.2—2013)。

(三) 培训教学设计

本设计以完成"停电更换1000kV交流输电线路子导线"为工作任务，按工作任务完成的标准化作业流程来设计各个培训阶段，每个阶段包括了具体的培训目标、培训内容、培训学时、培训方法（培训资源）、培训环境和考核评价等内容，如表2-16-1所示。

表2-16-1 停电更换1000kV交流输电线路子导线培训内容设计

培训流程	培训目标	培训内容	培训学时	培训方法与资源	培训环境	考核评价
1.理论教学	1.熟悉1000kV交流输电线路子导线的结构和安装方式。2.熟悉停电更换1000kV交流输电线路子导线工作的操作流程	讲授停电更换1000kV交流输电线路子导线工作的操作流程	2	培训方法:讲授法。培训资源:PPT、相关规程规范	多媒体教室	考勤、课堂提问和作业
2.准备工作	能完成作业前准备工作	1.作业现场查勘。2.编制培训标准化作业卡。3.填写培训操作工作票。4.完成本操作的工器具及材料准备	1	培训方法:1.现场查勘和工器具及材料清理采用现场实操方法。2.编写作业卡和填写工作票采用讲授方法。培训资源:1.1000kV实训线路。2.特高压工器具库房。3.空白工作票	1.特高压输电实训线路;2.多媒体教室	
3.作业现场准备	能完成作业现场准备工作	1.作业现场复勘。2.工作申请。3.作业现场布置。4.班前会。5.工器具及材料检查	2	培训方法:演示与角色扮演法。资源:1000kV实训线路	1000kV实训线路	
4.培训师演示	通过现场观摩,使学员初步领会本任务操作流程	1.登塔。2.验电、挂接地线。3.导线上的操作。4.渡线及附件安装。5.拆除工器具	8	培训方法:演示法。资源:1000kV实训线路	1000kV实训线路	
5.学员分组训练	能够分组完成停电更换1000kV交流输电线路子导线工作	1.学员分组(22人一组)训练停电更换1000kV交流输电线路子导线。2.培训师对学员操作进行指导和安全监护	7	培训方法:角色扮演法。资源:1000kV实训线路	1000kV实训线路	采用技能考核评分细则对学员操作评分
6.工作终结	1.使学员进一步辨析操作过程不足之处,便于后期提升。2.培训学员安全文明生产的工作作风	1.作业现场清理。2.向调度汇报工作。3.班后会,对本次工作任务进行点评总结	1	培训方法:讲授和归纳法	1000kV实训线路	

（四）作业流程

1. 工作任务

完成停电更换 1000kV 交流输电线路子导线。

2. 天气及作业现场要求

（1）停电更换 1000kV 交流输电线路子导线应在良好的天气进行。

如遇雷电（听见雷声、看见闪电）、雪、雹、雨、雾等，风力大于 5 级，不得进行作业；恶劣天气下必须开展带电抢修时，应组织有关人员充分讨论并编制必要的安全措施，经本单位批准后方可进行。

（2）作业人员精神状态良好，工作班成员认真学习工作票和安全技术措施，所有人员做到"四清楚"（作业任务精楚、危险点清楚、作业程序清楚、安全措施清楚）。

（3）作业前停送电联系人必须与调度联系履行工作许可手续，严禁约时停送电。工作负责人必须在得到许可人的许可工作命令后，方可在需检修的线路上验电、挂设接地线和进行检修工作。

（4）停电后，工作负责人应认真做好记录。

（5）登杆前应检查塔上是否有蜂窝，发现蜂窝严禁登塔。

（6）塔上作业人员必须使用双保险安全带。

3. 准备工作

3.1　危险点及其预控措施

（1）危险点——误登带电线路。

预控措施：

①登杆塔作业前，工作负责人、工作班成员应共同认真核查双重名称和识别标记（色标、判别标志等）与停电线路名称相符。

②登杆塔前应检查铁塔根部、基础等，必须牢固可靠。

③登杆塔前应检查登高工器具和设施，如安全带、脚钉、塔材等必须完整牢靠。

④不涉及挂设接地线的中间作业人员，应认真核实线路相序、色标、名称、编号与停电线路相符，确认线路名称无刷错、刷反等情况后，方可登杆。

（2）危险点——登塔时、塔上作业时违反安规进行操作，可能引起高空坠落。

预控措施：

①攀爬过程中，为防止登杆人员串落，登杆作业人员间距不得小于 1.6m。

②攀爬铁塔前应将脚底泥土清除干净，检查工具包完整，攀爬过程中不得掉落物件伤人。

③作业人员攀登杆塔时应戴好安全帽，穿软底鞋，动作不能过大，匀步攀登。

④攀爬过程中，安全带应收拾妥当，长尾绳放置在工具包内，主带应挂在肩上，防止攀爬过程中安全带勾挂脚钉和塔材，致使作业人员高空坠落。

⑤杆塔上移位时，不得失去安全带保护，做到踩稳抓牢。

⑥到达作业点位置，系好安全带（绳），应牢固可靠，不得低挂高用。

⑦未验电前，人体、无头绳等与导线的安全距离必须不小于9.5m，工作中应设专人监护。

(3) 危险点——高处坠物伤人

预控措施：

①地面人员不得站在作业点垂直下方。塔上人员应防止落物伤人，使用的工具、材料应用绳索传递。

②在高处作业应使用工具袋，较大的工具应固定在牢固的构件上，不准随便乱放。

③使用绞磨起吊过程中，应设专人指挥，统一配合，绝缘子串刚离地后应进行冲击检查。

(4) 危险点——防止感应电伤人

①为防感应电伤人，应在验电、挂设接地线时使用绝缘手套

②如需接触架空地线，在架空地线接触前应进行可靠接地。

(5) 危险点——现场作业安全监护。

①自作业开始至作业终结，安全监护人必须始终在现场对作业人员进行不间断的安全监护。

②工作负责人，监护人必须穿安全监护背心。

(6) 危险点——交通安全。

出车时应注意车辆行驶安全，谨慎驾驶车辆，禁止违法行车。

3.2 工器具及材料选择

停电更换1000kV交流输电线路子导线所需工器具及材料见表2-16-2。工器具出库前，应认真核对工器具的使用电压等级和试验周期，并检查确认外观良好、连接牢固、转动灵活，且符合本次工作任务的要求；工器具出库后，应防止脏污、受潮；金属工具和绝缘工器具应分开装运，防止因混装运输导致工器具变形、损伤等现象发生。

表2-16-2 停电更换1000kV交流输电线路子导线所需工器具及材料表

序号	名 称	规格型号	单位	数量	备 注
1	传递绳		根	2	
2	滑车		个	2	
3	安全帽		顶	22	
4	风速风向仪		块	1	
5	防潮帆布	2m×4m	块	2	
6	绳套		根	4	
7	防坠器	与杆塔防坠器装置型号对应	只	7	
8	安全带		副	7	
9	安全围栏		套	若干	
10	警示标示牌	"在此工作""从此进出""从此上下"	套	1	
11	红马甲	"工作负责人""安全监护人"	件	2	
12	导电膏		盒	1	
13	砂纸		张	1	
14	清洁毛巾		条	1	
15	对讲机		台	4	
16	操作杆		根	1	
17	验电器		根	1	
18	抛挂式接地线		组	1	
19	软梯		副	1	

3.3 作业人员分工

本任务作业人员分工如表2-16-3所示。

表2-16-3 停电更换1000kV交流输电线路子导线人员分工表

序号	工作岗位	数量(人)	工作职责
1	工作负责人	1	负责本次工作任务的人员分工、工作前的现场查勘、作业方案的制订、工作票的填写、现场复勘、办理工作许可手续、召开工作班前会、落实现场安全措施、负责作业过程中的安全监督、工作中突发情况的处理、工作质量的监督、工作后的总结
2	安全监护人	1	负责本次工作过程中的安全监护工作
3	塔上作业人员	7	负责本次停电更换1000kV交流输电线路子导线操作
4	地面辅助人员	10	负责本次作业过程的地面辅助工作
5	绞磨操作人员	2	负责本次作业过程的绞磨操作工作
6	信号指挥人员	2	负责2台绞磨启停的指挥工作

4. 工作程序

本任务工作流程如表2-16-4所示。

表2-16-4 停电更换1000kV交流输电线路子导线工作流程表

序号	作业内容	作业标准	安全注意事项	责任人
1	工作许可	(1)作业前停送电联系人必须与调度联系履行工作许可手续。	(1)不得未经工作许可人许可即开始工作 (2)严禁约时停送电	
2	现场布置	正确装设安全围栏并悬挂标示牌: (1)安全围栏范围应充分考虑高处坠物,以及对道路交通的影响。 (2)安全围栏出入口设置合理。 (3)妥当布置"从此进出""在此工作""车辆慢行"或"车辆绕行"等标示	对道路交通安全影响不可控时,应及时联系交通管理部门强化现场交通安全管控	
3	召开班前会	(1)全体工作成员列队。 (2)工作负责人宣读工作票,明确工作任务及人员分工;讲解工作中的安全措施和技术措施;查(问)全体工作成员精神状态;告知工作中存在的危险点及采取的预控措施。 (3)全体工作成员在工作票上签名确认	(1)工作票填写、签发和许可手续规范,签名完整。 (2)全体工作成员精神状态良好。 (3)全体工作成员明确任务分工、安全措施和技术措施	
4	检查工器具	(1)在防潮苫布上,将工器具按作业要求准备齐备,并分类定置摆放整齐。检查工器具外观和试验合格证,无遗漏。 (2)检查人员向工作负责人汇报各项检查结果符合作业要求	(1)防潮苫布数量足够,设置位置合理,保持清洁、干燥。 (2)工器具外观检查合格,无损伤、受潮、变形、失灵现象,合格证在有效期内	
5	登杆塔 (验电、挂接地线杆塔)	(1)登杆塔作业前,必须先核对线路名称及编号,所登杆塔应为作业地段前后两基杆塔,工作负责人、工作班成员应共同认真核查双重名称和识别标记(色标、判别标志等)。 (2)登杆塔前应检查铁塔根部、基础等,必须牢固可靠。 (3)攀爬过程,为防止登杆人员串落,登杆作业人员间距不得小于1.6m,安全带收拾妥当,长尾绳放置在工具包内,主带应挂在肩上,防止攀爬过程中安全带勾攀脚钉和塔材,致使作业人员高空坠落。 (4)登杆塔至横担处时,监护人和作业人员应再次核对停电线路的识别标记和双重名称	(1)作业人员攀登杆塔时应戴好安全帽,穿软底鞋,动作不能过大,匀步攀登。 (2)攀爬过程中,安全带应收拾妥当,长尾绳放置在工具包内,主带应挂在肩上,防止攀爬过程中安全带勾挂脚钉和塔材,致使作业人员高空坠落。 (3)杆塔上移位时,不得失去安全带保护,做到踩稳抓牢。 (4)到达作业点位置,系好安全带(绳),应牢固可靠,不得低挂高用。 (5)未验电前,人体、无头绳等与导线的安全距离必须不小于9.5m,工作中应设专人监护	

续表

序号	作业内容	作业标准	安全注意事项	责任人
6	验电、装设接地线	(1)登塔就位后,将安全带系在牢固可靠构件上,必须检查扣环是否正确就位。验电杆(笔)等工器具必须使用绳索传递。 (2)由专人用合格验电器、穿戴合格绝缘手套对作业段前后塔验电。 (3)验明线路确无电压后,开始装设接地线,检查接地线完好,按程序(先接接地端,后接导线端)装设好接地线。 (4)装设接地线时,必须使用绝缘绳或绝缘手柄进行操作,禁止直接用手操作接地线的金属部分的方式装设接地线。确认接地线的夹头与导线连接紧密可靠。	(1)验电杆在领用时和使用前应检查是否正常。 (2)禁止以缠绕导线的方式装设接地线。 (3)验电、装设接地线时,必须穿戴绝缘手套	
7	登杆塔(作业点)	(1)登杆塔作业前,必须先核对线路名称及编号,工作负责人、工作班成员应共同认真核查双重名称和识别标记(色标、判别标志等)。 (2)登杆塔前应检查铁塔根部、基础等,必须牢固可靠。 (3)攀爬过程,为防止登杆人员串落,登杆作业人员间距不得小于1.6m,安全带收拾妥当,长尾绳放置在工具包内,主带应挂在肩上,防止攀爬过程中安全带勾攀脚钉和塔材,致使作业人员高空坠落。 (4)登杆塔至横担时,监护人和作业人员应再次核对停电线路的识别标记和双重名称,确实无误后方可进入停电线路侧的横担。	(1)作业人员攀登杆塔时应戴好安全帽,穿软底鞋,动作不能过大,匀步攀登。 (2)攀爬过程中,安全带应收拾妥当,长尾绳放置在工具包内,主带应挂在肩上,防止攀爬过程中安全带勾挂脚钉和塔材,致使作业人员高空坠落。 (3)杆塔上移位时,不得失去安全带保护,做到踩稳抓牢。 (4)到达作业点位置,系好安全带(绳),应牢固可靠,不得低挂高用。 (5)未验电前,人体、无头绳等与导线的安全距离必须不小于9.5m,工作中应设专人监护	
8	导线上的操作	(1)作业人员进入导线,并拆除导线上的间隔棒,并在拆除位置做好标记。 (2)作业人员在直线塔沿软梯进入导线,并在安装单轮滑车,然后拆除防振锤,使用提线器和6T手扳葫芦提升子导线,取脱金具,将子导线翻进单轮滑车。 (3)将连接部分金具取下后,再将地线翻进固定于地线横担上的单轮滑车中。 (4)作业人员利用卡线器将5T绞磨连接子导线,再利用9T手扳葫芦收紧子导线,取脱导线,两端耐张塔利用绞磨将两端导线松至地面。 (5)作业人员拆除导线端的耐张线夹,新旧导线安装网套,并绑扎牢固,使用旋转接头连接新旧导线		

续表

序号	作业内容	作业标准	安全注意事项	责任人
9	渡线及附件安装	(1)牵引场与张力场同时2台绞磨同时启动,并有专人统一指挥。 (2)待新导线到达牵引场后张力场压接耐张线夹,压接后利用绞磨收线将压接后的导线连接到绝缘子金具上。 (3)牵引场收紧地线,待导线弧垂与原弧垂一致后作记号。 (4)导线松至地面压接另一端耐张线夹并收紧磨绳,将地线连接到绝缘子串金具上,直线塔将新导线连接到线夹里,紧固螺栓,按照之前记号位置安装防振锤。	(1)放线施工过程中,临时拉线、交叉跨越、直线塔、受力工器具应有专人看守,并检查受力情况良好。 (2)放线时,应保持通讯畅通,统一信号、统一指挥。 (3)施工过程中,严禁任何人在地线下方穿越或逗留。 (4)渡线过程中,应有专人随时检查地锚、转向滑车、磨绳的受力情况,并适时进行调整。 (5)放线过程中,收线绞磨操作应平稳,保持地线牵引平衡,预防地线跳槽	
10	拆除工器具	依序拆磨绳、手扳葫芦、滑车、卡线器、后备保护绳等工器具		
11	拆除接地线	在装设接地线的位置拆除接地线,拆除接地线时必须使用绝缘绳或绝缘手柄进行操作,禁止直接用手操作接地线的金属部分的方式装设接地线	拆除接地线时应佩戴绝缘手套	
12	下塔	塔上电工检查塔上无遗留物后,向工作负责人汇报,得到工作负责人同意后携带传递绳下塔	(1)下塔过程应使用防坠装置。 (2)下塔时禁止抓塔钉	
13	工作结束	(1)工作负责人组织全体工作成员整理工器具和材料,清理现场,做到"工完料尽场地清"。 (2)召开班后会,工作负责人进行工作总结和点评工作。点评本次工作的施工质量;点评全体工作成员的安全措施落实情况。 (3)工作负责人向工作许可人汇报工作结束,恢复停电线路送电,终结工作票		

二、考核标准

特高压交流输电线路运检技能考核评分细则

考生填写栏	编号：	姓名：	所在岗位：	单位：	日期：	年 月 日			
考评员填写栏	成绩：	考评员：	考评组长：	开始时间：		结束时间：		操作时长：	
考核模块	停电更换1000kV交流输电线路子导线			考核对象	特高压1000kV交流输电线路检修人员		考核方式	操作	考核时限 360min
任务描述	停电更换1000kV交流输电线路耐张段X相X#子导线								
工作规范及要求	1. 整个过程主要操作流程由小组成员共同完成,小组成员包括:塔上电工7人、地面辅助工10人、绞磨操作人员2人,协助参考人员完成工器具、材料的上、下传递工作,以及其他非技术性工作。 2. 操作前参考人员应作必要的安全检查。 3. 更换子导线时采用的工具应满足受力要求。 4. 工作开始应口头提出申请,工作结束时应口头汇报。 给定条件： 1. 培训基地：特高压交流1000kV交流实训线路耐张段A相7#子导线,导线型号：8×JL/G1A-630/45。 2. 工作票已办理,安全措施已经完备,工作开始、工作终结时应口头提出申请(调度或考评员)。 3. 必须按工作程序进行操作,工序错误扣除应做项目分值,出现重大人身、器材和操作安全隐患,考评员可下令终止操作(考核)。								
考核情景准备	1. 工器具：Φ16磨绳一捆、9T手扳葫芦2把、6T手扳葫芦1把、5T绞磨3台、吊绳滑车3套、绳套3根、旋转接头2套、网套2副、卡线器2个、提线器1套、翻线器1套、间隔棒专用工具2套、单轮滑车3个、放线架2套、放线盘2套、Φ20钢丝套8根、15T铁滑车8个、18T卸扣8个、10T卸扣6个、液压机(含液压钳)2套、对讲机若干、软梯1副、双保险安全带7套、垫木若干。 2. 材料：同型号导线3盘、耐张线夹2套。 3. 在培训线路上操作								
备注	1. 各项目得分均扣完为止,出现重大人身、器材和操作安全隐患,考评员可下令终止操作。 2. 设备、作业环境、安全带、安全帽、工器具、屏蔽服等不符合作业条件考评员可下令终止操作								

续表

序号	项目名称	质量要求	分值	扣分标准	扣分原因	扣分	得分
1	现场复勘	1)工作负责人到作业现场核对线路名称和杆塔编号、现场工作条件、缺陷部位等。2)现场气象条件符合作业要求。3)检查工作票填写完整，无涂改，检查是否所列安全措施与现场实际情况相符，必要时予以补充	5	1)未进行核对双重称号扣1分。2)未核实现场工作条件(气象)、缺陷部位扣1分。3)工作票填写出现涂改，每项扣0.5分，工作票编号有误，扣1分。工作票填写不完整，扣1.5分			
2	现场布置	正确装设安全围栏并悬挂标示牌：1)安全围栏范围应充分考虑高处坠物，以及对道路交通的影响。2)安全围栏出入口设置合理。3)妥当布置"从此进出""在此工作""从此上下"等标示	2	1)作业现场未装设围栏扣0.5分。2)未设立警示牌扣0.5分。3)未悬挂登塔作业标志扣0.5分			
3	召开班前会	1)全体工作成员全体人员正确佩戴安全帽、工作服。2)工作负责人穿红色背心，宣读工作票，明确工作任务及人员分工；讲解工作中的安全措施和技术措施；查(问)全体工作成员精神状态；告知工作中存在的危险点及采取的预控措施。3)全体工作成员在工作票上签名确认	3	1)工作人员着装不整齐扣0.5分，工作人员着装不整齐每人次扣0.5分。2)未进行分工本项不得分，分工不明扣1分。3)现场工作负责人未穿安全监护背心扣0.5分。4)工作票上工作班成员未签字或签字不全的扣1分			
4	工器具检查	1)工作人员按要求将工器具放在防防潮苫布上；防潮苫布应清洁、干燥。2)检查个人工具，活动扳手、平口钳、拔销钳、工具包符合质量要求。3)绞磨试机，检查档位是否正常；手扳葫芦是否符合质量要求。4)对钢丝绳、滑车进行检查，确认连接可靠，受力满足要求。5)检查安全带、个人保安线是否符合质量要求，在试验周期内。6)检查提线器、卡线器、翻线器是否符合要求，在试验周期内。7)确认更换的导线型号，长度，外观检查符合要求	10	1)未使用防潮布并定置摆放工器具扣1分。2)未检查工器具试验合格标签及外观检查每项扣1分。3)错、漏1项扣1分。4)未报告检查结果每项扣0.5分			

第二部分 技能模块培训及考核标准

续表

序号	项目名称	质量要求	分值	扣分标准	扣分原因	扣分	得分
5	场地布置	1）牵引场绞磨布置，正确完成绞磨位置布置、转角滑车位置布置。2）张力场绞磨布置，正确完成绞磨位置布置、转角滑车位置布置、线盘布置	4	错、漏1项扣2分。			
6	登塔	1）塔上电工核对线路双重名称、杆号、相别，检查铁塔是否满足登塔条件，并将结果向工作负责人汇报。2）塔上电工系好安全带，并正确对安全带、后备保护绳以及防坠器进行做冲击试验，向工作负责人汇报以后，方可登塔。3）登塔过程中系好防坠落保护装置，匀速登塔，脚踩脚钉，手抓主材。到达横担工作点后，将安全带系在牢固可靠构件上，必须检查扣环是否正确就位，选择合适位置布置滑车传递绳	5	1）未系安全带或安全带及后备保护绳未进行冲击试验各扣2分。2）手抓脚钉扣2分。3）滑车传递绳悬挂位置不便工具取用扣1分。4）传递时高空落物扣2分。5）传递过程工具与塔身磕碰扣2分。6）传递工具绳索打结混乱扣1分。7）工作负责人监护不到位扣2分。8）塔上电工操作不正确扣2分			
7	验电、装设接地线	1）地面电工将验电杆（笔）等工器具使用绳索传递给塔上电工。2）验电、并装设接地线。使用合格验电器，穿戴合格绝缘手套对作业段前后塔验电，验明线路确无电压后装设接地线，装设接地线时必须使用绝缘绳或绝缘手柄进行操作，禁止直接用手操作接地线的金属部分的方式装设接地线。确认接地线的夹头与导线连接紧密可靠	4	1）验电、装设接地线未佩戴绝缘手套，每项扣4分。2）使用缠绕导线的方式装设接地线扣4分			
8	导线上操作	1）塔上电工检查金具锈蚀情况，对绝缘子串进行冲击，经工作负责人许可后绝缘子串进入导线。2）塔上电工进入子导线，并拆除导线上的间隔棒，并在拆除位置做好标记。	18	1）未使用双保险安全带或失去安全保护3分。2）未对绝缘子串进行冲击扣2分。3）未检查金具扣1分。4）未使用传递绳传递间隔棒扣2分。5）未做记号扣1分。6）提升导线时未进行保护导线和塔材扣2分。			

续表

序号	项目名称	质量要求	分值	扣分标准	扣分原因	扣分	得分
8	导线上操作	3）塔上电工在直线塔沿软梯进入导线,并在安装单轮滑车,然后拆除防振锤,使用提线器和6T手扳葫芦提升子导线,取脱金具,将子导线翻进单轮滑车。 4）将连接部分金具取下后,再将地线翻进固定于地线横担上的单轮滑车中。 5）作业人员利用卡线器将5T绞磨连接子导线,再利用9T手扳葫芦收紧子导线,取脱导线,两端耐张塔利用绞磨将两端导线松至地面。 6）作业人员拆除导线端的耐张线夹,新旧导线安装网套,并绑扎牢固,使用旋转接头连接新旧导线。	18	7）上下过程中软梯使用不规范扣2分。 8）未先将绞磨与导线连接扣2分。 9）手扳葫芦安装位置不正确扣2分。 10）转向滑车安装未对塔材保护,1处扣1分。 11）新导线走向布置不规范扣3分 12）网套未进行绑扎处理,扣3分 13）网套长度不够扣2分 14）未正确使用旋转接头,扣1分			
9	渡线及附件安装	1）牵引场与张力场同时2台绞磨同时启动,并有专人统一指挥。 2）待新导线到达牵引场后张力场压接耐张线夹,压接后利用绞磨收线将压接后的导线连接到绝缘子金具上。 3）牵引场收紧地线,待线弧垂与原弧垂一致后作记号。 4）导线松至地面压接另一端耐张线夹并收紧磨绳,将地线连接到绝缘子串金具上,直线塔将新导线连接到线夹里,紧固螺栓,按照之前记号位置安装防振锤。 5）依序拆磨绳、手扳葫芦、滑车、卡线器、后备保护绳等工器具。	35	1）渡线是导线掉落地面1次扣2分 2）渡线过程中张力过大扣1分 3）未派专人指挥导致卡线扣2分 4）旧导线未及时整理回收扣2分 5）未指派专人测量弧垂扣2分 6）弧垂与其他导线弧垂不一致扣2分 7）耐张线夹型号与导线不符扣3分 8）连接完毕后未进行冲击试验扣2分 9）压接工艺不满足要求扣2分 10）未检查螺栓是否紧固到位扣2分 11）间隔棒安装位置不正确1处扣1分 12）间隔棒安装不规范,安装方向不正确扣2分 13）未检查金具连接情况扣3分 14）传递绳缠绕1次扣2分 15）未检查导线弧垂扣2分 16）塔材损伤1处扣2分			

续表

序号	项目名称	质量要求	分值	扣分标准	扣分原因	扣分	得分
10	拆除接地线	在装设接地线的位置拆除接地线，拆除接地线时必须使用绝缘绳或绝缘手柄进行操作，禁止直接用手操作接地线的金属部分的方式装设接地线。	4	1）拆除接地线未佩戴绝缘手套，每项扣4分。			
11	返回地面	塔上电工检查塔上无遗留物后，向工作负责人汇报，得到工作负责人同意后携带传递绳下塔	5	1）下塔过程未使用防坠装置扣2分。 2）塔上移位失去安全带保护的扣2分。 3）下塔抓塔钉，每处扣1分。 4）塔上有遗留物的，扣2分。			
12	工作结束	1）工作负责人组织全体工作成员整理工器具和材料，将工器具清洁后放入专用的箱（袋）中；清理现场，做到"工完料尽场地清"。 2）召开班后会，工作负责人进行工作总结和点评工作。点评本次工作的施工质量；点评全体工作成员的安全措施落实情况，并向调度汇报，恢复线路送电状态	5	1）工器具未清理扣2分。 2）工器具有遗漏扣2分。 3）未开班后会不得2分。 4）未拆除围栏扣2分。 5）未向调度汇报不得2分。			
	合计		100				

模块17 停电更换1000kV交流输电线路架空地线培训及考核标准

一、培训标准

(一) 培训要求

模块名称	停电更换1000kV交流输电线路架空地线	培训类别	操作类
培训方式	实操培训	培训学时	21学时
培训目标	1.了解架空地线的型号、结构及安装方式。 2.能在1000kV特高压线路的地线上验电、装设接地线。 3.能完成停电更换1000kV交流输电线路架空地线		
培训场地	特高压交流实训线路		
培训内容	停电更换1000kV交流输电线路架空地线		
适用范围	特高压交流输电线路检修人员		

(二) 引用规程规范

(1)《电工术语 架空线路》(GB/T 2900.51—1998)。

(2)《1000kV架空输电线路设计规范》(GB 50665—2011)。

(3)《1000kV交流输电线路检修规范》(DL/T 209—2008)。

(4)《1000kV交流输电线路运行规范》(DL/T 307—2010)。

(5)《国家电网公司电力安全工作规程(线路部分)》(Q/GDW1799.2—2013)。

(6)《架空送电线路运行规程》(DL/T 741—2010)。

(三) 培训教学设计

本设计以完成"停电更换1000kV交流输电线路架空地线"为工作任务,按工作任务完成的标准化作业流程来设计各个培训阶段,每个阶段包括了具体的培训目标、培训内容、培训学时、培训方法(培训资源)、培训环境和考核评价等内容,如表2-17-1所示。

表2-17-1　停电更换1000kV交流输电线路架空地线培训内容设计

培训流程	培训目标	培训内容	培训学时	培训方法与资源	培训环境	考核评价
1.理论教学	1.熟悉架空地线的型号、结构及安装方式。 2.熟悉验电、装设接地线的操作流程。 3.熟悉停电更换1000kV交流输电线路架空地线的工作流程	1.架空地线的型号、结构及安装方式。 2.验电、装设接地线的操作流程。 3.停电更换1000kV交流输电线路架空地线的工作流程	2	培训方法：讲授法。 培训资源：PPT、相关规程规范	多媒体教室	考勤、课堂提问和作业
2.准备工作	能完成作业前准备工作	1.作业现场查勘。 2.编制培训标准化作业卡。 3.填写培训操作工作票。 4.完成本操作的工器具及材料准备	1	培训方法： 1.现场查勘和工器具及材料清理采用现场实操方法。 2.编写作业卡和填写工作票采用讲授方法。 培训资源： 1.1000kV实训线路。 2.特高压工器具库房。 3.空白工作票	1.特高压输电实训线路； 2.多媒体教室	
3.作业现场准备	能完成作业现场准备工作	1.作业现场复勘。 2.工作许可。 3.作业现场布置。 4.班前会。 5.工器具及材料检查	2	培训方法：演示与角色扮演法。 培训资源：1000kV实训线路	1000kV实训线路	
4.培训师演示	通过现场观摩，使学员初步领会本任务操作流程	1.验电、装设接地线。 2.直线塔翻线。 3.耐张塔松线及连接新旧地线。 4.渡线及附件安装	8	培训方法：演示法。 培训资源：1000kV实训线路	1000kV实训线路	
5.学员分组训练	1.能完成验电、装设接地线操作。 2.能完成停电更换1000kV交流输电线路架空地线作业	1.学员分组（22人一组）训练验电、装设接地线和更换架空地线技能操作。 2.培训师对学员操作进行指导和安全监护	7	培训方法：角色扮演法。 培训资源：1000kV实训线路	1000kV实训线路	采用技能考核评分细则对学员操作评分
6.工作终结	1.使学员进一步辨析操作过程不足之处，便于后期提升。 2.培训学员安全文明生产的工作作风	1.作业现场清理。 2.向调度汇报工作。 3.班后会，对本次工作任务进行点评总结	1	培训方法：讲授和归纳法	1000kV实训线路	

（四）作业流程

1. 工作任务

完成停电更换1000kV交流输电线路架空地线。

2. 天气及作业现场要求

（1）停电更换1000kV交流输电线路架空地线应在良好的天气进行。

如遇雷电（听见雷声、看见闪电）、雪、雹、雨、雾等，风力大于5级，不得进行作业；恶劣天气下必须开展停电抢修时，应组织有关人员充分讨论并编制必要的安全措施，经本单位批准后方可进行。

（2）作业人员精神状态良好，工作班成员认真学习工作票和安全技术措施，所有人员做到"四清楚"（作业任务清楚、危险点清楚、作业程序清楚、安全措施清楚）。

（3）作业前停送电联系人必须与调度联系履行工作许可手续，严禁约时停送电。工作负责人必须在得到许可人的许可工作命令后，方可在需检修的线路上验电、挂设接地线和进行检修工作。

（4）停电后，工作负责人应认真做好记录。

（5）登杆前应检查塔上是否有蜂窝，发现蜂窝严禁登塔。

（6）塔上作业人员必须使用双保险安全带。

3. 准备工作

3.1 危险点及其预控措施

（1）危险点——误登带电线路

预控措施：

①登杆塔作业前，工作负责人、工作班成员应共同认真核查双重名称和识别标记（色标、判别标志等）与停电线路名称相符。

②登杆塔前应检查铁塔根部、基础等，必须牢固可靠。

③登杆塔前应检查登高工器具和设施，如安全带、脚钉、塔材等必须完整牢靠。

④不涉及挂设接地线的中间作业人员，应认真核实线路相序、色标、名称、编号与停电线路相符，确认线路名称无刷错、刷反等情况后，方可登杆。

（2）危险点——高处坠落

预控措施：

①攀爬过程中，为防止登杆人员串落，登杆作业人员间距不得小于1.6m。

②攀爬铁塔前应将脚底泥土清除干净，检查工具包完整，攀爬过程中不得掉落物件伤人。

③作业人员攀登杆塔时应戴好安全帽，穿软底鞋，动作不能过大，匀步攀登。

④攀爬过程中，安全带应收拾妥当，长尾绳放置在工具包内，主带应挂在肩上，防止攀爬过程中安全带勾挂脚钉和塔材，致使作业人员高空坠落。

⑤杆塔上移位时，不得失去安全带保护，做到踩稳抓牢。

⑥到达作业点位置，系好安全带（绳），应牢固可靠，不得低挂高用。

⑦未验电前，人体、无头绳等与导线的安全距离必须不小于9.5m，工作中应设专人监护。

（3）危险点——高处坠物伤人。

预控措施：

①地面人员不得站在作业点垂直下方。塔上人员应防止落物伤人，使用的工具、材料应用绳索传递。

②在高处作业应使用工具袋，较大的工具应固定在牢固的构件上，不准随便乱放。

③使用绞磨起吊过程中，应设专人指挥，统一配合，绝缘子串刚离地后应进行冲击检查。

（4）危险点——防止感应电伤人。

①为防感应电伤人，需在检修的线路上验电、挂设接地线。

②如需接触架空地线，在架空地线接触前应进行可靠接地。

（5）危险点——现场作业安全监护。

①自作业开始至作业终结，安全监护人必须始终在现场对作业人员进行不间断的安全监护。

②工作负责人，监护人必须穿安全监护背心。

（6）危险点——交通安全。

出车时应注意车辆行驶安全，谨慎驾驶车辆，禁止违法行车。

3.2 工器具及材料选择

停电更换1000kV交流输电线路架空地线所需工器具及材料见表2-17-2。工器具出库前，应认真核对工器具的使用电压等级和试验周期，并检查确认外观良好、连接牢固、转动灵活，且符合本次工作任务的要求；工器具出库后，应存放在工具袋或工具箱内进行运输，防止脏污、受潮；金属工具和绝缘工器具应分开装运，防止因混装运输导致工器具变形、损伤等现象发生。

表2-17-2　停电更换1000kV交流输电线路架空地线所需工器具及材料表

序号	名称	规格型号	单位	数量	备注
1	作业飞车		副	1	其他工具
2	磨绳	$\phi 16$	盘	1	其他工具
3	手板葫芦	6T	把	3	金属工具
4	绞磨	5T	台	3	机动工具

续表

序号	名称	规格型号	单位	数量	备注
5	吊绳滑车	1T	套	3	金属工具
6	绳套		根	3	其他工具
7	卡线器		个	2	金属工具
8	提线器		套	1	金属工具
9	单轮滑车		个	3	金属工具
10	放线架		套	2	金属工具
11	放线盘		套	2	金属工具
12	钢丝套	φ18	根	8	其他工具
13	铁滑车	10T	个	8	金属工具
14	卸扣	10T	个	8	金属工具
15	卸扣	8T	个	6	金属工具
16	液压机(含液压钳)		套	2	机动工具
17	对讲机		个	10	其他工具
18	全身式安全带	含带缓冲包长20M的后保绳	套	7	个人防护用具
19	安全背心		件	3	其他工具
20	安全围栏		卷	5	其他工具
21	垫木		块	若干	其他工具
22	地线	铝包钢绞线 JLB20A,150	盘	3	材料
23	耐张线夹	同型号	套	2	材料

3.3 作业人员分工

本任务作业人员分工如表2-17-3所示。

表2-17-3　停电更换1000kV交流输电线路架空地线人员分工表

序号	工作岗位	数量(人)	工作职责
1	工作负责人	1	负责本次工作任务的人员分工、工作前的现场查勘、作业方案的制订、工作票的填写、现场复勘、办理工作许可手续、召开工作班前会、落实现场安全措施、负责作业过程中的安全监督、工作中突发情况的处理、工作质量的监督、工作后的总结
2	安全监护人	1	负责本次工作过程中的安全监护工作
3	高空作业人员	7	负责本次停电更换1000kV交流输电线路架空地线操作
4	地面辅助人员	10	负责本次作业过程的地面辅助工作
5	绞磨操作人员	2	负责本次作业过程的绞磨操作工作
6	绞磨信号指挥人员	2	负责2台绞磨启停的指挥工作

4. 工作程序

本任务工作流程如表2-17-4所示。

表2-17-4　停电更换1000kV交流输电线路架空地线工作流程表

序号	作业内容	作业步骤及标准	安全措施及注意事项	责任人
1	现场复勘	工作负责人负责完成以下工作： (1)现场核对线路名称、杆塔编号，相别无误；基础及杆塔完好无异常；交叉跨越距离符合安全要求；确认缺陷情况及导地线规格型号等。 (2)检测风速等现场气象条件符合作业要求。 (3)检查地形环境符合作业要求。 (4)检查工作票所列安全措施与现场实际情况相符，必要时予以补充	(1)正确穿戴安全帽、工作服、工作鞋、劳保手套。 (2)不得在危及作业人员安全的气象条件下作业。 (3)严禁非工作人员、车辆进入作业现场	
2	工作许可	作业前停送电联系人必须与调度联系履行工作许可手续	不得未经工作许可人许可即开始工作	
3	现场布置	正确装设安全围栏并悬挂标示牌： (1)安全围栏范围应充分考虑高处坠物，以及对道路交通的影响。 (2)安全围栏出入口设置合理。 (3)妥当布置"从此进出""在此工作""车辆慢行"或"车辆绕行"等标示	对道路交通安全影响不可控时，应及时联系交通管理部门强化现场交通安全管控	
4	召开班前会	(1)全体工作成员列队。 (2)工作负责人宣读工作票，明确工作任务及人员分工；讲解工作中的安全措施和技术措施；查(问)全体工作成员精神状态；告知工作中存在的危险点及采取的预控措施。 (3)全体工作成员在工作票上签名确认	(1)工作票填写、签发和许可手续规范，签名完整。 (2)全体工作成员精神状态良好。 (3)全体工作成员明确任务分工、安全措施和技术措施	
5	检查工具	(1)在防潮苫布上，将工器具按作业要求准备齐备，并分类定置摆放整齐。检查工器具外观和试验合格证，无遗漏。 (2)检查人员向工作负责人汇报各项检查结果符合作业要求	(1)防潮苫布数量足够，设置位置合理，保持清洁、干燥。 (2)工器具外观检查合格，无损伤、受潮、变形、失灵现象，合格证在有效期内	

续表

序号	作业内容	作业步骤及标准	安全措施及注意事项	责任人
6	登塔	(1)登杆塔作业前,必须先核对线路名称及编号。对同塔多回线路,工作负责人、工作班成员应共同认真核查双重名称和识别标记(色标、判别标志等)。 (2)登杆塔前应检查铁塔根部、基础等,必须牢固可靠。 (3)攀爬过程,为防止登杆人员串落,登杆作业人员间距不得小于1.6m,安全带收拾妥当,长尾绳放置在工具包内,主带应挂在肩上,防止攀爬过程中安全带勾攀脚钉和塔材,致使作业人员高空坠落。 (4)登杆塔至横担处时,监护人和作业人员应再次核对停电线路的识别标记和双重名称,确实无误后方可进入停电线路侧的横担。	(1)作业人员攀登杆塔时应戴好安全帽,穿软底鞋,动作不能过大,匀步攀登。 (2)攀爬过程中,安全带应收拾妥当,长尾绳放置在工具包内,主带应挂在肩上,防止攀爬过程中安全带勾挂脚钉和塔材,致使作业人员高空坠落。 (3)杆塔上移位时,不得失去安全带保护,做到踩稳抓牢。 (4)到达作业点位置,系好安全带(绳),应牢固可靠,不得低挂高用。 (5)未验电前,人体、无头绳等与导线的安全距离必须不小于9.5m,工作中应设专人监护	
7	验电、装设接地线	(1)登杆就位后,将安全带系在牢固可靠构件或电杆上,必须检查扣环是否正确就位。验电杆(笔)等工器具必须使用绳索传递。 (2)验电时站位正确:宜工作且无触电危险的位置;验电时必须带绝缘手套。对线路的验电应逐相进行,按照先近后远,先验下层、后验上层的顺序。 (3)检查接地线完好,按程序(先接地端,后接导线端)装设好接地线。 (4)装设接地线时,必须使用绝缘绳或绝缘手柄进行操作,禁止直接用手操作接地线的金属部分的方式装设设接地线。确认接地线的夹头与导线连接紧密可靠	(1)验电前,应先在有电设备上进行试验,验证验电器良好,无法在有电设备上试验时可用高压发生器等确证验电器良好。 (2)高压验电必须戴绝缘手套、验电器的伸缩绝缘杆长度应拉足,验电时手应握在手柄处不得超过护环,人体与验电设备保持安全距离,雨雪天气时不得进行室外直接验电。 (3)装设接地线必须先接接地端,后接导体端,必须接触良好,拆除时顺序相反。禁止以缠绕导线的方式装设设接地线	
8	直线塔翻线	(1)作业人员在直线塔地线横担合适位置安装单轮滑车,并拆除地线上的防振锤。 (2)作业人员使用提线器和6T手板葫芦提升地线。 (3)将连接部分金具取下后,再将地线翻进固定于地线横担上的单轮滑车中		

续表

序号	作业内容	作业步骤及标准	安全措施及注意事项	责任人
9	安装卡线器	(1)作业人员在地线上量出防振锤位置并记录后,拆除防振锤 (2)作业人员乘坐飞车将卡线器卡至地线上合适位置后,并收紧磨绳。 (3)在地线两端通过6T手板葫芦收紧地线后,取脱地线连接金具	收紧地线后,应做冲击试验,确认无误后方可进行下一步操作。	
10	耐张塔松线及连接新旧地线	(1)作业人员松出两端的6T手扳葫芦,使磨绳受力后,拆除手扳葫芦。 (2)启动绞磨,利用绞磨将两端的地线松至地面。 (3)松至地面后,地面人员开断地线端的耐张线夹,打磨并压接新旧地线	(1)放线过程中,收线绞磨操作应平稳,保持地线牵引平衡,预防地线跳槽。 (2)使用绞磨时,绞磨盘上缠绕圈数不得少于5圈,且从下方卷入,并排列整齐,尾绳控制人员不得少于2人。 (3)放、紧线时,人员不得站在或跨在已受力的牵引绳、地线的内角侧和展放的地线圈内以及牵引绳或架空线的垂直下方,防止意外跑线时伤人。 (4)地线压接时人员禁止将身体任何部位放在液压机正上方,压接机具应有固定设施,操作时放置平稳,两侧扶线人员对准位置,手指不得伸入压模内	
11	渡线及附件安装	(1)牵引场与张力场同时2台绞磨同时启动,并有专人统一指挥。 (2)待新地线到达牵引场后张力场压接耐张线夹,压接后利用绞磨收线将压接后的地线连接到绝缘子金具上。 (3)牵引场收紧地线,待地线弧垂与原弧垂一致后作记号。 (4)地线松至地面压接另一端耐张线夹并收紧磨绳,将地线连接到绝缘子串金具上,直线塔将新地线连接到线夹里,紧固螺栓,按照之前记号位置安装防振锤	(1)放线施工过程中,临时拉线、交叉跨越、直线塔、受力工器具应有专人看守,并检查受力情况良好。 (2)放线时,应保持通信畅通,统一信号、统一指挥。 (3)施工过程中,严禁任何人在地线下方穿越或逗留。 (4)渡线过程中,应有专人随时检查地锚、转向滑车、磨绳的受力情况,并适时进行调整。 (5)放线过程中,收线绞磨操作应平稳,保持地线牵引平衡,预防地线跳槽	
12	拆除工器具	(1)依序拆磨绳、手扳葫芦、卡线器、后备保护绳等工器具。 (2)拆除接地线	(1)拆除接地线的顺序为先导体端,后接地端	
13	工作结束	(1)工作负责人组织全体工作成员整理工器具和材料,清理现场,做到"工完料尽场地清"。 (2)召开班后会,工作负责人进行工作总结和点评工作。点评本次工作的施工质量;点评全体工作成员的安全措施落实情况。 (3)工作负责人向工作许可人汇报工作结束,恢复停电线路送电,终结工作票		

二、考核标准

特高压交流输电线路运检技能考核评分细则

考生填写栏	编号:	姓 名:		所在岗位:		单 位:		日 期:	年 月 日	
考评员填写栏	成绩:	考评员:		考评组长:		开始时间:		结束时间:	操作时长:	
考核模块	停电更换1000kV交流输电线路架空地线		考核对象	特高压交流输电线路检修人员			考核方式	操作	考核时限	360min
任务描述	停电更换1000kV输电线路耐张段左架空地线									
工作规范及要求	1. 给定条件:1000kV交流实训线路003#-005#塔耐张段左架空线需要更换。线路已经停电、验电、挂接地线,所使用工器具、材料已经过测试,工作票已办理,安全措施已经完备。 2. 本作业工作应在良好天气下进行。如遇雷电(听见雷声、看见闪电)、雪、雹、雨、雾等,风力大于5级,不得进行作业。 3. 本项作业需工作负责人1名,安全监护人1人,塔上电工1人,塔上辅助工6人,地面辅助工10人,绞磨操作人员2人,协助人员完成工器具、材料的上、下传递工作,以及其他非技术性工作。 4. 工作负责人职责:负责本次工作任务的人员分工、工作前的现场查勘、作业方案的制订、工作票的填写、现场复勘、办理工作许可手续、召开工作班前会、落实现场安全措施、负责作业过程中的安全监督、工作中突发情况的处理、工作质量的监督、工作后的总结。 5. 专责监护人:负责作业现场的安全把控。 6. 高空作业人员职责:负责停电更换1000kV交流输电线路架空地线。 7. 地面辅助人员职责:负责传递工具、材料配合塔上电工。 8. 绞磨操作人员职责:负责本次作业过程的绞磨操作工作。 9. 在作业过程中,如遇雷、雨、大风或其他任何情况威胁到工作人员的安全时,工作负责人或监护人可根据情况,临时停止工作。 给定条件: 1. 培训基地:特高压1000kV交流实训线路003#~005#塔耐张段左架空线,地线型号:铝包钢绞线JLB20A,150。 2. 工作票已办理,安全措施已经完备,工作开始、工作终结时应口头提出申请(调度或考评员)。 3. 安全、正确地使用仪器对绝缘工具进行检测。 4. 必须按工作程序进行操作,工序错误扣除应做项目分值,出现重大人身、器材和操作安全隐患,考评员可下令终止操作(考核)									
考核情景准备	1. 线路:特高压交流1000kV线路003#~005#塔耐张段左架空线,工作内容:停电更换1000kV交流输电线路架空地线,地线型号:JLB20A,150。 2. 所需作业工器具:作业飞车1副、φ16磨绳一捆、6T手扳葫芦3把、5T绞磨3台、吊绳滑车3套、绳套3根、卡线器2个、提线器1套、单轮滑车3个、放线架2套、放线盘2套、φ18钢丝套8根、10T铁滑车8个、10T卸扣8个、8T卸扣6个、液压机(含液压钳)2套、对讲机若干、双保险安全带7套、垫木若干。 3. 材料:同型号地线1盘,接续管1套,耐张线夹2套。 4. 作业现场做好监护工作,作业现场安全措施(围栏等)已全部落实;禁止非作业人员进入现场,工作人员进入作业现场必须戴安全帽。 5. 考生自备工作服,安全帽,线手套,安全带(含二保绳)									
备注	1. 各项目得分均扣完为止,出现重大人身、器材和操作安全隐患,考评员可下令终止操作。 2. 设备、作业环境、安全带、安全帽、工器具、等不符合作业条件考评员可下令终止操作									

续表

序号	项目名称	质量要求	分值	扣分标准	扣分原因	扣分	得分
1	现场复勘	1)工作负责人到作业现场核对线路名称和杆塔编号、现场工作条件、缺陷部位等。 2)确认现场气象条件符合作业要求。 3)检查工作票填写完整,无涂改,检查是否所列安全措施与现场实际情况相符,必要时予以补充	5	1)未进行核对双重称号扣1分。 2)未核实现场工作条件(气象)、缺陷部位扣1分。 3)工作票填写出现涂改,每项扣0.5分,工作票编号有误,扣1分。工作票填写不完整,扣1.5分			
2	工作许可	1)工作负责人联系值班调控人员,按工作票内容履行工作许可手续。 2)汇报内容规范、完整	2	1)未联系调度部门(裁判)停用重合闸扣2分。 2)汇报专业用语不规范或不完整的各扣0.5分			
3	现场布置	正确装设安全围栏并悬挂标示牌: 1)安全围栏范围应充分考虑高处坠物,以及对道路交通的影响。 2)安全围栏出入口设置合理。 3)妥当布置"从此进出""在此工作""从此上下"等标示	3	1)作业现场未装设围栏扣0.5分。 2)未设立警示牌扣0.5分。 3)未悬挂登塔作业标志扣0.5分			
4	召开班前会	1)全体工作成员全体人员正确佩戴安全帽、工作服。 2)工作负责人穿红色背心,宣读工作票,明确工作任务及人员分工;讲解工作中的安全措施和技术措施;查(问)全体工作成员精神状态;告知工作中存在的危险点及采取的预控措施。 3)全体工作成员在工作票上签名确认	3	1)工作人员着装不整齐扣0.5分,工作人员着装不整齐每人次扣0.5分。 2)未进行分工本项不得分,分工不明扣1分。 3)现场工作负责人未穿安全监护背心扣0.5分。 4)工作票上工作班成员未签字或签字不全的扣1分			
5	工器具检查	1)工作人员按要求将工器具放在防防潮苫布上;防潮苫布应清洁、干燥。 2)工器具应按定置管理要求分类摆放;检查工器具外观和试验合格证,无遗漏。 3)活动扳手、平口钳、拔销钳、工具包符合质量要求。 4)绞磨试机,检查档位是否正常;手扳葫芦检查	10	1)未使用防潮布并定置摆放工器具扣1分。 2)未检查工器具试验合格标签及外观检查扣每项扣0.5分。 3)未正确对工器具进行检测每项扣1分。 4)未正确绞磨试机扣2分,未正确检查手扳葫芦扣2分			

续表

序号	项目名称	质量要求	分值	扣分标准	扣分原因	扣分	得分
5	工器具检查	5)对钢丝绳、滑车进行检查,确认连接可靠,受力满足要求。 6)安全带、个人保安线符合质量要求,在试验周期内。 7)提线器、卡线器检查符合要求,在试验周期内。 8)确认地线、接续管型号、长度、外观检查符合要求。 9)登塔人员再次核对双重名称、杆号、相别并报告	10	5)错漏1项扣1分。 6)未正确检查每项扣1分。 7)错漏1项扣0.5分。 8)错漏1项扣1分。 9)未核对双重名称扣2分			
6	登塔	1)检查杆塔基础无异常。 2)塔上电工穿好将安全带做冲击试验后,系好安全带后正确携带吊绳(吊绳头折叠、打死结、斜挎肩上)相继登塔。 3)登塔过程中应系好防坠落保护装置,匀速登塔,手抓主材,将安全带挂在肩上,登塔作业人员间距不得小于1.6m,工作负责人加强作业监护。 4)由塔身到横担上工作点不得失去安全带保护。 5)传递滑车安装位置正确,方便操作。	8	1)未检查1项扣1分。 2)未系安全带或安全带及后备保护绳未进行冲击试验各扣扣2分,未携带吊绳扣1分,吊绳携带方式不规范扣1分。 3)手抓脚钉扣2分,未沿脚钉侧主材登塔扣1分。 4)未正确使用安全带扣2分。 5)滑车传递绳悬挂位置不便工具取用扣1分。 6)传递时金属工具难以保证安全距离扣2分;工具绑扎不牢扣2分。 7)传递时高空落物扣2分。 8)传递过程工具与塔身磕碰扣2分。 9)传递工具绳索打结混乱扣1分			
7	验电、装设接地线	1)登杆就位后,将安全带系在牢固可靠构件或电杆上,必须检查扣环是否正确就位。验电杆(笔)等工器具必须使用绳索传递。 2)检查接地线完好,按程序(先接接地端,后接导线端)装设好接地线,接地夹头在横担上连接牢固。 3)装设接地线时,必须使用绝缘绳或绝缘手柄进行操作,禁止直接用手操作接地线的金属部分的方式装设接地线。确认接地线的夹头与导线连接紧密可靠	11	1)未进行验电、装设接地线扣5分。 2)装设接地线顺序错误扣3分。 3)接地线连接不牢固一处扣3分。 4)未使用绝缘手套一次扣2分			

续表

序号	项目名称	质量要求	分值	扣分标准	扣分原因	扣分	得分
8	直线塔翻线	1）作业人员在直线塔地线横担合适位置安装单轮滑车，并拆除地线上的防振锤。 2）作业人员使用提线器和6T手板葫芦提升地线。 3）将连接部分金具取下后，再将地线翻进固定于地线横担上的单轮滑车中	6	1）提升地线时未进行保护地线和塔材扣2分。 2）操作手板葫芦不当扣2分。 3）上下过程中软梯使用不规范扣2分			
9	安装卡线器	1）作业人员在地线上量出防振锤位置并记录后，拆除防振锤。 2）作业人员乘坐飞车将卡线器卡至地线上合适位置后，并收紧磨绳。 3）在地线两端通过6T手板葫芦收紧地线，做冲击试验，确认无误后取脱地线连接金具	8	1）未记录防振锤位置扣3分。 2）未正确使用作业飞车扣3分。 3）取脱连接金具前未冲击试验扣2分			
10	耐张塔松线及连接新旧地线	1）作业人员松出两端的6T手板葫芦，使磨绳受力后，冲击磨绳，拆除手板葫芦。 2）启动绞磨，利用绞磨将两端的地线松至地面。 3）布置张力场塔上新地线的转角滑车及地线走向。 4）地面人员开断地线端的耐张线夹，打磨并压接新旧地线，接续管压接工艺满足要求。	15	1）未冲击试验扣2分。 2）转向滑车安装未对塔材保护，1处扣1分。 3）新地线走向布置不规范扣3分。 4）地线未进行打磨处理，1处扣2分。 5）压接工艺不满足要求扣3分			
11	渡线及附件安装	1）牵引场与张力场同时2台绞磨同时启动，并有专人统一指挥。 2）渡线过程中档中央应派专人看守，防止地线落地。3）塔上作业人员看好接头，防止卡线。 4）绞磨操作人员控制好渡线速度，防止张力过大。 5）待新地线到达牵引场后张力场压接耐张线夹，压接后利用绞磨收线将压接后的地线连接到绝缘子金具上。 6）牵引场收紧地线，待地线弧垂与原弧垂一致后做记号。	12	1）渡线时地线掉落地面1次扣2分。 2）渡线过程中张力过大扣1分。 3）未派专人指挥导致卡线扣2分。 4）旧地线未及时整理回收扣2分。 5）未指派专人测量弧垂扣2分，弧垂与原弧垂不一致扣2分。 6）耐张线夹型号与地线不符扣3分。			

续表

序号	项目名称	质量要求	分值	扣分标准	扣分原因	扣分	得分
11	渡线及附件安装	7)地线松至地面压接另一端耐张线夹并收紧磨绳,将地线连接到绝缘子串金具上,直线塔将新地线连接到线夹里,紧固螺栓,按照之前记号位置安装防振锤。8)直线塔将新地线连接到线夹里,紧固螺栓,按照原要求安装防振锤。		7)连接完毕后未进行冲击试验扣2分。8)压接工艺不满足要求扣2分。9)未检查螺栓是否紧固到位扣2分。10)间隔棒安装位置不正确1处扣1分,间隔棒安装不规范,安装方向不正确扣2分			
12	返回地面	1)检查金具、附件是否齐全到位,连接部位是否牢固,松出磨绳,检查塔材是否损伤,检查地线弧垂,连接金具是否符合要求。2)依序拆磨绳、手扳葫芦、接地线、滑车、卡线器、后备保护绳等工器具,传递至地面。拆除接地线时,应先拆导线端,后拆接地端。先拆上层,后拆下层,先拆远端,后拆近端。3)塔上电工检查塔上无遗留物后,向工作负责人汇报,得到工作负责人同意后携带绝缘传递绳下塔。4)必须沿脚钉侧主材正确下塔,正确携带吊绳(吊绳头折双、打死结,斜挎肩上)。5)确认无遗留物	7	1)未检查金具连接情况扣3分,未检查地线弧垂扣2分。2)传递绳缠绕1次扣2分,塔材损伤1处扣2分。3)接地线拆除顺序不正确扣2分。4)未携带吊绳扣4分,吊绳携带方式不规范扣2分。5)手抓脚钉1次扣1分,未沿脚钉侧主材登塔扣4分。6)有遗留物扣4分			
13	工作结束	1)工作负责人组织全体工作成员整理工器具和材料,将工器具清洁后放入专用的箱(袋)中;清理现场,做到"工完料尽场地清"。2)召开班后会,工作负责人进行工作总结和点评工作。点评本次工作的施工质量;点评全体工作成员的安全措施落实情况。3)工作负责人向工作许可人汇报工作结束,恢复停电线路送电,终结工作票。4)在规定时间内按要求完成	10	1)未清理作业现场漏一项扣3分。2)未开班后会扣2分。3)未向工作许可人汇报扣2分。4)每超过1分钟,扣1分,超过5分钟后,未完成项不计分			
	合计		100				

Part I

Standard for Professional Training and Assessment for Operation Maintenance of UHV AC Transmission Line

I. General

In order to implement the principle of "safety first, preventive focus, and comprehensive management", standardize the professional training and assessment of 1000kV AC transmission line operation and inspection, and comprehensively improve the skill level of UHV AC transmission line maintenance personnel, this standard is formulated.

This standard is suitable for the training and examination of special skills of UHV AC transmission line maintenance personnel.

II. Preparation Basis

Post Ability Training Specification for Skilled Personnel of State Grid Company-Part 4: Operation and Inspection for Transmission Lines (330kV and Above) (Q/GDW11372.4-2015)

Post Ability Training Specification for Skilled Personnel of State Grid Company-Part 7: Live Working on Power Transmission Lines (Q/GDW11372.7-2015)

State Grid Corporation of China Working Regulations of Power Safety (Transmission Line Section) (Q/GDW1799.2-2013)

III. Training Objects

Maintenance personnel of UHV AC transmission line.

IV. Training Standards

This standard covers a total of 17 training modules, including 11 live maintenance operation training modules and 6 interruption maintenance operation modules.

(I) **Module Setting**

The training modules set up in this standard are shown in Table 1-1.

Table 1-1 Training Module Setting

S/N	Designation of module	Training hours
1	Live replacement of Type I composite insulator for 1000kV AC transmission line tangent tower	21
2	Live replacement of single-V composite insulator for 1000kV AC transmission line tangent tower	21

Part I Standard for Professional Training and Assessment for
Operation Maintenance of UHV AC Transmission Line

Table (Cont'd)

S/N	Designation of module	Training hours
3	The live replacement of 1 ~ 3 glass insulators on the cross arm side of 1000kV AC transmission line resisting-tensile tower	21
4	The live replacement of 1 ~ 3 glass insulators on the conductor side of 1000kV AC transmission line resisting-tensile tower	21
5	Live replacement of arbitrary single insulator of tensile glass insulator string of 1000kV AC transmission line	21
6	Live replacement of arbitrary section insulator of tensile glass insulator string of 1000kV AC transmission line	21
7	Live replacement of 1000kV AC transmission line conductor spacer	21
8	Live replacement of 1000kV AC transmission line conductor vibration damper	21
9	Live repair of 1000kV AC transmission line conductor	14
10	Live treatment of heating defects of 1000kV AC transmission line conductor drainage plate	14
11	Live replacement of ground wire connecting fitting for 1000kV AC transmission line tangent tower	14
12	Power-cut replacement of Type I composite insulator for 1000kV AC transmission line tangent tower	21
13	Power-cut replacement of single-V composite insulator for 1000kV AC transmission line tangent tower	21
14	Power-cut replacement of the whole strain insulator string on 1000kV AC transmission line	28
15	Power-cut repair of 1000kV AC transmission line overhead ground wire	14
16	Power-cut replacement of 1000kV AC transmission line sub-conductor	21
17	Power-cut replacement of 1000kV AC transmission line overhead ground wire	21

(Ⅱ) Module Description

The training content for each module of this standard is described in Table 1-2 ~ Table 1-18.

Table 1-2 Training Content for Live Replacement of Type I Composite Insulator for 1000kV AC Transmission Line Tangent Tower

Designation of module	Live replacement of Type I composite insulator for 1000kV AC transmission line tangent tower
Description	Learn to master the standardized operation flow of live replacement of Type I composite insulator for 1000kV AC transmission line tangent tower, train the special skills of trainees in entering equipotential through "basket method" and adopting equipotential operation method to replace Type I composite insulator of tangent tower in group, and improve the operation capability of maintenance personnel in live replacement project of Type I composite insulator for 1000kV AC transmission line tangent tower

Table 1-3 Training Content for Live Replacement of Single-V Composite Insulator for 1000kV AC Transmission Line Tangent Tower

Designation of module	Live replacement of single-V composite insulator for 1000kV AC transmission line tangent tower
Description	Learn to master the standardized operation flow of live replacement of single-V composite insulator for 1000kV AC transmission line tangent tower, train the special skills of trainees in entering equipotential through "basket method" and adopting equipotential operation method to replace single-V composite insulator of tangent tower in group, and improve the operation capability of maintenance personnel in live replacement project of single-V composite insulator for 1000kV AC transmission line tangent tower

Table 1-4 Training Content for Live Replacement of 1 ~ 3 Glass Insulators on the Cross Arm Side of 1000kV AC Transmission Line Resisting-tensile Tower

Designation of module	The live replacement of 1~3 glass insulators on the cross arm side of 1000kV AC transmission line resisting-tensile tower
Description	Learn to master the standardized operation flow of live replacement of 1~3 glass insulators on the cross arm side of 1000kV AC transmission line resisting-tensile tower, train the special skills of trainees in adopting ground potential operation method to replace 1~3 glass insulators on the cross arm side of resisting-tensile tower in group, and improve the operation capability of maintenance personnel in live replacement project of 1~3 glass insulators on the cross arm side of 1000kV AC transmission line resisting-tensile tower

Table 1-5 Training Content for Live Replacement of 1 ~ 3 Glass Insulators on the Conductor Side of 1000kV AC Transmission Line Resisting-tensile Tower

Designation of module	The live replacement of 1~3 glass insulators on the conductor side of 1000kV AC transmission line resisting-tensile tower
Description	Learn to master the standardized operation flow of live replacement of 1~3 glass insulators on the conductor side of 1000kV AC transmission line resisting-tensile tower, train the special skills of trainees in entering the equipotential along the strain insulator string through "two-span and three-short-circuit" operation mode and adopting equipotential operation method to replace 1~3 glass insulators on the conductor side of resisting-tensile tower in group, and improve the operation capability of maintenance personnel in live replacement project of 1~3 glass insulators on the conductor side of 1000kV AC transmission line resisting-tensile tower

Part I Standard for Professional Training and Assessment for
Operation Maintenance of UHV AC Transmission Line

Table 1-6 Training Content for Live Replacement of Arbitrary Single Insulator of Tensile Glass Insulator String of 1000kV AC Transmission Line

Designation of module	Live replacement of arbitrary single insulator of tensile glass insulator string of 1000kV AC transmission line
Description	Learn to master the standardized operation flow of live replacement of arbitrary single insulator of tensile glass insulator string of 1000kV AC transmission line, train the special skills of trainees in entering the intense electric field along the strain insulator string through "two-span and three-short-circuit" operation mode and adopting intermediate potential operation method to replace arbitrary single insulator of tensile glass insulator string in group, and improve the operation capability of maintenance personnel in live replacement project of arbitrary single insulator of tensile glass insulator string of 1000kV AC transmission line

Table 1-7 Training Content for Live Replacement of Arbitrary Section Insulator of Tensile Glass Insulator String of 1000kV AC Transmission Line

Designation of module	Live replacement of arbitrary section insulator of tensile glass insulator string of 1000kV AC transmission line
Description	Learn to master the standardized operation flow of live replacement of arbitrary section insulator of tensile glass insulator string of 1000kV AC transmission line, train the special skills of trainees in entering the intense electric field along the strain insulator string through "two-span and three-short-circuit" operation mode and adopting intermediate potential operation method to replace arbitrary section insulator of tensile glass insulator string in group, and improve the operation capability of maintenance personnel in live replacement project of arbitrary section insulator of tensile glass insulator string of 1000kV AC transmission line

Table 1-8 Training Content for Live Replacement of 1000kV AC Transmission Line Conductor Spacer

Designation of module	Live replacement of 1000kV AC transmission line conductor spacer
Description	Learn to master the standardized operation flow of live replacement of 1000kV AC transmission line conductor spacer, train the special skills of trainees in entering the equipotential along the strain insulator string through "two-span and three-short-circuit" operation mode and adopting equipotential operation method to replace eight-bundle conductor spacer in group, and improve the operation capability of maintenance personnel in live replacement project of 1000kV AC transmission line conductor spacer

Table 1-9 Training Content for Live Replacement of 1000kV AC Transmission Line Conductor Vibration Damper

Designation of module	Live replacement of 1000kV AC transmission line conductor vibration damper
Description	Learn to master the standardized operation flow of live replacement of 1000kV AC transmission line conductor vibration damper, train the special skills of trainees in entering the equipotential along the strain insulator string through "two-span and three-short-circuit" operation mode and adopting equipotential operation method to replace conductor vibration damper in group, and improve the operation capability of maintenance personnel in live replacement project of 1000kV AC transmission line conductor vibration damper

Table 1-10 Training Content for Live Repair of 1000kV AC Transmission Line Conductor

Designation of module	Live repair of 1000kV AC transmission line conductor
Description	Learn to master the standardized operation flow of live repair of 1000kV AC transmission line conductor, train the special skills of trainees in entering the equipotential along the strain insulator string through "two-span and three-short-circuit" operation mode and adopting equipotential operation method and use the patch rods of preformed armor rods to repair conductor in group, and improve the operation capability of maintenance personnel in live repair project of 1000kV AC transmission line conductor

Table 1-11 Training Content for Live Treatment of Heating Defects of 1000kV AC Transmission Line Conductor Drainage Plate

Designation of module	Live treatment of heating defects of 1000kV AC transmission line conductor drainage plate
Description	Learn to master the standardized operation flow of live treatment of heating defects of 1000kV AC transmission line conductor drainage plate, train the special skills of trainees in entering the equipotential along the strain insulator string through "two-span and three-short-circuit" operation mode and adopting equipotential operation method to tighten the connecting bolts of conductor drainage plate in group, and improve the operation capability of maintenance personnel in live treatment project of heating defects of 1000kV AC transmission line conductor drainage plate

Part I Standard for Professional Training and Assessment for Operation Maintenance of UHV AC Transmission Line

Table 1-12 Training Content for Live Replacement of Ground Wire Connecting Fitting for 1000kV AC Transmission Line Tangent Tower

Designation of module	Live replacement of ground wire connecting fitting for 1000kV AC transmission line tangent tower
Description	Learn to master the standardized operation flow of Live replacement of ground wire connecting fitting for 1000kV AC transmission line tangent tower, train the special skills of trainees in adopting ground potential operation method to replace ground wire connecting fitting for tangent tower in group, and improve the operation capability of maintenance personnel in live replacement project of ground wire connecting fitting for 1000kV AC transmission line tangent tower

Table 1-13 Training Content for Power-cut Replacement of Type I Composite Insulator for 1000kV AC Transmission Line Tangent Tower

Designation of module	Power-cut replacement of Type I composite insulator for 1000kV AC transmission line tangent tower
Description	Learn to master the standardized operation flow of power-cut replacement of Type I composite insulator for 1000kV AC transmission line tangent tower, train the special skills of trainees in adopting power-cut operation method to replace Type I composite insulator of tangent tower in group, and improve the operation capability of maintenance personnel in power-cut replacement project of Type I composite insulator for 1000kV AC transmission line tangent tower

Table 1-14 Training Content for Power-cut Replacement of Single-V Composite Insulator for 1000kV AC Transmission Line Tangent Tower

Designation of module	Power-cut replacement of single-V composite insulator for 1000kV AC transmission line tangent tower
Description	Learn to master the standardized operation flow of power-cut replacement of single-V composite insulator for 1000kV AC transmission line tangent tower, train the special skills of trainees in adopting power-cut operation method to replace single-V composite insulator of tangent tower in group, and improve the operation capability of maintenance personnel in power-cut replacement project of single-V composite insulator for 1000kV AC transmission line tangent tower

Table 1-15 Training Content for Power-cut Replacement of Whole Strain Insulator String on 1000kV AC Transmission Line

Designation of module	Power-cut replacement of the whole strain insulator string on 1000kV AC transmission line
Description	Learn to master the standardized operation flow of power-cut replacement of whole strain insulator string on 1000kV AC transmission line, train the special skills of trainees in adopting power-cut operation method to replace whole strain insulator string in group, and improve the operation capability of maintenance personnel in power-cut replacement project of whole strain insulator string on 1000kV AC transmission line

Table 1-16　Training Content for Power-cut Repair of 1000kV AC Transmission Line Overhead Ground Wire

Designation of module	Power-cut repair of 1000kV AC transmission line overhead ground wire
Description	Learn to master the standardized operation flow of power-cut repair of 1000kV AC transmission line overhead ground wire, train the special skills of trainees in adopting power-cut operation method and the hanging wheels to the operation position to repair the overhead ground wire with the patch rods of preformed armor rods in group, and improve the operation capability of maintenance personnel in power-cut repair project of 1000kV AC transmission line overhead ground wire

Table 1-17　Training Content for Power-cut Replacement of 1000kV AC Transmission Line Sub-conductor

Designation of module	Power-cut replacement of 1000kV AC transmission line sub-conductor
Description	Learn to master the standardized operation flow of power-cut replacement of 1000kV AC transmission line sub-conductor, train the special skills of trainees in adopting power-cut operation method to replace sub-conductor in group, and improve the operation capability of maintenance personnel in power-cut replacement project of 1000kV AC transmission line sub-conductor

Table 1-18　Training Content for Power-cut Replacement of 1000kV AC Transmission Line Overhead Ground Wire

Designation of module	Power-cut replacement of 1000kV AC transmission line overhead ground wire
Description	Learn to master the standardized operation flow of power-cut replacement of 1000kV AC transmission line overhead ground wire, train the special skills of trainees in adopting power-cut operation method to replace overhead ground wire in group, and improve the operation capability of maintenance personnel in power-cut replacement project of 1000kV AC transmission line overhead ground wire

(Ⅲ) Teaching Design

The training and teaching implementation process of each module in this standard is divided into 6 training stages: theoretical teaching, preparations, work site preparation, trainer demonstration, group training of trainees and end of work. Each training stage is designed according to the standardized operation process to complete the work task, including training objectives, training contents, training hours, training methods (training resources), training environment, assessment and evaluation, etc.

V. Assessment Standards

The assessment module in this standard corresponds to the training module. After completing the module training, the assessment can be implemented according to the corresponding detailed rules for assessment and scoring. The detailed rules for assessment and scoring mainly set the contents and scope, including examinee information, assessor information, assessment time and duration, assessment module, assessee, assessment mode, assessment time limit, task description, work specification and requirements, preparation of assessment scenario, etc. According to the operation standardization process of each module, the preparation stage, work implementation stage and work end stage are worked out, the quality requirements, assessment score and deduction standard of each specific process section are formulated, and the deduction reason column, deduction column and scoring column filled out by the assessor are provided. The assessment results of each module shall be the sum of the evaluation scores of each process section obtained to complete the whole process of the work tasks set out in the detailed rules for assessment and scoring. In the assessment module, the group or the individual can be assessed and scored. If it is necessary to assess several roles involved in the operation, the role turn method can be used to complete the assessment and scoring of the examinees. The assessment module set up in this standard are shown in Table 1-19.

Table 1-19 Assessment Module Setting

S/N	Designation of module	Assessment method	Assessment time limit/min	Number of assessment roles
1	Live replacement of Type I composite insulator for 1000kV AC transmission line tangent tower	Operation	120	3
2	Live replacement of single-V composite insulator for 1000kV AC transmission line tangent tower	Operation	120	3
3	The live replacement of 1 ~ 3 glass insulators on the cross arm side of 1000kV AC transmission line resisting-tensile tower	Operation	90	2
4	The live replacement of 1 ~ 3 glass insulators on the conductor side of 1000kV AC transmission line resisting-tensile tower	Operation	90	2

Table (Cont'd)

S/N	Designation of module	Assessment method	Assessment time limit/min	Number of assessment roles
5	Live replacement of arbitrary single insulator of tensile glass insulator string of 1000kV AC transmission line	Operation	90	2
6	Live replacement of arbitrary section insulator of tensile glass insulator string of 1000kV AC transmission line	Operation	90	3
7	Live replacement of 1000kV AC transmission line conductor spacer	Operation	90	3
8	Live replacement of 1000kV AC transmission line conductor vibration damper	Operation	90	3
9	Live repair of 1000kV AC transmission line conductor	Operation	60	3
10	Live treatment of heating defects of 1000kV AC transmission line conductor drainage plate	Operation	60	3
11	Live replacement of ground wire connecting fitting for 1000kV AC transmission line tangent tower	Operation	60	2
12	Power-cut replacement of Type I composite insulator for 1000kV AC transmission line tangent tower	Operation	150	3
13	Power-cut replacement of single-V composite insulator for 1000kV AC transmission line tangent tower	Operation	150	3
14	Power-cut replacement of the whole strain insulator string on 1000kV AC transmission line	Operation	360	4
15	Power-cut repair of 1000kV AC transmission line overhead ground wire	Operation	100	3
16	Power-cut replacement of 1000kV AC transmission line sub-conductor	Operation	360	4
17	Power-cut replacement of 1000kV AC transmission line overhead ground wire	Operation	360	4

VI. Conditions of Guarantee

(I) Teacher Allocation

The teacher allocation and requirements of each training module in this standard are shown in Table 1-20.

Part I Standard for Professional Training and Assessment for Operation Maintenance of UHV AC Transmission Line

Table 1-20 Teacher Allocation and Requirements for Training Module

Designation of module	Lecturer		Skill operation demonstration trainer		Ground auxiliary personnel	
	Qty.	Requirements	Qty.	Requirements	Qty.	Requirements
Live replacement of Type I composite insulator for 1000kV AC transmission line tangent tower	1	Clarify the standardized operation flow of live replacement of Type I composite insulator for 1,000kV AC transmission line tangent tower, be familiar with the development trend of UHV AC transmission line, have solid high voltage technology and theoretical knowledge of live working, be familiar with key technical points and hazard precontrol in operation, basket method entering and exiting the intense electric field and use method for potential transfer rod, can fully explain the action essentials, reasons and methods of each operation step, can flexibly use appropriate teaching methods, and have good teaching organization ability.	5	Be familiar with the working content and standardized operation flow of live replacement of Type I composite insulator for 1,000kV AC transmission line tangent tower, participate in the live replacement of Type I composite insulator for 1,000kV AC transmission line tangent tower many times, and have rich working experience and good operation skills.The role of the Responsible Person can correctly organize the whole operation, determine whether the work site conditions and safety measures of the work order are complete and supplemented, make clear all kinds of safety measures and hazard precontrol, organize and implement the safety measures listed in the work order, and effectively supervise the operators to implement the on-site safety measures in accordance with the rules and regulations; the role of electrician on tower shall be familiar with the work content and work flow, master the safety measures, clarify the scope of work, hazard and precontrol measures, and be able to use tools and protective equipment skillfully.	5	Make clear the working content and standardized operation flow of live replacement of Type I composite insulator for 1, 000kV AC transmission line tangent tower, have corresponding electrician operation skills, can timely understand the orders issued by the role of the Responsible Person, correctly operate the related tools, correctly lift and transfer the tools and insulators, etc.

· 275 ·

Table (Cont'd)

Designation of module	Lecturer		Skill operation demonstration trainer		Ground auxiliary personnel	
	Qty.	Requirements	Qty.	Requirements	Qty.	Requirements
Live replacement of single-V composite insulator for 1000kV AC transmission line tangent tower	1	Clarify the standardized operation flow of live replacement of Type I composite insulator for 1,000kV AC transmission line tangent tower, be familiar with the development trend of UHV AC transmission line, have solid high voltage technology and theoretical knowledge of live working, be familiar with key technical points and hazard precontrol in operation, basket method entering and exiting the intense electric field and use method for potential transfer rod, can fully explain the action essentials, reasons and methods of each operation step, can flexibly use appropriate teaching methods, and have good teaching organization ability.	5	Be familiar with the working content and standardized operation flow of live replacement of single-V composite insulator for 1,000kV AC transmission line tangent tower, participate in the live replacement of single-V composite insulator for 1,000kV AC transmission line tangent tower many times, and have rich working experience and good operation skills. The role of the Responsible Person can correctly organize the whole operation, determine whether the work site conditions and safety measures of the work order are complete and supplemented, make clear all kinds of safety measures and hazard precontrol, organize and implement the safety measures listed in the work order, and effectively supervise the operators to implement the on-site safety measures in accordance with the rules and regulations; the role of electrician on tower shall be familiar with the work content and work flow, master the safety measures, clarify the scope of work, hazard and precontrol measures, and be able to use tools and protective equipment skillfully.	6	Make clear the working content and standardized operation flow of live replacement of single-V composite insulator for 1,000kV AC transmission line tangent tower, have corresponding electrician operation skills, can timely understand the orders issued by the Responsible Person, correctly operate the related tools, correctly lift and transfer the tools and insulators, etc.

Part I Standard for Professional Training and Assessment for Operation Maintenance of UHV AC Transmission Line

Table (Cont'd)

Designation of module	Lecturer		Skill operation demonstration trainer		Ground auxiliary personnel	
	Qty.	Requirements	Qty.	Requirements	Qty.	Requirements
The live replacement of 1~3 glass insulators on the cross arm side of 1000kV AC transmission line resisting-tensile tower	1	Clarify the standardized operation flow of live replacement of 1~3 glass insulators on the cross arm side of 1,000kV AC transmission line resisting-tensile tower, be familiar with the development trend of UHV AC transmission line, have solid high voltage technology and theoretical knowledge of live working, be familiar with key technical points and hazard precontrol in operation, and replacement methods and quality standards for single insulator, can fully explain the action essentials, reasons and methods of each operation step, can flexibly use appropriate teaching methods, and have good teaching organization ability.	2	Be familiar with the working content and standardized operation flow of live replacement of 1~3 glass insulators on the cross arm side of 1,000kV AC transmission line resisting-tensile tower, participate in the live replacement of 1~3 glass insulators on the cross arm side of 1,000kV AC transmission line resisting-tensile tower many times, and have rich working experience and good operation skills. The role of the Responsible Person can correctly organize the whole operation, determine whether the work site conditions and safety measures of the work order are complete and supplemented, make clear all kinds of safety measures and hazard precontrol, organize and implement the safety measures listed in the work order, and effectively supervise the operators to implement the on-site safety measures in accordance with the rules and regulations; the role of electrician on tower shall be familiar with the work content and work flow, master the safety measures, clarify the scope of work, hazard and precontrol measures, and be able to use tools and protective equipment skillfully.	3	Make clear the working content and standardized operation flow of live replacement of 1~3 glass insulators on the cross arm side of 1,000kV AC transmission line resisting-tensile tower, have corresponding electrician operation skills, can timely understand the orders issued by the role of the Responsible Person, correctly operate the related tools, and correctly transfer the tools and insulators, etc.

Standard for Professional Training and Assessment for Operation Maintenance of UHV AC Transmission Line

Table (Cont'd)

Designation of module	Lecturer		Skill operation demonstration trainer		Ground auxiliary personnel	
	Qty.	Requirements	Qty.	Requirements	Qty.	Requirements
The live replacement of 1~3 glass insulators on the conductor side of 1000kV AC transmission line resisting-tensile tower	1	Clarify the standardized operation flow of live replacement of 1~3 glass insulators on the conductor side of 1,000kV AC transmission line resisting-tensile tower, be familiar with the development trend of UHV AC transmission line, have solid high voltage technology and theoretical knowledge of live working, be familiar with key technical points and hazard precontrol in operation, principle of operation mode in entering and exiting electric field along insulator and use method for potential transfer rod and replacement methods and quality standards for single insulator, can fully explain the action essentials, reasons and methods of each operation step, can flexibly use appropriate teaching methods, and have good teaching organization ability.	2	Be familiar with the working content and standardized operation flow of live replacement of 1~3 glass insulators on the conductor side of 1,000kV AC transmission line resisting-tensile tower, participate in the live replacement of 1~3 glass insulators on the conductor side of 1,000kV AC transmission line resisting-tensile tower many times, and have rich working experience and good operation skills. The role of the Responsible Person can correctly organize the whole operation, determine whether the work site conditions and safety measures of the work order are complete and supplemented, make clear all kinds of safety measures and hazard precontrol, organize and implement the safety measures listed in the work order, and effectively supervise the operators to implement the on-site safety measures in accordance with the rules and regulations; the role of electrician on tower shall be familiar with the work content and work flow, master the safety measures, clarify the scope of work, hazard and precontrol measures, and be able to use tools and protective equipment skillfully.	3	Make clear the working content and standardized operation flow of live replacement of 1~3 glass insulators on the conductor side of 1,000kV AC transmission line resisting-tensile tower, have corresponding electrician operation skills, can timely understand the orders issued by the role of the Responsible Person, correctly operate the related tools, and correctly transfer the tools and insulators, etc.

Part I Standard for Professional Training and Assessment for Operation Maintenance of UHV AC Transmission Line

Table (Cont'd)

Designation of module	Lecturer		Skill operation demonstration trainer		Ground auxiliary personnel	
	Qty.	Requirements	Qty.	Requirements	Qty.	Requirements
Live replacement of arbitrary single insulator of tensile glass insulator string of 1000kV AC transmission line	1	Clarify the standardized operation flow of live replacement of arbitrary single insulator of tensile glass insulator string of 1,000kV AC transmission line, be familiar with the development trend of UHV AC transmission line, have solid high voltage technology and theoretical knowledge of live working, be familiar with key technical points and hazard precontrol in operation, principle of operation mode in entering and exiting electric field along insulator and replacement methods and quality standards for single insulator, can fully explain the action essentials, reasons and methods of each operation step, can flexibly use appropriate teaching methods, and have good teaching organization ability.	2	Be familiar with the working content and standardized operation flow of live replacement of arbitrary single insulator of tensile glass insulator string of 1,000kV AC transmission line, participate in the live replacement of arbitrary single insulator of tensile glass insulator string of 1,000kV AC transmission line many times, and have rich working experience and good operation skills. The role of the Responsible Person can correctly organize the whole operation, determine whether the work site conditions and safety measures of the work order are complete and supplemented, make clear all kinds of safety measures and hazard precontrol, organize and implement the safety measures listed in the work order, and effectively supervise the operators to implement the on-site safety measures in accordance with the rules and regulations; the role of electrician on tower shall be familiar with the work content and work flow, master the safety measures, clarify the scope of work, hazard and precontrol measures, and be able to use tools and protective equipment skillfully.	3	Make clear the working content and standardized operation flow of live replacement of arbitrary single insulator of tensile glass insulator string of 1,000kV AC transmission line, have corresponding electrician operation skills, can timely understand the orders issued by the role of the Responsible Person, correctly operate the related tools, and correctly transfer the tools and insulators, etc.

Standard for Professional Training and Assessment for Operation Maintenance of UHV AC Transmission Line

Table (Cont'd)

Designation of module	Lecturer		Skill operation demonstration trainer		Ground auxiliary personnel	
	Qty.	Requirements	Qty.	Requirements	Qty.	Requirements
Live replacement of arbitrary section insulator of tensile glass insulator string of 1000kV AC transmission line	1	Clarify the standardized operation flow of live replacement of arbitrary section insulator of tensile glass insulator string of 1,000kV AC transmission line, be familiar with the development trend of UHV AC transmission line, have solid high voltage technology and theoretical knowledge of live working, be familiar with key technical points and hazard precontrol in operation and principle of operation mode in entering and exiting electric field along insulator, can fully explain the action essentials, reasons and methods of each operation step, can flexibly use appropriate teaching methods, and have good teaching organization ability.	5	Be familiar with the working content and standardized operation flow of live replacement of arbitrary section insulator of tensile glass insulator string of 1,000kV AC transmission line, have carried out the live replacement of arbitrary section insulator of tensile glass insulator string of 1,000kV AC transmission line, and have rich working experience and good operation skills. The role of the Responsible Person can correctly organize the whole operation, determine whether the work site conditions and safety measures of the work order are complete and supplemented, make clear all kinds of safety measures and hazard precontrol, organize and implement the safety measures listed in the work order, and effectively supervise the operators to implement the on-site safety measures in accordance with the rules and regulations; the role of electrician on tower shall be familiar with the work content and work flow, master the safety measures, clarify the scope of work, hazard and precontrol measures, and be able to use tools and protective equipment skillfully.	6	Make clear the working content and standardized operation flow of live replacement of arbitrary section insulator of tensile glass insulator string of 1,000kV AC transmission line, have corresponding electrician operation skills, can timely understand the orders issued by the role of the Responsible Person, correctly operate the related tools, and correctly transfer the tools and insulators, etc.

Part I Standard for Professional Training and Assessment for Operation Maintenance of UHV AC Transmission Line

Table (Cont'd)

Designation of module	Lecturer		Skill operation demonstration trainer		Ground auxiliary personnel	
	Qty.	Requirements	Qty.	Requirements	Qty.	Requirements
Live replacement of 1000kV AC transmission line conductor spacer	1	Clarify the standardized operation flow of live replacement of 1,000kV AC transmission line conductor spacer, be familiar with the development trend of UHV AC transmission line, have solid high voltage technology and theoretical knowledge of live working, be familiar with key technical points and hazard precontrol in operation, principle of operation mode in entering and exiting electric field along insulator and use method for potential transfer rod, can fully explain the action essentials, reasons and methods of each operation step, can flexibly use appropriate teaching methods, and have good teaching organization ability.	3	Be familiar with the working content and standardized operation flow of live replacement of 1,000kV AC transmission line conductor spacer, participate in the live replacement of 1,000kV AC transmission line conductor spacer many times, and have rich working experience and good operation skills. The role of the Responsible Person can correctly organize the whole operation, determine whether the work site conditions and safety measures of the work order are complete and supplemented, make clear all kinds of safety measures and hazard precontrol, organize and implement the safety measures listed in the work order, and effectively supervise the operators to implement the on-site safety measures in accordance with the rules and regulations; the role of electrician on tower shall be familiar with the work content and work flow, master the safety measures, clarify the scope of work, hazard and precontrol measures, and be able to use tools and protective equipment skillfully.	2	Make clear the working content and standardized operation flow of live replacement of 1,000kV AC transmission line conductor spacer, have corresponding electrician operation skills, can timely understand the orders issued by the role of the Responsible Person, correctly operate the related tools, and correctly transfer the tools and spacers, etc.

Table (Cont'd)

Designation of module	Lecturer		Skill operation demonstration trainer		Ground auxiliary personnel	
	Qty.	Requirements	Qty.	Requirements	Qty.	Requirements
Live replacement of 1000kV AC transmission line conductor vibration damper	1	Clarify the standardized operation flow of live replacement of 1,000kV AC transmission line conductor vibration damper, have solid high voltage technology and theoretical knowledge of live working, be familiar with key technical points and hazard precontrol in operation, principle of operation mode in entering and exiting electric field along insulator and use method for potential transfer rod, can fully explain the action essentials, reasons and methods of each operation step, can flexibly use appropriate teaching methods, and have good teaching organization ability.	3	Be familiar with the working content and standardized operation flow of live replacement of 1,000kV AC transmission line conductor vibration damper, participate in the live replacement of 1,000kV AC transmission line conductor vibration damper many times, and have rich working experience and good operation skills. The role of the Responsible Person can correctly organize the whole operation, determine whether the work site conditions and safety measures of the work order are complete and supplemented, make clear all kinds of safety measures and hazard precontrol, organize and implement the safety measures listed in the work order, and effectively supervise the operators to implement the on-site safety measures in accordance with the rules and regulations; the role of electrician on tower shall be familiar with the work content and work flow, master the safety measures, clarify the scope of work, hazard and precontrol measures, and be able to use tools and protective equipment skillfully.	2	Make clear the working content and standardized operation flow of live replacement of 1,000kV AC transmission line conductor vibration damper, have corresponding electrician operation skills, can timely understand the orders issued by the role of the Responsible Person, correctly operate the related tools, and correctly transfer the tools and vibration dampers, etc.

Part I Standard for Professional Training and Assessment for Operation Maintenance of UHV AC Transmission Line

Table (Cont'd)

Designation of module	Lecturer		Skill operation demonstration trainer		Ground auxiliary personnel	
	Qty.	Requirements	Qty.	Requirements	Qty.	Requirements
Live repair of 1000kV AC transmission line conductor	1	Clarify the standardized operation flow of live repair of 1,000kV AC transmission line conductor, be familiar with the development trend of UHV AC transmission line, have solid high voltage technology and theoretical knowledge of live working, be familiar with key technical points and hazard precontrol in operation principle of operation mode in entering and exiting electric field along insulator, use method for potential transfer rod and methods and quality standards for conductor repair, can fully explain the action essentials, reasons and methods of each operation step, can flexibly use appropriate teaching methods, and have good teaching organization ability.	3	Be familiar with the working content and standardized operation flow of live repair of 1,000kV AC transmission line conductor, participate in the live repair of 1,000kV AC transmission line conductor many times, and have rich working experience and good operation skills. The role of the Responsible Person can correctly organize the whole operation, determine whether the work site conditions and safety measures of the work order are complete and supplemented, make clear all kinds of safety measures and hazard precontrol, organize and implement the safety measures listed in the work order, and effectively supervise the operators to implement the on-site safety measures in accordance with the rules and regulations; the role of electrician on tower shall be familiar with the work content and work flow, master the safety measures, clarify the scope of work, hazard and precontrol measures, and be able to use tools and protective equipment skillfully.	2	Make clear the working content and standardized operation flow of live repair of 1,000kV AC transmission line conductor, have corresponding electrician operation skills, can timely understand the orders issued by the Responsible Person, correctly operate the related tools, and correctly transfer the tools and preformed armor rods, etc.

Standard for Professional Training and Assessment for Operation Maintenance of UHV AC Transmission Line

Table (Cont'd)

Designation of module	Lecturer		Skill operation demonstration trainer		Ground auxiliary personnel	
	Qty.	Requirements	Qty.	Requirements	Qty.	Requirements
Live treatment of heating defects of 1000kV AC transmission line conductor drainage plate	1	Clarify the standardized operation flow of live treatment of heating defects of 1000kV AC transmission line conductor drainage plate, have solid high voltage technology and theoretical knowledge of live working, be familiar with key technical points and hazard precontrol in operation, principle of operation mode in entering and exiting electric field along insulator and use method for potential transfer rod, can fully explain the action essentials, reasons and methods of each operation step, can flexibly use appropriate teaching methods, and have good teaching organization ability.	3	Be familiar with the working content and standardized operation flow of live treatment of heating defects of 1000kV AC transmission line conductor drainage plate, participate in the live treatment of heating defects of 1000kV AC transmission line conductor drainage plate many times, and have rich working experience and good operation skills. The role of the Responsible Person can correctly organize the whole operation, determine whether the work site conditions and safety measures of the work order are complete and supplemented, make clear all kinds of safety measures and hazard precontrol, organize and implement the safety measures listed in the work order, and effectively supervise the operators to implement the on-site safety measures in accordance with the rules and regulations; the role of electrician on tower shall be familiar with the work content and work flow, master the safety measures, clarify the scope of work, hazard and precontrol measures, and be able to use tools and protective equipment skillfully.	2	Make clear the working content and standardized operation flow of live treatment of heating defects of 1000kV AC transmission line conductor drainage plate, have corresponding electrician operation skills, can timely understand the orders issued by the role of the Responsible Person, correctly operate the related tools, and correctly transfer the tools, etc.

Part I Standard for Professional Training and Assessment for Operation Maintenance of UHV AC Transmission Line

Table (Cont'd)

Designation of module	Lecturer		Skill operation demonstration trainer		Ground auxiliary personnel	
	Qty.	Requirements	Qty.	Requirements	Qty.	Requirements
Live replacement of ground wire connecting fitting for 1000kV AC transmission line tangent tower	1	Clarify the standardized operation flow of live replacement of ground wire connecting fitting for 1000kV AC transmission line tangent tower, be familiar with the development trend of UHV AC rarnsmission line, have solid high voltage technology and theoretical knowledge of live working, be familiar with the basic principle of ground potential operation and equivalent circuit diagram, the use method for ground wire lifter, and the replacement methods and quality standards for ground wire fitting, can fully explain the action essentials, reasons and methods of each operation step, can flexibly use appropriate teaching methods, and have good teaching organization ability.	2	Be familiar with the working content and standardized operation flow of live replacement of ground wire connecting fitting for 1000kV AC transmission line tangent tower, have carried out the live replacement of ground wire connecting fitting for 1000kV AC transmission line tangent tower, and have rich working experience and good operation skills.The role of the Responsible Person can correctly organize the whole operation, determine whether the work site conditions and safety measures of the work order are complete and supplemented, make clear all kinds of safety measures and hazard precontrol, organize and implement the safety measures listed in the work order, and effectively supervise the operators to implement the on-site safety measures in accordance with the rules and regulations; the role of electrician on tower shall be familiar with the work content and work flow, master the safety measures, clarify the scope of work, hazard and pre-control measures, and be able to use tools and protective equipment skillfully.	2	Make clear the working content and standardized operation flow of live replacement of ground wire connecting fitting for 1000kV AC transmission line tangent tower, have corresponding electrician operation skills, can timely understand the orders issued by the role of the Responsible Person, correctly operate the related tools, and correctly transfer the tools, etc.

Standard for Professional Training and Assessment for Operation Maintenance of UHV AC Transmission Line

Table (Cont'd)

Designation of module	Lecturer		Skill operation demonstration trainer		Ground auxiliary personnel	
	Qty.	Requirements	Qty.	Requirements	Qty.	Requirements
Power-cut replacement of Type I composite insulator for 1000kV AC transmission line tangent tower	1	Clarify the standardized operation flow of power-cut replacement of Type I composite insulator for 1000kV AC transmission line tangent tower, be familiar with the development trend of UHV AC transmission line, have knowledge of UHV AC transmission line structure and mechanics, be familiar with the process of interruption maintenance and the use of tools, and the replacement methods and quality standards for Type I composite insulator for tangent tower, can fully explain the action essentials, reasons and methods of each operation step, can flexibly use appropriate teaching methods, and have good teaching organization ability.	4	Be familiar with the working content and standardized operation flow of power-cut replacement of Type I composite insulator for 1000kV AC transmission line tangent tower, participate in the power-cut replacement of Type I composite insulator for 1000kV AC transmission line tangent tower many times, and have rich working experience and good operation skills. The role of the Responsible Person can correctly organize the whole operation, determine whether the work site conditions and safety measures of the work order are complete and supplemented, make clear all kinds of safety measures and hazard precontrol, organize and implement the safety measures listed in the work order, and effectively supervise the operators to implement the on-site safety measures in accordance with the rules and regulations; the role of electrician on tower shall be familiar with the work content and work flow, master the safety measures, clarify the scope of work, hazard and precontrol measures, and be able to use various tools skillfully.	5	Make clear the working content and standardized operation flow of power-cut replacement of Type I composite insulator for 1000kV AC transmission line tangent tower, have corresponding electrician operation skills, can timely understand the orders issued by the role of the Responsible Person, correctly operate the related tools, correctly transfer the tools and insulators, etc.

Part I Standard for Professional Training and Assessment for Operation Maintenance of UHV AC Transmission Line

Table (Cont'd)

Designation of module	Lecturer		Skill operation demonstration trainer		Ground auxiliary personnel	
	Qty.	Requirements	Qty.	Requirements	Qty.	Requirements
Power-cut replacement of single-V composite insulator for 1000kV AC transmission line tangent tower	1	Clarify the operation flow of power-cut replacement of single-V composite insulator for 1000kV AC transmission line tangent tower, be familiar with the development trend of UHV AC transmission line, have knowledge of UHV AC transmission line structure and mechanics, be familiar with the process of interruption maintenance and the use of tools, and the replacement methods and quality standards for single-V composite insulator for tangent tower, can fully explain the action essentials, reasors and methods of each operation step, can flexibly use appropriate teaching methods, and have good teaching organization ability.	4	Be familiar with the working content and standardized operation flow of power-cut replacement of single-V composite insulator for 1000kV AC transmission line tangent tower, participate in the power-cut replacement of single-V composite insulator for 1000kV AC transmission line tangent tower many times, and have rich working experience and good operation skills. The role of the Responsible Person can correctly organize the whole operation, determine whether the work site conditions and safety measures of the work order are complete and supplemented, make clear all kinds of safety measures and hazard precontrol, organize and implement the safety measures listed in the work order, and effectively supervise the operators to implement the on-site safety measures in accordance with the rules and regulations; the role of electrician on tower shall be familiar with the work content and work flow, master the safety measures, clarify the scope of work, hazard and precontrol measures, and be able to use various tools skillfully.	5	Make clear the working content and standardized operation flow of power-cut replacement of single-V composite insulator for 1000kV AC transmission line tangent tower, have corresponding electrician operation skills, can timely understand the orders issued by the Responsible Person, correctly operate the related tools, correctly transfer the tools and insulators, etc.

Table (Cont'd)

Designation of module	Lecturer		Skill operation demonstration trainer		Ground auxiliary personnel	
	Qty.	Requirements	Qty.	Requirements	Qty.	Requirements
Power-cut replacement of the whole strain insulator string on 1000kV AC transmission line	1	Clarify the operation flow of power-cut replacement of whole strain insulator string on 1000kV AC transmission line, be familiar with the development trend of UHV AC transmission line, have knowledge of UHV AC transmission line structure and mechanics, be familiar with the installation and use of pulley block, and the process of interruption maintenance and the use of tools, can fully explain the action essentials, reasons and methods of each operation step, can flexibly use appropriate teaching methods, and have good teaching organization ability.	7	Be familiar with the working content and standardized operation flow of power-cut replacement of whole strain insulator string on 1000kV AC transmission line, have carried out the power-cut replacement of whole strain insulator string on 1000kV AC transmission line, and have rich working experience and good operation skills. The role of the Responsible Person can correctly organize the whole operation, determine whether the work site conditions and safety measures of the work order are complete and supplemented, make clear all kinds of safety measures and hazard precontrol, organize and implement the safety measures listed in the work order, and effectively supervise the operators to implement the on-site safety measures in accordance with the rules and regulations; the role of electrician on tower shall be familiar with the work content and work flow, master the safety measures, clarify the scope of work, hazard and precontrol measures, and be able to use various tools skillfully.	9	Make clear the working content and standardized operation flow of power-cut replacement of whole strain insulator string on 1000kV AC transmission line, have corresponding electrician operation skills, can timely understand the orders issued by the role of the Responsible Person, correctly operate the related tools, correctly control and operate the winching, and correctly transfer the tools and insulators, etc.

Part I Standard for Professional Training and Assessment for Operation Maintenance of UHV AC Transmission Line

Table (Cont'd)

Designation of module	Lecturer		Skill operation demonstration trainer		Ground auxiliary personnel	
	Qty.	Requirements	Qty.	Requirements	Qty.	Requirements
Power-cut repair of 1000kV AC transmission line overhead ground wire	1	Clarify the operation flow of power-cut repair of 1000kV AC transmission line overhead ground wire, be familiar with the development trend of UHV AC transmission line, have knowledge of UHV AC transmission line structure and mechanics, be familiar with the hanging wheel structure and the use, and the repair methods and quality standards for overhead ground wire, can fully explain the action essentials, reasons and methods of each operation step, can flexibly use appropriate teaching methods, and have good teaching organization ability.	3	Be familiar with the working content and standardized operation flow of power-cut repair of 1000kV AC transmission line overhead ground wire, participate in the power-cut repair of 1000kV AC transmission line overhead ground wire many times, and have rich working experience and good operation skills. The role of the Responsible Person can correctly organize the whole operation, determine whether the work site conditions and safety measures of the work order are complete and supplemented, make clear all kinds of safety measures and hazard precontrol, organize and implement the safety measures listed in the work order, and effectively supervise the operators to implement the on-site safety measures in accordance with the rules and regulations; the role of electrician on tower shall be familiar with the work content and work flow, master the safety measures, clarify the scope of work, hazard and precontrol measures, and be able to use various tools skillfully.	2	Make clear the working content and standardized operation flow of power-cut repair of 1000kV AC transmission line overhead ground wire, have corresponding electrician operation skills, can timely understand the orders issued by the Responsible Person, correctly operate the related tools, and correctly transfer the tools and preformed armor rods, etc.

Table (Cont'd)

Designation of module	Lecturer		Skill operation demonstration trainer		Ground auxiliary personnel	
	Qty.	Requirements	Qty.	Requirements	Qty.	Requirements
Power-cut replacement of 1000kV AC transmission line sub-conductor	1	Clarify the operation flow of power-cut replacement of 1000kV AC transmission line sub-conductor, be familiar with the development trend of UHV AC transmission line, have knowledge of UHV AC transmission line structure and mechanics, be familiar with the work site layout of sub-conductor replacement operation, can fully explain the action essentials, reasons and methods of each operation step, can flexibly use appropriate teaching methods, and have good teaching organization ability.	8	Be familiar with the working content and standardized operation flow of power-cut replacement of 1000kV AC transmission line sub-conductor, have carried out the power-cut replacement of 1000kV AC transmission line sub-conductor, and have rich working experience and good operation skills. The role of the Responsible Person can correctly organize the whole operation, be familiar with the work site layout, determine whether the work site conditions and safety measures of the work order are complete and supplemented, make clear all kinds of safety measures and hazard precontrol, organize and implement the safety measures listed in the work order, and effectively supervise the operators to implement the on-site safety measures in accordance with the rules and regulations; the role of electrician on tower shall be familiar with the work content and work flow, master the safety measures, clarify the scope of work, hazard and precontrol measures, and be able to use various tools skillfully.	14	Make clear the working content and standardized operation flow of power-cut replacement of 1000kV AC transmission line sub-conductor, have corresponding electrician operation skills, can timely understand the orders issued by the Responsible Person, correctly operate the related tools, correctly control and operate the winching, and correctly transfer the tools, etc.

Part I Standard for Professional Training and Assessment for Operation Maintenance of UHV AC Transmission Line

Table (Cont'd)

Designation of module	Lecturer		Skill operation demonstration trainer		Ground auxiliary personnel	
	Qty.	Requirements	Qty.	Requirements	Qty.	Requirements
Power-cut replacement of 1000kV AC transmission line overhead ground wire	1	Clarify the operation flow of power-cut replacement of 1000kV AC transmission line overhead ground wire, be familiar with the development trend of UHV AC transmission line, have knowledge of UHV AC transmission line structure and mechanics, be familiar with the work site layout of overhead ground wire replacement operation, can fully explain the action essentials, reasons and methods of each operation step, can flexibly use appropriate teaching methods, and have good teaching organization ability.	8	Be familiar with the working content and standardized operation flow of power-cut replacement of 1000kV AC transmission line overhead ground wire, and have rich working experience and good operation skills. The role of the Responsible Person can correctly organize the whole operation, be familiar with the work site layout, determine whether the work site conditions and safety measures of the work order are complete and supplemented, make clear all kinds of safety measures and hazard precontrol, organize and implement the safety measures listed in the work order, and effectively supervise the operators to implement the on-site safety measures in accordance with the rules and regulations; the role of electrician on tower shall be familiar with the work content and work flow, master the safety measures, clarify the scope of work, hazard and pre-control measures, and be able to use various tools skillfully.	14	Make clear the working content and standardized operation flow of power-cut replacement of 1000kV AC transmission line overhead ground wire, have corresponding electrician operation skills, can timely understand the orders issued by the Responsible Person, correctly operate the related tools, correctly control and operate the winching, and correctly transfer the tools, etc.

(Ⅱ) Practical Training Conditions

The training conditions of each training module in this standard are shown in Table 1-21.

Table 1-21 Training Conditions of Training Module

Designation of module	Facilities and equipment	Tools and materials
Live replacement of Type I composite insulator for 1000kV AC transmission line tangent tower	1000kV AC transmission line tangent tower, Type I composite insulator	Table 2-1-3
Live replacement of single-V composite insulator for 1000kV AC transmission line tangent tower	1000kV AC transmission line tangent tower, single-V composite insulator	Table 2-2-3
The live replacement of 1~3 glass insulators on the cross arm side of 1000kV AC transmission line resisting-tensile tower	1000kV AC transmission line resisting-tensile tower, glass insulator	Table 2-3-4
The live replacement of 1~3 glass insulators on the conductor side of 1000kV AC transmission line resisting-tensile tower	1000kV AC transmission line resisting-tensile tower, glass insulator	Table 2-4-4
Live replacement of arbitrary single insulator of tensile glass insulator string of 1000kV AC transmission line	1000kV AC transmission line resisting-tensile tower, glass insulator	Table 2-5-4
Live replacement of arbitrary section insulator of tensile glass insulator string of 1000kV AC transmission line	1000kV AC transmission line resisting-tensile tower, glass insulator	Table 2-6-3
Live replacement of 1000kV AC transmission line conductor spacer	1000kV AC transmission line resisting-tensile tower, eight-bundle conductor	Table 2-7-3
Live replacement of 1000kV AC transmission line conductor vibration damper	1000kV AC transmission line resisting-tensile tower	Table 2-8-3
Live repair of 1000kV AC transmission line conductor	1000kV AC transmission line resisting-tensile tower	Table 2-9-3
Live treatment of heating defects of 1000kV AC transmission line conductor drainage plate	1000kV AC transmission line resisting-tensile tower	Table 2-10-3
Live replacement of ground wire connecting fitting for 1000kV AC transmission line tangent tower	1000kV AC transmission line tangent tower, type of overhead ground wire: JLB20A-150	Table 2-11-2
Power-cut replacement of Type I composite insulator for 1000kV AC transmission line tangent tower	1000kV AC transmission line tangent tower, Type I composite insulator	Table 2-12-2

Table (Cont'd)

Designation of module	Facilities and equipment	Tools and materials
Power-cut replacement of single-V composite insulator for 1000kV AC transmission line tangent tower	1000kV AC transmission line tangent tower, single-V composite insulator	Table 2-13-2
Power-cut replacement of the whole strain insulator string on 1000kV AC transmission line	1000kV AC transmission line resisting-tensile tower	Table 2-14-2
Power-cut repair of 1000kV AC transmission line overhead ground wire	1000kV AC transmission line 1, type of overhead ground wire:JLB20A-150	Table 2-15-2
Power-cut replacement of 1000kV AC transmission line sub-conductor	1000kV AC transmission line tensile section 1, type of conductor:8×JL/G1A-630/45	Table 2-16-2
Power-cut replacement of 1000kV AC transmission line overhead ground wire	1000kV AC transmission line tensile section 1, type of overhead ground wire:JLB20A-150	Table 2-17-2

VII. Description of Preparation

(I) Background of Preparation

State Grid Corporation of China began to start the research and construction of UHV power grid in 2004. By September 2016, 21 UHV projects have been put into operation, including 7 UHV AC projects and 14 UHV DC projects. At the same time, State Grid Corporation of China has further studied the operation law and safety mechanism of the large power grid, scientifically planned and built a reasonable grid structure, and focused on solving the problem of "strong DC and weak AC" of UHV, speeding up the formation of a strong UHV backbone grid frame and constantly strengthening the ability to resist serious faults. With the continuous construction and operation of UHV AC transmission line, the workload of line operation and inspection will be greatly increased, the skill level of UHV AC maintenance personnel will be improved, and the reserve of this kind of talents is imminent, but there is no professional training and assessment standards for UHV AC transmission line operation and inspection at present. For this reason, with the strong support of State Grid Sichuan Electric Power Corporation, the Skills Training Center of State Grid Sichuan Electric Power Corporation, State Grid Sichuan Electric Power Corporation, has prepared this standard, summed up 17 UHV AC transmission line maintenance modules, developed the training standards and detailed rules for assessment and scoring, including 11 live maintenance operations and 6 interruption maintenance operations.

(II) **Main Principles of Preparation**

The preparation idea of this standard is to improve the ability of UHV AC maintenance personnel to complete the common maintenance work projects, to follow the individual practical, self-selected system, training and evaluation integrated independent training and assessment framework, and to establish a diversified training and assessment model.

This standard is prepared in accordance with the following principles.

1. Focus on improving the special skill level of personnel, and follow the "sufficient knowledge and necessary skills".

2. Summarize typical work tasks, select representative projects, adopt modular training and assessment.

3. Realize the special promotion training and assessment of a single module, and can also select the required modules for systematic training and assessment.

4. Synchronize the assessment standard with the training standard.

5. Realize the assessment standard applicable to the performance evaluation of a role, single person and working group.

(III) **Standard Structure**

This standard is divided into two parts. The first part is the general description of the professional training and assessment standards for UHV AC transmission line operation and inspection, which is divided into 7 parts, including general principles, preparation basis, training object, training standard, assessment standard, guarantee condition and preparation description.

The second part is the training and assessment standards for each skill module, with a total of 17 modules, including module 1 to module 11 for live maintenance operation, module 12 to module 17 for interruption maintenance operation work, in no special order for difficulty. Each module is composed of two parts: training standard and assessment standard. The training standard is divided into 4 parts, including training requirements, referenced rules and specifications, teaching design for training and operation process; the assessment standard is the detailed rules for the assessment and scoring of the operation and inspection skills of UHV AC transmission line.

When selecting multiple modules to constitute the systematic training content, the duplicate theoretical teaching and training contents shall be removed.

Part II

Skill Module Training and Assessment Standards

Module 1 Standards for Training and Assessment on Live Replacement of Type I Composite Insulator for 1000kV AC Transmission Line Tangent Tower

I. Training Standard

(I) Training Requirements

Designation of module	Live replacement of Type I composite insulator for 1000kV AC transmission line tangent tower	Type of training	Operation
Training method	Practical operation training	Hours of training	21 training hours
Training objectives	1. Master the electrical significance of "basket method" operation mode when the tangent tower enters and exits 1000kV intense electric field. 2. Can enter 1000kV equipotential operation point by adopting the "basket method". 3. Can independently complete the replacement of Type I composite insulator for 1000kV AC transmission line tangent tower (equipotential operation method)		
Training venue	UHV AC training line		
Training content	With the cooperation of equipotential with ground potential, enter the equipotential for operation through the "basket method", and replace the Type I composite insulator of for tangent tower by the equipotential operation method		
Scope of application	Maintenance personnel of UHV AC transmission line		

(II) Referenced Rules and Specifications

(1) Electrotechnical Terminology (GB/T2900.55-2002).

(2) Insulated Tackles for Live Working (GB/T13034-2008).

(3) Live Working-Insulating Ropes (GB/T 13035-2008).

(4) Calculation Method of Live Working Minimum Approach Distance on AC Transmission Line (GB/T 18037-2000)

(5) Screen Clothes for Live Working on 1000kV AC (GB/T25726-2010).

(6) Code for Design of 1000kV Overhead Transmission Line (GB50665-2011).

Part II Skill Module Training and Assessment Standards

(7) Maintenance Code for 1000kV AC Transmission Line (DL/T209-2008).

(8) Operation Code for 1000kV AC Transmission Line (DL/T307-2010).

(9) Technical Guide for Live Working on 1000kV AC Transmission Line (DL/T392-2015).

(10) Guidelines of Insulation Coordination for Live Working (DL/T876-2004).

(11) Minimum Requirements for Utilization of Tools, Devices and Equipment for Live Working (DL/T 877-2004)

(12) Preventive Test Code of Tools, Devices and Equipment for Live Working (DL/T 976-2005).

(13) State Grid Corporation of China Working Regulations of Power Safety (Transmission Line Section) (Q/GDW1799.2-2013).

(Ⅲ) Teaching Design for Training

To complete the work task of "live replacement of Type I composite insulator for 1000kV AC transmission line tangent tower", each training stage shall be designed according to the standard operation procedure for work task completion. Each stage includes specific training objectives, training content, hours of training, training methods (training resources), training environment, assessment and evaluation, etc, as shown in the Table 2-1-1.

(Ⅳ) Operation Flow

1. Work Task

With the cooperation of equipotential with ground potential, enter the equipotential for operation by adopting the "basket method", and replace the Type I composite insulator of for tangent tower.

2. Requirements for Weather and Work Site

(1) The live replacement of Type I composite insulator for 1000kV AC transmission line tangent tower shall be carried out in good weather. In case of lightning (hearing thunder or seeing lightning), snow, hail, rain, fog and so on, live working is prohibited. When the wind force is greater than level 5, or the relative humidity of the air is greater than 80%, it is unsuitable for live working; when emergency live repair is required in bad weather, relevant personnel shall be organized to fully discuss and prepare necessary safety measures, which can be implemented after being approved by the unit.

(2) The operating personnel shall be in good spirits, and be familiar with the organiza-

Table 2-1-1 Training Content Design for Live Replacement of Type I Composite Insulator for 1000kV AC Transmission Line Tangent Tower

Training schedule	Training objectives	Training content	Hours of training	Training methods and resources	Preparation of training conditions	Assessment and evaluation
1. Theoretical teaching	1. Preliminarily master the basic method for entering and exiting 1000kV intense electric field in basket method. 2. Be familiar with the methods for potential transfer. 3. Be familiar with replacement method for Type I composite insulator for transmission line tangent tower	1. The electrical significance of operation mode of entering and exiting the intense electric field in basket method; 2. The use method for the potential transfer rod during entering and exiting the UHV intense electric field. 3. Replacement method and quality standard for Type I composite insulator for transmission line tangent tower	2	Training methods: Lecture. Training resources: PPT, relevant regulations and specifications	Multimedia classroom	Attendance, classroom questions and assignments
2. Preparations	Be able to complete the preparation before operation	1. Work site survey; 2. Preparation of the standardized operation card; 3. Filling of the work order; 4. Preparation of tools and materials for this operation	1	Training methods: 1. Work site survey and cleaning of tools and materials shall be practiced at site; 2. Preparation of operation card and the filling of work order shall adopt lecture method; Training resources: 1. 1000kV training line; 2. UHV tools warehouse; 3. Blank work order	1. UHV training transmission line 2. Multimedia classroom	

· 298 ·

Part II Skill Module Training and Assessment Standards

Table (Cont'd)

Training schedule	Training objectives	Training content	Hours of training	Training methods and resources	Preparation of training conditions	Assessment and evaluation
3. Work site preparation	Be able to complete the preparations of work site	1. Work site re-survey; 2. Job application; 3. Work site layout; 4. Pre-shift meeting; 5. Inspection of tools and materials	1	Training methods: demonstration and role play. Resources: 1000kV training line	1000kV training line	
4. Trainer's demonstration	The trainees can preliminarily understand the operation process of the task through inspecting and learning from each other's work	1. The arrangement and installation of the tools on tower; 2. The equipotential electrician enters and exits the intense electric field by adopting the basket method; 3. The equipotential electrician and the equipotential electrician cooperate with each other to complete the replacement of Type I composite insulator by using the load transfer device.	2	Training methods: Demonstration. Resources: 1000kV training line	1000kV training line	

Standard for Professional Training and Assessment for Operation Maintenance of UHV AC Transmission Line

Table (Cont'd)

Training schedule	Training objectives	Training content	Hours of training	Training methods and resources	Preparation of training conditions	Assessment and evaluation
5. Group training of trainees	1. Can complete the operation of entering and exiting 1000kV intense electric field; 2. Can complete the replacement of Type I composite insulator for 1000kV transmission line tangent tower	1. Trainees are grouped (11 persons per group) to train the skill of entering and exiting 1000kV intense electric field and replacing Type I composite insulator for tangent tower; 2. Trainers guide the operation of trainees and conduct safety supervision	14	Training methods: Role play. Resources: 1000kV training line	1000kV training line	Score the operation of trainees according to the detailed rules for skill assessment and scoring
6. End of the work	1. Trainees can further identify the shortcomings during the operation process for later improvement; 2. Train the trainees in the working style of safe and civilized production	1. Cleaning up the work site; 2. Report to dispatcher; 3. Comment and summarize the work task this time at post-shift meeting	1	Training method: Lecture and inductive method	1000kV training line	

tional and technical measures to ensure safety in work; they shall hold the live working qualification within the validity period.

(3) The Responsible Person shall organize relevant personnel to complete field investigation in advance, determine the operating methods, required working apparatus and necessary measures according to the results, and handle the work orders for live working.

(4) The work site shall be reasonably set up with fence and warning signs. Non-operating personnel is forbidden to enter.

(5) The line re-closing device shall be deactivated in the Project.

(6) Safety working distance and effective insulation length are shown in Table 2-1-2.

Table 2-1-2 Safe Distance for Live Replacement of Type I Composite Insulator for 1000kV AC Transmission Line Tangent Tower

Altitude/m	Minimum safe distance/m		Minimum combined clearance/m		Minimum effective insulation length of insulating tools/m	Minimum distance between exposed part of human body and electrified body when potential is transferred /m
	Middle phase	Side phase	Middle phase	Side phase		
H≤1000	6.8	6.0	6.9	6.7	6.8	0.5
1000<H≤2000	7.4	6.6	7.6	7.3	7.2	0.5

3. Preparations

3.1 Hazards and precontrol measures

(1) Hazard - precontrol measures for electric shock injury include the following points:

① Before work, Responsible Person should contact the control personnel on duty, deactivate the line re-closing, and perform the licensing procedures.

② The Responsible Person must disclose the safety measures, technical measures and work tasks. Before climbing the tower, ground potential operators on tower must carefully check the dual names of the line, the number of the pole and tower, and phase, and then the tower can be climbed after all have been confirmed correct.

③ If lines lose power suddenly during work, operators shall still deem it to be charged. The Responsible Person shall contact the control personnel as soon as possible, and no forced energization is allowed before the on-duty control personnel getting in touch with the Responsible Person.

④ Insulating tool and insulating ropes shall be free of damage, moisture, deformation, and failure. It is not allowed to use non-insulating ropes (such as cotton rope, manila rope, and steel wire rope).

⑤ The equipotential operator shall wear flame-retardant underwear and full set of screen clothes outside (including hat, dresses & trousers, gloves, socks and shoes). All parts shall be in excellent connection conditions, and the resistance of the whole suit of shielding clothes shall not exceed 20Ω.

⑥ Prior to potential transfer, equipotential operators shall obtain permission from the Responsible Person, and the minimum distance between the exposed part of the human body and the electrified body shall not be less than 0.5 m; equipotential operators must use potential transfer rods for potential transfer; the safe distance between the ground potential operator and the charged body is less than that specified in Table 2-1-2.

⑦ When transmitting large metal objects by using insulating ropes, ground potential operators shall not touch them before ground connection.

⑧ The Responsible Person and the Special Supervisor shall continuously monitor the operators and correct their nonstandard operation or actions in violation at any time. Special attention shall be paid to operators work at heights to ensure that there is enough safety distance (meeting the requirements in Table 2-1-2). It is forbidden to contact two non-connected electrified bodies or make contact with electrified body and grounding body at the same time.

(2) Hazard - precontrol measures for falling accident:

① Before climbing, operators work at heights must satisfy the requirements of this operation, such as physical condition, mental state and skill and quality.

② After climbing the tower to the operation point, the ground electrician shall fasten the safety rope, check and confirm that it is firm, arrange the track rope of the basket, and the human body backup protection rope of the equipotential electrician shall be reasonable and reliably anchored. The ground electrician shall reliably connects the insulated hoisting rope with the basket using a closed hook. The equipotential electrician shall fasten the insulated and main protection ropes, and enter the basket after passing the impact test on the basket. After the equipotential electrician fastens the safety belt, the relevant work can be carried out. The equipotential electrician shall fasten the protection rope at the operation point.

③ Supervisors shall correct the nonstandard or illegal actions at any time. Special attention shall be paid to operators to prevent them from losing the protection of safety belt or insulated backup protection rope during transposition, and it is forbidden to fasten the safety belt or insulated backup protection rope in a position lower than the operating personnel.

④ Personnel shall inspect the shackles and tower materials for fastening condition before climbing the tower, and shall grasp the main materials by hand but not grasp the shackles by hand when climbing the tower.

(3) Hazard - injury caused by objects falling from heights. Precontrol measures:

① Operator working at heights should put personal tools and fragmentary materials into the tools bag. It is strictly forbidden to hang objects in heights or keep in the mouth.

② Ground operator should correctly wear a helmet and use the knots. The vertical distance from the work site should not be less than the falling radius.

③ The work site shall be set up with fence and warning signs. It shall be noted at any time that Supervisor shall prohibit irrelevant personnel and vehicles from entering operation area.

3.2 Selection of tools and instruments and materials

Tools and materials required for live replacement of Type I composite insulator for 1000kV AC transmission line tangent tower can be seen in Table 2-1-3. Before delivering tools and instruments out of warehouse, application voltage class and test period shall be carefully checked and they shall be inspected to ensure that appearance is intact, connection is firm, rotation is flexible and meet the working task requirements. After delivering tools and instruments out of warehouse, they shall be stored in tools bag or tool kit for transportation to avoid contamination and damp. Metal tools and insulated tools shall be separately loaded and transported to avoid deformation, damage or other defects caused by mixed loading and transportation.

Table 2-1-3 Tools and Materials Required for Live Replacement of Type I Composite Insulator for 1000kV AC Transmission Line Tangent Tower

S/N	Name of tool and appliance	Specification	Unit	Qty.	Remarks
1	A whole suit of shielding clothes	Shielding efficiency ≥ 60 dB (shield efficiency of shielding mask ≥ 20 dB)	Set	5	
2	Conductive shoes	The size depends on the wearer	Nos.	5	

Table (Cont'd)

S/N	Name of tool and appliance	Specification	Unit	Qty.	Remarks
3	Flame retardant underwear	Pure mulberry silk	Set	2	
4	Safety belt of double insurance	Suspender	Nos.	4	
5	Safety helmet		Nos.	11	
6	Falling protector	Corresponding to the type of pole and tower falling protector	Nos.	4	
7	Insulated protection rope		Nos.	4	
8	Eight-bundle wire lifter		Nos.	2	
9	Hydraulic leading screw		Nos.	2	
10	Plane leading screw		Nos.	2	
11	Special joint		Nos.	4	
12	Motor winching	3T	Set	1	
13	Potential transfer rod		Nos.	2	
14	Wire rope sleeve		Nos.	4	
15	Insulated transmission rope	TJS-14	Nos.	3	
16	Insulated track rope	TJS-16	Nos.	1	
17	2-2 insulated tackle	JH20-2	Nos.	2	
18	Insulated tackle	JH10-1	Nos.	6	
19	Insulated suspender	φ53mm	Group	2	
20	Insulated noose	1T	Nos.	3	
21	Cradle		Set	1	
22	Insulation resistance meter	5000V	Piece	1	
23	Temperature and humidity indicator		Piece	1	
24	Anemometer		Piece	1	
25	Interphone		Nos.	Several	
26	Multimeter		Piece	1	
27	Moisture-proof tarpaulin		Piece	4	
28	Guardianship clothes of Responsible Person		Piece	1	
29	Guardianship clothes of Special Supervisor		Piece	1	

Part II Skill Module Training and Assessment Standards

Table (Cont'd)

S/N	Name of tool and appliance	Specification	Unit	Qty.	Remarks
30	Tool bag (kit)		Nos.	Several	
31	Security fence		Set	Several	
32	Warning sign	"Work Here", "Access from Here", "Access from Here"	Set	1	
33	Composite insulators	FXBZ-1000/420	pcs	1	

3.3 Division of labor for operators

Division of labor for operators of the task is shown in Table 2-1-4.

Table 2-1-4 Division of Labor for Live Replacement of Type I Composite Insulator for 1000kV AC Transmission Line Tangent Tower

S/N	Post	Qty. (person)	Responsibilities
1	Responsible Person	1	Be responsible for the labor division of operating personnel of the work task, field investigation before work, preparation of the operation plan, work order filling, field re-investigation, performing work permit procedures, holding pre-shift meeting, implementation of the site safety measures, safety supervision in the operation process, dealing with emergency situations in work, quality surveillance of the work, and the summary after the work
2	Special Supervisor	1	Be responsible for the safety control of the work site
3	Equipotential electrician	2	Cooperate with ground electrician in installing wire lifting system (planar leading screw, special joint, insulating suspender, hydraulic screw rod), operating hydraulic leading screw to transfer wire load, and disassembling and assembling insulator string, etc.
4	Ground potential electrician	2	Be responsible for installation of basket, wire lifting system (planar leading screw, special joint, insulating suspender, eight-bundle wire lifter, hydraulic screw rod), insulating grinding rope and cooperating with equal potential electrician in entering and exiting the potential, disassembly and assembly of composite insulator string, etc.
5	Ground electrician	5	Be responsible for transferring tools and composite insulator string, etc.

4. Work Procedure

The workflow of this task is shown in Table 2-1-5.

Table 2-1-5 Workflow of Live Replacement of Type I Composite Insulator for 1000kV AC Transmission Line Tangent Tower

S/N	Work Content	Operation Steps and Standards	Safety Measures and Precautions	Responsible Person
1	Site re-survey	The Responsible Person shall complete the following work: (1) Check the line title, the number of the pole and tower and ensure the phases are correct; guarantee that the foundation and the pole and tower are intact and in normal condition; ensure that the cross and span distance meets the safety requirements; confirm the defect conditions and the specifications and models of earth wires. (2) Check that the site meteorological conditions such as wind speed and humidity should meet the operation requirements. (3) Check that the terrain and environment shall meet the operation requirements. (4) Check that the safety measures listed in the work order are in line with the actual situations on site, and the measures will be supplemented if necessary	(1) Correctly wear helmet, working clothes, work shoes and protective gloves. (2) Operation under meteorological conditions that may endanger the safety of operators is forbidden. (3) Non-operation personnel and vehicles are strictly prohibited from entering the working site.	
2	Work Permit	(1) The Responsible Person is responsible for contacting the on-duty control personnel and applying for stopping the line re-closing as per the contents of the work order. (2) Live working could be started only after being approved by the on-duty control personnel.	Live working shall not be started without the permission of the on-duty control personnel	
3	Site layout	Install the security fence and hang the signboards correctly: (1) The security fence should take full account of falling objects from the heights and the influence on road traffic. (2) The entrance and exit of the security fence shall be set reasonably. (3) Signs such as "Access from Here", "Work Here", "Access from Here" shall be properly and well arranged	When the influence on road traffic safety is uncontrollable, the traffic management department should be contacted in time to strengthen the on-site control of traffic safety.	

Part II Skill Module Training and Assessment Standards

Table (Cont'd)

S/N	Work Content	Operation Steps and Standards	Safety Measures and Precautions	Responsible Person
4	Hold a pre-shift meeting	(1) All working personnel shall line up. (2) The Responsible Person shall wear red waistcoat and read out the work order and be clear with work task and division of personnel; explain safety measures and technical measures in work; check (inquire after) mental state of all working personnel; inform of hazards in work and precontrol measures. (3) All working personnel shall sign on the work order for confirmation.	(1) The work order shall be filled in, issued and approved in a standardized manner, and the signature shall be complete. (2) All working personnel shall be in good mental states. (3) All working personnel shall be clear with task division of works, safety measures and technical measures	
5	Inspection tool	(1) The ground electrician and equipotential electrician on tower shall wear the shielding clothes in a right way and pass the inspection, which shall be supervised and inspected by the Responsible Person. (2) Wear personal safety equipment correctly (proper size and easy lock), and the Responsible Person shall supervise and inspect it. (3) Measure the wind speed, wind direction and humidity, check the insulation performance of insulating tools, and make records	(1) Check carefully for damage, deformation and failure before using metal and insulating tools. Carry out segment insulation detection with such insulated tools with insulation resistance meter of 2,500V or above and with the resistance no less than 700MΩ, and wipe it off with a clean dry towel. (2) Use a multimeter to measure the resistance between the farthest ends of the shielding clothes and trousers, which shall not be greater than 20Ω. The Responsible Person shall check the connection of the electrician's shielding clothing. (3) Check the tool assembly and make sure the connection is reliable. (4) The live working tools used at site shall be placed on the moisture-proof canvas	

Table (Cont'd)

S/N	Work Content	Operation Steps and Standards	Safety Measures and Precautions	Responsible Person
6	Climbing the tower	(1) After checking the line name and pole and tower number, the ground electrician on tower shall check the stress of the safety belt and the falling protector in the equipotential shock test. (2) The ground electrician on tower carries the insulated transmission rope to climb the tower, and the equipotential electrician then climbs the tower. When they reach the operation point of the cross arm, they choose the appropriate position to fasten the safety belt, and the ground electrician on tower installs the insulated tackle and the insulated transmission rope in the appropriate position of the cross arm. Then he shall cooperate with the ground electrician to separate the insulated transmission rope for lifting preparation	(1) After checking the correct line name and pole and tower number, the tower can be climbed for operation. (2) The anti-falling device installed on tower shall be used in the process of climbing the tower; when moving and transposition on the pole and tower, the safety protection shall not be lost, and the operators must climb and grasp the components securely. (3) The working electrician must wear a whole suit of qualified shielding clothes which must be connected reliably. Before the cross arm enters the equipotential, the equipotential electrician shall check and confirm that each part of the shielding clothes are connected reliably before the next operation	
7	Installation of tackle and basket	(1) The ground electrician transfers the basket, basket track rope, insulated protection rope, 2-2 insulated tackle block and potential transfer rod to the cross arm by means of insulated transmission rope. (2) The electrician on tower installs the 2-2 insulated tackle block and the basket in the appropriate position of the upper plane of the cross arm, and the basket track rope in the appropriate position of the cross arm	(1) The lifting of insulating sling shall be smooth, free of collision and winding in transfer. (2) After the basket is installed, the electrician on tower shall carry out the careful inspection and check to the basket situation. (3) 2-2 tackle block and basket shall be installed reliably in a suitable position on the cross arm	

Part II Skill Module Training and Assessment Standards

Table (Cont'd)

S/N	Work Content	Operation Steps and Standards	Safety Measures and Precautions	Responsible Person
8	Entering the intense electric field	(1) No. 1 equipotential electrician shall fasten the insulated protection rope to enter the basket, and the ground electrician slowly releases the control rope of 2-2 insulated tackle block and slows down when the basket is about 2m away from the live conductor. (2) During the continuous movement of the basket to the conductor, the equipotential electrician holds the potential transfer rod in hand and faces the live conductor, and applies for potential transfer to the Responsible Person. After obtaining the approval, the equipotential electrician quickly extends the potential transfer rod when the basket is 0.5m away from the conductor and hooks it onto the nearest sub-conductor for potential transfer. (3) The equipotential electrician shall fasten the safety belt after entering the intense electric field, and decide whether to release the insulated protection rope according to the operation requirements. At the same time, the equipotential electrician shall control the head not to exceed the grading ring on the conductor side. (4) The ground electrician tightens the control rope of 2-2 insulated tackle block, and transfers the basket up to the cross-arm. No. 2 equipotential electrician fastens the insulated protection rope and enters the basket, and then enter the intense electric field by same method	(1) Before entering the equipotential, the equipotential electrician shall re-check and confirm that each part of the shielding clothes, the potential transfer rod and the insulated shielding clothes are connected reliably before the next operation. The Special Supervisor is responsible for inspecting the connection of the equipotential electrician's shielding clothes, and the reliable connection of the potential transfer rod. (2) The equipotential electrician must obtain the permission of the Responsible Person before entering the potential. (3) The equipotential electrician must fasten the protective rope before entering the basket. (4) When the ground electrician cooperates with the equipotential electrician to enter the equipotential, the control rope of the tackle block shall be pulled and released stably. (5) The combined gap composed of the gaps between the equipotential electrician and the grounding body and the electrified body in the process of entering the potential shall not be less than 6.9m (middle phase)/ 6.7m (side phase). (6) The Special Supervisor shall be responsible for monitoring the safety precautions for the equipotential electrician entering the intense electric field. The Special Supervisor shall remind the dangerous and irregular actions of the operators on tower in a timely manner and stop them if necessary.	

Table (Cont'd)

S/N	Work Content	Operation Steps and Standards	Safety Measures and Precautions	Responsible Person
9	Install tools and transfer conductor load	(1) The ground electrician transfers tools such as planar screw rod, insulated suspender, eight-bundle wire lifter and hydraulic stringing system to the working position, and the equipotential electrician cooperates with the ground electrician to install the insulator replacement tool on both sides of the composite insulator string to be replaced (installed vertically along the insulator string, with the planar screw rod and hydraulic stringing system installed on the cross arm side). (2) After checking that each part of the load-bearing tools is installed reliably and obtaining the consent of the Responsible Person, the ground electrician shall tighten the planar screw rod first, and then tighten the hydraulic stringing system after the planar screw rod is properly stressed, so as to relax the insulator string. (3) After checking that the load-bearing tools are stressed normally and obtaining the consent of the Responsible Person, the equipotential electrician dismantles the socket hanging plate bolts on the conductor side. (4) The ground electrician transfers the control rope of the composite insulator string to the equipotential electrician who installs it at the tail of the composite insulator string. (5) The ground electrician ties the insulated transmission rope to the upper end of the composite insulator string, and then takes out the fitting pin connected between the composite insulator string and the ball head hanging ring. The ground electrician starts the motor winching, and cooperates with the ground electrician to disconnect the composite insulator string with the ball head hanging ring	(1) The upper and lower working electricians shall cooperate closely, and all working electricians shall obey the unified command of the Responsible Person. (2) The minimum safe distance of the ground electrician to the electrified body and the equipotential electrician to the grounding body shall not be less than 6.8m (medium phase)/6.0m (side phase). The effective insulated length of insulated suspender and insulating rope shall not be less than 6.8m. (3) The binding rope buckles of the upper and lower transfer tools of the pole and tower shall be correct and reliable, and the electrician on tower shall not drop objects from high place. (4) After the tool is stressed and passes the impulse inspection, it shall be reported to the Responsible Person, and then the operation can be continued only with the permission of the Responsible Person. (5) The Special Supervisor shall remind the dangerous and irregular actions of the operators on tower in a timely manner and stop them if necessary.	

Part II Skill Module Training and Assessment Standards

Table (Cont'd)

S/N	Work Content	Operation Steps and Standards	Safety Measures and Precautions	Responsible Person
10	Replacement of insulator string	(1) The ground electrician controls the control rope of the composite insulator string, and slowly puts the composite insulator string to the ground by using the motor winching. Attention shall be paid to the control rope of composite insulator string, which shall not collide with the load-bearing tools, conductor and pole and tower. (2) The ground electrician transfers the insulated transmission rope and the control rope of the composite insulator string to the new composite insulator respectively. (3) The ground electrician starts the motor winching to transfer the new composite insulator string to the working position on tower. The ground potential electrician restores the connection between the new composite insulator string and ball head hanging ring, and restore the fitting pin. (4) The ground electrician slowly loosens the motor winching to make the composite insulator string naturally vertical, and the ground electrician restores the connection between the socket hanging plate and the metal yoke plate, and installs the cotter pin	(1) When hoisting the insulator string, care shall be taken not to collide with the pole and tower, and the ground electrician shall pull the tail rope of the insulator string. (2) The rope shall not rub against the pole and tower. The binding rope buckle shall be correct and reliable. (3) When hoisting the insulator string with the winching, the winching shall be arranged stably. The tail rope shall be controlled by an electrician with working experience of live work, shall be tightened at any time and shall not be negligently loosened. (4) When hoisting the insulator string with the motor winching, the stress conditions of the winching and steering tackle must be checked before operation. (5) The Special Supervisor shall remind the dangerous and irregular actions of the operators on tower in a timely manner and stop them if necessary.	
11	Remove tools	(1) After checking the reliable connection of the composite insulator string and obtaining the consent of the Responsible Person, the ground electrician loosens the hydraulic stringing system and the planar screw rod.	(1) After the composite insulator is installed and reset, the connection of each part shall be checked in detail for correctness, and the wire lifting tools shall not be removed until the consent of the Responsible Person is obtained.	

Table (Cont'd)

S/N	Work Content	Operation Steps and Standards	Safety Measures and Precautions	Responsible Person
11	Remove tools	(2) After checking that the stress of the composite insulator string is normal and obtaining the consent of the Responsible Person, the ground electrician shall cooperate with the equipotential electrician to remove the planar screw rod, the insulated suspender, the eight-bundle wire lifter and the hydraulic stringing system, and transfer them to the ground.	(2) The tools shall not collide with each other in the process of transfer, and the rope buckle shall be bound correctly and reliably. (3) The Special Supervisor shall remind the dangerous and irregular actions of the operators on tower in a timely manner and stop them if necessary.	
12	Exiting the potential	(1) No. 1 equipotential electrician shall fasten the insulated protective rope, hook the metal end of the potential transfer rod to the sub-conductor, enter the basket with one hand holding the insulated handle, and then keep the arm straight so that the basket is 0.5m away from the sub-conductor. (2) The equipotential electrician applies to the Responsible Person for exiting the potential. After obtaining the consent, the equipotential electrician quickly removes the connection between the potential transfer rod and the sub-conductor, and puts the potential transfer rod back in the basket. (3) At the same time, the ground electrician quickly tightens the control rope of the 2-2 insulated tackle block, pulls the basket upward to the cross arm and stops it. Then, the equipotential electrician climbs the cross arm and fastens the safety belt. (4) The ground electrician transfers the basket to the No. 2 equipotential electrician by means of insulated transmission rope. Then, the equipotential electrician enters the basket after checking that there is no object left on the conductor, and exits the potential by the same method.	(1) The upper and lower working electricians shall cooperate closely and follow the command of the Responsible Person. (2) The equipotential electrician must obtain the permission of the Responsible Person before exiting the potential. (3) The equipotential electrician must fasten the protective rope before entering the basket. (4) When the ground electrician cooperates with the equipotential electrician to enter the equipotential, the control rope of the tackle block shall be pulled and released stably. (5) The combined gap composed of the gaps between the equipotential electrician and the grounding body and the electrified body in the process of exiting the potential shall not be less than 6.9m (middle phase)/ 6.7m (side phase). (6) The Special Supervisor shall be responsible for monitoring the safety precautions for the equipotential electrician exiting the intense electric field. The Special Supervisor shall remind the dangerous and irregular actions of the operators on tower in a timely manner and stop them if necessary.	

Part II Skill Module Training and Assessment Standards

Table (Cont'd)

S/N	Work Content	Operation Steps and Standards	Safety Measures and Precautions	Responsible Person
13	Remove the basket and return to the ground	(1) The electrician on tower shall cooperate the removal of the basket track rope, the insulated protection rope, the 2-2 insulated tackle block and the basket and transfer of them to the ground. (2) After checking that there is no object left on tower, the equipotential electricians shall report it to the Responsible Person and then climb down the tower with insulated transmission rope after obtaining the consent of the Responsible Person.	(1) The tools shall not collide with each other in the process of transfer, and the rope buckle shall be bound correctly and reliably. (2) The anti-falling device installed on tower shall be used in the process of climbing the tower; when moving and transposition on the pole and tower, the safety protection shall not be lost, and the operators must grasp the components securely.	
14	End of the work	(1) The site and tools shall be cleaned up, and any left objects on the pole (tower) shall be carefully checked. The Responsible Person shall comprehensively inspect the completion of the work, count the number of people, declare the end of the work if no error is found and evacuate from the construction site. (2) The dispatcher shall be notified of the end of work, the completion formalities for the work order shall be handled	It is forbidden to restore the line re-closing in the appointed time	

Standard for Professional Training and Assessment for Operation Maintenance of UHV AC Transmission Line

II. Assessment Standard

Detailed Rules for Assessment and Scoring of Operation and Inspection Skills of UHV AC Transmission Line

Fill-in Column of Examinee	No.:	Name:	Position:	Unit:	Date:	MM/DD/YYYY	
Fill-in Column of Assessor	Grade:	Assessor:	Assessment Team Leader:	Starting time:	Closing time:	Operation Duration:	
Assessment module	Live replacement of Type I composite insulator for 1000kV AC transmission line tangent tower		Assessee	Maintenance personnel of UHV AC transmission line	Assessment method	Operation	Assessment Time Limit
							120min
Job Description	Live replacement of Type I composite insulator for 1000kV AC transmission line tangent tower						
Work Specifications and Requirements	1. Live working shall be carried out in good weather. In case of thunder, rain, snow or fog, no live working shall be carried out. When the wind force is greater than Level 5 and the humidity is greater than 80%, live working should not be carried out. 2. This operation requires 1 Responsible Person, 1 Special Supervisor, 2 ground electricians, 5 ground electricians and 2 equipotential electricians. They shall adopt the basket transfer method to enter the intense electric field for insulator replacement. 3. Responsibilities of Responsible Person: Be responsible for division of operating personnel of the task, work order reading, handling re-closing deactivation of the line and the formalities of work permit, getting work permits, holding pre-shift meeting, dealing with emergency situations in work, quality surveillance, and the summary after work. 4. Special Supervisor: Be responsible for safety control of the work site. 5. Responsibilities of ground electrician: Cooperate with ground electrician in installing wire lifting system (planar leading screw, special joint, insulating suspender, hydraulic screw rod), operating hydraulic leading screw to transfer wire load, and disassembling and assembling insulator string. 6. Responsibilities of ground electrician on tower: Be responsible for installation of basket, wire lifting system (planar leading screw, special joint, insulating suspender, eight-bundle wire lifter, hydraulic screw rod), insulating grinding rope and cooperating with equal potential electrician in entering and exiting the potential, disassembly and assembly of composite insulator string, etc. 7. Responsibilities of ground electrician: Be responsible for transferring tools and materials and cooperating with equipotential electrician in entering and exiting the equipotential. 8. During the live working, if thunder, rain, strong wind or any other circumstance threaten the safety of the staff, the Responsible Person or Supervisor may stop working temporarily according to the circumstances.						

Part II Skill Module Training and Assessment Standards

Table (Cont'd)

Work Specifications and Requirements	Given conditions: 1. Training base: Phase A Type I composite insulator of UHV AC 1000kV training line tangent tower, type of insulator: FXBZ-1000/420. 2. Work orders have been handled, safety measures have been completed (re-closing has been deactivated), and oral application (dispatcher or assessor) shall be made at the beginning and end of the work. 3. The instrument shall be used safely and correctly to test the insulating tool. 4. The operation must be carried out according to the working procedures. The scores of the items to be carried out shall be deducted for the process error. In case of major hidden dangers of personal, equipment and operational safety, the assessor may order the termination of the operation (assessment)
Assessment scenario preparation	1. Line: Phase A of UHV AC 10000kV practical training line tangent tower, work content: live replacement of Phase A Type I composite insulator of UHV AC 1000kV training line tangent tower, type of insulator: FXBZ-1000/420. 2. Required work tools: 3 insulated transmission ropes (TJS-14), 1 insulated track rope (TJS-16), 6 insulated tackles (JH10-1), 2 2-2 insulated tackles (JH20-2), 1 basket, 2 hydraulic stringing systems, 2 eight-bundle wire lifters, 2 groups of insulated suspenders, 1 motor winching, insulation detector, 1 potential transfer rod, insulation resistance meter (5000V), 5 suits of shielding clothes (shielding efficiency ≥ 60dB), 1 multimeter, 4 tarpaulins, 1 insulation tester, 1 temperature and humidity meter, 1 anemometer, 2 cotton towels. 3. The work site shall be monitored, and the safety measures (fence, etc.) on the work site have been fully implemented; non-operation personnel are prohibited from entering the site, and the staff must wear safety helmets when entering the work site. 4. Examinees shall bring their own work clothes, flame retardant cotton underwear, safety helmets, gloves, safety belts (including double-protective ropes)
Remarks	1. The deduction shall be done until the scores of each item are deducted completely. In case of major hidden dangers of personal, equipment and operational safety, the assessor may order the termination of the operation. 2. When equipment, working environment, safety belt, safety helmet, tool, shielding clothes, etc., do not conform to the operation condition, the assessor may order the termination of the operation

Table (Cont'd)

S/N	Project name	Quality requirements	Score	Deduction standard	Reasons for deduction	Deduction	Scoring
1	Site re-survey	1) The Responsible Person shall go to the work site to check the line name, pole and tower number, on-site working conditions, defective parts and so on. 2) Check that the site meteorological conditions such as wind speed and humidity should meet the operation requirements. 3) Check whether the work order is complete and unmodified, check whether the safety measures listed are consistent with the actual situation on site, and supplement it if necessary	5	1) Deduct 1 point for failure to check the double title. 2) Deduct 1 point for failure to verify on-site working conditions (meteorology), defective parts. 3) Deduct 0.5 points/item for any alteration in the work order filling, and deduct 1 point for incorrect work order number. Deduct 1.5 points for each incomplete work order			
2	Work Permit	1) The Responsible Person is responsible for contacting the on-duty control personnel and applying for stopping the line re-closing as per the contents of the work order. 2) Reporting content is standardized and complete	2	1) Deduct 2 points for failure to contact the dispatching department (referee) for deactivating the re-closing. 2) Deduct 0.5 points for non-standard or incomplete terminology reporting respectively			

Part II Skill Module Training and Assessment Standards

Table (Cont'd)

S/N	Project name	Quality requirements	Score	Deduction standard	Reasons for deduction	Deduction	Scoring
3	Site layout	Install the security fence and hang the signboards correctly: 1) The security fence should take full account of falling objects from the heights and the influence on road traffic. 2) The entrance and exit of the security fence shall be set reasonably. 3) Signs such as "Access from Here", "Work Here", "Access from Here" shall be properly and well arranged	3	1) Deduct 0.5 points for failure to arrange the fence at the work site. 2) Deduct 0.5 points for failure to arrange the warning board. 3) Deduct 0.5 points for failure to hang the tower climbing operation sign			
4	Hold a pre-shift meeting	1) All staff and personnel shall wear safety helmets and work clothes correctly. 2) Responsible Person shall wear red vest and read out the work order and be clear with work task and division of personnel; explain safety measures and technical measures in work; check (inquire after) mental state of all working personnel; inform of hazards in work and precontrol measures. 3) All working personnel shall sign on the work order for confirmation.	3	1) Deduct 0.5 points/person for the staff not dressing uniformly. Deduct 0.5 points/person for the staff not dressing uniformly 2) Give no points to this item for no division of labor, and deduct 1 point for unclear division of labor. 3) Deduct 0.5 points for the on-site Responsible Person not wearing a safety monitoring vest. 4) Deduct 1 point for the work shift member failing to sign or signing incompletely on the work order.			

· 317 ·

Table (Cont'd)

S/N	Project name	Quality requirements	Score	Deduction standard	Reasons for deduction	Deduction	Scoring
5	Inspection of tools	1) The staff shall place the tools on the moisture-proof tarpaulin as required; the moisture-proof tarpaulin shall be clean and dry. 2) The tools shall be placed in category according to the requirements of the fixed management; the insulated tools shall not be mixed with metal tools and materials; and the appearance inspection shall be done on the tools. 3) The surface of insulated tools shall not be worn, deformed and damaged, and the operation shall be flexible. Carry out segment insulation detection with such insulated tools with insulation resistance meter of 2,500V or above and with the resistance no less than 700M Ω, and wipe it off with a clean dry towel. 4) The ground potential and equipotential personnel on tower shall correctly wear a whole suit of qualified shield clothes and conductive shoes as required, with each part connected well, shall not wear chemical fiber clothes next to the skin in the shielding clothes and shall fasten safety belts; the Responsible Person shall carefully check whether they wears it correctly. 5) Tower climbing personnel shall check the double name, pole number and phase again and report them	7	1) Deduct 1 point for failure to use moisture-proof cloth and place tools to designed positions. 2) Deduct 0.5 points/item for failure to check qualified label of tool test and appearance inspection. 3) Deduct 1 point/item for failure to use testing instrument for testing the tools. 4) Deduct 2 points/person time for the operator failing to wear the shielding clothes correctly and each part connected well. 5) Deduct 1 point for the on-site Responsible Person failing to check the safety protective equipment of the tower climbing operators. 6) Deduct 2 points/person for the tower climbing personnel failing to check the double name of the line, pole number and phase. 7) Deduct 2 points/person for the tower climbing personnel failing to report the check results			

Part II Skill Module Training and Assessment Standards

Table (Cont'd)

S/N	Project name	Quality requirements	Score	Deduction standard	Reasons for deduction	Deduction	Scoring
6	Climbing the tower	1) The ground electrician and the equipotential electrician on tower shall wear a whole suit of qualified shielding clothes, fasten the safety belt after performing the impulse test on the safety belt, and carry the insulated transmission rope to climb the tower one after another. 2) During the tower climbing process, they shall fasten the anti-fall protection device, climb the tower to an appropriate position, fasten the safety belt, arrange the insulated transmission rope, and then cooperate with the ground electrician to make lifting preparation of the insulated transmission rope separately.	5	1) Deduct 2 points for failure to fasten the safety belt or for failure to perform the impulse test on the safety belt and the backup protection rope. 2) Deduct 2 points for grasping the shackles with hands. 3) Deduct 1 point for inconvenient suspension position of tackle transmission rope for taking tools. 4) Deduct 2 points for metal tools that are difficult to ensure safe distance during transfer; deduct 2 points for tools that are not bound securely. 5) Deduct 2 points for falling object at high place.			
6	Climbing the tower	3) During tower climbing, the electrician shall fasten the anti-fall protection device, climb the tower at a uniform speed, grasp the main material by hand, hang the safety belt on the shoulder and keep the safety distance of more than 6.8m (medium phase)/6.0m (medium phase) away from the electrified body, and the Responsible Person shall strengthen the operation monitoring	5	6) Deduct 2 points for tools colliding with the tower body during the transfer process. 7) Deduct 1 point for knotting and disordered rope in tool transfer. 8) Deduct 2 points for the Responsible Person failing to monitor the operation in place. 9) Deduct 2 points for incorrect operation of electrician on tower			

· 319 ·

Standard for Professional Training and Assessment for Operation Maintenance of UHV AC Transmission Line

Table (Cont'd)

S/N	Project name	Quality requirements	Score	Deduction standard	Reasons for deduction	Deduction	Scoring
7	Installation of tackle block and basket	1) The lifting of insulating sling shall be smooth, free of collision and winding in transfer. 2) After the basket is installed, the electrician on tower shall carry out the careful inspection and check to the basket situation. 3) 2-2 tackle block and basket shall be installed reliably in a suitable position on the cross arm	5	1) Deduct 0.5 points for rope winding of 2-2 tackle block. 2) Deduct 1 point for unreasonable installation position of track rope. 3) Deduct 1 point for unsuitable length of insulated protection rope. 4) Deduct 1 point for unsteadiness and collision of transfer tools.			
8	Entering the intense electric field	1) After re-checking and confirming that each part of the shielding clothes is connected reliably, the equal potential electrician carries out the impulse test on the basket, and fastens the protection rope to climb the basket after reporting it to the Responsible Person. 2) The ground electrician slowly loosens the control rope of the 2-2 insulated tackle block, and slows down when it is about 2m away from the conductor. The equipotential electrician applies to the Responsible Person for potential transfer at a distance of 0.5m from the conductor, and quickly extends the potential transfer rod to hook it onto the nearest sub-conductor for potential transfer after obtaining the consent.	10	1) Deduct 1 point for the equipotential electrician failing to conduct the impulse on the basket. 2) Deduct 1 point for failure to fasten the insulated protection rope. 3) Deduct 2 points for the ground electrician failing to check the safety measures of equipotential electrician. 4) Deduct 1 point for the ground electrician failing to control the tail rope of the tackle steadily. 5) Deduct 2 points for the equipotential electrician failing to apply to the Responsible Person before entering the intense electric field; deduct 1 point for starting to enter the intense electric field without consent after application.			

Part II Skill Module Training and Assessment Standards

Table (Cont'd)

S/N	Project name	Quality requirements	Score	Deduction standard	Reasons for deduction	Deduction	Scoring
8	Entering the intense electric field	3) After entering the intense electric field, the equipotential electrician shall protect the human body, and control the head not to exceed the grading ring on the conductor side. 4) The combined gap composed of the gaps between the electrician and the grounding body and the electrified body in the process of entering the potential shall not be less than 6.9m (middle phase)/6.7m (side phase), and the potential transfer rod must be used for potential transfer when entering the intense electric field.	10	6) Deduct 1 point for unskillful potential transfer and 5 points for potential transfer without using the potential transfer rod. 7) Deduct 2 points for the equipotential electrician failing to fasten the safety belt after entering the intense electric field. 8) Deduct 2 points for the equipotential electrician with head exceeding the grading ring on the conductor side after entering the intense electric field.			
9	Install tools and transfer conductor load	1) The ground electrician transfers tools such as planar screw rod, insulated suspender, eight-bundle wire lifter and hydraulic stringing system to the working position, and the equipotential electrician cooperates with the ground electrician to install the insulator replacement tool on both sides of the composite insulator string to be replaced. 2) After checking that each part of the load-bearing tools is installed reliably and obtaining the consent of the Responsible Person, the ground electrician shall tighten the planar screw rod first, and then tighten the hydraulic stringing system after the planar screw rod is properly stressed, so as to relax the insulator string.	15	1) Deduct 1 point for unsteady transfer of tools; deduct 1 point for each collision. 2) Deduct 2 points for failure to check the reliable installation and good stress of load-bearing tools, and 2 points for failure to report it and obtain the consent of the Responsible Person. 3) Deduct 1 point for ineffective communication between the ground electrician and the equipotential electrician.			

· 321 ·

Table (Cont'd)

S/N	Project name	Quality requirements	Score	Deduction standard	Reasons for deduction	Deduction	Scoring
9	Install tools and transfer conductor load	3) After checking that the load-bearing tools are stressed normally and obtaining the consent of the Responsible Person, the equipotential electrician dismantles the socket hanging plate bolts on the conductor side. 4) The ground electrician transfers the control rope of the composite insulator string to the equipotential electrician who installs it at the tail of the composite insulator string. 5) The ground electrician ties the insulated transmission rope to the upper end of the composite insulator string, and then takes out the fitting pin connected between the composite insulator string and the ball head hanging ring. The ground electrician starts the motor winching, and cooperates with the ground electrician to disconnect the composite insulator string with the ball head hanging ring	15	4) Deduct 5 points for failure to check the load-bearing tools before removing the insulator string, and 2 points for failure to report the check results. 5) Deduct 2 points for inappropriate position of the insulating rope fastening the composite insulator. 6) Deduct 2 points for each collision during the transfer of the insulator string. 7) Deduct 1 point for the ground electrician failing to control the motor winching in place.			
10	Replacement of insulator string	1) The ground electrician controls the control rope of the composite insulator string, and slowly puts the composite insulator string to the ground by using the motor winching. Attention shall be paid to the control rope of composite insulator string, which shall not collide with the load-bearing tools, conductor and pole and tower.	17	1) Deduct 1 point for the ground electrician failing to control the tail rope of the insulator well. 2) Deduct 1 point for unreasonable rope buckle binding.			

Part II Skill Module Training and Assessment Standards

Table (Cont'd)

S/N	Project name	Quality requirements	Score	Deduction standard	Reasons for deduction	Deduction	Scoring
10	Replacement of insulator string	2) The ground electrician transfers the insulated transmission rope and the control rope of the composite insulator string to the new composite insulator respectively. 3) The ground electrician starts the motor winching to transfer the new composite insulator string to the working position on tower. The ground potential electrician restores the connection between the new composite insulator string and ball head hanging ring, and restore the fitting pin. 4) The ground electrician slowly loosens the motor winching to make the composite insulator string naturally vertical, and the ground electrician restores the connection between the socket hanging plate and the metal yoke plate, and installs the cotter pin.	17	3) Deduct 2 points for failure to check the steering of winching and the stress of tackle. 5) Deduct 2 points for each collision during the transfer of the insulator string. 4) Deduct 5 points for failing to install the insulator string in place. 5) Deduct 5 points for the operator failing to check the connection of insulator string. 6) Deduct 2 points for the operator failing to check the installation of pin in place. 7) Deduct 2 points for the Special Supervisor failing to fulfill the supervision responsibility.			
11	Remove tools	1) After checking the reliable connection of the composite insulator string and obtaining the consent of the Responsible Person, the ground electrician loosens the hydraulic stringing system and the planar screw rod. 2) After obtaining the consent of the Responsible Person, the ground electrician shall co-operate with the equipotential electrician to remove the planar screw rod, the insulated suspender, the eight-bundle wire lifter and the hydraulic stringing system, and transfer them to the ground.	5	1) Deduct 1 point for failing to check the connection of each part for correctness in detail, and for failing to obtain the consent of the Responsible Person. 2) Deduct 1 point for unsteady transfer of tools; deduct 1 point for each collision.			

· 323 ·

Table (Cont'd)

S/N	Project name	Quality requirements	Score	Deduction standard	Reasons for deduction	Deduction	Scoring
12	Exiting the potential	1) An equipotential electrician shall fasten the insulated protective rope, hook the metal end of the potential transfer rod to the sub-conductor, enter the basket with one hand holding the insulated handle, and then keep the arm straight so that the basket is 0.5m away from the sub-conductor. 2) The equipotential electrician applies to the Responsible Person for exiting the potential. After obtaining the consent, the equipotential electrician quickly removes the connection between the potential transfer rod and the sub-conductor, and puts the potential transfer rod back in the basket. 3) At the same time, the ground electrician quickly tightens the control rope of the 2-2 insulated tackle block, pulls the basket upward to the cross arm and stops it. Then, the equipotential electrician climbs the cross arm and fastens the safety belt. 4) The ground electrician transfers the basket to the another equipotential electrician by means of insulated transmission rope. Then, the equipotential electrician enters the basket after checking that there is no object left on the conductor, and exits the potential by the same method.	8	1) Deduct 2 points for failing to report the end of work, and 1 point for each item left in the intense electric field. 2) Deduct 1 point for the equipotential electrician failing to fasten the insulated protection rope. 3) Deduct 1 point for the ground electrician failing to control the tail rope of the tackle steadily. 4) Deduct 2 points for the equipotential electrician failing to apply to the Responsible Person before entering the intense electric field, and 1 point for starting to enter the intense electric field without consent after application. 5) Deduct 1 point for unskillful potential transfer and 5 points for potential transfer without using the potential transfer rod.			

Part II Skill Module Training and Assessment Standards

Table (Cont'd)

S/N	Project name	Quality requirements	Score	Deduction standard	Reasons for deduction	Deduction	Scoring
13	Remove the basket and return to the ground	1) The electrician on tower shall cooperate the removal of the basket track rope, the insulated protection rope, the 2-2 insulated tackle block and the basket and transfer of them to the ground. 2) After checking that there is no object left on tower, the equipotential electricians shall report it to the Responsible Person and then climb down the tower with insulated transmission rope after obtaining the consent of the Responsible Person.	5	1) Deduct 2 points for failure to use the falling protector when climbing down the tower. 2) Deduct 2 points for loosing the protection of the safety belt when moving on the tower. 3) Deduct 1 point for grasping the tower nail when climbing down the tower. 4) Deduct 2 points for any objects left on tower.			
14	End of the work	1) The Responsible Person shall organize all working members to put working apparatus and materials in order and put them in a special kit (bag) after cleaning; clean the site to ensure that "the materials are removed and the site is cleaned after construction". 2) In the post-shift meeting, Responsible Person will give work summaries and comments. Comments include the construction quality of this work and the implementation of safety measures from all working personnel. 3) Responsible Person shall report to the on-duty control personnel that the work is over, apply for the restoration of circuit re-closing and terminate the work order.	5	1) Deduct 2 points for failure to clean the tools. 2) Deduct 2 points for missing tools. 3) Deduct 2 points for failure to hold the post-shift meeting. 4) Deduct 2 points for failure to remove the fence. 5) Deduct 2 points for failure to report to dispatcher			
	Total		100				

Module 2　Standards for Training and Assessment on Live Replacement of Single-V Composite Insulator for 1000kV AC Transmission Line Tangent Tower

I. Training Standard

(I) Training Requirements

Designation of module	Live replacement of single-V composite insulator for 1000kV AC transmission line tangent tower	Type of training	Operation
Training method	Practical operation training	Hours of training	21 training hours
Training objectives	1. Master the electrical significance of "basket method" operation mode when the tangent tower enters and exits 1000kV intense electric field. 2. Can enter 1000kV equipotential operation point by adopting the "basket method". 3. Be able to independently complete the replacement of single-V composite insulator for 1000kV AC transmission line tangent tower (equipotential operation method)		
Training venue	UHV AC training line		
Training content	With the cooperation of equipotential with ground potential, enter the equipotential for operation through the "basket method", and replace the single-V composite insulator of for tangent tower by the equipotential operation method		
Scope of application	Maintenance personnel of UHV AC transmission line		

(II) Referenced rules and specifications

(1) Electrotechnical Terminology (GB/T2900.55-2002).

(2) Insulated Tackles for Live Working (GB/T13034-2008).

(3) Live Working-Insulating Ropes (GB/T 13035-2008).

(4) Calculation Method of Live Working Minimum Approach Distance on AC Transmission Line (GB/T 18037-2000)

(5) Screen Clothes for Live Working on 1000kV AC (GB/T25726-2010).

Part II Skill Module Training and Assessment Standards

(6) Code for Design of 1000kV Overhead Transmission Line (GB50665-2011).

(7) Maintenance Code for 1000kV AC Transmission Line (DL/T209-2008).

(8) Operation Code for 1000kV AC Transmission Line (DL/T307-2010).

(9) Technical Guide for Live Working on 1000kV AC Transmission Line (DL/T392-2015).

(10) Guidelines of Insulation Coordination for Live Working (DL/T876-2004).

(11) Minimum Requirements for Utilization of Tools, Devices and Equipment for Live Working (DL/T 877-2004)

(12) Preventive Test Code of Tools, Devices and Equipment for Live Working (DL/T 976-2005).

(13) State Grid Corporation of China Working Regulations of Power Safety (Transmission Line Section) (Q/GDW1799.2-2013).

(Ⅲ) Teaching Design for Training

To complete the work task of "live replacement of single-V composite insulator for 1000kV AC transmission line tangent tower", each training stage shall be designed according to the standard operation procedure for work task completion. Each stage includes specific training objectives, training content, hours of training, training methods (training resources), training environment, assessment and evaluation, etc, as shown in the Table 2-2-1.

(Ⅳ) Operation Flow

1. Work Task

With the cooperation of equipotential with ground potential, enter the equipotential for operation by adopting the "basket method", and replace the single-V composite insulator of for tangent tower.

2. Requirements for Weather and Work Site

(1) The live replacement of single-V composite insulator for 1000kV AC transmission line tangent tower shall be carried out in good weather. In case of lightning (hearing thunder or seeing lightning), snow, hail, rain, fog and so on, live working is prohibited. When the wind force is greater than level 5, or the relative humidity of the air is greater than 80%, it is unsuitable for live working; when emergency live repair is required in bad weather, relevant personnel shall be organized to fully discuss and prepare necessary safety measures, which can be implemented after being approved by the unit.

Standard for Professional Training and Assessment for Operation Maintenance of UHV AC Transmission Line

Table 2-2-1 Training Content Design for Live Replacement of Single-V Composite Insulator for 1000kV AC Transmission Line Tangent Tower

Training schedule	Training objectives	Training content	Hours of training	Training methods and resources	Preparation of training conditions	Assessment and evaluation
1. Theoretical teaching	1. Preliminarily master the basic method for entering and exiting 1000kV intense electric field in basket method. 2. Be familiar with the methods for potential transfer. 3. Be familiar with replacement method for single-V composite insulator for transmission line tangent tower	1. The electrical significance of operation mode of entering and exiting the intense electric field in basket method. 2. The use method for the potential transfer rod during entering and exiting the UHV intense electric field. 3. Replacement method and quality standard for single-V composite insulator for transmission line tangent tower	2	Training methods: Lecture. Training resources: PPT, relevant regulations and specifications	Multimedia classroom	Attendance, classroom questions and assignments
2. Preparations	Be able to complete the preparation before operation	1. Work site survey. 2. Preparation of the standardized operation card. 3. Filling of the work order. 4. Preparation of tools and materials for this operation	1	Training methods: 1. Site survey and cleaning of tools and materials shall be practiced at site. 2. Preparation of operation card and the filling of work order shall adopt lecture method. Training resources: 1. 1000kV practical training line. 2. UHV tools warehouse. 3. Blank work order	1. UHV training transmission line 2. Multi-Media classroom	

· 328 ·

Table (Cont'd)

Training schedule	Training objectives	Training content	Hours of training	Training methods and resources	Preparation of training conditions	Assessment and evaluation
3. Work site preparation	Be able to complete the preparations of work site	1. Work site re-survey. 2. Job application. 3. Work site layout. 4. Pre-shift meeting. 5. Inspection of tools and materials	1	Training methods: demonstration and role play. Resources: 1000kV training line	1000kV training line	
4. Trainer's demonstration	The trainees can preliminarily understand the operation process of the task through inspecting and learning from each other's work	1. The arrangement and installation of the tools on tower. 2. The equipotential electrician enters and exits the intense electric field by adopting the basket method. 3. The equipotential electrician and the equipotential electrician cooperate with each other to complete the replacement of single-V composite insulator by using the load transfer device.	2	Training methods: Demonstration. Resources: 1000kV training line	1000kV training line	

Table (Cont'd)

Training schedule	Training objectives	Training content	Hours of training	Training methods and resources	Preparation of training conditions	Assessment and evaluation
5. Group training of trainees	1. Can complete the operation of entering and exiting 1000kV intense electric field. 2. Can complete the replacement of single-V composite insulator for 1000kV transmission line tangent tower	1. Trainees are grouped (12 persons per group) to train the skill of entering and exiting 1000kV intense electric field and replacing single-V composite insulator for tangent tower. 2. Trainers guide the operation of trainees and conduct safety supervision	14	Training methods: Role play. Resources: 1000kV training line	1000kV training line	Score the operation of trainees according to the detailed rules for skill assessment and scoring
6. End of the work	1. Enable the trainees to further distinguish the shortcomings of the operation process and facilitate the promotion in the later stage. 2. Train the trainees in the working style of safe and civilized production atmosphere	1. Cleaning up the work site. 2. Report to dispatcher. 3. Comment and summarize the work task this time at post-shift meeting	1	Training method: Lecture and inductive method	1000kV training line	

(2) The operating personnel shall be in good spirits, and be familiar with the organizational and technical measures to ensure safety in work; they shall hold the live working qualification within the validity period.

(3) The Responsible Person shall organize relevant personnel to complete field investigation in advance, determine the operating methods, required working apparatus and necessary measures according to the results, and handle the work orders for live working.

(4) The work site shall be reasonably set up with fence and warning signs. Non-operating personnel is forbidden to enter.

(5) The line re-closing device shall be deactivated in the Project.

Safe working distance and effective insulation length during operation are shown in Table 2-2-2.

Table 2-2-2 Safe Distance for Live Replacement of Single-V Composite Insulator for 1000kV AC Transmission Line Tangent Tower

Altitude/m	Minimum safe distance/m		Minimum combined clearance /m		Minimum effective insulation length of insulating tools /m	Minimum distance between exposed part of human body and electrified body when potential is transferred /m
	Middle phase	Side phase	Middle phase	Side phase		
$H \leqslant 1000$	6.8	6.0	6.9	6.7	6.8	0.5
$1000 < H \leqslant 2000$	7.4	6.6	7.6	7.3	6.8	0.5

3. Preparations

3.1 Hazards and precontrol measures

(1) Hazard - electric shock injury

Precontrol measures:

① Before work, Responsible Person should contact the control personnel on duty, deactivate the line re-closing, and perform the licensing procedures.

② The Responsible Person must disclose the safety measures, technical measures and work tasks. Before climbing the tower, ground potential operators on tower must carefully check the dual names of the line, the number of the pole and tower, and phase, and then the tower can be climbed after all have been confirmed correct.

③ If lines lose power suddenly during work, operators shall still deem it to be charged. The Responsible Person shall contact the control personnel as soon as possible, and no forced energization is allowed before the on-duty control personnel getting in touch with the

Responsible Person.

④ Insulating tool and insulating ropes shall be free of damage, moisture, deformation, and failure. It is not allowed to use non-insulating ropes (such as cotton rope, manila rope, and steel wire rope).

⑤ The equipotential operator shall wear flame-retardant underwear and full set of screen clothes outside (including hat, dresses & trousers, gloves, socks and shoes). All parts shall be in excellent connection conditions, and the resistance of the whole suit of shielding clothes shall not exceed 20Ω.

⑥ Prior to potential transfer, equipotential operators shall obtain permission from the Responsible Person, and the minimum distance between the exposed part of the human body and the electrified body shall not be less than 0.5m; equipotential operators must use potential transfer rods for potential transfer; the safe distance between the ground potential operator and the charged body is less than that specified in Table 2-2-2.

⑦ When transmitting large metal objects by using insulating ropes, ground potential operators shall not touch them before ground connection.

⑧ The Responsible Person and the Special Supervisor shall continuously monitor the operators and correct their nonstandard operation or actions in violation at any time. Special attention shall be paid to operators work at heights to ensure that there is enough safety distance (meeting the requirements in Table 2-2-2). It is forbidden to contact two non-connected electrified bodies or make contact with electrified body and grounding body at the same time.

(2) Hazard - precontrol measures for falling accident:

① Before climbing, operators work at heights must satisfy the requirements of this operation, such as physical condition, mental state and skill and quality.

② After climbing the tower to the operation point, the ground electrician shall fasten the safety rope, check and confirm that it is firm, arrange the track rope of the basket, and the human body backup protection rope of the equipotential electrician shall be reasonable and reliably anchored. The ground electrician shall reliably connects the insulated hoisting rope with the basket using a closed hook. The equipotential electrician shall fasten the insulated and main protection ropes, and enter the basket after passing the impact test on the basket. After the equipotential electrician fastens the safety belt, the relevant work can be carried out. The equipotential electrician shall fasten the protection rope at the operation point.

Part II Skill Module Training and Assessment Standards

③Supervisors shall correct the nonstandard or illegal actions at any time. Special attention shall be paid to operators to prevent them from losing the protection of safety belt or insulated backup protection rope during transposition, and it is forbidden to fasten the safety belt or insulated backup protection rope in a position lower than the operating personnel.

④ Personnel shall inspect the shackles and tower materials for fastening condition before climbing the tower, and shall grasp the main materials by hand but not grasp the shackles by hand when climbing the tower.

(3) Hazard - injury caused by objects falling from heights.

Precontrol measures:

① Operator working at heights should put personal tools and fragmentary materials into the tools bag. It is strictly forbidden to hang objects in heights or keep in the mouth.

② Ground operator should correctly wear a helmet and use the knots. The vertical distance from the work site should not be less than the falling radius.

③ The work site shall be set up with fence and warning signs. It shall be noted at any time that Supervisor shall prohibit irrelevant personnel and vehicles from entering operation area.

3.2 Selection of tools and instruments and materials

Tools and materials required for live replacement of Type I composite insulator for 1000kV AC transmission line tangent tower can be seen in Table 2-2-3. Before delivering tools and instruments out of warehouse, application voltage class and test period shall be carefully checked and they shall be inspected to ensure that appearance is intact, connection is firm, rotation is flexible and meet the working task requirements. After delivering tools and instruments out of warehouse, they shall be stored in tools bag or tool kit for transportation to avoid contamination and damp. Metal tools and insulated tools shall be separately loaded and transported to avoid deformation, damage or other defects caused by mixed loading and transportation.

Table 2-2-3 Tools and Materials Required for Live Replacement of Single-V Composite Insulator for 1000kV AC Transmission Line Tangent Tower

S/N	Name of tool and appliance	Specification	Unit	Qty.	Remarks
1	A whole suit of shielding clothes	Shielding efficiency $\geqslant$ 60 dB (shield efficiency of shielding mask $\geqslant$ 20 dB)	Set	5	
2	Conductive shoes	The size depends on the wearer	Nos.	5	

Table (Cont'd)

S/N	Name of tool and appliance	Specification	Unit	Qty.	Remarks
3	Insulated pull-rod		Set	2	
4	Tension transfer device		Set	1	
5	Safety belt of double insurance	Suspender	Nos.	5	
6	Cradle		Nos.	1	
7	Basket track rope	TJS-φ16mm	Nos.	1	
8	Conductor backup protection rope	φ32mm	Nos.	1	
9	Insulated transmission rope	TJS-14	Nos.	2	
10	Insulated guide rail rope	TJS-16	Nos.	1	
11	Insulated backup protection rope	TJS-16	Nos.	4	
12	Insulated tackle	JH10-1	Nos.	2	
13	Insulated tackle	JH20-2	Nos.	1	
14	2-2 insulated tackle	JH20-2	Group	1	
15	Potential transfer rod		Nos.	1	
16	Hydraulic screw rod		Nos.	2	
17	Conductor airfoil clamp		Nos.	1	
18	Multimeter		Set	1	
19	Wind speed, temperature and humidity tester	HT-8321	Nos.	1	
20	Safety fence		Set	Several	
21	Warning sign	"Work Here", "Access from Here"	Set	1	
22	Red waistcoat	"Responsible Person"	Piece	1	
23	Moisture-proof tarpaulin	2m×4m	Piece	5	
24	Motor winching	1T	Set	1	
25	U-ring	3T、5T、8T	Nos.	3	
26	Insulation resistance meter	5000V	Piece	1	
27	Falling protector	Corresponding to the type of pole and tower falling protector	Nos.	4	
28	Special joint		Nos.	4	
29	Insulation jacks		Nos.	6	
30	Steel wire jack		Nos.	4	
31	Interphone		Nos.	5	
32	Personal tools and instruments		Set	4	
33	Safety helmet		Nos.	12	
34	Composite insulators	FXBZ-1000/420	pcs	1	

Part II Skill Module Training and Assessment Standards

3.3 Division of labor for operators

Division of labor for operators of the task is shown in Table 2-2-4.

Table 2-2-4 Division of Labor for Live Replacement of Single-V Composite Insulator for 1000kV AC Transmission Line Tangent Tower

S/N	Post	Qty. (person)	Responsibilities
1	Responsible Person	1	Be responsible for division of operating personnel of the task, field investigation before work, implementation of the operation plan, work order filling, field re-investigation, performing work permit procedures, holding pre-shift meeting, the implementation of safety measures on site, safety supervision in the operation process, dealing with emergency situations in work, quality surveillance, and the summary after work.
2	Special Supervisor	1	Be responsible for the safety control of the work site
3	Equipotential electrician	2	Cooperate with ground electrician in installing wire lifting system (planar leading screw, special joint, insulating suspender, hydraulic screw rod), operating hydraulic leading screw to transfer wire load, and disassembling and assembling insulator string, etc.
4	Ground potential electrician	2	Be responsible for installation of basket, wire lifting system (planar leading screw, special joint, insulating suspender, hydraulic screw rod), insulating grinding rope and cooperating with equal potential electrician in entering and exiting the potential, disassembly and assembly of composite insulator string, etc.
5	Ground electrician	6	Be responsible for transferring tools and composite insulator string, etc.

4. Work Procedure

The workflow of this task is shown in Table 2-2-5.

Table 2-2-5 Workflow of Live Replacement of Single-V Composite Insulator for 1000kV AC Transmission Line Tangent Tower

S/N	Work Content	Operation Steps and Standards	Safety Measures and Precautions	Responsible Person
1	Site re-survey	The Responsible Person shall complete the following work: (1) Check the line title, the number of the pole and tower and ensure the phases are correct; guarantee that the foundation and the pole and tower are intact and in normal condition; ensure that the cross and span distance meets the safety requirements; confirm the defect conditions and the specifications and models of earth wires. (2) Check that the site meteorological conditions such as wind speed and humidity should meet the operation requirements. (3) Check that the terrain and environment shall meet the operation requirements. (4) Check that the safety measures listed in the work order are in line with the actual situations on site, and the measures will be supplemented if necessary	(1) Correctly wear helmet, working clothes, work shoes and protective gloves. (2) Operation under meteorological conditions that may endanger the safety of operators is forbidden. (3) Non-operation personnel and vehicles are strictly prohibited from entering the working site.	
2	Work Permit	(1) The Responsible Person is responsible for contacting the on-duty control personnel and applying for stopping the line re-closing as per the contents of the work order. (2) Live working could be started only after being approved by the on-duty control personnel.	Live operation shall not be started without the permission of the on-duty control personnel.	
3	Site layout	Install the security fence and hang the signboards correctly: (1) The security fence should take full account of falling objects from the heights and the influence on road traffic. (2) The entrance and exit of the security fence shall be set reasonably. (3) Signs such as "Access from Here", "Work Here", "Access from Here" shall be properly and well arranged	When the influence on road traffic safety is uncontrollable, the traffic management department should be contacted in time to strengthen the on-site control of traffic safety.	

Part II Skill Module Training and Assessment Standards

Table (Cont'd)

S/N	Work Content	Operation Steps and Standards	Safety Measures and Precautions	Responsible Person
4	Hold a pre-shift meeting	(1) All working personnel shall line up. (2) The Responsible Person shall wear red waistcoat and read out the work order and be clear with work task and division of personnel; explain safety measures and technical measures in work; check (inquire after) mental state of all working personnel; inform of hazards in work and precontrol measures. (3) All working personnel shall sign on the work order for confirmation.	(1) The work order shall be filled in, issued and approved in a standardized manner, and the signature shall be complete. (2) All working personnel shall be in good mental states. (3) All working personnel shall be clear with task division of works, safety measures and technical measures	
5	Inspect tools	(1) The ground electrician and equipotential electrician on tower shall wear the shielding clothes in a right way and pass the inspection, which shall be supervised and inspected by the Responsible Person. (2) Wear personal safety equipment correctly (proper size and easy lock), and the Responsible Person shall supervise and inspect it. (3) Measure the wind speed, wind direction and humidity, check the insulation performance of insulating tools, and make records	(1) Check carefully for damage, deformation and failure before using metal and insulating tools. Carry out segment insulation detection with such insulated tools with insulation resistance meter of 2,500V or above and with the resistance no less than 700MΩ, and wipe it off with a clean dry towel. (2) Use a multimeter to measure the resistance between the farthest ends of the shielding clothes and trousers, which shall not be greater than 20Ω. The Responsible Person shall check the connection of the electrician's shielding clothing. (3) Check the tool assembly and make sure the connection is reliable. (4) The live working tools used at site shall be placed on the moisture-proof canvas	

Table (Cont'd)

S/N	Work Content	Operation Steps and Standards	Safety Measures and Precautions	Responsible Person
6	Climbing the tower	(1) After checking the line name and pole and tower number, the ground electrician on tower shall check the stress of the safety belt and the falling protector in the equipotential shock. (2) The ground electrician on tower carries the insulated transmission rope to climb the tower, and the equipotential electrician then climbs the tower. When they reach the operation point of the cross arm, they choose the appropriate position to fasten the safety belt, and the ground electrician on tower installs the insulated tackle and the insulated transmission rope in the appropriate position of the cross arm. Then he shall cooperate with the ground electrician to separate the insulated transmission rope for lifting preparation	(1) After checking the correct line name and pole and tower number, the tower can be climbed for operation. (2) The anti-falling device installed on tower shall be used in the process of climbing the tower; when moving and transposition on the pole and tower, the safety protection shall not be lost, and the operators must climb and grasp the components securely. (3) The working electrician must wear a whole suit of qualified shielding clothes which must be connected reliably. Before the cross arm enters the equipotential, the equipotential electrician shall check and confirm that each part of the shielding clothes are connected reliably before the next operation	
7	Installation of tackle block, basket and grinding rope	(1) The ground electrician transfers the basket, basket track rope, insulated protection rope, 2-2 insulated tackle block and potential transfer rod to the cross arm by means of insulated transmission rope. (2) The electrician on tower installs the 2-2 insulated tackle block and the basket in the appropriate position of the upper plane of the cross arm, and the basket track rope in the appropriate position of the cross arm	(1) The lifting of insulating sling shall be smooth, free of collision and winding in transfer. (2) After the basket is installed, the electrician on tower shall carry out the careful inspection and check to the basket situation. (3) 2-2 tackle block and basket shall be installed reliably in a suitable position on the cross arm	

Part II Skill Module Training and Assessment Standards

Table (Cont'd)

S/N	Work Content	Operation Steps and Standards	Safety Measures and Precautions	Responsible Person
8	Entering the intense electric field	(1) No. 1 equipotential electrician shall fasten the insulated protection rope to enter the basket, and the ground electrician slowly releases the control rope of 2-2 insulated tackle block and slows down when the basket is about 2m away from the live conductor. (2) During the continuous movement of the basket to the conductor, the equipotential electrician holds the potential transfer rod in hand and faces the live conductor, and applies for potential transfer to the Responsible Person. After obtaining the approval, the equipotential electrician quickly extends the potential transfer rod when the basket is 0.5m away from the conductor and hooks it onto the nearest sub-conductor for potential transfer. (3) The equipotential electrician shall fasten the safety belt after entering the intense electric field, and decide whether to release the insulated protection rope according to the operation requirements. At the same time, the equipotential electrician shall control the head not to exceed the grading ring on the conductor side. (4) The ground electrician tightens the control rope of 2-2 insulated tackle block, and transfers the basket up to the cross-arm. No. 2 equipotential electrician fastens the insulated protection rope and enters the basket, and then enter the intense electric field by same method	(1) Before entering the equipotential, the equipotential electrician shall re-check and confirm that each part of the shielding clothes, the potential transfer rod and the insulated shielding clothes are connected reliably before the next operation. The Special Supervisor is responsible for inspecting the connection of the equipotential electrician's shielding clothes, and the reliable connection of the potential transfer rod. (2) The equipotential electrician must obtain the permission of the Responsible Person before entering the potential. (3) The equipotential electrician must fasten the protective rope before entering the basket. (4) When the ground electrician cooperates with the equipotential electrician to enter the equipotential, the control rope of the tackle block shall be pulled and released stably. (5) The combined gap composed of the gaps between the equipotential electrician and the grounding body and the electrified body in the process of entering the potential shall not be less than 6.9m (middle phase)/6.7m (side phase). (6) The Special Supervisor shall be responsible for monitoring the safety precautions for the equipotential electrician entering the intense electric field. The Special Supervisor shall remind the dangerous and irregular actions of the operators on tower in a timely manner and stop them if necessary.	

Table (Cont'd)

S/N	Work Content	Operation Steps and Standards	Safety Measures and Precautions	Responsible Person
9	Install tools and transfer conductor load	(1) The ground electrician carries the insulating rope to the cross arm operation point above the medium phase conductor, and installs the insulated tackle and the insulated transmission rope to the appropriate position. (2) The ground electrician transfers the airfoil clamp, insulated pull-rod, hydraulic leading screw and conductor backup protection rope to the working position, and the ground electrician cooperates with the equipotential electrician to install the tools correctly. (3) After checking the reliability of each component, the equipotential electrician tightens the hydraulic leading screw to make it bear a little force and check the stress point. (4) With the consent of the Responsible Person, the equipotential electrician continues to tighten the hydraulic leading screw evenly to relax the composite insulator.	(1) The upper and lower working electricians shall cooperate closely, and all working electricians shall obey the unified command of the Responsible Person. (2) The minimum safe distance of the ground electrician to the electrified body and the equipotential electrician to the grounding body shall not be less than 6.8m (medium phase)/6.0m (side phase). The effective insulated length of insulated suspender and insulating rope shall not be less than 6.8m. (3) The binding rope buckles of the upper and lower transfer tools of the pole and tower shall be correct and reliable, and the electrician on tower shall not drop objects from high place. (4) After the tool is stressed and passes the impulse inspection, it shall be reported to the Responsible Person, and then the operation can be continued only with the permission of the Responsible Person. (5) The Special Supervisor shall remind the dangerous and irregular actions of the operators on tower in a timely manner and stop them if necessary. (6) The conductor backup protection rope must be installed reliably, and all 8 sub-conductors must be bound securely.	

Part II Skill Module Training and Assessment Standards

Table (Cont'd)

S/N	Work Content	Operation Steps and Standards	Safety Measures and Precautions	Responsible Person
10	Remove the insulator string	(1) The ground electrician tightens the tension transfer device, disconnects the connecting bolt at the parallel hanging plate, and relaxes the tension transfer device about 300mm. (2) The equipotential electrician installs the insulated transmission rope to the tail of the composite insulator. The reverse beam tackle is installed, and the ground electrician tightens the reverse beam rope. (3) The equipotential electrician remove the bolts of the socket hanging plate. The ground electrician cooperates to make the insulator string naturally perpendicular. (4) The ground electrician installs the grinding rope of hoisting insulator string, starts the motor winching, and cooperates to remove the tension transfer device. (5) The ground electrician controls the tail rope of the composite insulator string, and cooperates with the motor winching to slowly put the composite insulator string to the ground.	(1) When the insulator string is withdrawn from operation, the receiving parts shall be checked in detail and may be removed with the consent of the Responsible Person. (2) They shall be arranged steadily when the winching is used, and the tail rope control shall not be negligently loosened. (3) The stress conditions of the winching and steering tackle must be checked before operation. (4) The tail rope of the insulator string shall be pulled at any time to ensure that it does not collided with the pole and tower	
11	Replaced with the new insulator string	(1) The ground electrician uses the motor winching to transfer the new composite insulator string to the tower. The ground electrician restores the connection between the new composite insulator and the tension transfer device. (2) The ground electrician slowly loosens the motor winching to make the composite insulator naturally perpendicular.	(1) The insulator string shall not collide with the pole and tower in the process of hoisting. (2) The rope shall not rub against the pole and tower, and shall be bound securely and reliably. (3) The lifting tail rope shall be tightened at any time and shall not be relaxed. (4) The stress conditions of the winching and steering tackle must be checked before operation.	

Table (Cont'd)

S/N	Work Content	Operation Steps and Standards	Safety Measures and Precautions	Responsible Person
11	Replaced with the new insulator string	(3) The ground electrician tightens the reverse beam rope of composite insulator to pull the tail of the composite insulator string to the working position of the ground electrician on the conductor side. The equipotential electrician restores the connection between the socket hanging plate and the conductor yoke plate, and installs the cotter pin. (4) The ground electrician tightens the tension transfer device, restores the connection between the parallel hanging plate and the new composite insulator, and installs the cotter pin.	(5) The tension transfer device is reliably connected with the composite insulator string	
12	Remove tools	(1) After checking the reliable connection of the composite insulator string and obtaining the consent of the Responsible Person, the ground electrician loosens the hydraulic stringing system and the planar screw rod. (2) After checking that the stress of the composite insulator string is normal and obtaining the consent of the Responsible Person, the ground electrician shall cooperate with the equipotential electrician to remove the insulated pull-rod, hydraulic screw rod, airfoil clamp, conductor backup protection rope, and transfer them to the ground.	(1) After the composite insulator is installed and reset, the connection of each part shall be checked in detail for correctness, and the wire lifting tools shall not be removed until the consent of the Responsible Person is obtained. (2) The tools shall not collide with each other in the process of transfer, and the rope buckle shall be bound correctly and reliably. (3) The Special Supervisor shall remind the dangerous and irregular actions of the operators on tower in a timely manner and stop them if necessary.	

Part II Skill Module Training and Assessment Standards

Table (Cont'd)

S/N	Work Content	Operation Steps and Standards	Safety Measures and Precautions	Responsible Person
13	Exiting the potential	(1) No. 1 equipotential electrician shall fasten the insulated protective rope, hook the metal end of the potential transfer rod to the sub-conductor, enter the basket with one hand holding the insulated handle, and then keep the arm straight so that the basket is 0.5m away from the sub-conductor. (2) No. 1 equipotential electrician applies to the Responsible Person for exiting the potential. After obtaining the consent, the equipotential electrician quickly removes the connection between the potential transfer rod and the sub-conductor, and puts the potential transfer rod back in the basket. (3) At the same time, the ground electrician quickly tightens the control rope of the 2-2 insulated tackle block, pulls the basket upward to the cross arm and stops it. Then, the equipotential electrician climbs the cross arm and fastens the safety belt. (4) The ground electrician transfers the basket to the No. 2 equipotential electrician by means of insulated transmission rope. Then, the equipotential electrician enters the basket after checking that there is no object left on the conductor, and exits the potential by the same method.	(1) The upper and lower working electricians shall cooperate closely and follow the command of the Responsible Person. (2) The equipotential electrician must obtain the permission of the Responsible Person before exiting the potential. (3) The equipotential electrician must fasten the protective rope before entering the basket. (4) When the ground electrician cooperates with the equipotential electrician to enter the equipotential, the control rope of the tackle block shall be pulled and released stably. (5) The combined gap composed of the gaps between the equipotential electrician and the grounding body and the electrified body in the process of exiting the potential shall not be less than 6.9m (middle phase)/6.7m (side phase). (6) The Special Supervisor shall be responsible for monitoring the safety precautions for the equipotential electrician exiting the intense electric field. The Special Supervisor shall remind the dangerous and irregular actions of the operators on tower in a timely manner and stop them if necessary.	

Table (Cont'd)

S/N	Work Content	Operation Steps and Standards	Safety Measures and Precautions	Responsible Person
14	Remove the basket and return to the ground	(1) The electrician on tower shall co-operate the removal of the basket track rope, the insulated protection rope, the 2-2 insulated tackle block and the basket and transfer of them to the ground. (2) After checking that there is no object left on tower, the equipotential electricians shall report it to the Responsible Person and then climb down the tower with insulated transmission rope after obtaining the consent of the Responsible Person.	(1) The tools shall not collide with each other in the process of transfer, and the rope buckle shall be bound correctly and reliably. (2) The anti-falling device installed on tower shall be used in the process of climbing the tower; when moving and transposition on the pole and tower, the safety protection shall not be lost, and the operators must grasp the components securely.	
15	End of the work	(1) The site and tools shall be cleaned up, and any left objects on the pole (tower) shall be carefully checked. The Responsible Person shall comprehensively inspect the completion of the work, count the number of people, declare the end of the work if no error is found and evacuate from the construction site. (2) The dispatcher shall be notified of the end of work, the completion formalities for the work order shall be handled	It is forbidden to restore the line re-closing in the appointed time	

II. Assessment Standard

Detailed Rules for Assessment and Scoring of Operation and Inspection Skills of UHV AC Transmission Line

Fill-in Column of Examinee	No.:		Unit:		Name:		Position:		Date:	MM/DD/YYYY
Fill-in Column of Assessor	Grade:		Assessor:		Assessment Team Leader:		Starting time:		Closing time:	Operation Duration:
Assessment module	Live replacement of single-V composite insulator for 1000kV AC transmission line tangent tower			Assessee	Maintenance personnel of UHV AC transmission line		Assessment method	Operation	Assessment Time Limit	120min
Job Description	Live replacement of single-V composite insulator for 1000kV AC transmission line tangent tower									
Work Specifications and Requirements	1. Live working shall be carried out in good weather. In case of thunder, rain, snow or fog, no live working shall be carried out. When the wind force is greater than Level 5 and the humidity is greater than 80%, live working should not be carried out. 2. This operation requires 1 Responsible Person, 1 Special Supervisor, 2 ground electricians, 6 ground electricians and 2 equipotential electricians. They shall adopt the basket transfer method to enter the intense electric field for insulator replacement. 3. Responsibilities of Responsible Person: Be responsible for division of operating personnel of the task, work order reading, handling re-closing deactivation of the line and the formalities of work permit, getting work permits, holding pre-shift meeting, dealing with emergency situations in work, quality surveillance, and the summary after work. 4. Special Supervisor: Be responsible for safety control of the work site. 5. Responsibilities of ground electrician: Cooperate with ground electrician in installing wire lifting system (airfoil clamp, planar leading screw, special joint, insulating suspender, hydraulic screw rod), operating hydraulic leading screw to transfer wire load, and disassembling and assembling insulator string. 6. Responsibilities of ground electrician on tower: Be responsible for installation of basket, wire lifting system (plane leading screw, special joint, insulating suspender, hydraulic screw rod), insulating grinding rope and cooperating with equal potential electrician in entering and exiting the potential, disassembly and assembly of composite insulator string, etc.									

Part II Skill Module Training and Assessment Standards

	Table (Cont'd)
Work Specifications and Requirements	7. Responsibilities of ground electrician: Be responsible for transferring tools and materials and cooperating with equipotential electrician in entering and exiting the equipotential. 8. During the live working, if thunder, rain, strong wind or any other circumstance threaten the safety of the staff, the Responsible Person or Supervisor may stop working temporarily according to the circumstances. Given conditions: 1. Training base: Phase B Single-V composite insulator of UHV AC 1000kV training line tangent tower, type of insulator: FXBZ-1000/420. 2. Work orders have been handled, safety measures have been completed (re-closing has been deactivated), and oral application (dispatcher or assessor) shall be made at the beginning and end of the work. 3. The instrument shall be used safely and correctly to test the insulating tool. 4. The operation must be carried out according to the working procedures. The scores of the items to be carried out shall be deducted for the process error. In case of major hidden dangers of personal, equipment and operational safety, the assessor may order the termination of the operation (assessment)
Assessment scenario preparation	1. Line: Phase B of UHV AC 10000kV training line tangent tower, work content: live replacement of Phase A Type I composite insulator of UHV AC 1000kV training line tangent tower, type of insulator: FXBZ-1000/420. 2. Required work tools: 3 insulated transmission ropes (TJS-14), 1 insulated track rope (TJS-16), 6 insulated tackles (JH10-1), 2 2-2 insulated tackles (JH20-2), 1 basket, 2 hydraulic stringing systems, 2 groups of insulated suspenders, 1 motor winching, insulation detector, 1 potential transfer rod, insulation resistance meter (5000V), 5 suits of shielding clothes (shielding efficiency ≥ 60dB), 1 multimeter, 4 tarpaulins, 1 insulation tester, 1 temperature and humidity meter, 1 anemometer, 2 cotton towels. 3. The work site shall be monitored, and the safety measures (fence, etc.) on the work site have been fully implemented; non-operation personnel are prohibited from entering the site, and the staff must wear safety helmets when entering the work site. 4. Examinees shall bring their own work clothes, flame retardant cotton underwear, safety helmets, gloves, safety belts (including double-protective ropes)
Remarks	1. The deduction shall be done until the scores of each item are deducted completely. In case of major hidden dangers of personal, equipment and operational safety, the assessor may order the termination of the operation. 2. When equipment, working environment, safety belt, safety helmet, tool, shielding clothes, etc., do not conform to the operation condition, the assessor may order the termination of the operation

Part II Skill Module Training and Assessment Standards

Table (Cont'd)

S/N	Project name	Quality requirements	Score	Deduction standard	Reasons for deduction	Deduction	Scoring
1	Site re-survey	1) The Responsible Person shall go to the work site to check the line name, pole and tower number, on-site working conditions, defective parts and so on. 2) Check that the site meteorological conditions such as wind speed and humidity should meet the operation requirements. 3) Check whether the work order is complete and unmodified, check whether the safety measures listed are consistent with the actual situation on site, and supplement it if necessary	5	1) Deduct 1 point for failure to check the double title. 2) Deduct 1 point for failure to verify on-site working conditions (meteorology), defective parts. 3) Deduct 0.5 points/item for any alteration in the work order filling, and deduct 1 point for incorrect work order number. Deduct 1.5 points for each incomplete work order			
2	Work Permit	1) The Responsible Person is responsible for contacting the on-duty control personnel and applying for stopping the line re-closing as per the contents of the work order. 2) Reporting content is standardized and complete	2	1) Deduct 2 points for failure to contact the dispatching department (referee) for deactivating the re-closing. 2) Deduct 0.5 points for non-standard or incomplete terminology reporting respectively			
3	Site layout	Install the security fence and hang the signboards correctly: 1) The security fence should take full account of falling objects from the heights and the influence on road traffic. 2) The entrance and exit of the security fence shall be set reasonably. 3) Signs such as "Access from Here", "Work Here", "Access from Here" shall be properly and well arranged	3	1) Deduct 0.5 points for failure to arrange the fence at the work site. 2) Deduct 0.5 points for failure to arrange the warning board. 3) Deduct 0.5 points for failure to hang the tower climbing operation sign			

Table (Cont'd)

S/N	Project name	Quality requirements	Score	Deduction standard	Reasons for deduction	Deduction	Scoring
4	Hold a pre-shift meeting	1) All staff and personnel shall wear safety helmets and work clothes correctly. 2) Responsible Person shall wear red vest and read out the work order and be clear with work task and division of personnel; explain safety measures and technical measures in work; check (inquire after) mental state of all working personnel; inform of hazards in work and precontrol measures. 3) All working personnel shall sign on the work order for confirmation.	3	1) Deduct 0.5 points/person for the staff not dressing uniformly. Deduct 0.5 points/person for the staff not dressing uniformly 2) Give no points to this item for no division of labor, and deduct 1 point for unclear division of labor. 3) Deduct 0.5 points for the on-site Responsible Person not wearing a safety monitoring vest. 4) Deduct 1 point for the work shift member failing to sign or signing incompletely on the work order.			
5	Inspection of tools	1) The staff shall place the tools on the moisture-proof tarpaulin as required; the moisture-proof tarpaulin shall be clean and dry. 2) The tools shall be placed in category according to the requirements of the fixed management; the insulated tools shall not be mixed with metal tools and materials; and the appearance inspection shall be done on the tools. 3) The surface of insulated tools shall not be worn, deformed and damaged, and the operation shall be flexible. Carry out segment insulation detection with such insulated tools with insulation resistance meter of 2,500V or above and with the resistance no less than 700 $M\Omega$, and wipe it off with a clean dry towel.	7	1) Deduct 1 point for failure to use moisture-proof cloth and place tools to designed positions. 2) Deduct 0.5 points/item for failure to check qualified label of tool test and appearance inspection. 3) Deduct 1 point/item for failure to use testing instrument for testing the tools. 4) Deduct 2 points/person time for the operator failing to wear the shielding clothes correctly and each part connected well.			

Part II Skill Module Training and Assessment Standards

Table (Cont'd)

S/N	Project name	Quality requirements	Score	Deduction standard	Reasons for deduction	Deduction	Scoring
5	Inspection of tools	4) The ground potential and equipotential personnel on tower shall correctly wear a whole suit of qualified shield clothes and conductive shoes as required, with each part connected well, shall not wear chemical fiber clothes next to the skin in the shielding clothes and shall fasten safety belts; the Responsible Person shall carefully check whether they wears it correctly. 5) Tower climbing personnel shall check the double name, pole number and phase again and report them		5) Deduct 1 point for the on-site Responsible Person failing to check the safety protective equipment of the tower climbing operators. 6) Deduct 2 points/person for the tower climbing personnel failing to check the double name of the line, pole number and phase. 7) Deduct 2 points/person for the tower climbing personnel failing to report the check results			
6	Climbing the tower	1) The ground electrician and the equipotential electrician on tower shall wear a whole suit of qualified shielding clothes, fasten the safety belt after performing the impulse test on the safety belt, and carry the insulated transmission rope to climb the tower one after another. 2) During the tower climbing process, they shall fasten the anti-fall protection device, climb the tower to an appropriate position, fasten the safety belt, arrange the insulated transmission rope, and then cooperate with the ground electrician to make lifting preparation of the insulated transmission rope separately.	5	1) Deduct 2 points for failure to fasten the safety belt or for failure to perform the impulse test on the safety belt and the backup protection rope. 2) Deduct 2 points for grasping the shackles with hands. 3) Deduct 1 point for inconvenient suspension position of tackle transmission rope for taking tools. 4) Deduct 2 points for metal tools that are difficult to ensure safe distance during transfer; deduct 2 points for tools that are not bound securely.			

Table (Cont'd)

S/N	Project name	Quality requirements	Score	Deduction standard	Reasons for deduction	Deduction	Scoring
6	Climbing the tower	3) During tower climbing, the electrician shall fasten the anti-fall protection device, climb the tower at a uniform speed, grasp the main material by hand, hang the safety belt on the shoulder and keep the safety distance of more than 6.8m (medium phase)/ 6.0m (medium phase) away from the electrified body, and the Responsible Person shall strengthen the operation monitoring		5) Deduct 2 points for falling object at high place. 6) Deduct 2 points for tools colliding with the tower body during the transfer process. 7) Deduct 1 point for knotting and disordered rope in tool transfer. 8) Deduct 2 points for the Responsible Person failing to monitor the operation in place. 9) Deduct 2 points for incorrect operation of electrician on tower			
7	Installation of tackle block and Cradle	1) The lifting of insulating sling shall be smooth, free of collision and winding in transfer. 2) After the basket is installed, the electrician on tower shall carry out the careful inspection and check to the basket situation. 3) 2-2 tackle block and basket shall be installed reliably in a suitable position on the cross arm	5	1) Deduct 0.5 points for rope winding of 2-2 tackle block. 2) Deduct 1 point for unreasonable installation position of track rope. 3) Deduct 1 point for unsuitable length of insulated protection rope. 4) Deduct 1 point for unsteadiness and collision of transfer tools.			

Part II Skill Module Training and Assessment Standards

Table (Cont'd)

S/N	Project name	Quality requirements	Score	Deduction standard	Reasons for deduction	Deduction	Scoring
8	Entering the intense electric field	1) After re-checking and confirming that each part of the shielding clothes is connected reliably, the equal potential electrician carries out the impulse test on the basket, and fastens the protection rope to climb the basket after reporting it to the Responsible Person. 2) The ground electrician slowly loosens the control rope of the 2-2 insulated tackle block, and slows down when it is about 2m away from the conductor. The equipotential electrician applies to the Responsible Person for potential transfer at a distance of 0.5m from the conductor, and quickly extends the potential transfer rod to hook it onto the nearest sub-conductor for potential transfer after obtaining the consent. 3) After entering the intense electric field, the equipotential electrician shall protect the human body, and control the head not to exceed the grading ring on the conductor side. (4) The combined gap composed of the gaps between the electrician and the grounding body and the electrified body in the process of entering the intense electric field shall not be less than 6.9m (middle phase)/6.7m (side phase), and the potential transfer rod must be used for potential transfer when entering the intense electric field.	10	1) Deduct 1 point for the equipotential electrician failing to conduct the impulse on the basket. 2) Deduct 1 point for failure to fasten the insulated protection rope. 3) Deduct 2 points for the ground electrician failing to check the safety measures of equipotential electrician. 4) Deduct 1 point for the ground electrician failing to control the tail rope of the tackle steadily. 5) Deduct 2 points for the equipotential electrician failing to apply to the Responsible Person before entering the intense electric field; deduct 1 point for starting to enter the intense electric field without consent after application. 6) Deduct 1 point for unskillful potential transfer and 5 points for potential transfer without using the potential transfer rod. 7) Deduct 2 points for the equipotential electrician failing to fasten the safety belt after entering the intense electric field. 8) Deduct 2 points for the equipotential electrician with head exceeding the grading ring on the conductor side after entering the intense electric field.			

· 351 ·

Table (Cont'd)

S/N	Project name	Quality requirements	Score	Deduction standard	Reasons for deduction	Deduction	Scoring
9	Install tools and transfer conductor load	1) The ground electrician carries the insulating rope to the cross arm operation point above the medium phase conductor, and installs the insulated tackle and the insulated transmission rope to the appropriate position. 2) The ground electrician transfers the airfoil clamp, insulated pull-rod, hydraulic leading screw and conductor backup protection rope to the working position, and the ground electrician cooperates with the equipotential electrician to install the tools correctly. 3) After checking the reliability of each component, the equipotential electrician tightens the hydraulic leading screw to make it bear a little force and check the stress point. 4) With the consent of the Responsible Person, the equipotential electrician continues to tighten the hydraulic leading screw evenly to relax the composite insulator.	15	1) Deduct 1 point for unsteady transfer of tools; deduct 1 point for each collision. 2) Deduct 2 points for failure to check the reliable installation and good stress of load-bearing tools, and 2 points for failure to report it and obtain the consent of the Responsible Person. 3) Deduct 1 point for ineffective communication between the ground electrician and the equipotential electrician. 4) Deduct 5 points for failure to check the load-bearing tools before removing the insulator string, and 2 points for failure to report the check results. 5) Deduct 2 points for inappropriate position of the insulating rope fastening the composite insulator. 6) Deduct 2 points for each collision during the transfer of the insulator string. 7) Deduct 1 point for the ground electrician failing to control the motor winching in place.			

Part II Skill Module Training and Assessment Standards

Table (Cont'd)

S/N	Project name	Quality requirements	Score	Deduction standard	Reasons for deduction	Deduction	Scoring
10	Replacement of insulator string	1) The ground electrician controls the control rope of the composite insulator string, and slowly puts the composite insulator string to the ground by using the motor winching.Attention shall be paid to the fact that the control rope of the composite insulator string shall not collide with the bearing tools, conductor and pole and tower. 2) The ground electrician transfers the insulated transmission rope and the control rope of the composite insulator string to the new composite insulator string respectively. 3) The ground electrician starts the motor winching to transfer the new composite insulator string to the working position on tower. The ground potential electrician restores the connection between the new composite insulator string and ball head hanging ring, and restore the fitting pin. 4) The ground electrician slowly loosens the motor winching to make the composite insulator string naturally vertical, and the ground electrician restores the connection between the socket hanging plate and the metal yoke plate, and installs the cotter pin.	17	1) Deduct 1 point for the ground electrician failing to control the tail rope of the insulator well. 2) Deduct 1 point for unreasonable rope buckle binding. 3) Deduct 2 points for failure to check the steering of winching and the stress of tackle.5) Deduct 2 points for each collision during the transfer of the insulator string. 4) Deduct 5 points for failing to install the insulator string in place.5) Deduct 5 points for the operator failing to check the connection of insulator string. 6) Deduct 2 points for the operator failing to check the installation of pin in place. 7) Deduct 2 points for the Special Supervisor failing to fulfill the supervision responsibility.			

Table (Cont'd)

S/N	Project name	Quality requirements	Score	Deduction standard	Reasons for deduction	Deduction	Scoring
11	Remove tools	1) After checking the reliable connection of the composite insulator string and obtaining the consent of the Responsible Person, the ground electrician loosens the hydraulic stringing system and the planar screw rod. 2) After checking that the stress of the composite insulator string is normal and obtaining the consent of the Responsible Person, the ground electrician shall cooperate with the equipotential electrician to remove the insulated pull-rod, hydraulic screw rod, airfoil clamp, conductor backup protection rope, and transfer them to the ground.	5	1) Deduct 1 point for failing to check the connection of each part for correctness in detail, and for failing to obtain the consent of the Responsible Person. 2) Deduct 1 point for unsteady transfer of tools; deduct 1 point for each collision.			
12	Exiting the potential	1) No. 1 equipotential electrician shall fasten the insulated protective rope, hook the metal end of the potential transfer rod to the sub-conductor, enter the basket with one hand holding the insulated handle, and then keep the arm straight so that the basket is 0.5m away from the sub-conductor. 2) The equipotential electrician applies to the Responsible Person for exiting the potential. After obtaining the consent, the equipotential electrician quickly removes the connection between the potential transfer rod and the sub-conductor, and puts the potential transfer rod back in the basket.	8	1) Deduct 2 points for failing to report the end of work, and 1 point for each item left in the intense electric field. 2) Deduct 1 point for the equipotential electrician failing to fasten the insulated protection rope. 3) Deduct 1 point for the ground electrician failing to control the tail rope of the tackle steadily. 4) Deduct 2 points for the equipotential electrician failing to apply to the Responsible Person before entering the intense electric field, and 1 point for starting to enter the intense electric field without consent after application.			

Table (Cont'd)

S/N	Project name	Quality requirements	Score	Deduction standard	Reasons for deduction	Deduction	Scoring
12	Exiting the potential	3) At the same time, the ground electrician quickly tightens the control rope of the 2-2 insulated tackle block, pulls the basket upward to the cross arm and stops it. Then, the equipotential electrician climbs the cross arm and fastens the safety belt. 4) The ground electrician transfers the basket to the another equipotential electrician by means of insulated transmission rope. Then, the equipotential electrician enters the basket after checking that there is no object left on the conductor, and exits the potential by the same method.	8	5) Deduct 1 point for unskillful potential transfer and 5 points for potential transfer without using the potential transfer rod.			
13	Remove the basket and return to the ground	1) The electrician on tower shall cooperate the removal of the basket track rope, the insulated protection rope, the 2-2 insulated tackle block and the basket and transfer of them to the ground. 2) After checking that there is no object left on tower, the equipotential electricians shall report it to the Responsible Person and then climb down the tower with insulated transmission rope after obtaining the consent of the Responsible Person.	5	1) Deduct 2 points for failure to use the falling protector when climbing down the tower. 2) Deduct 2 points for loosing the protection of the safety belt when moving on the tower. 3) Deduct 1 point for grasping the tower nail when climbing down the tower. 4) Deduct 2 points for any objects left on tower.			

Table (Cont'd)

S/N	Project name	Quality requirements	Score	Deduction standard	Reasons for deduction	Deduction	Scoring
14	End of the work	1) The Responsible Person shall organize all working members to put working apparatus and materials in order and put them in a special kit (bag) after cleaning; clean the site to ensure that "the materials are removed and the site is cleaned after construction". 2) In the post-shift meeting, Responsible Person will give work summaries and comments. Comments include the construction quality of this work and the implementation of safety measures from all working personnel. 3) Responsible Person shall report to the on-duty control personnel that the work is over, apply for the restoration of circuit re-closing and terminate the work order.	10	1) Deduct 2 points for failure to clean the tools. 2) Deduct 2 points for missing tools. 3) Deduct 2 points for failure to hold the post-shift meeting. 4) Deduct 2 points for failure to remove the fence. 5) Deduct 2 points for failure to report to dispatcher			
	Total		100				

Module 3 Standards for Training and Assessment on Live Replacement of 1 ~ 3 Glass Insulators on the Cross Arm Side of 1000kV AC Transmission Line Resisting-tensile Tower

I. Training Standard

(I) Training Requirements

Designation of module	The live replacement of 1~3 glass insulators on the cross arm side of 1000kV AC transmission line resisting-tensile tower	Type of training	Operation
Training method	Practical operation training	Hours of training	21 training hours
Training objectives	1. Master the electrical significance of electromagnetic protection in ground potential operation method. 2. Can independently complete the live replacement of 1~3 glass insulators on the cross arm side of 1000kV AC transmission line resisting-tensile tower (intermediate potential operation method)		
Training venue	UHV AC training line		
Training content	Operation of live replacement of 1~3 glass insulators on the cross arm side of 1000kV AC transmission line resisting-tensile tower by adopting the ground potential operation method		
Scope of application	Maintenance personnel of UHV AC transmission line		

(II) Referenced Rules and Specifications

(1) Code for Design of 1000kV Overhead Transmission Line (GB50665-2011).

(2) Maintenance Code for 1000kV AC Transmission Line (DL/T209-2008).

(3) Operation Code for 1000kV AC Transmission Line (DL/T307-2010).

(4) Technical Guide for Live Working on 1000kV AC Transmission Line (DL/T392-2015).

(5) Calculation Method of Live Working Minimum Approach Distance on AC Transmission Line (GB/T 19185-2008)

(6) Guidelines of Insulation Coordination for Live Working (DL/T867-2004).

(7) Live Working-Guidelines for the Installation of Power Transmission Line Conductors and Earthwires-stringing Equipment and Accessory Items (DL/T 1007-2006).

(8) State Grid Corporation of China on the Management Regulations of Live Working (Trial Implementation) (SGCC [2007] No.751).

(9) State Grid Corporation of China Working Regulations of Power Safety (Transmission Line Section) (Q/GDW1799.2-2013).

(10) Electrotechnical Terminology - Overhead Line (GB/T 2900.51-1998).

(11) Electrotechnical Terminology- Live Working (GB/T2900.55-2002).

(12) Live Working - Terminology for Tools, Equipment and Devices (GB/T 14286-2002).

(13) Minimum Requirements for Utilization of Tools, Devices and Equipment for Live Working (DL/T 877-2004).

(14) Preventive Test Code of Tools, Devices and Equipment for Live Working (DL/T 976-2005).

(15) Insulated Tackles for Live Working (GB/T13034-2008).

(16) Live Working-Insulating Ropes (GB 13035-2008).

(17) Screen Clothes for Live Working on 1000kV AC (GB/T25726-2010).

(Ⅲ) Teaching Design for Training

To complete the work task of "live replacement of 1~3 glass insulators on the cross arm side of 1000kV AC transmission line resisting-tensile tower", each training stage shall be designed according to the standard operation procedure for work task completion. Each stage includes specific training objectives, training content, hours of training, training methods (training resources), training environment, assessment and evaluation, etc, as shown in the Table 2-3-1.

Part II Skill Module Training and Assessment Standards

Table 2-3-1 Live Replacement of 1 ~ 3 Glass Insulators on the Cross Arm Side of 1000kV AC Transmission Line Resisting-tensile Tower

Training schedule	Training objectives	Training content	Hours of training	Training methods and resources	Preparation of training conditions	Assessment and evaluation
1. Theoretical teaching	1. Preliminarily master the basic method for electromagnetic protection in ground potential operation method. 2. Be familiar with the replacement method of strain single insulator of transmission line	1. Master the electrical significance of electromagnetic protection in ground potential operation method. 2. Method and quality standard for replacement of strain single insulator of transmission line	2	Training methods: Lecture. Training resources: PPT, relevant regulations and specifications	Multimedia classroom,	Attendance, classroom questions and assignments
2. Preparations	Be able to complete the preparation before operation	1. Work site survey. 2. Preparation of the standardized operation card. 3. Filling of the work order. 4. Preparation of tools and materials for this operation	1	Training methods: 1. Site survey and cleaning of tools and materials shall be practiced at site. 2. Preparation of operation card and the filling of work order shall adopt lecture method. Training resources: 1. 1000kV practical training line. 2. UHV tools warehouse. 3. Blank work order	1. UHV training transmission line. 2. Multi-Media classroom	
3. Work site preparation	Be able to complete the preparations of work site	1. Work site re-survey. 2. Job application. 3. Work site layout. 4. Pre-shift meeting. 5. Inspection of tools and materials	1	Training methods: demonstration and role play. Resources: 1000kV training line	1000kV training line	

Standard for Professional Training and Assessment for Operation Maintenance of UHV AC Transmission Line

Table (Cont'd)

Training schedule	Training objectives	Training content	Hours of training	Training methods and resources	Preparation of training conditions	Assessment and evaluation
4. Trainer's demonstration	The trainees can preliminarily understand the operation process of the task through inspecting and learning from each other's work	1. The ground electrician assembles tools. 2. The ground electrician completes the replacement of single glass insulator	2	Training methods: Demonstration. Resources: 1000kV training line	1000kV training line	
5. Group training of trainees	1. Can complete the replacement of strain single glass insulator on the cross arm side	1. The trainees are grouped (6 persons per group) to train the skill operation of replacing insulator. 2. Trainers guide the operation of trainees and conduct safety supervision	14	Training methods: Role play. Resources: 1000kV training line	1000kV training line	Score the operation of trainees according to the detailed rules for skill assessment and scoring
6. End of the work	1. Enable the trainees to further distinguish the shortcomings of the operation process and facilitate the promotion in the later stage. 2. Train the trainees in the working style of safe and civilized production	1. Cleaning up the work site. 2. Report to dispatcher the end of work. 3. Comment and summarize the work task this time at post-shift meeting	1	Training method: Lecture and inductive method	1000kV training line	

(IV) Operation Flow

1. Work Task

The live replacement of 1 ~ 3 glass insulators on the cross arm side of 1000kV AC transmission line resisting-tensile tower.

2. Requirements for Weather and Work Site

(1)The live replacement of 1 ~ 3 glass insulators on the cross arm side of 1000kV AC transmission line resisting-tensile tower shall be carried out in good weather. In case of lightning (hearing thunder or seeing lightning), snow, hail, rain, fog and so on, live working is prohibited. When the wind force is greater than level 5, it is unsuitable for live working; when the relative humidity of the air is greater than 80%, insulating tools with moisture-proof properties shall be used if live working needs to be carried out. When emergency live repair is required in bad weather, relevant personnel shall be organized to fully discuss and prepare necessary safety measures, which can be implemented after being approved by the unit.

(2)The operating personnel should be in good mental states, be familiar with the organizational and technical measures to ensure safety in work and master the methods for emergency rescue and first aid for electric shock in heights; they should hold the qualification certificate for live working within the validity period.

(3)Responsible Person should organize the relevant personnel to complete field investigation in advance, determine the operating methods, required working apparatus and necessary measures according to the results, and handle the work order for live working.

(4)The work site shall be reasonably set up with fence and warning signs. Non-operating personnel is forbidden to enter.

(5)The line re-closing device shall be deactivated in the Project.

(6)Mode of operation: Ground potential operation.

(7)Safe working distance and effective insulation length during operation are shown in Table 2-3-2.

Table 2-3-2 Safe Distance for Live Replacement of Any Single Glass Insulator of Tensile Glass Insulator String of 1000kV AC Transmission Line

Altitude/m	Minimum safe distance between the intermediate potential operator and the electrified body /m		Minimum effective insulation length of insulating tools /m
	Middle phase	Side phase	
$H \leqslant 1000$	6.8	6.0	6.8
$1000 < H \leqslant 2000$	7.4	6.6	7.2

Note: The values in the table do not include the human body occupying gap which shall not be less than 0.5 m during operation.

(8) When the ground potential electrician enters the cross arm side of strain insulator string, the number of insulators shorted by the human body shall not be more than 4. The minimum number of good insulators shall meet the requirements of Table 2-3-3 after deducting the number of insulators shorted by human body and defective insulators from the strain insulator string.

Table 2-3-3 Minimum Combined Gap and Minimum Number of Good Insulators

Altitude	Structural height of single glass insulator (mm)	Minimum total length of good insulator string (m)	Minimum number of good insulators
$H \leqslant 1000$	170	7.2	43
	195		37
	205		36
$1000 < H \leqslant 2000$	170	8.0	47
	195		41
	205		39

Note: The values in the table do not include the human body occupying gap which shall not be less than 0.5 m during operation.

3. Preparations

3.1 Hazards and precontrol measures

(1) Hazard - precontrol measures for electric shock injury:

① Before work, Responsible Person should contact the control personnel on duty, deactivate the line re-closing, and perform the licensing procedures.

② Before climbing the tower, operators on tower must carefully check the dual names of the line, the number of the pole and tower, and phase, and then the tower can be climbed after all have been confirmed correct.

③ If lines lose power suddenly during work, operators shall still deem it to be charged. The Responsible Person shall contact the control personnel as soon as possible, and no forced energization is allowed before the on-duty control personnel getting in touch with the Responsible Person.

④ Insulating tool and insulating ropes shall be free of damage, moisture, deformation, and failure. It is not allowed to use non-insulating ropes (such as cotton rope, manila rope, and steel wire rope).

⑤ The ground electrician shall wear clean and dry gloves when operating the insulating tools. When entering the work site, the live working tools shall be placed on damp-proof canvas or insulating mat to prevent dirt and dampness of the insulating tools in use.

⑥ The ground electrician shall wear flame-retardant underwear and full set of qualified shielding clothes outside (including hat, dresses & trousers, gloves, socks and shoes). All parts shall be in excellent connection conditions.

⑦ When the ground electrician enters the cross arm side of the strain insulator string, the positions of the hands and feet shall be correspondingly consistent, and the number of insulators shorted by the human body and the tools shall conform to the provisions of Table 2-3-3.

⑦ When transmitting large metal objects by using insulating rope, the ground potential electricians and ground electricians shall ground the metal objects before contacting.

⑨ During the live working, the Responsible Person (Supervisor) shall continuously monitor the operators and correct their nonstandard operation or actions in violation at any time. Special attention shall be paid to operators work at heights to ensure that there is enough safety distance (meeting the requirements in Table 2-3-2). It is forbidden to contact two non-connected electrified bodies or make contact with electrified body and grounding body at the same time.

(2) Hazard - precontrol measures for falling accident:

① Before climbing, operators work at heights must satisfy the requirements of this operation, such as physical condition, mental state and skill and quality.

② Double safety belts shall be used by the personnel for Work at Heights. When climbing up and down the tower, the main material shall be grasped with hands, with shackles trodden and moving at a uniform speed.

③ Before the operation, the ground electrician shall carefully inspect hydraulic lead screw, closed clamp and so on to ensure that the load-bearing tools are qualified; when moving along the insulator string, the positions of hands and feet must be correspondingly consistent, and the safety belt shall be fastened to the insulator string supported by hands and moved synchronously; when replacing insulators, the load-bearing tools shall be installed reliably. Before and after load transfer, impulse test shall be conducted to determine their reliability, and it shall be reported to the Responsible Person in a timely manner. The implementation can only be carried out with the permission of the Responsible Person.

④ Supervisor shall correct irregularities and violations at any time. Special attention shall be paid to operators for Work at Heights to prevent them from losing the protection of the safety belt or insulated backup protection rope during transposition. The safety belt or insulated backup protection rope shall not be fastened lower than operating personnel.

(3) Hazard - injury caused by objects falling from heights.

Precontrol measures:

① Operator working at heights should put personal tools and fragmentary materials into the tools bag. It is strictly forbidden to hang objects in heights or keep in the mouth.

② Ground electrician should correctly wear a helmet and use the knots. The vertical distance from the work site should not be less than the falling radius.

③ The work site shall be set up with fence and warning signs. It shall be noted at any time that Supervisor shall prohibit irrelevant personnel and vehicles from entering operation area.

3.2 Selection of tools and instruments and materials

Tools and materials required for live replacement of 1 ~ 3 glass insulators on the cross arm side of 1000kV AC transmission line resisting-tensile tower can be seen in Table 2-3-4. Before delivering tools and instruments out of warehouse, application voltage class and test period shall be carefully checked and they shall be inspected to ensure that appearance is intact, connection is firm, rotation is flexible and meet the working task requirements. After delivering tools and instruments out of warehouse, they shall be stored in tools bag or tool kit for transportation to avoid contamination and damp. Metal tools and insulated tools shall be separately loaded and transported to avoid deformation, damage or other defects caused by mixed loading and transportation.

Table 2-3-4 Tools and Materials Required for Live Replacement of 1 ~ 3 Glass Insulators on the Cross Arm Side of 1000kV AC Transmission Line Resisting-tensile Tower

S/N	Name	Specification	Unit	Qty.	Remarks
1	Shielding clothes	Type I	Set	2	PPE
2	Conductive shoes	The size depends on the wearer	Nos.	2	PPE
3	Flame retardant underwear	Pure mulberry silk	Set	2	PPE
4	Safety belt of double insurance	Suspender	Nos.	2	PPE
5	Safety helmet		Nos.	7	PPE

Part II Skill Module Training and Assessment Standards

Table (Cont'd)

S/N	Name	Specification	Unit	Qty.	Remarks
6	Goggles		Nos.	2	PPE
7	Insulated transmission rope	φ14mm, with the length matching the lifting height	Nos.	1	Insulating tool
8	Insulated backup protection rope	φ16mm	Nos.	2	Insulating tool
9	Insulated noose	φ14mm	Nos.	2	Insulating tool
10	Insulated tackle	1T	Nos.	1	Insulating tool
11	Strain end clamp		Nos.		Metal tool
12	Hydraulic leading screw		Nos.		Metal tool
13	Closed clamp (rear clamp)		Nos.		Metal tool
14	Insulation resistance tester	2,500V, with electrode width 2cm and pole width 2cm	Set	1	Other tools
15	Multimeter		Set	1	Other tools
16	Wind speed, temperature and humidity tester		Nos.	1	Other tools
17	Safety fence		Set	Several	Other tools
18	Warning sign	"Work Here", "Access from Here" "Slow Down" and "Blocking"	Set	1	Other tools
19	Red waistcoat	"Responsible Person" "Special Supervisor"	Piece	1	Other tools
20	Moisture-proof tarpaulin	3m×3m	Piece	2	Other tools
21	Personal tools	Wrench, vice	Set	1	Other tools
22	Pin puller		Nos.	1	Other tools
23	Falling protector	Corresponding to the type of pole and tower anti-fall device	Nos.	2	Other tools
24	Towel	Cotton	Nos.	1	Other tools
25	Insulator		Piece	1	Material

3.3 Division of labor for operators

Division of labor for operators of the task is shown in Table 2-3-5.

Table 2-3-5 Division of Labor for Live Replacement of 1 ~ 3 Glass Insulators on the Cross Arm Side of 1000kV AC Transmission Line Resisting-tensile Tower

S/N	Post	Qty. (person)	Responsibilities
1	Responsible Person	1	Be responsible for division of operating personnel of the task, work order reading, handling re-closing deactivation of the line and the formalities of work permit, getting work permits, holding pre-shift meeting, dealing with emergency situations in work, quality surveillance, and the summary after work
2	Special Supervisor	1	Be responsible for the safety control of the work site
3	Ground potential electrician	1	Be responsible for tool installation and insulator replacement
4	Ground electrician	3	Be responsible for ground auxiliary works during the operation.

4. Work Procedure

The workflow of this task is shown in Table 2-3-6.

Table 2-3-6 Workflow of Live Replacement of 1 ~ 3 Glass Insulators on the Cross Arm Side of 1000kV AC Transmission Line Resisting-tensile Tower

S/N	Work Content	Operation Steps and Standards	Safety Measures and Precautions	Responsible Person
1	Site re-survey	The Responsible Person shall complete the following work: (1) Check the line name, the number of the pole and tower and dual number on spot to ensure they are correct; guarantee that the base and the pole and tower are intact and in normal condition, and ensure that the cross and span distance meets the safety requirements; confirm the defect conditions and the specifications and models of earth wires. (2) Check that the site meteorological conditions such as wind speed and humidity should meet the operation requirements. (3) Check that the terrain and environment shall meet the operation requirements. (4) Check that the safety measures listed in the work order are in line with the actual situations on site, and the measures will be supplemented if necessary	(1) Correctly wear helmet, working clothes, work shoes and protective gloves. (2) Operation under meteorological conditions that may endanger the safety of operators is forbidden. (3) Non-operation personnel and vehicles are strictly prohibited from entering the working site.	

Part II Skill Module Training and Assessment Standards

Table (Cont'd)

S/N	Work Content	Operation Steps and Standards	Safety Measures and Precautions	Responsible Person
2	Work Permit	(1) The Responsible Person is responsible for contacting the on-duty control personnel and applying for stopping the line re-closing as per the contents of the work order. (2) Live working could be started only after being approved by the on-duty control personnel.	Live working shall not be started without the permission of the on-duty control personnel Beginning of the work	
3	Site layout	Install the security fence and hang the signboards correctly: (1) The security fence should take full account of falling objects from the heights and the influence on road traffic. (2) The entrance and exit of the security fence shall be set reasonably. (3) Signs such as "Access from Here", "Work Here", "Slow Down" or "Blocking" shall be properly arranged.	When the influence on road traffic safety is uncontrollable, the traffic management department should be contacted in time to strengthen the on-site control of traffic safety	
4	Hold a pre-shift meeting	(1) All working personnel shall line up. (2) The Responsible Person shall wear red waistcoat and read out the work order and be clear with work task and division of personnel; explain safety measures and technical measures in work; check (inquire after) mental state of all working personnel; inform of hazards in work and precontrol measures. (3) All working personnel shall sign on the work order for confirmation.	(1) The work order shall be filled in, issued and approved in a standardized manner, and the signature shall be complete. (2) All working personnel shall be in good mental states. (3) All working personnel shall be clear with task division of works, safety measures and technical measures	

Table (Cont'd)

S/N	Work Content	Operation Steps and Standards	Safety Measures and Precautions	Responsible Person
5	Inspect tools	(1) All necessary tools and instruments shall be prepared as per the operation requirements and placed on the moisture-proof tarpaulin regularly according to the category and location. The appearance and test certificate of tools and instruments shall be checked to ensure there is no omission. (2) The surface insulation resistance of insulating tools and insulating ropes shall be tested with insulation resistance tester in correct methods, and the value shall be not less than 700MΩ. (3) The new insulator is wiped up, and it is intact in appearance inspection, without rust, cracks and breakage. The surface insulation resistance shall be tested with insulation resistance tester in correct methods, and the value shall be not less than 500MΩ. (4) The internal resistance of full shielding clothes shall be tested with a multimeter in correct methods, and the value shall be not more than 20Ω. (5) The inspector shall report to the Responsible Person that all inspection results are in conformity with the operation requirements	(1) The waterproof tarpaulin shall be enough in quantity and reasonable in position, and be clean and dry. (2) Before using metal and insulating tools, they shall be carefully checked for damage, dampness, deformation and failure, and the certificate of conformity is within the validity period. (3) Insulating tools and insulating ropes are tested as acceptable.	
6	Climbing the tower	(1) The ground electrician shall check the double name and phase of the line again, inspect and confirm that the shackles are complete and firm; fasten safety belts and attach falling protectors; perform impact inspection on safety belt, backup protection rope and falling protector in correct method; the Responsible Person shall check and confirm that the connection conditions of each connection point of the double safety belt and shielding clothes worn by them are in good condition, including shoulder harness, pectoral harness, belt, back rope sling, buckle and ring.	(1) Safety belt and falling protector shall pass the impulse test. (2) The safety belt and insulated transmission rope shall be prevented from hooking up tower materials. (3) The minimum safe distance between the human body and the conductor shall conform to the regulations shown in Table 2-3-2. (4) It is forbidden to grasp shackles with hands.	

Part II Skill Module Training and Assessment Standards

Table (Cont'd)

S/N	Work Content	Operation Steps and Standards	Safety Measures and Precautions	Responsible Person
6	Climbing the tower	(2) The electrician shall carry toolkit and insulated transmission rope (including insulated tackle) in correct methods. (3) Main band of safety belt and backup protection rope shall obliquely across the shoulder. (4) The soles shall be cleaned, and climbing the tower shall be done in turn with the permission of the Responsible Person. (5) The electrician shall tread on shackles, grasp the main materials, and climb the tower uniformly to the proper position of the cross arm with safety belt well fastened and then break away from the falling protector.	(5) Falling protector shall be properly used. (6) Protection of safety belt shall not be lost during transposition.	
7	Install tools and transfer conductor tension	(1) The ground electrician transfers the closed clamp (front clamp), hydraulic leading screw, strain end clamp and so on to the ground electrician by using the insulated transmission rope. The lifting process shall be smooth, free from collision and winding, and the knot shall be used correctly. (2) The ground electrician first installs the strain end clamp on the towing plate and then the closed clamp (rear clamp) on the third insulator on the cross arm side, and connects the hydraulic leading screw. Each part of the load-bearing tools is securely and reliably installed. (3) Equipotential electricians shall check and confirm that all parts of the load-bearing tools are installed in good condition. With the permission of the Responsible Person, they can operate the hydraulic leading screw so that it is gradually stressed and the insulators to be replaced are relaxed. The two hydraulic leading screws shall be uniformly stressed.	(1) The safe distance between the human body and the electrified body shall not be less than that specified in Table 2-3-2. (2) Objects falling from high place shall be avoided. (3) After deducting inferior insulators, insulators shorted by human body and tools, the number of good insulators shall conform to that specified in Table 2-3-3.	

Table (Cont'd)

S/N	Work Content	Operation Steps and Standards	Safety Measures and Precautions	Responsible Person
8	Replacement of insulator	(1) The ground electrician shall perform impulse test, check and confirm that the load-bearing tools are stressed normally. With the permission of the Responsible Person, they shall tie the old insulator with the insulated transmission rope, take out the fitting pins at both ends of the old insulator, continue to operate and tighten the hydraulic leading screw until the old insulator is removed. The two hydraulic leading screws shall be stressed uniformly, and the operating handle shall not knock on the insulator. (2) The ground electrician uses the other end of the insulated transmission rope to fasten the new insulator, and lifts the new insulator to the ground electrician by means of the old lower and the new upper. The lifting process shall be smooth, free from collision and winding, and the knot shall be used correctly. (3) The ground electrician installs the new insulator, resets the fitting pin at its two ends, and installs it in place	(1) The safe distance between the human body and the electrified body shall not be less than that specified in Table 2-3-2. (2) Objects falling from high place shall be avoided	
9	Remove tools	(1) The ground electricians check and confirm that the connection of the new insulator is reliable and, with the permission of the Responsible Person, they operate and loosen the hydraulic leading screw so that the replaced insulator is gradually stressed. (2) After load transfer, the ground electricians shall perform the impulse test, check and confirm that the new insulator is in good condition. With the permission of the Responsible Person, they can remove the insulated transmission rope tied to the insulator, fasten it to the proper position of the load-bearing tools, remove the closed clamp (front clamp), hydraulic leading screw, strain end clamp and other load-bearing tools, and transfer them to the ground with the cooperation of the ground electrician. The transfer process shall be smooth, free from collision and winding, and the knot shall be used correctly.	(1) The safe distance between the human body and the electrified body shall not be less than that specified in Table 2-3-2. (2) Objects falling from high place shall be avoided	

Part II Skill Module Training and Assessment Standards

Table (Cont'd)

S/N	Work Content	Operation Steps and Standards	Safety Measures and Precautions	Responsible Person
10	Evacuation from the pole and tower	The ground potential electricians shall remove the insulated transmission rope after checking that there is no object left on the tower. With the permission from Responsible Person, the electrician shall properly hang the falling protector, unfasten and put straight the safety belt, carry the insulated transmission rope correctly and step down to the ground from the tower in uniform steps by treading on shackles and grasping the main material	(1) The safe distance between the human body and the electrified body shall not be less than that specified in Table 2-3-2. (2) Protection of safety belt shall not be lost during transposition. (3) Hand slippage and step missing shall be avoided and it's forbidden to grasp the shackles with hands. (4) Falling protector shall be properly used. (5) The insulated transmission rope and safety belt shall be prevented from hooking up the tower materials or shackles	
11	End of the work	(1) The Responsible Person shall organize all working members to put working apparatus and materials in order and put them in a special kit (bag) after cleaning; clean the site to ensure that "the materials are removed and the site is cleaned after construction". (2) In the post-shift meeting, Responsible Person will give work summaries and comments. Comments include the construction quality of this work and the implementation of safety measures from all working personnel. (3) Responsible Person shall report to the on-duty control personnel that the work is over, apply for the restoration of circuit re-closing and terminate the work order.		

Standard for Professional Training and Assessment for Operation Maintenance of UHV AC Transmission Line

II. Assessment Standard

Detailed Rules for Assessment and Scoring of Operation and Inspection Skills of UHV AC Transmission Line

Fill-in Column of Examinee	No.:		Name:	Position:	Date:	MM/DD/YYYY
Fill-in Column of Assessor	Grade:		Assessor:	Assessment Team Leader:	Starting time:	Closing time:	Operation Duration:
Assessment module	The live replacement of 1 ~ 3 glass insulators on the cross arm side of 1000kV AC transmission line resisting-tensile tower	Assessee	Maintenance personnel of UHV AC transmission line	Assessment method	Operation	Assessment Time Limit	90min
Job Description	The live replacement of 1 ~ 3 glass insulators on the cross arm side of 1000kV AC transmission line resisting-tensile tower						
Work Specifications and Requirements	1. Live working shall be carried out in good weather.In case of thunder, rain, snow or fog, no live working shall be carried out.When the wind force is greater than Level 5, live working should not be carried out. When the humidity is greater than 80%, insulating tools with moisture-proof properties shall be used if live working needs to be carried out. 2. Six persons are required for this operation, including 1 Responsible Person, 1 special Supervisor, 1 ground electrician and 3 ground electricians. 3. Responsibilities of Responsible Person (Supervisor): Be responsible for division of operating personnel of the task, work order reading, handling re-closing deactivation of the line and the formalities of work permit, getting work permits, holding pre-shift meeting, safety supervision in the operation process, dealing with emergency situations in work, quality surveillance, and the summary after work 4. During the live working, if thunder, rain, strong wind or any other circumstance threaten the safety of the staff, the Responsible Person or Supervisor may stop working temporarily according to the circumstances Given conditions: 1. Work orders have been handled, safety measures have been completed (re-closing has been deactivated), and oral application (dispatcher or assessor) shall be made at the beginning and end of the work.						

Table (Cont'd)

Work Specifications and Requirements	2. The instrument shall be used safely and correctly to test the insulating tool. 3. The operation must be carried out according to the working procedures. The scores of the items to be carried out shall be deducted for the process error. In case of major hidden dangers of personal, equipment and operational safety, the assessor may order the termination of the operation (assessment)
Assessment scenario preparation	1. Tower type: 1000kV AC resisting-tensile tower. 2. Required operation tools: 2 safety belts (including double-protective rope), 2 sets of Type I shielding clothes, 2 sheets of moisture-proof cloth, 1 multimeter, 1 insulation resistance detector, 1 anemometer, 1 2-in-one temperature and humidity detector, 2 hydraulic leading screws, 2 insulating ropes, 1 set of closed clamp, 1 tackle, 2 goggles, 1 pin puller and 1 set of hand-operated tools. 3. The work site shall be monitored, and the safety measures (fence, etc.) on the work site have been fully implemented; non-operation personnel are prohibited from entering the site, and the staff must wear safety helmets when entering the work site. 4. Examinees shall bring their own work clothes, helmets and gloves
Remarks	1. The deduction shall be done until the scores of each item are deducted completely. In case of major hidden dangers of personal, equipment and operational safety, the assessor may order the termination of the operation. 2. When equipment, working environment, safety belt, safety helmet, tool, shielding clothes, etc., do not conform to the operation condition, the assessor may order the termination of the operation

Table (Cont'd)

S/N	Project name	Quality requirements	Score	Deduction standard	Reasons for deduction	Deduction	Scoring
1	Site re-survey	1) The Responsible Person shall go to the work site to check the line name, pole and tower number, on-site working conditions, defective parts and so on. 2) Check that the site meteorological conditions such as wind speed and humidity should meet the operation requirements. 3) Check whether the work order is complete and unmodified, check whether the safety measures listed are consistent with the actual situation on site, and supplement it if necessary	5	1) Deduct 1 point/item for failure to check the line name, pole and tower number, on-site working conditions and defective parts. 2) Deduct 1 point/item for failure to detect wind speed, humidity and other on-site meteorological conditions. 3) Deduct 0.5 points/part for any alteration in the work order filling; deduct 1 point for incorrect work order number; deduct 1.5 points for incomplete work order filling			
2	Work Permit	1) The Responsible Person is responsible for contacting the on-duty control personnel (referee) and applying for stopping the line re-closing as per the contents of the work order. 2) Reporting content is standardized and complete	2	1) Deduct 2 points for failure to contact the dispatching department (referee) for deactivating the re-closing. 2) Deduct 1 point for non-standard or incomplete terminology reporting			
3	Site layout	Install the security fence and hang the signboards correctly: 1) The security fence should take full account of falling objects from the heights and the influence on road traffic. 2) The entrance and exit of the security fence shall be set reasonably. 3) Signs such as "Access from Here", "Work Here", "Access from Here" shall be properly and well arranged	3	1) Deduct 1 point for failure to arrange the fence at the work site. 2) Deduct 1 point for failure to arrange the warning board. 3) Deduct 1 point for failure to hang the tower climbing operation sign			

Part II Skill Module Training and Assessment Standards

Table (Cont'd)

S/N	Project name	Quality requirements	Score	Deduction standard	Reasons for deduction	Deduction	Scoring
4	Hold a pre-shift meeting	1) All staff and personnel shall wear safety helmets and work clothes correctly. 2) Responsible Person shall wear red vest and read out the work order and be clear with work task and division of personnel; explain safety measures and technical measures in work; check (inquire after) mental state of all working personnel; inform of hazards in work and precontrol measures. 3) All working personnel shall sign on the work order for confirmation.	3	1) Deduct 0.5 points/person for the staff not dressing uniformly. 2) Deduct 3 points for no division of labor; deduct 1 point for unclear division of labor. 3) Deduct 1 point for the on-site Responsible Person not wearing a safety monitoring vest. 4) Deduct 1 point for the work shift member failing to sign or signing incompletely on the work order.			
5	Inspection of tools	1) All necessary tools and instruments shall be prepared as per the operation requirements and placed on the moisture-proof tarpaulin regularly according to the category and location. The appearance and test certificate of tools and instruments shall be checked to ensure there is no omission. 2) The surface insulation resistance of insulating tools and insulating ropes shall be tested with insulation resistance tester in correct methods, and the value shall be not less than 700MΩ.	7	1) Deduct 1 point for failure to use moisture-proof tarpaulin and place tools to designed positions. 2) Deduct 0.5 points/item for failure to check the appearance of tools and pass the test certificate. 3) Deduct 1 point/item for failure to use testing instrument for testing the tools. 4) Deduct 1 point for non-standard reporting of test results; deduct 0.5 points/item for incomplete reporting			

· 375 ·

Standard for Professional Training and Assessment for Operation Maintenance of UHV AC Transmission Line

Table (Cont'd)

S/N	Project name	Quality requirements	Score	Deduction standard	Reasons for deduction	Deduction	Scoring
5	Inspection of tools	3) The new insulator is wiped up, and it is intact in appearance inspection, without rust, cracks and breakage. The surface insulation resistance shall be tested with insulation resistance tester in correct methods, and the value shall be not less than 500MΩ. 4) The internal resistance of full shielding clothes shall be tested with a multimeter in correct methods, and the value shall be not more than 20Ω. 5) The inspector shall report to the Responsible Person that all inspection results are in conformity with the operation requirements					
6	Climbing the tower	1) The ground electrician shall check the double name and phase of the line again, inspect and confirm that the shackles are complete and firm; fasten safety belts and attach falling protectors; perform impulse test on double safety belts, backup protection rope and falling protector in correct method;The Responsible Person shall check and confirm that the connection conditions of each connection point of the double safety belt and shielding clothes worn by the ground electricians are in good condition, including shoulder harness, pectoral harness, belt, back rope sling, buckle and ring.	5	1) Deduct 1 point/item for the intermediate potential electrician failing to check the double name of the line, pole number, phase and tower material; deduct 1 point for non-report after checking. 2) Deduct 2 points/item for failure to perform the impulse test on double safety belts and falling protector. 3) Deduct 1 point for the on-site Responsible Person failing to check the safety protection equipment of the intermediate potential electrician. 4) Deduct 0.5 points/time for hands grasping the shackles.			

Part II Skill Module Training and Assessment Standards

Table (Cont'd)

S/N	Project name	Quality requirements	Score	Deduction standard	Reasons for deduction	Deduction	Scoring
		2) The electrician shall carry toolkit and insulated transmission rope (including insulated tackle) in correct methods. 3) During the process of climbing the tower, the electrician shall fasten the anti-fall device, tread on shackles, grasp the main materials, and climb the tower uniformly to the proper position with safety belt well fastened and then break away from the falling protector		5) Deduct 1 point for the unreasonable suspension position of the tackle transmission rope. 6) Deduct 5 points for loosing the protection of the safety belt during transmission position			
7	Installation tool	1) The ground electrician lifts the closed clamp (front clamp), hydraulic leading screw, strain end clamp and so on to the ground electrician by using the insulated transmission rope. The lifting process shall be smooth, free from collision and winding, and the knot shall be used correctly. 2) The ground electrician first installs the strain end clamp on the towing plate and then the closed clamp (rear clamp) on the third insulator on the cross arm side, and connects the hydraulic leading screw. Each part of the load-bearing tools is securely and reliably installed. 3) Equipotential electricians shall check and confirm that all parts of the load-bearing tools are installed in good condition. With the permission of the Responsible Person, they can operate the hydraulic leading screw so that it is gradually stressed and the insulators to be replaced are relaxed. The two hydraulic leading screws shall be uniformly stressed.	20	1) Deduct 1 point/time for unstable lifting process, colliding and winding. 2) Deduct 2 points/time for falling objects from heights. 3) Deduct 2 points for improper installation and fixing of clamp. 4) Deduct 3 points for failure to check the installation of load-bearing tools; deduct 1 point for failure to report after checking; deduct 1 point for starting tightening the leading screw without the consent of Responsible Person after reporting. 5) Deduct 3 points/time for the number of shorted insulators exceeding 4 during the operation. 6) Deduct 2 points for collision and breakage of insulators in the clamp installation. 7) Deduct 2 points for failure to tighten leading screw in a balanced way			

Table (Cont'd)

S/N	Project name	Quality requirements	Score	Deduction standard	Reasons for deduction	Deduction	Scoring
8	Replacement of insulator	1) The ground electrician shall perform impulse test, check and confirm that the load-bearing tools are stressed normally. With the permission of the Responsible Person, they shall tie the old insulator with the insulated transmission rope, take out the fitting pins at both ends of the old insulator, continue to operate and tighten the hydraulic leading screw until the old insulator is removed. The two hydraulic leading screws shall be stressed uniformly, and the operating handle shall not knock on the insulator. 2) The ground electrician uses the other end of the insulated transmission rope to fasten the new insulator, and lifts the new insulator to the ground electrician by means of the old lower and the new upper. The lifting process shall be smooth, free from collision and winding, and the knot shall be used correctly. 3) The ground electrician installs the new insulator, resets the fitting pin at its two ends	20	1) Deduct 3 points for failure to check the stress of load-bearing tools; deduct 2 points for failure to report after checking; deduct 1 point for removing fitting pins at both ends of the old insulator without the consent of Responsible Person after reporting. 2) Deduct 2 points for failure to tighten leading screw in a balanced way. 3) Deduct 1 point/time for the operating handle hitting the insulator. 4) Deduct 1 point for wrong knot. 5) Deduct 2 points/time for falling objects from heights. 6) Deduct 1 point for new and old insulators colliding with each other. 7) Deduct 1 point for the transmission insulator and tower body colliding with each other. 8) Deduct 2 points for winding insulating ropes			

Part II Skill Module Training and Assessment Standards

Table (Cont'd)

S/N	Project name	Quality requirements	Score	Deduction standard	Reasons for deduction	Deduction	Scoring
9	Remove tools	1) The ground electricians check that the connection of the new insulator is reliable and, with the permission of the Responsible Person, they operate and loosen the hydraulic leading screw so that the replaced insulator is gradually stressed. 2) After load transfer, the equipotential electricians shall perform the impulse test, check and confirm that the new insulator is in good condition. With the permission of the Responsible Person, they can remove the insulated transmission rope tied to the insulator, fasten it to the proper position of the load-bearing tools, remove the closed clamp (front clamp), hydraulic leading screw, strain end clamp and other load-bearing tools, and transfer them to the ground with the cooperation of the ground electrician. The transfer process shall be smooth, free from collision and winding, and the knot shall be used correctly.	20	1) Deduct 3 points for failure to check the connection of new insulator; deduct 2 points for failure to report after checking; deduct 1 point for starting loosening the hydraulic leading screw without the consent of Responsible Person after reporting. 2) Deduct 3 points for failure to check the stress of new insulator; deduct 2 points for failure to report after checking; deduct 1 point for starting removing the hydraulic leading screw without the consent of Responsible Person after reporting. 3) Deduct 1 point for incorrect use of knot when binding the tools. 4) Deduct 2 points/time for falling objects from heights. 5) Deduct 1 point/time for the tools colliding with each other; deduct 1 point/time for the tool colliding with the electrified body or tower body; deduct 2 points for winding insulating ropes			

Table (Cont'd)

S/N	Project name	Quality requirements	Score	Deduction standard	Reasons for deduction	Deduction	Scoring
10	Return to the ground	The ground potential electricians shall remove the insulated transmission rope after checking that there is no object left on the tower. With the permission from Responsible Person, the electrician shall properly hang the falling protector, unfasten and put straight the safety belt, carry the insulated transmission rope correctly and step down to the ground from the tower in uniform steps by treading on shackles and grasping the main material	5	1) Deduct 5 points for failure to use the falling protector when climbing down the tower. 2) Deduct 5 points for loosing the protection of the safety belt when moving on the tower. 3) Deduct 1 point/time for hands grasping the shackles when climbing down the tower. 4) Deduct 2 points for any objects left on tower.			
11	End of the work	1) The Responsible Person shall organize all working members to put working apparatus and materials in order and put them in a special kit (bag) after cleaning; clean the site to ensure that "the materials are removed and the site is cleaned after construction". 2) In the post-shift meeting, Responsible Person will give work summaries and comments. Comments include the construction quality of this work and the implementation of safety measures from all working personnel. 3) Responsible Person shall report to the on-duty control personnel that the work is over, apply for the restoration of circuit re-closing and terminate the work order.	10	1) Deduct 2 points for failure to clean the tools. 2) Deduct 2 points for missing tools. 3) Deduct 10 points for failure to hold the post-shift meeting. 4) Deduct 2 points for failure to remove the fence. 5) Deduct 2 points for failure to report to dispatcher			
	Total		100				

Module 4 Standards for Training and Assessment on Live Replacement of 1 ~ 3 Glass Insulators on the Conductor Side of 1000kV AC Transmission Line Resisting-tensile Tower

Ⅰ. Training Standard

(Ⅰ) Training Requirements

Designation of module	The live replacement of 1~3 glass insulators on the conductor side of 1000kV AC transmission line resisting-tensile tower	Type of training	Operation
Training method	Practical operation training	Hours of training	21 training hours
Training objectives	1. Master the "two-span and three-short-circuit" operation mode in entering and exiting 1000kV intense electric field along strain insulator string and electrical significance of potential transfer. 2. Can complete the entry of 1000kV equipotential operation point along the strain insulator string. 3. Be able to independently complete the live replacement of 1~3 glass insulators on the conductor side of 1000kV AC transmission line resisting-tensile tower (equipotential operation method)		
Training venue	UHV AC training line		
Training content	The "two-span and three-short-circuit" operation mode is adopted to enter the equipotential along the tensile glass insulator string, and the equipotential operation method is adopted for live replacement of 1~3 glass insulators on the conductor side of 1000kV AC transmission line resisting-tensile tower		
Scope of application	Maintenance personnel of UHV AC transmission line		

(Ⅱ) Referenced rules and specifications

(1) Code for Design of 1000kV Overhead Transmission Line (GB50665-2011).

(2) Maintenance Code for 1000kV AC Transmission Line (DL/T209-2008).

(3) Operation Code for 1000kV AC Transmission Line (DL/T307-2010).

(4)Technical Guide for Live Working on 1000kV AC Transmission Line (DL/T392-2015).

(5)Calculation Method of Live Working Minimum Approach Distance on AC Transmission Line (GB/T 19185-2008)

(6)Guidelines of Insulation Coordination for Live Working (DL/T867-2004).

(7)Live Working-Guidelines for the Installation of Power Transmission Line Conductors and Earthwires-stringing Equipment and Accessory Items (DL/T 1007-2006).

(8)State Grid Corporation of China on the Management Regulations of Live Working (Trial Implementation) (SGCC [2007] No.751).

(9)State Grid Corporation of China Working Regulations of Power Safety (Transmission Line Section) (Q/GDW1799.2-2013).

(10)Electrotechnical Terminology- Live Working (GB/T2900.55-2002).

(11)Live Working - Terminology for Tools, Equipment and Devices (GB/T 14286-2002).

(12)Minimum Requirements for Utilization of Tools, Devices and Equipment for Live Working (DL/T 877-2004).

(13)Preventive Test Code of Tools, Devices and Equipment for Live Working (DL/T 976-2005).

(14)Insulated Tackles for Live Working (GB/T13034-2008).

(15)Live Working-Insulating Ropes (GB 13035-2008).

(16)Screen Clothes for Live Working on 1000kV AC (GB/T25726-2010).

(Ⅲ) Teaching Design for Training

To complete the work task of "live replacement of 1 ~ 3 glass insulators on the conductor side of 1000kV AC transmission line resisting-tensile tower", each training stage shall be designed according to the standard operation procedure for work task completion. Each stage includes specific training objectives, training content, hours of training, training methods (training resources), training environment, assessment and evaluation, etc, as shown in the Table 2-4-1.

Part II Skill Module Training and Assessment Standards

Table 2-4-1 Live Replacement of 1~3 Glass Insulators on the Conductor Side of 1000kV AC Transmission Line Resisting-tensile Tower

Training schedule	Training objectives	Training content	Hours of training	Training methods and resources	Preparation of training conditions	Assessment and evaluation
1. Theoretical teaching	Preliminarily master the basic method for entering and exiting 1000kV intense electric field along the insulator string. 2. Be familiar with the methods for potential transfer. 3. Be familiar with the replacement method of strain single insulator of transmission line	1. The electrical significance of the "two-span and three-short-circuit" operation mode when entering and leaving the intense electric field along the insulator. 2. The use methods for potential transfer rod. 3. Method and quality standard for replacement of strain single insulator of transmission line	2	Training methods: Lecture. Training resources: PPT, relevant regulations and specifications	Multimedia classroom	Attendance, classroom questions and assignments
2. Preparations	Be able to complete the preparation before operation	1. Work site survey. 2. Preparation of the standardized operation card. 3. Filling of the work order. 4. Preparation of tools and materials for this operation	1	Training methods: 1. Site survey and cleaning of tools and materials shall be practiced at site. 2. Preparation of operation card and the filling of work order shall adopt lecture method. Training resources: 1. 1000kV training line. 2. UHV tools warehouse. 3. Blank work order	1. UHV training transmission line. 2. Multi-Media classroom	
3. Work site preparation	Be able to complete the preparations of work site	1. Work site re-survey. 2. Job application. 3. Work site layout. 4. Pre-shift meeting. 5. Inspection of tools and materials	1	Training methods: Demonstration and role play. Resources: 1000kV training line	1000kV training line	

· 383 ·

Table (Cont'd)

Training schedule	Training objectives	Training content	Hours of training	Training methods and resources	Preparation of training conditions	Assessment and evaluation
4. Trainer's demonstration	The trainees can preliminarily understand the operation process of the task through inspecting and learning from each other's work	1. The equipotential electrician enters and exits the intense electric field along the strain insulator string. 2. The equipotential electrician assembles tools. 3. The equipotential electrician completes the replacement of single glass insulator	2	Training methods: Demonstration method. Resources; 1000kV training line	1000kV training line	
5. Group training of trainees	1. Can complete the operation of entering and exiting 1000kV intense electric field. 2. Be able to complete the replacement of single glass insulator	1. The trainees are grouped (6 persons per group) to train the skill operation of entering and exiting 1000kV intense electric field and replacing insulator. 2. Trainers guide the operation of trainees and conduct safety supervision	14	Training methods: Role play. Resources; 1000kV training line	1000kV training line	Score the operation of trainees according to the detailed rules for skill assessment and scoring
6. End of the work	1. Enable the trainees to further distinguish the shortcomings of the operation process and facilitate the promotion in the later stage. 2. Train the trainees in the working style of safe and civilized production	1. Cleaning up the work site. 2. Report to dispatcher. 3. Comment and summarize the work task this time at post-shift meeting	1	Training methods: Lecture and inductive method	1000kV training line	

Part II Skill Module Training and Assessment Standards

(IV) Operation Flow

1. Work Task

The live replacement of 1~3 glass insulators on the conductor side of 1000kV AC transmission line resisting-tensile tower.

2. Requirements for Weather and Work Site

(1) The live replacement of 1~3 glass insulators on the conductor side of 1000kV AC transmission line resisting-tensile tower should be carried out in good weather. In case of lightning (hearing thunder or seeing lightning), snow, hail, rain, fog and so on, live working is prohibited. When the wind force is greater than level 5, it is unsuitable for live working; when the relative humidity of the air is greater than 80%, insulating tools with moisture-proof properties shall be used if live working needs to be carried out. When emergency live repair is required in bad weather, relevant personnel shall be organized to fully discuss and prepare necessary safety measures, which can be implemented after being approved by the unit.

(2) The operating personnel should be in good mental states, be familiar with the organizational and technical measures to ensure safety in work and master the methods for emergency rescue and first aid for electric shock in heights; they should hold the qualification certificate for live working within the validity period.

(3) The Responsible Person shall organize relevant personnel to complete field investigation in advance, determine the operating methods, required working apparatus and necessary measures according to the results, and handle the work orders for live working.

(4) The work site shall be reasonably set up with fence and warning signs. Non-operating personnel is forbidden to enter.

(5) The line re-closing device shall be deactivated in the Project.

(6) Mode of operation: Equipotential operation method

(7) Safety working distance and effective insulation length are shown in Table 2-4-2.

Table 2-4-2 Safe Distance for Live Replacement of 1~3 Glass Insulators on the Conductor Side of 1000kV AC Transmission Line Resisting-tensile Tower

Altitude/m	Minimum safe distance between equipotential electrician and grounding frame /m		Minimum effective insulation length of insulating tools /m	Minimum combined clearance /m	
	Middle phase	Side phase		Middle phase	Side phase
H≤1000	6.8	6.0	6.8	6.9	6.7
1000<H≤2000	7.4	6.6	7.2	7.6	7.3

Note: The values in the table do not include the human body occupying gap which shall not be less than 0.5 m during operation.

(8) When the equipotential electrician enters the equipotential along the strain insulator string, the number of insulators shorted by the human body shall not be more than 4. The minimum number of good insulators shall meet the requirements of Table 2-4-3 after deducting the number of insulators shorted by human body and defective insulators from the strain insulator string.

Table 2-4-3 Minimum Combined Gap and Minimum Number of Good Insulators

Altitude	Structural height of single glass insulator (mm)	Minimum total length of good insulator string (m)	Minimum number of good insulators
H≤1000	170	7.2	43
	195		37
	205		36
1000<H≤2000	170	8.0	47
	195		41
	205		39

Note: The values in the table do not include the human body occupying gap which shall not be less than 0.5 m during operation.

3. Preparations

3.1 Hazards and precontrol measures

(1) Hazard - precontrol measures for electric shock injury:

① Before work, Responsible Person should contact the control personnel on duty, deactivate the line re-closing, and perform the licensing procedures.

② Before climbing the tower, operators on tower must carefully check the dual names of the line, the number of the pole and tower, and phase, and then the tower can be climbed after all have been confirmed correct.

③ If lines lose power suddenly during work, operators shall still deem it to be charged. The Responsible Person shall contact the control personnel as soon as possible, and no forced energization is allowed before the on-duty control personnel getting in touch with the Responsible Person.

④ Insulating tool and insulating ropes shall be free of damage, moisture, deformation, and failure. It is not allowed to use non-insulating ropes (such as cotton rope, manila rope,

Part II　Skill Module Training and Assessment Standards

and steel wire rope).

⑤ The ground electrician shall wear clean and dry gloves when operating the insulating tools. When entering the work site, the live working tools shall be placed on damp-proof canvas or insulating mat to prevent dirt and dampness of the insulating tools in use.

⑥ The equipotential electrician shall wear flame-retardant underwear and full set of qualified shielding clothes outside (including hat, dresses & trousers, gloves, socks and shoes). All parts shall be in excellent connection conditions.

⑦ When the equipotential electrician moves along the insulator string, the positions of the hands and feet shall be correspondingly consistent, and the number of intact insulators shorted by the human body and the tools shall conform to the provisions of Table 2-4-3.

⑧ The equipotential electrician enters the intense electric field by means of "two-span and three-short-circuit" operation mode (also called free operation method). When the operator moves in parallel to the three insulators of the grading ring on the conductor side, the movement shall be stopped, and the potential transfer shall be carried out by using the potential transfer rod which shall be 0.4m long. At the time of potential transfer, the distance between the human face and the charged body shall not be less than 0.5m.

⑨ When transmitting large metal objects by using insulating ropes, ground operators shall not touch them before ground connection.

⑩ During the live working, the Responsible Person (Supervisor) shall continuously monitor the operators and correct their nonstandard operation or actions in violation at any time. Special attention shall be paid to high-place operators to ensure that there is enough safety distance and combined gap (meeting the requirements in Table 2-4-2). It is forbidden to touch two non-connected electrified bodies or electrified body and grounding body.

(2) Hazard - precontrol measures for falling accident:

① Before climbing, operators work at heights must satisfy the requirements of this operation, such as physical condition, mental state and skill and quality.

② Double safety belts shall be used by the personnel for Work at Heights. When climbing up and down the tower, the main material shall be grasped with hands, with shackles trodden and moving at a uniform speed.

③ Before the operation, the equipotential electrician shall carefully inspect hydraulic lead screw, closed clamp, wire end clamp and so on to ensure that the load-bearing tools are qualified; when moving along the insulator string, the positions of hands and feet must be correspondingly consistent, and the safety belt shall be fastened to the insulator string supported by hands and moved synchronously; when replacing insulators, the load-bearing tools shall be installed reliably. Before and after load transfer, impulse test shall be conducted to determine their reliability, and it shall be reported to the Responsible Person in a timely manner. The implementation can only be carried out with the permission of the Responsible Person.

④ Supervisor shall correct irregularities and violations at any time. Special attention shall be paid to operators for Work at Heights to prevent them from losing the protection of the safety belt or insulated backup protection rope during transposition. The safety belt or insulated backup protection rope shall not be fastened lower than operating personnel.

(3) Hazard - injury caused by objects falling from heights. Precontrol measures:

① Operator working at heights should put personal tools and fragmentary materials into the tools bag. It is strictly forbidden to hang objects in heights or keep in the mouth.

② Ground operator should correctly wear a helmet and use the knots. The vertical distance from the work site should not be less than the falling radius.

③ The work site shall be set up with fence and warning signs. It shall be noted at any time that Supervisor shall prohibit irrelevant personnel and vehicles from entering operation area.

3.2 Selection of tools and instruments and materials

Tools and materials required for live replacement of 1 ~ 3 glass insulators on the conductor side of 1000kV AC transmission line resisting-tensile tower can be seen in Table 2-4-4. Before delivering tools and instruments out of warehouse, application voltage class and test period shall be carefully checked and they shall be inspected to ensure that appearance is intact, connection is firm, rotation is flexible and meet the working task requirements. After delivering tools and instruments out of warehouse, they shall be stored in tools bag or tool kit for transportation to avoid contamination and damp. Metal tools and insulated tools shall be separately loaded and transported to avoid deformation, damage or other defects caused by mixed loading and transportation.

Table 2-4-4 Tools and Materials Required for Live Replacement of 1 ~ 3 Glass Insulators on the Conductor Side of 1000kV AC Transmission Line Resisting-tensile Tower

S/N	Name	Specification	Unit	Qty.	Remarks
1	Shielding clothes	Type I	Set	2	PPE
2	Conductive shoes	The size depends on the wearer	Nos.	2	PPE
3	Flame retardant underwear	Pure mulberry silk	Set	2	PPE
4	Safety belt of double insurance	Suspender	Nos.	2	PPE
5	Safety helmet		Nos.	7	PPE
6	Goggles		Nos.	2	PPE
7	Insulated transmission rope	φ14mm, with the length matching the lifting height	Nos.	1	Insulating tool
8	Insulated backup protection rope	φ16mm	Nos.	3	Insulating tool
9	Insulated noose	φ14mm	Nos.	2	Insulating tool
10	Insulated tackle	1T	Nos.	1	Insulating tool
11	Conductor end clamp		Nos.	1	Metal tool
12	Hydraulic leading screw	8T	Nos.	2	Metal tool
13	Closed clamp	Tc4	Set	1	Metal tool
14	Potential transfer rod		Nos.	2	Other tools
15	Pin puller		Nos.	1	Other tools
16	Insulation resistance tester	2,500V, with electrode width 2cm and pole width 2cm	Set	1	Other tools
17	Multimeter		Set	1	Other tools
18	Wind speed, temperature and humidity tester		Nos.	1	Other tools
19	Safety fence		Set	Several	Other tools
20	Warning sign	"Work Here", "Access from Here" "Slow Down" and "Blocking"	Set	1	Other tools
21	Red waistcoat	"Responsible Person" "Special Supervisor"	Piece	1	Other tools
22	Moisture-proof tarpaulin	3m×3m	Piece	2	Other tools
23	Personal tools	Wrench, vice	Set	1	Other tools
24	Falling protector	Corresponding to the type of pole and tower anti-fall device	Nos.	2	Other tools
25	Towel	Cotton	Nos.	1	Other tools
26	Insulator		Piece	1	Material

3.3 Division of labor for operators

Division of labor for operators of the task is shown in Table 2-4-5.

Table 2-4-5 Division of Labor for Live Replacement of 1 ~ 3 Glass Insulators on the Conductor Side of 1000kV AC Transmission Line Resisting-tensile Tower

S/N	Post	Qty. (person)	Responsibilities
1	Responsible Person	1	Be responsible for division of operating personnel of the task, work order reading, handling re-closing deactivation of the line and the formalities of work permit, getting work permits, holding pre-shift meeting, dealing with emergency situations in work, quality surveillance, and the summary after work
2	Special Supervisor	1	Be responsible for the safety control of the work site
3	Equipotential electrician	1	Be responsible for tool installation and insulator replacement
4	Ground electrician	3	Be responsible for ground auxiliary works during the operation.

4. Work Procedure

The workflow of this task is shown in Table 2-4-6.

Table 2-4-6 Workflow of Live Replacement of 1 ~ 3 Glass Insulators on the Conductor Side of 1000kV AC Transmission Line Resisting-tensile Tower

S/N	Work Content	Operation Steps and Standards	Safety Measures and Precautions	Responsible Person
1	Site re-survey	The Responsible Person shall complete the following work: (1) Check the line name, the number of the pole and tower and dual number on spot to ensure they are correct; guarantee that the base and the pole and tower are intact and in normal condition, and ensure that the cross and span distance meets the safety requirements; confirm the defect conditions and the specifications and models of earth wires. (2) Check that the site meteorological conditions such as wind speed and humidity should meet the operation requirements. (3) Check that the terrain and environment shall meet the operation requirements. (4) Check that the safety measures listed in the work order are in line with the actual situations on site, and the measures will be supplemented if necessary	(1) Correctly wear helmet, working clothes, work shoes and protective gloves. (2) Operation under meteorological conditions that may endanger the safety of operators is forbidden. (3) Non-operation personnel and vehicles are strictly prohibited from entering the working site.	

Part II Skill Module Training and Assessment Standards

Table (Cont'd)

S/N	Work Content	Operation Steps and Standards	Safety Measures and Precautions	Responsible Person
2	Work Permit	(1) The Responsible Person is responsible for contacting the on-duty control personnel and applying for stopping the line re-closing as per the contents of the work order. (2) Live working could be started only after being approved by the on-duty control personnel.	Live working shall not be started without the permission of the on-duty control personnel	
3	Site layout	Install the security fence and hang the signboards correctly: (1) The security fence should take full account of falling objects from the heights and the influence on road traffic. (2) The entrance and exit of the security fence shall be set reasonably. (3) Signs such as "Access from Here", "Work Here", "Slow Down" or "Blocking" shall be properly arranged.	When the influence on road traffic safety is uncontrollable, the traffic management department should be contacted in time to strengthen the on-site control of traffic safety	
4	Hold a pre-shift meeting	(1) All working personnel shall line up. (2) The Responsible Person shall wear red waistcoat and read out the work order and be clear with work task and division of personnel; explain safety measures and technical measures in work; check (inquire after) mental state of all working personnel; inform of hazards in work and pre-control measures. (3) All working personnel shall sign on the work order for confirmation.	(1) The work order shall be filled in, issued and approved in a standardized manner, and the signature shall be complete. (2) All working personnel shall be in good mental states. (3) All working personnel shall be clear with task division of works, safety measures and technical measures	

Table (Cont'd)

S/N	Work Content	Operation Steps and Standards	Safety Measures and Precautions	Responsible Person
5	Inspect tools	(1) All necessary tools and instruments shall be prepared as per the operation requirements and placed on the moisture-proof tarpaulin regularly according to the category and location. The appearance and test certificate of tools and instruments shall be checked to ensure there is no omission. (2) The surface insulation resistance of insulating tools and insulating ropes shall be tested with insulation resistance tester in correct methods, and the value shall be not less than 700MΩ. (3) The new insulator is wiped up, and it is intact in appearance inspection, without rust, cracks and breakage. The surface insulation resistance shall be tested with insulation resistance tester in correct methods, and the value shall be not less than 500MΩ. (4) The internal resistance of full shielding clothes shall be tested with a multimeter in correct methods, and the value shall be not more than 20Ω. (5) The inspector shall report to the Responsible Person that all inspection results are in conformity with the operation requirements	(1) The waterproof tarpaulin shall be enough in quantity and reasonable in position, and be clean and dry. (2) Before using metal and insulating tools, they shall be carefully checked for damage, dampness, deformation and failure, and the certificate of conformity is within the validity period. (3) Insulating tools and insulating ropes are tested as acceptable. Qualified	
6	Climbing the tower	(1) The equipotential electrician shall check the double name and phase of the line again, inspect and confirm that the shackles are complete and firm; fasten safety belts and attach falling protectors; perform impact inspection on safety belt, backup protection rope and falling protector in correct method; the Responsible Person shall check and confirm that the connection conditions of each connection point of the double safety belt and shielding clothes worn by the equipotential electrician are in good condition, including shoulder harness, pectoral harness, belt, back rope sling, buckle and ring.	(1) Safety belt and falling protector shall pass the impulse test. (2) The safety belt and insulated transmission rope shall be prevented from hooking up tower materials. (3) The minimum safe distance between the human body and the conductor shall conform to the regulations shown in Table 2-4-2. (4) It is forbidden to grasp shackles with hands.	

Part II Skill Module Training and Assessment Standards

Table (Cont'd)

S/N	Work Content	Operation Steps and Standards	Safety Measures and Precautions	Responsible Person
6	Climbing the tower	(2) The electrician shall carry toolkit and insulated transmission rope (including insulated tackle) in correct methods. (3) Main band of safety belt and backup protection rope shall obliquely across the shoulder. (4) The soles shall be cleaned, and climbing the tower shall be done with the permission of the Responsible Person. (5) The electrician shall tread on shackles, grasp the main materials, and climb the tower uniformly to the proper position of the cross arm with safety belt well fastened and then break away from the falling protector.	(5) Falling protector shall be properly used. (6) Protection of safety belt shall not be lost during transposition.	
7	Entering the intense electric field	(1) The equipotential electrician carries the insulated transmission rope, transposes it to the hanging point of the operation phase strain insulator string, hangs the main safety belt on the insulator string connection fittings, and leaves the safety belt backup protection rope at the proper position of the cross arm. (2) The good connection of all parts of the shielding clothes and the insulator string and the location of the faulty insulators shall be re-checked and confirmed. With the permission of the Responsible Person, the electrician shall grasp one string and tread the other one, and enter the intense electric field along the insulator string in the operation mode of "two-span and three-short-circuit". When reaching the three pieces of insulators outside the grading ring on the conductor side, the operator shall stop the movement, and use the potential transfer rod for potential transfer, so as to realize the equipotential between the operator and the conductor. (3) The main safety belt shall be transferred and fastened to the appropriate position of the conductor, cross the grading ring to reach the operation point on the conductor, and the insulated transmission rope is arranged in the appropriate position of the conductor, which shall be secure, reliable and convenient for operation.	(1) The safety belt and insulated transmission rope shall be prevented from hooking up tower materials. (2) Protection of safety belt shall not be lost during transposition. (3) The safe distance between the human body and the grounding body, and the combined gap between the human body and the grounding body and the electrified body shall not be less than that specified in Table 2-4-2. (4) When the potential is transferred, the distance between the human face and the electrified body shall not be less than 0.5m.	

Table (Cont'd)

S/N	Work Content	Operation Steps and Standards	Safety Measures and Precautions	Responsible Person
8	Installation of load-bearing tools	(1) The ground electrician lifts the closed clamp (front clamp), hydraulic leading screw, conductor end clamp and so on to the equipotential electrician by using the insulated transmission rope. The lifting process shall be smooth, free from collision and winding, and the knot shall be used correctly. (2) The equipotential electrician shall install the conductor end clamp at the appropriate position on the conductor side, install the closed clamp (front clamp) on the third insulator on the conductor side, and connect the hydraulic leading screw. Each part of the load-bearing tools is securely and reliably installed. (3) Equipotential electricians shall check and confirm that all parts of the load-bearing tools are installed in good condition. With the permission of the Responsible Person, they can operate the hydraulic leading screw so that it is gradually stressed and the insulators to be replaced are relaxed. The two hydraulic leading screws shall be uniformly stressed.	(1) The safe distance between the human body and the grounding body shall not be less than that specified in Table 2-4-2. (2) Objects falling from high place shall be avoided. (3) After deducting inferior insulators, insulators shorted by human body and tools, the number of good insulators shall conform to that specified in Table 2-4-3.	
9	Replacement of insulator	(1) The equipotential electrician shall perform impulse test, check and confirm that the load-bearing tools are stressed normally. With the permission of the Responsible Person, they shall tie the old insulator securely with the insulated transmission rope, take out the fitting pins at both ends of the old insulator, continue to operate and tighten the hydraulic leading screw until the old insulator is removed. The two hydraulic leading screws shall be stressed uniformly, and the operating handle shall not knock on the insulator. (2) The ground electrician uses the other end of the insulated transmission rope to fasten the new insulator, and transfers the new insulator to the intermediate potential electrician by means of the old lower and the new upper. The lifting process shall be smooth, free from collision and winding, and the knot shall be used correctly. (3) They shall install new insulator and reset fitting pins at both ends.	(1) The safe distance between the human body and the grounding body shall not be less than that specified in Table 2-4-2. (2) Objects falling from high place shall be avoided	

Part II Skill Module Training and Assessment Standards

Table (Cont'd)

S/N	Work Content	Operation Steps and Standards	Safety Measures and Precautions	Responsible Person
10	Remove tools	(1) The equipotential electricians check that the connection of the new insulator is reliable and, with the permission of the Responsible Person, they operate and loosen the hydraulic leading screw so that the replaced insulator is gradually stressed. (2) After load transfer, the equipotential electricians shall perform the impulse test, check and confirm that the new insulator is in good condition. With the permission of the Responsible Person, they can remove the insulated transmission rope tied to the insulator, fasten it to the proper position of the load-bearing tools, remove the hydraulic leading screw, closed clamp, conductor end clamp and other load-bearing tools, and transfer them to the ground with the cooperation of the ground electrician. The transfer process shall be smooth, free from collision and winding, and the knot shall be used correctly.	(1) The safe distance between the human body and the grounding body shall not be less than that specified in Table 2-4-2. (2) Objects falling from high place shall be avoided	
11	Exit the intense electric field	(1) The equipotential electrician shall remove and put the insulated transmission rope in order after checking that there is no object left in the operation area. (2) The main safety belt shall be transferred and fastened at an appropriate position of the insulator string, returned to the insulator string through the grading ring, hooked at the appropriate position of the grading ring by the potential transfer rod, and stopped when the main belt is moved along the insulator string toward the lateral side of the cross arm to three pieces of insulators outside the grading ring; the equipotential electrician shall grasps the insulator with one hand and the potential transfer rod with the other, so as to quickly disengage from the equipotential by using the potential transfer rod. (3) They can reach the cross arm along the insulator string in the "two-span and three-short-circuit" operation mode.	(1) The combined gap between the human body and the grounding body and the charged body shall not be less than that specified in Table 2-4-2. (2) Protection of safety belt shall not be lost during transposition.	

Table (Cont'd)

S/N	Work Content	Operation Steps and Standards	Safety Measures and Precautions	Responsible Person
12	Evacuation from the pole and tower	The equipotential electrician shall check that there is no object left on the tower. After being approved by the Responsible Person, the falling protector shall be properly hung and the safety belt shall be unfastened and put straight, the electrician shall carry the insulated transmission rope correctly and step down to the ground from the tower in uniform steps by treading on shackles and grasping the main materials	(1) Protection of safety belt shall not be lost during transposition. (2) Hand slippage and step missing shall be avoided and it's forbidden to grasp the shackles with hands. (3) Falling protector shall be properly used. (4) The insulated transmission rope and safety belt shall be prevented from hooking up the tower materials or shackles	
13	End of the work	(1) The Responsible Person shall organize all working members to put working apparatus and materials in order and put them in a special kit (bag) after cleaning; clean the site to ensure that "the materials are removed and the site is cleaned after construction". (2) In the post-shift meeting, Responsible Person will give work summaries and comments.Comments include the construction quality of this work and the implementation of safety measures from all working personnel. (3) Responsible Person shall report to the on-duty control personnel the end of the work and terminate the work order.		

II. Assessment Standard

Part II Skill Module Training and Assessment Standards

Detailed Rules for Assessment and Scoring of Operation and Inspection Skills of UHV AC Transmission Line

Fill-in Column of Examinee	No.:		Name:		Position:		Date:		MM/DD/YYYY
Fill-in Column of Assessor	Unit:		Assessor:		Assessment Team Leader:		Starting time:	Closing time:	Operation Duration:
Assessment module	Grade:		The live replacement of 1~3 glass insulators on the conductor side of 1000kV AC transmission line resisting-tensile tower	Assessee	Maintenance personnel of UHV AC transmission line	Assessment method	Operation	Assessment Time Limit	90min
Job Description	The live replacement of 1~3 glass insulators on the conductor side of 1000kV AC transmission line resisting-tensile tower								
Work Specifications and Requirements	1. Live working shall be carried out in good weather. In case of thunder, rain, snow or fog, no live working shall be carried out. When the wind force is greater than Level 5, live working should not be carried out. When the humidity is greater than 80%, insulating tools with moisture-proof properties shall be used if live working needs to be carried out. 2. Six persons are required for this operation, including 1 Responsible Person, 1 Special Supervisor, 1 equipotential electrician and 4 ground electricians. 3. Responsibilities of Responsible Person (Supervisor): Be responsible for division of operating personnel of the task, work order reading, handling re-closing deactivation of the line and the formalities of work permit, getting work permits, holding pre-shift meeting, safety supervision in the operation process, dealing with emergency situations in work, quality surveillance, and the summary after work. 4. During the live working, if thunder, rain, strong wind or any other circumstance threaten the safety of the staff, the Responsible Person or Supervisor may stop working temporarily according to the circumstances Given conditions: 1. Work orders have been handled, safety measures have been completed (re-closing has been deactivated), and oral application (dispatcher or assessor) shall be made at the beginning and end of the work. 2. The instrument shall be used safely and correctly to test the insulating tool. 3. The operation must be carried out according to the working procedures. The scores of the items to be carried out shall be deducted for the process error. In case of major hidden dangers of personal, equipment and operational safety, the assessor may order the termination of the operation (assessment)								

Standard for Professional Training and Assessment for Operation Maintenance of UHV AC Transmission Line

Table (Cont'd)

Assessment scenario preparation	1. Tower type: 1000kV AC resisting-tensile tower; 2. Required operation tools: 2 safety belts (including double-protective rope), 2 sets of Type 1 shielding clothes, 2 sheets of moisture-proof cloth, 1 multimeter, 1 insulation resistance detector, 1 anemometer, 1 2-in-one temperature and humidity detector, 2 hydraulic leading screws, 2 insulating ropes, 1 set of closed clamp, 1 tackle, 2 potential transfer rods, 2 goggles, 1 pin puller and 1 set of hand-operated tools. 3. The work site shall be monitored, and the safety measures (fence, etc.) on the work site have been fully implemented; non-operation personnel are prohibited from entering the site, and the staff must wear safety helmets when entering the work site. 4. Examinees shall bring their own work clothes, helmets and gloves
Remarks	1. The deduction shall be done until the scores of each item are deducted completely. In case of major hidden dangers of personal, equipment and operational safety, the assessor may order the termination of the operation. 2. When equipment, working environment, safety belt, safety helmet, tool, shielding clothes, etc., do not conform to the operation condition, the assessor may order the termination of the operation

S/N	Project name	Quality requirements	Score	Deduction standard	Reasons for deduction	Deduction	Scoring
1	Site re-survey	1) The Responsible Person shall go to the work site to check the line name, pole and tower number, on-site working conditions, defective parts and so on. 2) Check that the site meteorological conditions such as wind speed and humidity should meet the operation requirements. 3) Check whether the work order is complete and unmodified, check whether the safety measures listed are consistent with the actual situation on site, and supplement it if necessary	5	1) Deduct 1 point/item for failure to check the line name, pole and tower number, on-site working conditions and defective parts. 2) Deduct 1 point/item for failure to detect wind speed, humidity and other on-site meteorological conditions. 3) Deduct 0.5 points/part for any alteration in the work order filling; deduct 1 point for incorrect work order number; deduct 1.5 points for incomplete work order filling			

Part II Skill Module Training and Assessment Standards

Table (Cont'd)

S/N	Project name	Quality requirements	Score	Deduction standard	Reasons for deduction	Deduction	Scoring
2	Work Permit	1) The Responsible Person is responsible for contacting the on-duty control personnel (referee) and applying for stopping the line re-closing as per the contents of the work order. 2) Reporting content is standardized and complete	2	1) Deduct 2 points for failure to contact the dispatching department (referee) for deactivating the re-closing. 2) Deduct 1 point for non-standard or incomplete terminology reporting			
3	Site layout	Install the security fence and hang the signboards correctly: 1) The security fence should take full account of falling objects from the heights and the influence on road traffic. 2) The entrance and exit of the security fence shall be set reasonably. 3) Signs such as "Access from Here", "Work Here", "Access from Here" shall be properly and well arranged	3	1) Deduct 1 point for failure to arrange the fence at the work site. 2) Deduct 1 point for failure to arrange the warning board. 3) Deduct 1 point for failure to hang the tower climbing operation sign			
4	Hold a pre-shift meeting	1) All staff and personnel shall wear safety helmets and work clothes correctly. 2) Responsible Person shall wear red vest and read out the work order and be clear with work task and division of personnel; explain safety measures and technical measures in work; check (inquire after) mental state of all working personnel; inform of hazards in work and precontrol measures. 3) All working personnel shall sign on the work order for confirmation.	3	1) Deduct 0.5 points/person for the staff not dressing uniformly. 2) Deduct 3 points for no division of labor; deduct 1 point for unclear division of labor. 3) Deduct 1 point for the on-site Responsible Person not wearing a safety monitoring vest. 4) Deduct 1 point for the work shift member failing to sign or signing incompletely on the work order.			

· 399 ·

Table (Cont'd)

S/N	Project name	Quality requirements	Score	Deduction standard	Reasons for deduction	Deduction	Scoring
5	Inspection of tools	1) All necessary tools and instruments shall be prepared as per the operation requirements and placed on the moisture-proof tarpaulin regularly according to the category and location. The appearance and test certificate of tools and instruments shall be checked to ensure there is no omission. 2) The surface insulation resistance of insulating tools and insulating ropes shall be tested with insulation resistance tester in correct methods, and the value shall be not less than 700MΩ. 3) The new insulator is wiped up, and it is intact in appearance inspection, without rust, cracks and breakage. The surface insulation resistance shall be tested with insulation resistance tester in correct methods, and the value shall be not less than 500MΩ. 4) The internal resistance of full shielding clothes shall be tested with a multimeter in correct methods, and the value shall be not more than 20Ω. 5) The inspector shall report to the Responsible Person that all inspection results are in conformity with the operation requirements	7	1) Deduct 1 point for failure to use moisture-proof tarpaulin and place tools to designed positions. 2) Deduct 0.5 points/item for failure to check the appearance of tools and pass the test certificate. 3) Deduct 1 point/item for failure to use testing instrument for testing the tools. 4) Deduct 1 point for non-standard reporting of test results; deduct 0.5 points/item for incomplete reporting			

Part II Skill Module Training and Assessment Standards

Table (Cont'd)

S/N	Project name	Quality requirements	Score	Deduction standard	Reasons for deduction	Deduction	Scoring
6	Climbing the tower	1) The equipotential electrician shall check the double name and phase of the line again, inspect and confirm that the shackles are complete and firm; fasten safety belts and attach falling protectors; perform impulse test on double safety belts, backup protection rope and falling protector in correct method;The Responsible Person shall check and confirm that the connection conditions of each connection point of the double safety belt and shielding clothes worn by the equipotential electricians are in good condition, including shoulder harness, pectoral harness, belt, back rope sling, buckle and ring. 2) The electrician shall carry toolkit and insulated transmission rope (including insulated tackle) in correct methods. 3) During the process of climbing the tower, the electrician shall fasten the anti-fall device, tread on shackles, grasp the main materials, and climb the tower uniformly to the proper position with safety belt well fastened and then break away from the falling protector	5	1) Deduct 1 point/item for the equipotential electrician failing to check the double name of the line, pole number, phase and tower material; deduct 1 point for non-report after checking. 2) Deduct 2 points/item for failure to perform the impulse test on double safety belts and falling protector. 3) Deduct 1 point for the on-site Responsible Person failing to check the safety protection equipment of the intermediate potential electrician. 4) Deduct 0.5 points/time for hands grasping the shackles. 5) Deduct 1 point for the unreasonable suspension position of the tackle transmission rope. 6) Deduct 5 points for loosing the protection of the safety belt during transposition			

Table (Cont'd)

S/N	Project name	Quality requirements	Score	Deduction standard	Reasons for deduction	Deduction	Scoring
7	Entering the intense electric field	1) The equipotential electrician carries the insulated transmission rope, transposes it to the hanging point of the operation phase strain insulator string, hangs the main safety belt on the insulator string connection fittings, and leaves the safety belt backup protection rope at the proper position of the cross arm. 2) The good connection of all parts of the shielding clothes and the insulator string and the location of the faulty insulators shall be re-checked and confirmed. With the permission of the Responsible Person, the electrician shall grasp one string and tread the other one, and enter the intense electric field along the insulator string in the operation mode of "two-span and three-short-circuit". When reaching the three pieces of insulators outside the grading ring on the conductor side, the operator shall stop the movement, and use the potential transfer rod for potential transfer, so as to realize the equipotential between the operator and the conductor. 3) The main safety belt shall be transferred and fastened to the appropriate position of the conductor, cross the grading ring to reach the operation point on the conductor, and the insulated transmission rope is arranged in the appropriate position of the conductor, which shall be secure, reliable and convenient for operation.	5	1) Deduct 2 points for irrational fastening position and irregular use of safety belt back protection rope. 2) Deduct 1 point/item for the equipotential electrician failing to check the connection of the shielding clothes and insulator string and the location of the faulty insulator. 3) Deduct 5 points for entering the intense electric field without permission of the Responsible Person. 4) Deduct 2 points/time for the incorrect action of equipotential electrician entering the equipotential and repeat discharging. 5) Deduct 3 points for the incorrect action of potential transfer; deduct 2 points for failure to report and 1 point for potential transfer without the consent of the Responsible Person after reporting. 6) Deduct 1 point for the unreasonable installation position of the insulated transmission rope. 7) Deduct 2 points/time for falling objects from heights. 8) Deduct 5 points for loosing the protection of the safety belt during transposition			

Part II Skill Module Training and Assessment Standards

Table (Cont'd)

S/N	Project name	Quality requirements	Score	Deduction standard	Reasons for deduction	Deduction	Scoring
8	Installation tool	1) The ground electrician transfers the closed clamp (front clamp), hydraulic leading screw, conductor end clamp and so on to the equipotential electrician by using the insulated transmission rope. The lifting process shall be smooth, free from collision and winding, and the knot shall be used correctly. 2) The equipotential electrician shall first install the conductor end clamp at the appropriate position on the conductor side, install the closed clamp (front clamp) on the third insulator on the conductor side, and connect the hydraulic leading screw. Each part of the load-bearing tools is securely and reliably installed. 3) Equipotential electricians shall check and confirm that all parts of the load-bearing tools are installed in good condition. With the permission of the Responsible Person, they can operate the hydraulic leading screw so that it is gradually stressed and the insulators to be replaced are relaxed. The two hydraulic leading screws shall be uniformly stressed.	15	1) Deduct 1 point/time for unstable lifting process, colliding and winding. 2) Deduct 2 points/time for falling objects from heights. 3) Deduct 2 points for improper installation and fixing of clamp. 4) Deduct 3 points for failure to check the installation of load-bearing tools; deduct 1 point for failure to report after checking; deduct 1 point for starting tightening the leading screw without the consent of Responsible Person after reporting. 5) Deduct 3 points/time for the number of shorted insulators exceeding 4 during the operation. 6) Deduct 2 points for collision and breakage of insulators in the clamp installation. 7) Deduct 2 points for failure to tighten leading screw in a balanced way			

Standard for Professional Training and Assessment for Operation Maintenance of UHV AC Transmission Line

Table (Cont'd)

S/N	Project name	Quality requirements	Score	Deduction standard	Reasons for deduction	Deduction	Scoring
9	Replacement of insulator	1) The equipotential electrician shall perform impulse test, check and confirm that the load-bearing tools are stressed normally. With the permission of the Responsible Person, they shall tie the old insulator with the insulated transmission rope, take out the fitting pins at both ends of the old insulator, continue to operate and tighten the hydraulic leading screw until the old insulator is removed. The two hydraulic leading screws shall be stressed uniformly, and the operating handle shall not knock on the insulator. 2) The ground electrician uses the other end of the insulated transmission rope to fasten the new insulator, and transfers the new insulator to the intermediate potential electrician by means of the old lower and the new upper. The lifting process shall be smooth, free from collision and winding, and the knot shall be used correctly. 3) They shall install new insulator and reset fitting pins at both ends.	20	1) Deduct 3 points for failure to check the stress of load-bearing tools; deduct 2 points for failure to report after checking; deduct 1 point for removing fitting pins at both ends of the old insulator without the consent of Responsible Person after reporting. 2) Deduct 2 points for failure to tighten leading screw in a balanced way. 3) Deduct 1 point/time for the operating handle hitting the insulator. 4) Deduct 1 point for wrong knot. 5) Deduct 2 points/time for falling objects from heights. 6) Deduct 1 point for new and old insulators colliding with each other. 7) Deduct 1 point for the transmission insulator and tower body colliding with each other. 8) Deduct 2 points for winding insulating ropes			

Table (Cont'd)

S/N	Project name	Quality requirements	Score	Deduction standard	Reasons for deduction	Deduction	Scoring
10	Remove tools	1) The equipotential electricians check that the connection of the new insulator is reliable and, with the permission of the Responsible Person, they operate and loosen the hydraulic leading screw so that the replaced insulator is gradually stressed. 2) After load transfer, the equipotential electricians shall perform the impulse test, check and confirm that the new insulator is in good condition. With the permission of the Responsible Person, they can remove the insulated transmission rope tied to the insulator, fasten it to the proper position of the load-bearing tools, remove the hydraulic leading screw, closed clamp, conductor end clamp and other load-bearing tools, and transfer them to the ground with the cooperation of the ground electrician. The transfer process shall be smooth, free from collision and winding, and the knot shall be used correctly.	15	1) Deduct 3 points for failure to check the connection of new insulator; deduct 2 points for failure to report after checking; deduct 1 point for starting loosening the hydraulic leading screw without the consent of Responsible Person after reporting. 2) Deduct 3 points for failure to check the stress of new insulator; deduct 2 points for failure to report after checking; deduct 1 point for starting removing the hydraulic leading screw without the consent of Responsible Person after reporting. 3) Deduct 1 point for incorrect use of knot when binding the tools. 4) Deduct 2 points/time for falling objects from heights. 5) Deduct 1 point/time for the tools colliding with each other; deduct 1 point/time for the tool colliding with the electrified body or tower body; deduct 2 points for winding insulating ropes			

Table (Cont'd)

S/N	Project name	Quality requirements	Score	Deduction standard	Reasons for deduction	Deduction	Scoring
11	Exit the intense electric field	1) The equipotential electrician shall remove and put the insulated transmission rope in order after checking that there is no object left in the operation area. 2) The main safety belt shall be transferred and fastened at an appropriate position of the insulator string, returned to the insulator string through the grading ring, hooked at the appropriate position of the grading ring by the potential transfer rod, and stopped when the main belt is moved along the insulator string toward the lateral side of the cross arm to three pieces of insulators outside the grading ring; the equipotential electrician shall grasps the insulator with one hand and the potential transfer rod with the other, so as to quickly disengage from the equipotential by using the potential transfer rod. 3) They can reach the cross arm along the insulator string in the "two-span and three-short-circuit" operation mode.	5	1) Deduct 2 points for the potential transfer without applying to the Responsible Person; deduct 1 point for starting the work without the consent after applying. 2) Deduct 1 point for the inappropriate position applied for the potential transfer. 3) Deduct 2 points for the incorrect action of equipotential electrician exiting the intense electric field and repeat discharging. 4) Deduct 1 point for failure to effectively control backup protection rope			

Part II Skill Module Training and Assessment Standards

Table (Cont'd)

S/N	Project name	Quality requirements	Score	Deduction standard	Reasons for deduction	Deduction	Scoring
12	Return to the ground	After checking that there is no object left on the tower, the equipotential electrician on the tower shall report it to the Responsible Person and then climb down the tower with insulated transmission rope after obtaining the consent of the Responsible Person	5	1) Deduct 2 points for failure to use the falling protector when climbing down the tower. 2) Deduct 2 points for loosing the protection of the safety belt when moving on the tower. 3) Deduct 1 point for grasping the tower nail when climbing down the tower. 4) Deduct 2 points for any objects left on tower.			
13	End of the work	1) The Responsible Person shall organize all working members to put working apparatus and materials in order and put them in a special kit (bag) after cleaning; clean the site to ensure that "the materials are removed and the site is cleaned after construction". 2) In the post-shift meeting, Responsible Person will give work summaries and comments. Comments include the construction quality of this work and the implementation of safety measures from all working personnel. 3) Responsible Person shall report to the on-duty control personnel that the work is over, apply for the restoration of circuit re-closing and terminate the work order.	10	1) Deduct 2 points for failure to clean the tools. 2) Deduct 2 points for missing tools. 3) Deduct 2 points for failure to hold the post-shift meeting. 4) Deduct 2 points for failure to remove the fence. 5) Deduct 2 points for failure to report to dispatcher			
	Total		100				

Module 5 Standards for Training and Assessment on Live Replacement of Arbitrary Single Insulator of Tensile Glass Insulator String of 1000kV AC Transmission Line

I. Training Standard

(I) Training Requirements

Designation of module	Live replacement of arbitrary single insulator of tensile glass insulator string of 1000kV AC transmission line	Type of training	Operation
Training method	Practical operation training	Hours of training	21 training hours
Training objectives	1. Master the electrical significance of "two-span and three-short-circuit" operation mode in entering and exiting 1000kV intense electric field along strain insulator string. 2. Be able to complete the entry of 1000kV intermediate potential operation point along the strain insulator string. 3. Be able to independently complete the live replacement operation of any single glass insulator of tensile glass insulator string of 1000kV AC transmission line (intermediate potential operation method)		
Training venue	UHV AC training line		
Training content	The "two-span and three-short-circuit" operation mode is adopted to enter the intense electric field along the tensile glass insulator string, and the intermediate potential operation method is adopted for live replacement operation of any single glass insulator of tensile glass insulator string of 1000kV AC transmission line		
Scope of application	Maintenance personnel of UHV AC transmission line		

(II) Referenced Procedures and Specifications

(1) Code for Design of 1000kV Overhead Transmission Line (GB50665-2011).

(2) Maintenance Code for 1000kV AC Transmission Line (DL/T209-2008).

(3) Operation Code for 1000kV AC Transmission Line (DL/T307-2010).

Part II Skill Module Training and Assessment Standards

(4)Technical Guide for Live Working on 1000kV AC Transmission Line (DL/T392-2015).

(5)Calculation Method of Live Working Minimum Approach Distance on AC Transmission Line (GB/T 19185-2008)

(6)Guidelines of Insulation Coordination for Live Working (DL/T867-2004).

(7)Live Working-Guidelines for the Installation of Power Transmission Line Conductors and Earthwires-stringing Equipment and Accessory Items (DL/T 1007-2006).

(8)State Grid Corporation of China on the Management Regulations of Live Working (Trial Implementation) (SGCC [2007] No.751).

(9)State Grid Corporation of China Working Regulations of Power Safety (Transmission Line Section) (Q/GDW1799.2-2013).

(10)Electrotechnical Terminology - Overhead Line (GB/T 2900.51-1998).

(11)Electrotechnical Terminology- Live Working (GB/T2900.55-2002).

(12)Live Working - Terminology for Tools, Equipment and Devices (GB/T 14286-2002).

(13)Minimum Requirements for Utilization of Tools, Devices and Equipment for Live Working (DL/T 877-2004).

(14)Preventive Test Code of Tools, Devices and Equipment for Live Working (DL/T 976-2005).

(15)Insulated Tackles for Live Working (GB/T13034-2008).

(16)Live Working-Insulating Ropes (GB 13035-2008).

(17)Screen Clothes for Live Working on 1000kV AC (GB/T25726-2010).

(Ⅲ) Teaching Design for Training

To complete the work task of "live replacement of arbitrary single insulator of tensile glass insulator string of 1000kV AC transmission line", each training stage shall be designed according to the standard operation procedure for work task completion. Each stage includes specific training objectives, training content, hours of training, training methods (training resources), training environment, assessment and evaluation, etc, as shown in the Table 2-5-1.

Standard for Professional Training and Assessment for Operation Maintenance of UHV AC Transmission Line

Table 2-5-1 Live Replacement of Any Single Glass Insulator of Tensile Glass Insulator String of 1000kv AC Transmission Line

Training schedule	Training objectives	Training content	Hours of training	Training methods and resources	Preparation of training conditions	Assessment and evaluation
1. Theoretical teaching	1. Preliminarily master the basic method for entering and exiting 1000kV intense electric field along the insulator string. 2. Be familiar with the replacement method of strain single glass insulator of transmission line	1. The electrical significance of the "two-span and three-short-circuit" operation mode when entering and leaving the intense electric field along the insulator. 2. Method and quality standard for replacement of strain single glass insulator of transmission line	2	Training methods: Lecture. Training resources: PPT, relevant regulations and specifications	Multimedia classroom	Attendance, classroom questions and assignments
2. Preparations	Be able to complete the preparation before operation	1. Work site survey. 2. Preparation of the standardized operation card. 3. Filling of the work order. 4. Preparation of tools and materials for this operation	1	Training methods: 1. Site survey and cleaning of tools and materials shall be practiced at site. 2. Preparation of operation card and the filling of work order shall adopt lecture method. Training resources: 1. 1000kV practical training line. 2. UHV tools warehouse. 3. Blank work order	1. UHV training transmission line. 2. Multi-Media classroom	

· 410 ·

Part II Skill Module Training and Assessment Standards

Table (Cont'd)

Training schedule	Training objectives	Training content	Hours of training	Training methods and resources	Preparation of training conditions	Assessment and evaluation
3. Work site preparation	Be able to complete the preparations of work site	1. Work site re-survey. 2. Job application. 3. Work site layout. 4. Pre-shift meeting. 5. Inspection of tools and materials	1	Training methods: demonstration and role play. Resources: 1000kV training line	1000kV training line	
4. Trainer's demonstration	The trainees can preliminarily understand the operation process of the task through inspecting and learning from each other's work	1. The intermediate potential electrician enters and exits the intense electric field along the strain insulator string. 2. The intermediate potential electrician assembles tools. 3. The intermediate potential electrician completes the replacement of single glass insulator	2	Training methods: Demonstration. Resources: 1000kV training line	1000kV training line	

Table (Cont'd)

Training schedule	Training objectives	Training content	Hours of training	Training methods and resources	Preparation of training conditions	Assessment and evaluation
5. Group training of trainees	1. Can complete the operation of entering and exiting 1000kV intense electric field. 2. Be able to complete the replacement of single glass insulator	1. The trainees are grouped (6 persons per group) to train the skill operation of entering and exiting 1000kV intense electric field and replacing insulator. 2. Trainers guide the operation of trainees and conduct safety supervision	14	Training methods: Role play. Resources: 1000kV training line	1000kV training line	Score the operation of trainees according to the detailed rules for skill assessment and scoring
6. End of the work	1. Enable the trainees to further distinguish the shortcomings of the operation process and facilitate the promotion in the later stage. 2. Train the trainees in the working style of safe and civilized production	1. Cleaning up the work site. 2. Report to dispatcher the end of work. 3. Comment and summarize the work task this time at post-shift meeting	1	Training method: Lecture and inductive method	1000kV training line	

(Ⅳ) Operation Flow

1. Work Task

The live replacement of arbitrary single glass insulator of tensile glass insulator string of 1000kV AC transmission line.

2. Requirements for Weather and Work Site

(1) The live replacement operation of any single glass insulator of tensile glass insulator string of 1000kV AC transmission line should be carried out in good weather.

In case of lightning (hearing thunder or seeing lightning), snow, hail, rain, fog and so on, live working is prohibited. When the wind force is greater than level 5, it is unsuitable for live working; when the relative humidity of the air is greater than 80%, insulating tools with moisture-proof properties shall be used if live working needs to be carried out. When emergency live repair is required in bad weather, relevant personnel shall be organized to fully discuss and prepare necessary safety measures, which can be implemented after being approved by the unit.

(2) The operating personnel should be in good mental states, be familiar with the organizational and technical measures to ensure safety in work and master the methods for emergency rescue and first aid for electric shock in heights; they should hold the qualification certificate for live working within the validity period.

(3) The Responsible Person shall organize relevant personnel to complete field investigation in advance, determine the operating methods, required working apparatus and necessary measures according to the results, and handle the work orders for live working.

(4) The work site shall be reasonably set up with fence and warning signs. Non-operating personnel is forbidden to enter.

(5) The line re-closing device shall be deactivated in the Project.

(6) Mode of operation: Intermediate potential operation method.

(7) Safety working distance and effective insulation length are shown in Table 2-5-2.

Table 2-5-2 Safe Distance for Live Replacement of Any Single Glass Insulator of Tensile Glass Insulator String of 1000kV AC Transmission Line

Altitude/m	Minimum safe distance between the intermediate potential operator and the electrified body /m		Minimum effective insulation length of insulating tools /m	Minimum combined clearance /m	
	Middle phase	Side phase		Middle phase	Side phase
$H \leqslant 1000$	6.8	6.0	6.8	6.9	6.7
$1000 < H \leqslant 2000$	7.4	6.6	7.2	7.6	7.3

Note: The values in the table do not include the human body occupying gap which shall not be less than 0.5 m during operation.

(8) When the intermediate potential operator enters ±1000kV intense electric field along the strain insulator string, the number of insulators shorted by the human body shall not be more than 4. The minimum number of good insulators shall meet the requirements of Table 2-5-3 after deducting the number of insulators shorted by human body and defective insulators from the strain insulator string.

Table 2-5-3　Minimum Combined Gap and Minimum Number of Good Insulators

Altitude	Structural height of single glass insulator (mm)	Minimum total length of good insulator string (m)	Minimum number of good insulators
$H \leqslant 1000$	170	7.2	43
	195		37
	205		36
$1000 < H \leqslant 2000$	170	8.0	47
	195		41
	205		39

Note: The values in the table do not include the human body occupying gap which shall not be less than 0.5 m during operation.

3. Preparations

3.1 Hazards and precontrol measures

(1) Hazard - electric shock injury

Precontrol measures:

① Before work, Responsible Person should contact the control personnel on duty, deactivate the line re-closing, and perform the licensing procedures.

② Before climbing the tower, operators on tower must carefully check the dual names of the line, the number of the pole and tower, and phase, and then the tower can be climbed after all have been confirmed correct.

③ If lines lose power suddenly during work, operators shall still deem it to be charged. The Responsible Person shall contact the on-duty control personnel as soon as possible, and no forced energization is allowed before the on-duty control personnel getting in touch with the Responsible Person.

④ Insulating tool and insulating ropes shall be free of damage, moisture, deformation, and failure. It is not allowed to use non-insulating ropes (such as cotton rope, manila rope, and steel wire rope).

⑤ The ground electrician shall wear clean and dry gloves when operating the insulating tools. When entering the work site, the live working tools shall be placed on damp-proof can-

Part II Skill Module Training and Assessment Standards

vas or insulating mat to prevent dirt and dampness of the insulating tools in use.

⑥ The intermediate potential electrician shall wear flame-retardant underwear and full set of qualified shielding clothes outside (including hat, dresses & trousers, gloves, socks and shoes). All parts shall be in excellent connection conditions.

⑦ When the intermediate potential operator moves along the insulator string, the positions of the hands and feet shall be correspondingly consistent, and the number of insulators shorted by the human body and the tools shall conform to the provisions of Table 2-5-3.

⑧ The operator enters the intense electric field by means of "two-span and three-short-circuit" operation mode (also called free operation method).

⑨ When transmitting large metal objects by using insulating ropes, ground electrician shall not touch them before ground connection.

⑩ During the live working, the Responsible Person (Supervisor) shall continuously monitor the operators and correct their nonstandard operation or actions in violation at any time. Special attention shall be paid to high-place operators to ensure that there is enough safety distance and combined gap (meeting the requirements in Table 2-5-2). It is forbidden to touch two non-connected electrified bodies or electrified body and grounding body.

(2) Hazard - precontrol measures for falling accident:

① Before climbing, operators work at heights must satisfy the requirements of this operation, such as physical condition, mental state and skill and quality.

② Double safety belts shall be used by the personnel for Work at Heights. When climbing up and down the tower, the main material shall be grasped with hands, with shackles trodden and moving at a uniform speed.

③ Before the operation, the intermediate potential operator shall carefully inspect hydraulic lead screw, closed clamp and so on to ensure that the load-bearing tools are qualified; when moving along the insulator string, the positions of hands and feet must be correspondingly consistent, and the safety belt shall be fastened to the insulator string supported by hands and moved synchronously; when replacing insulators, the load-bearing tools shall be installed reliably. Before and after load transfer, impulse test shall be conducted to determine their reliability, and it shall be reported to the Responsible Person in a timely manner. The implementation can only be carried out with the permission of the Responsible Person.

④ Supervisor shall correct irregularities and violations at any time. Special attention shall be paid to operators for Work at Heights to prevent them from losing the protection of the safety belt or insulated backup protection rope during transposition. The safety belt or insulated backup protection rope shall not be fastened lower than operating personnel.

(3) Hazard - injury caused by objects falling from heights

Precontrol measures:

① Operator working at heights should put personal tools and fragmentary materials into the tools bag. It is strictly forbidden to hang objects in heights or keep in the mouth.

② Ground electrician should correctly wear a helmet and use the knots. The vertical distance from the work site should not be less than the falling radius.

③ The work site shall be set up with fence and warning signs. It shall be noted at any time that Supervisor shall prohibit irrelevant personnel and vehicles from entering operation area.

3.2 Selection of tools and instruments and materials

Tools and materials required for live replacement operation of any single glass insulator of tensile glass insulator string of 1000kV AC transmission line can be seen in Table 2-5-4. Before delivering tools and instruments out of warehouse, application voltage class and test period shall be carefully checked and they shall be inspected to ensure that appearance is intact, connection is firm, rotation is flexible and meet the working task requirements. After delivering tools and instruments out of warehouse, they shall be stored in tools bag or tool kit for transportation to avoid contamination and damp. Metal tools and insulated tools shall be separately loaded and transported to avoid deformation, damage or other defects caused by mixed loading and transportation.

Table 2-5-4　Tools and Materials Required for Live Replacement of Arbitrary Single Glass Insulator of Tensile Glass Insulator String of 1000kV AC Transmission Line

S/N	Name	Specification	Unit	Qty.	Remarks
1	Shielding clothes	Type I	Set	2	PPE
2	Conductive shoes	The size depends on the wearer	Nos.	2	PPE
3	Flame retardant underwear	Pure mulberry silk	Set	2	PPE
4	Safety belt of double insurance	Suspender	Nos.	2	PPE
5	Safety helmet		Nos.	7	PPE
6	Goggles		Nos.	2	PPE
7	Insulated transmission rope	φ14mm, with the length matching the lifting height	Nos.	1	Insulating tool
8	Insulated backup protection rope	φ16mm	Nos.	3	Insulating tool
9	Insulated noose	φ14mm	Nos.	2	Insulating tool

Part II Skill Module Training and Assessment Standards

Table (Cont'd)

S/N	Name	Specification	Unit	Qty.	Remarks
10	Insulated tackle	1T	Nos.	1	Insulating tool
11	Hydraulic leading screw	8T	Nos.	2	Metal tool
12	Closed clamp	Tc4	Set	1	Metal tool
13	Pin puller		Nos.	1	Other tools
14	Insulation resistance tester	2,500V, with electrode width 2cm and interelectrode width 2cm	Set	1	Other tools
15	Multimeter		Set	1	Other tools
16	Wind speed, temperature and humidity tester		Nos.	1	Other tools
17	Safety fence		Set	Several	Other tools
18	Warning sign	"Work Here", "Access from Here""Slow Down" and "Blocking"	Set	1	Other tools
19	Red waistcoat	"Responsible Person" "Special Supervisor"	Piece	1	Other tools
20	Moisture-proof tarpaulin	3m×3m	Piece	2	Other tools
21	Personal tools	Wrench, vice	Set	1	Other tools
22	Falling protector	Corresponding to the type of pole and tower anti-fall device	Nos.	2	Other tools
23	Towel	Cotton	Nos.	1	Other tools
24	Insulator	The same model with replaced insulators	Piece	1	Material

3.3 Division of labor for operators

Division of labor for operators of the task is shown in Table 2-5-5.

Table 2-5-5 Division of Labor for Live Replacement of Any Single Glass Insulator of Tensile Glass Insulator String of 1000kv AC Transmission Line

S/N	Post	Qty. (person)	Responsibilities
1	Responsible Person	1	Be responsible for division of operating personnel of the task, work order reading, handling re-closing deactivation of the line and the formalities of work permit, getting work permits, holding pre-shift meeting, dealing with emergency situations in work, quality surveillance, and the summary after work
2	Special Supervisor	1	Be responsible for the safety control of the work site
3	Intermediate potential operator	1	Be responsible for tool installation and insulator replacement
4	Ground electrician	3	Be responsible for ground auxiliary works during the operation.

4. Work Procedure

The workflow of this task is shown in Table 2-5-6.

Table 2-5-6　Workflow of Live Replacement of Any Single Glass Insulator of Tensile Glass Insulator String of 1000kV AC Transmission Line

S/N	Work Content	Operation Steps and Standards	Safety Measures and Precautions	Responsible Person
1	Site re-survey	The Responsible Person shall complete the following work: (1) Check the line name, the number of the pole and tower and dual number on spot to ensure they are correct; guarantee that the base and the pole and tower are intact and in normal condition, and ensure that the cross and span distance meets the safety requirements; confirm the defect conditions and the specifications and models of earth wires. (2) Check that the site meteorological conditions such as wind speed and humidity should meet the operation requirements. (3) Check that the terrain and environment shall meet the operation requirements. (4) Check that the safety measures listed in the work order are in line with the actual situations on site, and the measures will be supplemented if necessary	(1) Correctly wear helmet, working clothes, work shoes and protective gloves. (2) Operation under meteorological conditions that may endanger the safety of operators is forbidden. (3) Non-operation personnel and vehicles are strictly prohibited from entering the working site.	
2	Work Permit	(1) The Responsible Person is responsible for contacting the on-duty control personnel and applying for stopping the line re-closing as per the contents of the work order. (2) Live working could be started only after being approved by the on-duty control personnel.	Live working shall not be started without the permission of the on-duty control personnel	
3	Site layout	Install the security fence and hang the signboards correctly: (1) The security fence should take full account of falling objects from the heights and the influence on road traffic. (2) The entrance and exit of the security fence shall be set reasonably. (3) Signs such as "Access from Here", "Work Here", "Slow Down" or "Blocking" shall be properly arranged.	When the influence on road traffic safety is uncontrollable, the traffic management department should be contacted in time to strengthen the on-site control of traffic safety	,

Part II Skill Module Training and Assessment Standards

Table (Cont'd)

S/N	Work Content	Operation Steps and Standards	Safety Measures and Precautions	Responsible Person
4	Hold a pre-shift meeting	(1) All working personnel shall line up. (2) The Responsible Person shall wear red waistcoat and read out the work order and be clear with work task and division of personnel; explain safety measures and technical measures in work; check (inquire after) mental state of all working personnel; inform of hazards in work and precontrol measures. (3) All working personnel shall sign on the work order for confirmation.	(1) The work order shall be filled in, issued and approved in a standardized manner, and the signature shall be complete. (2) All working personnel shall be in good mental states. (3) All working personnel shall be clear with task division of works, safety measures and technical measures	
5	Inspect tools	(1) All necessary tools and instruments shall be prepared as per the operation requirements and placed on the moisture-proof tarpaulin regularly according to the category and location. The appearance and test certificate of tools and instruments shall be checked to ensure there is no omission. (2) The surface insulation resistance of insulating tools and insulating ropes shall be tested with insulation resistance tester in correct methods, and the value shall be not less than 700MΩ. (3) The new insulator is wiped up, and it is intact in appearance inspection, without rust, cracks and breakage. The surface insulation resistance shall be tested with insulation resistance tester in correct methods, and the value shall be not less than 500MΩ. (4) The internal resistance of full shielding clothes shall be tested with a multimeter in correct methods, and the value shall be not more than 20Ω. (5) The inspector shall report to the Responsible Person that all inspection results are in conformity with the operation requirements	(1) The waterproof tarpaulin shall be enough in quantity and reasonable in position, and be clean and dry. (2) Before using metal and insulating tools, they shall be carefully checked for damage, dampness, deformation and failure, and the certificate of conformity is within the validity period. (3) Insulating tools and insulating ropes are tested as acceptable.	

Table (Cont'd)

S/N	Work Content	Operation Steps and Standards	Safety Measures and Precautions	Responsible Person
6	Climbing the tower	(1) The intermediate potential electrician shall check the double name and phase of the line again, inspect and confirm that the shackles are complete and firm; fasten safety belts and attach falling protectors; perform impact inspection on safety belt, backup protection rope and falling protector in correct method; the Responsible Person shall check and confirm that the connection conditions of each connection point of the double safety belt and shielding clothes worn by the equipotential electricians are in good condition, including shoulder harness, pectoral harness, belt, back rope sling, buckle and ring. (2) The electrician shall carry toolkit and insulated transmission rope (including insulated tackle) in correct methods. (3) Main band of safety belt and backup protection rope shall obliquely across the shoulder. (4) The soles shall be cleaned, and climbing the tower shall be done with the permission of the Responsible Person. (5) The electrician shall tread on shackles, grasp the main materials, and climb the tower uniformly to the proper position of the cross arm with safety belt well fastened and then break away from the falling protector.	(1) Safety belt and falling protector shall pass the impulse test. (2) The safety belt and insulated transmission rope shall be prevented from hooking up tower materials. (3) The minimum safe distance between the human body and the conductor shall conform to the regulations shown in Table 2-5-2. (4) It is forbidden to grasp shackles with hands. (5) Falling protector shall be properly used. (6) Protection of safety belt shall not be lost during transposition.	
7	Entering the intense electric field	(1) The intermediate potential electrician carries the insulated transmission rope, transposes it to the hanging point of the operation phase strain insulator string, hangs the main safety belt on the insulator string connection fittings, and leaves the safety belt backup protection rope at the proper position of the cross arm. (2) The intermediate potential electrician shall recheck and confirm the good connection of all parts of the shielding clothes and the insulator string and the location of the faulty insulators. With the permission of	(1) The safety belt and insulated transmission rope shall be prevented from hooking up tower materials. (2) Protection of safety belt shall not be lost during transposition.	

… Part II Skill Module Training and Assessment Standards

Table (Cont'd)

S/N	Work Content	Operation Steps and Standards	Safety Measures and Precautions	Responsible Person
		the Responsible Person, the electrician shall grasp one string and tread the other one, and steadily move to the operation point along the insulator string in the operation mode of "two-span and three-short-circuit"; the positions of hands and feet must be correspondingly consistent, and the safety belt shall be fastened to the insulator string supported by hands and moved synchronously. (3) After reaching the operation point, the intermediate potential electrician shall fix the insulated tackle at a proper position of insulator string by insulating rope sleeve, and thread it into the insulated transmission rope. The installation shall be firm, stable and convenient for work.	(3) The safe distance between the human body and the electrified body, and the combined gap between the human body and the grounding body and the electrified body shall not be less than that specified in Table 2-5-2.	
8	Install tools and transfer conductor tension	(1) The ground electrician transfers the closed clamp, hydraulic leading screw and so on to the intermediate potential electrician by using the insulated transmission rope. The lifting process shall be smooth, free from collision and winding, and the knot shall be used correctly. (2) The intermediate potential electrician installs the closed clamp (front clamp) in the slot of the rear two insulators where insulators need be replaced, the rear clamp is installed on the steel cap of the front insulator where the insulator needs to be replaced, and the hydraulic leading screw is connected. Each part of the load-bearing tools is securely and reliably installed. (3) Equipotential electricians shall check and confirm that all parts of the load-bearing tools are installed in good condition. With the permission of the Responsible Person, they can operate the hydraulic leading screw so that it is gradually stressed and the insulators to be replaced are relaxed. The two hydraulic leading screws shall be uniformly stressed.	(1) The combined gap between the human body and the grounding body and the charged body shall not be less than that specified in Table 2-5-2. (2) Objects falling from high place shall be avoided. (3) After deducting inferior insulators, insulators shorted by human body and tools, the number of good insulators shall conform to that specified in Table 2-5-3.	

Table (Cont'd)

S/N	Work Content	Operation Steps and Standards	Safety Measures and Precautions	Responsible Person
9	Replacement of insulator	(1) The intermediate potential electrician shall perform impulse test, check and confirm that the load-bearing tools are stressed normally. With the permission of the Responsible Person, they shall tie the old insulator with the insulated transmission rope, take out the fitting pins at both ends of the old insulator, continue to operate and tighten the hydraulic leading screw until the old insulator is removed. The two hydraulic leading screws shall be stressed uniformly, and the operating handle shall not knock on the insulator. (2) The ground electrician uses the other end of the insulated transmission rope to fasten the new insulator, and transfers the new insulator to the intermediate potential electrician by means of the old lower and the new upper. The lifting process shall be smooth, free from collision and winding, and the knot shall be used correctly. (3) The intermediate potential electrician installs the new insulator, resets the fitting pin at its two ends, and confirms that it is installed in place	(1) The combined gap between the human body and the grounding body and the charged body shall not be less than that specified in Table 2-5-2. (2) Objects falling from high place shall be avoided	
10	Remove tools	(1) The intermediate potential electricians shall check that the connection of the new insulator is reliable and, with the permission of the Responsible Person, they operate and loosen the hydraulic leading screw so that the replaced insulator is gradually stressed. (2) After load transfer, the intermediate potential electricians shall perform the impulse test, check and confirm that the new insulator is in good condition. With the permission of the Responsible Person, they can remove the insulated transmission rope tied to the insulator, fasten it to the proper position of the load-bearing tools, remove the hydraulic leading screw, closed clamp and other load-bearing tools, and transfer them to the ground with the cooperation of the ground electrician. The transfer process shall be smooth, free from collision and winding, and the knot shall be used correctly.	(1) The combined gap between the human body and the grounding body and the charged body shall not be less than that specified in Table 2-5-2. (2) Objects falling from high place shall be avoided	

Part II Skill Module Training and Assessment Standards

Table (Cont'd)

S/N	Work Content	Operation Steps and Standards	Safety Measures and Precautions	Responsible Person
11	Exit the intense electric field	(1) The intermediate potential electricians shall remove the insulated transmission rope after checking that there is no object left in the operation area. (2) They shall return to the cross arm with the insulated transmission rope along the insulator string in the "two-span and three-short-circuit" operation mode	(1) The combined gap between the human body and the grounding body and the charged body shall not be less than that specified in Table 2-5-2. (2) Protection of safety belt shall not be lost during transposition.	
12	Evacuation from the pole and tower	The intermediate potential electricians shall check that there is no object left on the tower. After being approved by the Responsible Person, the falling protector shall be properly hung and the safety belt shall be unfastened and put straight, the electrician shall carry the insulated transmission rope correctly and step down to the ground from the tower in uniform steps by treading on shackles and grasping the main materials	(1) Protection of safety belt shall not be lost during transposition. (2) Hand slippage and step missing shall be avoided and it's forbidden to grasp the shackles with hands. (3) Falling protector shall be properly used. (4) The insulated transmission rope and safety belt shall be prevented from hooking up the tower materials or shackles	
13	End of the work	(1) The Responsible Person shall organize all working members to put working apparatus and materials in order and put them in a special kit (bag) after cleaning; clean the site to ensure that "the materials are removed and the site is cleaned after construction". (2) In the post-shift meeting, Responsible Person will give work summaries and comments. Comments include the construction quality of this work and the implementation of safety measures from all working personnel. (3) Responsible Person shall report to the on-duty control personnel that the work is over, apply for the restoration of circuit re-closing and terminate the work order.		

II. Assessment Standard

Detailed Rules for Assessment and Scoring of Operation and Inspection Skills of UHV AC Transmission Line

Fill-in Column of Examinee	No.:		Name:		Position:		Date:	MM/DD/YYYY
Fill-in Column of Assessor	Grade:		Assessor:		Team Assessment Leader:		Closing time:	Operation Duration:
Assessment module	Live replacement of arbitrary single insulator of tensile glass insulator string of 1000kV AC transmission line		Assessee	Maintenance personnel of UHV AC transmission line		Assessment method	Operation	Assessment Time Limit
								90min
Job Description	The live replacement of arbitrary single glass insulator of tensile tower glass insulator string of 1000kV AC transmission line.							
Work Specifications and Requirements	1. Live working shall be carried out in good weather. In case of thunder, rain, snow or fog, no live working shall be carried out. When the wind force is greater than Level 5, live working should not be carried out. When the humidity is greater than 80%, insulating tools with moisture-proof properties shall be used if live working needs to be carried out. 2. Six persons are required for this operation, including 1 Responsible Person, 1 special Supervisor, 1 intermediate potential electrician and 3 ground electricians. 3. Responsibilities of Responsible Person (Supervisor): Be responsible for division of operating personnel of the task, work order reading, handling re-closing deactivation of the line and the formalities of work permit, getting work permits, holding pre-shift meeting, safety supervision in the operation process, dealing with emergency situations in work, quality surveillance, and the summary after work 4. During the live working, if thunder, rain, strong wind or any other circumstance threaten the safety of the staff, the Responsible Person or Supervisor may stop working temporarily according to the circumstances. Given conditions: 1. Work orders have been handled, safety measures have been completed (re-closing has been deactivated), and oral application (dispatcher or assessor) shall be made at the beginning and end of the work. 2. The instrument shall be used safely and correctly to test the insulating tool. 3. The operation must be carried out according to the working procedures. The scores of the items to be carried out shall be deducted for the process error. In case of major hidden dangers of personal, equipment and operational safety, the assessor may order the termination of the operation (assessment)							

Part II Skill Module Training and Assessment Standards

Table (Cont'd)

Assessment scenario preparation	1. Tower type: 1000kV AC resisting-tensile tower. 2. Required operation tools: 2 double safety belts, 2 sets of 1-shaped shielding clothes, 2 sheets of moisture-proof cloth, 1 multimeter, 1 insulation resistance detector, 1 anemometer, 1 2-in-one temperature and humidity detector, 2 hydraulic leading screws, 2 insulating ropes, 1 set of closed clamp, 1 tackle, 1 goggles, 1 pin puller and 1 set of hand-operated tools. 3. The work site shall be monitored, and the safety measures (fence, etc.) on the work site have been fully implemented; non-operation personnel are prohibited from entering the site, and the staff must wear safety helmets when entering the work site. 4. Examinees shall bring their own work clothes, helmets and gloves
Remarks	1. The deduction shall be done until the scores of each item are deducted completely. In case of major hidden dangers of personal, equipment and operational safety, the assessor may order the termination of the operation. 2. When equipment, working environment, safety belt, safety helmet, tool, shielding clothes, etc., do not conform to the operation condition, the assessor may order the termination of the operation

S/N	Project name	Quality requirements	Score	Deduction standard	Reasons for deduction	Deduction	Scoring
1	Site re-survey	1) The Responsible Person shall go to the work site to check the line name, pole and tower number, on-site working conditions, defective parts and so on. 2) Check that the site meteorological conditions such as wind speed and humidity should meet the operation requirements. 3) Check whether the work order is complete and unmodified, check whether the safety measures listed are consistent with the actual situation on site, and supplement it if necessary	5	1) Deduct 1 point/item for failure to check the line name, pole and tower number, on-site working conditions and defective parts. 2) Deduct 1 point/item for failure to detect wind speed, humidity and other on-site meteorological conditions. 3) Deduct 0.5 points/part for any alteration in the work order filling; deduct 1 point for incorrect work order number; deduct 1.5 points for incomplete work order filling			

Standard for Professional Training and Assessment for Operation Maintenance of UHV AC Transmission Line

Table (Cont'd)

S/N	Project name	Quality requirements	Score	Deduction standard	Reasons for deduction	Deduction	Scoring
2	Work Permit	1) The Responsible Person is responsible for contacting the on-duty control personnel (referee) and applying for stopping the line re-closing as per the contents of the work order. 2) Reporting content is standardized and complete	2	1) Deduct 2 points for failure to contact the dispatching department (referee) for deactivating the re-closing. 2) Deduct 1 point for non-standard or incomplete terminology reporting			
3	Site layout	Install the security fence and hang the signboards correctly: 1) The security fence should take full account of falling objects from the heights and the influence on road traffic. 2) The entrance and exit of the security fence shall be set reasonably. 3) Signs such as "Access from Here", "Work Here", "Access from Here" shall be properly and well arranged	3	1) Deduct 1 point for failure to arrange the fence at the work site. 2) Deduct 1 point for failure to arrange the warning board. 3) Deduct 1 point for failure to hang the tower climbing operation sign			
4	Hold a pre-shift meeting	1) All staff and personnel shall wear safety helmets and work clothes correctly. 2) Responsible Person shall wear red vest and read out the work order and be clear with work task and division of personnel; explain safety measures and technical measures in work; check (inquire after) mental state of all working personnel; inform of hazards in work and precontrol measures. 3) All working personnel shall sign on the work order for confirmation.	3	1) Deduct 0.5 points/person for the staff not dressing uniformly. 2) Deduct 3 points for no division of labor; deduct 1 point for unclear division of labor. 3) Deduct 1 point for the on-site Responsible Person not wearing a safety monitoring vest. 4) Deduct 1 point for the work shift member failing to sign or signing incompletely on the work order.			

· 426 ·

Part II Skill Module Training and Assessment Standards

Table (Cont'd)

S/N	Project name	Quality requirements	Score	Deduction standard	Reasons for deduction	Deduction	Scoring
5	Inspection of tools	1) All necessary tools and instruments shall be prepared as per the operation requirements and placed on the moisture-proof tarpaulin regularly according to the category and location. The appearance and test certificate of tools and instruments shall be checked to ensure there is no omission. 2) The surface insulation resistance of insulating tools and insulating ropes shall be tested with insulation resistance tester in correct methods, and the value shall be not less than 700MΩ. 3) The new insulator is wiped up, and it is intact: in appearance inspection, without rust, cracks and breakage. The surface insulation resistance shall be tested with insulation resistance tester in correct methods, and the value shall be not less than 500MΩ. 4) The internal resistance of full shielding clothes shall be tested with a multimeter in correct methods, and the value shall be not more than 20Ω. 5) The inspector shall report to the Responsible Person that all inspection results are in conformity with the operation requirements	7	1) Deduct 1 point for failure to use moisture-proof tarpaulin and place tools to designed positions. 2) Deduct 0.5 points/item for failure to check the appearance of tools and pass the test certificate. 3) Deduct 1 point/item for failure to use testing instrument for testing the tools. 4) Deduct 1 point for non-standard reporting of test results; deduct 0.5 points/item for incomplete reporting			

Table (Cont'd)

S/N	Project name	Quality requirements	Score	Deduction standard	Reasons for deduction	Deduction	Scoring
6	Climbing the tower	1) The intermediate potential electrician shall check the double name and phase of the line again, inspect and confirm that the shackles are complete and firm; fasten safety belts and attach falling protectors; perform impulse test on double safety belts, backup protection rope and falling protector in correct method. The Responsible Person shall check and confirm that the connection conditions of each connection point of the double safety belt and shielding clothes worn by the intermediate potential electricians are in good condition, including shoulder harness, pectoral harness, belt, back rope sling, buckle and ring. 2) The electrician shall carry toolkit and insulated transmission rope (including insulated tackle) in correct methods. 3) During the process of climbing the tower, the electrician shall fasten the anti-fall device, tread on shackles, grasp the main materials, and climb the tower uniformly to the proper position with safety belt well fastened and then break away from the falling protector	5	1) Deduct 1 point/item for the intermediate potential electrician failing to check the double name of the line, pole number, phase and tower material; deduct 1 point for non-report after checking. 2) Deduct 2 points/item for failure to perform the impulse test on double safety belts and falling protector. 3) Deduct 1 point for the on-site Responsible Person failing to check the safety protection equipment of the intermediate potential electrician. 4) Deduct 0.5 points/time for hands grasping the shackles. 5) Deduct 1 point for the unreasonable suspension position of the tackle transmission rope. 6) Deduct 5 points for loosing the protection of the safety belt during transposition			

Part II Skill Module Training and Assessment Standards

Table (Cont'd)

S/N	Project name	Quality requirements	Score	Deduction standard	Reasons for deduction	Deduction	Scoring
7	Entering the intense electric field	1) The intermediate potential electrician carries the insulated transmission rope, transposes it to the hanging point of the operation phase strain insulator string, hangs the main safety belt on the insulator string connection fittings, and leaves the safety belt backup protection rope at the proper position of the cross arm. 2) The intermediate potential electrician shall recheck and confirm the good connection of all parts of the shielding clothes and the insulator string and the location of the faulty insulators. With the permission of the Responsible Person, the electrician shall grasp one string and tread the other one, and steadily move to the operation point along the insulator string in the operation mode of "two-span and three-short-circuit"; the positions of hands and feet must be correspondingly consistent, and the safety belt shall be fastened to the insulator string supported by hands and moved synchronously. 3) After reaching the operation point, the intermediate potential electrician shall fix the insulated tackle at a proper position of insulator string by insulating rope sleeve, and thread it into the insulated transmission rope. The installation shall be firm, stable and convenient for work.	5	1) Deduct 2 points for irrational fastening position and irregular use of safety belt back protection rope. 2) Deduct 1 point/item for the intermediate potential electrician failing to check the connection of the shielding clothes and insulator string and the location of the faulty insulator. 3) Deduct 5 points for entering the intense electric field without permission of the Responsible Person. 4) Deduct 2 points/time for the incorrect action of intermediate potential electrician entering the intense electric field and repeat discharging. 5) Deduct 1 point for the unreasonable installation position of the insulated transmission rope. 6) Deduct 2 points/time for falling objects from heights. 7) Deduct 5 points for loosing the protection of the safety belt during transposition			

Table (Cont'd)

S/N	Project name	Quality requirements	Score	Deduction standard	Reasons for deduction	Deduction	Scoring
8	Installation tool	1) The ground electrician shall use insulated transmission rope to transfer the closed clamp, hydraulic leading screw and other tools to the intermediate potential electrician respectively, and use the knots in a correct manner, which shall be smooth, free from collision and winding. 2) The intermediate potential electrician installs the closed clamp (front clamp) in the slot of the rear two insulators where insulators need be replaced, the rear clamp is installed on the steel cap of the front insulator where the insulator needs to be replaced, and the hydraulic leading screw is connected. Each part of the load-bearing tools is securely and reliably installed. 3) Equipotential electricians shall check and confirm that all parts of the load-bearing tools are installed in good condition. With the permission of the Responsible Person, they can operate the hydraulic leading screw so that it is gradually stressed and the insulators to be replaced are relaxed. The two hydraulic leading screws shall be uniformly stressed.	15	1) Deduct 1 point/time for unstable lifting process, colliding and winding. 2) Deduct 2 points/time for falling objects from heights. 3) Deduct 2 points for improper installation and fixing of clamp. 4) Deduct 3 points for failure to check the installation of load-bearing tools; deduct 1 point for failure to report after checking; deduct 1 point for starting tightening the leading screw without the consent of Responsible Person after reporting. 5) Deduct 3 points/time for the number of shorted insulators exceeding 4 during the operation. 6) Deduct 15 points for collision and breakage of insulators in the clamp installation. 7) Deduct 2 points for failure to tighten leading screw in a balanced way			

Part II   Skill Module Training and Assessment Standards

Table (Cont'd)

S/N	Project name	Quality requirements	Score	Deduction standard	Reasons for deduction	Deduction	Scoring
9	Replacement of insulator	1) The intermediate potential electrician shall perform impulse test, check and confirm that the load-bearing tools are stressed normally. With the permission of the Responsible Person, they shall tie the old insulator with the insulated transmission rope, take out the fitting pins at both ends of the old insulator, continue to operate and tighten the hydraulic leading screw until the old insulator is removed. The two hydraulic leading screws shall be stressed uniformly, and the operating handle shall not knock on the insulator. 2) The ground electrician uses the other end of the insulated transmission rope to fasten the new insulator, and transfers the new insulator to the intermediate potential electrician by means of the old lower and the new upper. The lifting process shall be smooth, free from collision and winding, and the knot shall be used correctly. 3) They shall install new insulator and reset fitting pins at both ends.	20	1) Deduct 3 points for failure to check the stress of load-bearing tools; deduct 2 points for failure to report after checking; deduct 1 point for removing fitting pins at both ends of the old insulator without the consent of Responsible Person after reporting. 2) Deduct 2 points for failure to tighten leading screw in a balanced way. 3) Deduct 1 point/time for the operating handle hitting the insulator. 4) Deduct 1 point for wrong knot. 5) Deduct 2 points/time for falling objects from heights. 6) Deduct 1 point for new and old insulators colliding with each other. 7) Deduct 1 point for the transmission insulator and tower body colliding with each other. 8) Deduct 2 points for winding insulating ropes			

Standard for Professional Training and Assessment for Operation Maintenance of UHV AC Transmission Line

Table (Cont'd)

S/N	Project name	Quality requirements	Score	Deduction standard	Reasons for deduction	Deduction	Scoring
10	Remove tools	1) The intermediate potential electricians shall check that the connection of the new insulator is reliable and, with the permission of the Responsible Person, they operate and loosen the hydraulic leading screw so that the replaced insulator is gradually stressed. 2) After load transfer, the intermediate potential electricians shall perform the impulse test, check and confirm that the new insulator is in good condition. With the permission of the Responsible Person, they can remove the insulated transmission rope tied to the insulator, fasten it to the proper position of the load-bearing tools, remove the hydraulic leading screw, closed clamp and other load-bearing tools, and transfer them to the ground with the cooperation of the ground electrician. The transfer process shall be smooth, free from collision and winding, and the knot shall be used correctly.	15	1) Deduct 3 points for failure to check the connection of new insulator; deduct 2 points for failure to report after checking; deduct 1 point for starting loosening the hydraulic leading screw without the consent of Responsible Person after reporting. 2) Deduct 3 points for failure to check the stress of new insulator; deduct 2 points for failure to report after checking; deduct 1 point for starting removing the hydraulic leading screw without the consent of Responsible Person after reporting. 3) Deduct 1 point for incorrect use of knot when binding the tools. 4) Deduct 2 points/time for falling objects from heights. 5) Deduct 1 point/time for the tools colliding with each other; deduct 1 point/time for the tool colliding with the electrified body or tower body; deduct 2 points for winding insulating ropes			

Part II Skill Module Training and Assessment Standards

Table (Cont'd)

S/N	Project name	Quality requirements	Score	Deduction standard	Reasons for deduction	Deduction	Scoring
11	Exit the intense electric field	1) The intermediate potential electricians shall fasten the insulated transmission rope after checking that there is no object left in the operation area to make preparation for exiting the potential. 2) The intermediate potential electrician exits the equipotential according to the operation mode of "two-span and three-short-cut".	5	1) Deduct 2 points for exiting the intense electric field without applying to the Responsible Person; deduct 1 point for starting exiting the intense electric field without the consent after applying. 2) Deduct 2 points/time for the incorrect action of intermediate potential electrician exiting the intense electric field and repeat discharging. 3) Deduct 1 point for failure to effectively control backup protection rope			
12	Return to the ground	After checking that there is no object left on the tower, the electricians can climb down the tower with insulated transmission rope after obtaining the consent of the Responsible Person.	5	1) Deduct 5 points for failure to use the falling protector when climbing down the tower. 2) Deduct 5 points for loosing the protection of the safety belt when moving on the tower. 3) Deduct 1 point/time for hands grasping the shackles when climbing down the tower. 4) Deduct 2 points for any objects left on tower.			

· 433 ·

Table (Cont'd)

S/N	Project name	Quality requirements	Score	Deduction standard	Reasons for deduction	Deduction	Scoring
13	End of the work	1) The Responsible Person shall organize all working members to put working apparatus and materials in order and put them in a special kit (bag) after cleaning; clean the site to ensure that "the materials are removed and the site is cleaned after construction". 2) In the post-shift meeting, Responsible Person will give work summaries and comments. Comments include the construction quality of this work and the implementation of safety measures from all working personnel. 3) Responsible Person shall report to the on-duty control personnel that the work is over, apply for the restoration of circuit re-closing and terminate the work order.	10	1) Deduct 2 points for failure to clean the tools. 2) Deduct 2 points for missing tools. 3) Deduct 10 points for failure to hold the post-shift meeting. 4) Deduct 2 points for failure to remove the fence. 5) Deduct 2 points for failure to report to dispatcher			
	Total		100				

Module 6 Standards for Training and Assessment on Live Replacement of Arbitrary Section Insulator of Tensile Glass Insulator String of 1000kV AC Transmission Line

I. Training Standard

(I) Training Requirements

Designation of module	Live replacement of arbitrary section insulator of tensile glass insulator string of 1000kV AC transmission line	Type of training	Operation
Training method	Practical operation training	Hours of training	21 training hours
Training objectives	1. Master the operating principle of live replacement of arbitrary section insulator for 1000kV current transmission line tension glass insulator string. 2. Grasp the layout methods of tools and instruments for live working of 1000kV AC transmission line, the method of "two-span and three-short-circuit" in and out along the insulator string, and the method of replacing arbitrary section insulator (note: arbitrary section insulator refers to single or multiple insulator(s) within continuous 7 and less than 7 insulator strings). 3. Be capable of live replacement of arbitrary section insulator of tensile glass insulator string of 1000kV AC transmission line		
Training venue	UHV AC training line		
Training content	The intermediate potential operation method is adopted for the live replacement of arbitrary section insulator of tensile glass insulator string of 1000kV AC transmission line, and the operation phase is the intermediate phase		
Scope of application	Maintenance personnel of UHV AC transmission line		

(II) Referenced Rules and Specifications

(1) Electrotechnical Terminology (GB/T2900.55-2002).

(2) Insulated Tackles for Live Working (GB/T13034-2008).

(3) Live Working-Insulating Ropes (GB/T 13035-2008).

(4) Calculation Method of Live Working Minimum Approach Distance on AC Transmis-

sion Line (GB/T 18037-2000)

(5)Screen Clothes for Live Working on 1000kV AC (GB/T25726-2010).

(6)Code for Design of 1000kV Overhead Transmission Line (GB50665-2011).

(7)Maintenance Code for 1000kV AC Transmission Line (DL/T209-2008).

(8)Operation Code for 1000kV AC Transmission Line (DL/T307-2010).

(9)Technical Guide for Live Working on 1000kV AC Transmission Line (DL/T392-2015).

(10)Minimum Requirements for Utilization of Tools, Devices and Equipment for Live Working (DL/T 877-2004)

(11)State Grid Corporation of China Working Regulations of Power Safety (Transmission Line Section) (Q/GDW1799.2-2013).

(Ⅲ) Teaching Design for Training

To complete the work task of "live replacement of arbitrary section insulator of tensile glass insulator string of 1000kV AC transmission line", each training stage shall be designed according to the standard operation procedure for work task completion. Each stage includes specific training objectives, training content, hours of training, training methods (training resources), training environment, assessment and evaluation, etc, as shown in the Table 2-6-1.

(Ⅳ) Operation Flow

1. Work Task

The "two-span and three-short-circuit" operation mode is adopted to enter the intense electric field and arrive at the operation site along the tensile glass insulator string, and the intermediate potential operation method is adopted for live replacement operation of any arbitrary section insulator of tensile glass insulator string of 1000kV AC transmission line.

2. Requirements for Weather and Work Site

(1) The live replacement operation of any arbitrary section insulator of tensile glass insulator string of 1000kV AC transmission line should be carried out in good weather. In case of lightning (hearing thunder or seeing lightning), snow, hail, rain, fog and so on, live working is prohibited. When the wind force is greater than level 5 or the relative humidity of the air is greater than 80%, it is unsuitable for live working. When electric emergency repair must be carried out in bad weather, relevant personnel shall be organized to fully discuss and prepare necessary safety measures, which can only be carried out after the approval of the

Part II Skill Module Training and Assessment Standards

Table 2-6-1 Design for Training Content for Live Replacement of Arbitrary Section Insulator of Tensile Glass Insulator String of 1000kV AC Transmission Line

Training schedule	Training objectives	Training content	Hours of training	Training methods and resources	Preparation of training conditions	Assessment and evaluation
1. Theoretical teaching	1. Preliminarily master the basic methods in and out of 1000kV intense electric field along insulator string; 2. Be familiar with the layout methods of arbitrary section insulator tool replacement and the methods of replacing multiple insulators.	1. Theoretical basis and principles for adopting the method of "two-span and three-short-circuit" in and out of intense electric field along the insulator string. 2. Methods of layout of arbitrary section insulator tool replacement for 1000kV AC transmission line	2	Training methods: Lecture. Training resources: PPT, relevant regulations and specifications, and video presentation	Multimedia classroom	Attendance, classroom questions and assignments
2. Preparations	Be able to complete the preparation before operation	1. Work site survey. 2. Preparation of the standardized operation card. 3. Filling of the work order. 4. Preparation of tools and materials for this operation	1	Training methods: 1. Site survey and cleaning of tools and materials shall be practiced at site. 2. Preparation of operation card and the filling of work order shall adopt lecture method. Training resources: 1. 1000kV practical training line. 2. UHV tools warehouse. 3. Blank work order	1. UHV training transmission line 2. Multi-Media classroom	

Table (Cont'd)

Training schedule	Training objectives	Training content	Hours of training	Training methods and resources	Preparation of training conditions	Assessment and evaluation
3. Work site preparation	Be able to complete the preparations of work site	1. Work site re-survey. 2. Work permit. 3. Work site layout. 4. Pre-shift meeting. 5. Inspection of tools and materials	1	Training methods: Demonstration and role play. Resources: 1000kV training line	1000kV training line	
4. Trainer's demonstration	The trainees can preliminarily understand the operation process of the task through inspecting and learning from each other's work	1. Detection of null-value insulator. 2. The equipotential electrician enters and exits the intense electric field along the strain insulator string. 3. The equipotential electrician (or ground potential electrician) cooperates with the intermediate potential electrician to complete the tool installation. 4. Replacement of arbitrary section insulator completed by the intermediate potential electrician	2	Training methods: Demonstration. Resources: 1000kV training line	1000kV training line	

Part II Skill Module Training and Assessment Standards

Table (Cont'd)

Training schedule	Training objectives	Training content	Hours of training	Training methods and resources	Preparation of training conditions	Assessment and evaluation
5. Group training of trainees	1. Be able to complete the layout operation of tools and instruments for live working. 2. Be able to complete the replacement of arbitrary section insulator for 1000kV transmission line	1. The trainees are grouped (11 persons per group) to be trained about the skills and operation of tool installation and insulator replacement. 2. Trainers guide the operation of trainees and conduct safety supervision	14	Training methods: Role play. Resources: 1000kV training line	1000kV training line	Score
6. End of the work	1. Enable the trainees to further distinguish the shortcomings of the operation process and facilitate the promotion in the later stage. 2. Train the trainees in the working style of safe and civilized production	1. Cleaning up the work site. 2. Report to dispatcher. 3. Comment and summarize the work task this time at post-shift meeting	1	Training method: Lecture and inductive method	1000kV training line	

production leader in charge of the unit.

(2) The operating personnel shall be in good spirits, and be familiar with the organizational and technical measures to ensure safety in work; they shall hold the live working qualification within the validity period.

(3) The Responsible Person shall organize relevant personnel to complete field investigation in advance, determine the operating methods, required working apparatus and necessary measures according to the results, and handle the work orders for live working.

(4) The work site shall be reasonably set up with fence and warning signs. Non-operating personnel is forbidden to enter.

(5) The line re-closing device shall be deactivated in the Project.

(6) Safety working distance and effective insulation length are shown in Table 2-6-2.

Table 2-6-2 Safe Distance for Live Replacement of Arbitrary Section Insulator of Tensile Glass Insulator String of 1000kV AC Transmission Line

Voltage class /m	Safe distance between human body and electrified body /m	Minimum effective insulation length/m		Minimum combined clearance /m	Minimum distance between exposed part of human body and electrified body when potential is transferred /m
		Insulating bar	Insulated load-bearing tools and insulating ropes		
1000kV	6.8	6.8	6.8	6.9	0.5

(7) For operation on 1000kV transmission line, the number of insulators in good operation phase shall be no less than 37 pieces.

3. Preparations

3.1 Hazards and precontrol measures

(1) Hazard - precontrol measures for electric shock injury:

① Before work, the Responsible Person shall contact the control personnel on duty, deactivate the working line re-closing, and perform the licensing procedures.

② Before climbing the tower, ground potential operators on tower must carefully check the name of the line, the number of the iron tower, and phase, and then the tower can be climbed after all have been confirmed correct.

③ If lines lose power suddenly during work, operators shall still deem it to be charged. The Responsible Person shall contact the control personnel as soon as possible, and no forced energization is allowed before the on-duty control personnel getting in touch with the Responsible Person.

④ Insulating tool and insulating ropes shall be free of damage, moisture, deformation, and failure. It is not allowed to use non-insulating ropes (such as cotton rope, manila rope, and steel wire rope).

⑤ The ground electrician shall wear clean and dry gloves when operating the insulating tools. When entering the work site, the live working tools shall be placed on moisture-proof tarpaulin to prevent dirt and dampness of the insulating tools in use.

⑥ The equipotential operator shall wear flame-retardant underwear and qualified full set of shielding clothes over the underwear (including shielding mask, hat, dresses & trousers, gloves, socks and shoes). All parts shall be in excellent connection conditions.

⑦ Before potential transfer, equipotential operator shall obtain the approval of Responsible Person, and the minimum distance between exposed part of human body and electrified body shall not be lower than 0.5m. During potential transfer, the operation shall be fast, and end shall not be used for power charging and discharging; During delivering tools and materials to ground potential operators, the effective length of insulating tools or insulating ropes shall not be lower than the requirements as specified in Table 2-6-2.

⑧ When transmitting large metal objects by using insulating ropes, ground potential operators shall not touch them before ground connection.

⑨ The Responsible Person shall continuously monitor the operators and correct their nonstandard operation or actions in violation of rules and regulations at any time. Operators working at heights shall be specially monitored to ensure that there is enough safe distance (meeting the requirements in Table 2-6-2).

(2) Hazard - falling accident

Precontrol measures:

① Before climbing the tower, operators working at heights must satisfy the requirements of this operation, such as physical condition, mental state, and skill and quality.

② The operators on the tower must correctly use the safety belt, which shall be tied on the firm parts and in reasonable location convenient for operation. While climbing the tower and during displacement on tower, the safety protection is required.

③ When the operator on the tower is moving along the insulator string, the locations of hands and feet must be kept consistent accordingly. The safety belt shall be tied on the handheld insulator string, and hands and feet shall move synchronously.

④Supervisors shall correct the nonstandard or illegal actions and behaviors at any time. Special monitoring shall be conducted to operators to prevent them from losing the protection of safety belt or insulated backup protection rope during transposition, and it is forbidden to fasten the safety belt or insulated backup protection rope in a position lower than the operating personnel.

(3) Hazard - injury caused by objects falling from heights.

Precontrol measures:

① Operator working at heights should put personal tools and fragmentary materials into the tools bag. It is strictly forbidden to hang objects in heights or keep in the mouth.

② Ground operator should correctly wear a helmet and use the knots. The vertical distance from the work site should not be less than the falling radius.

③ The work site shall be set up with fence and warning signs. It shall be noted at any time that Supervisor shall prohibit irrelevant personnel and vehicles from entering operation area.

(4) Hazard - clamps not installed properly. Precontrol measures:

① Before replacing the insulator, it is required to carefully check the clamps, hydraulic leading screws, closed clamps, etc. at the end in details, to ensure that the load-bearing parts will work normally. After the insulator is removed, the clamps, hydraulic leading screws, closed clamps, etc. at the end shall be able to bear the conductor load.

② When replacing the insulator by clamps, the clamps shall be reliably installed, and it is required to judge its reliability before the clamps are stressed and the conductor load is moved. In case of no error after check, the insulator can be replaced. While replacing the insulator, it is required to report to the Responsible Person, which can be started after the approval by the Responsible Person.

3.2 Selection of tools and instruments and materials

Tools and materials required for live replacement operation of arbitrary section insulator of tensile glass insulator string of 1000kV AC transmission line can be seen in Table 2-6-3. Before delivering tools and instruments out of warehouse, application voltage class and test period shall be carefully checked and they shall be inspected to ensure that appearance is intact, connection is firm, rotation is flexible and meet the working task requirements. After delivering tools and instruments out of warehouse, they shall be stored in tools

Part II Skill Module Training and Assessment Standards

bag or tool kit for transportation to avoid contamination and damp. Metal tools and insulated tools shall be separately loaded and transported to avoid deformation, damage or other defects caused by mixed loading and transportation.

Table 2-6-3 Tools and Materials Required for Live Replacement of Arbitrary Section Insulator of Tensile Glass Insulator String of 1000kV AC Transmission Line

S/N	Name	Specification	Unit	Qty.	Remarks
1	Shielding clothes	Type I	Set	4	
2	Conductive shoes	The size depends on the wearer	Nos.	2	
3	Flame retardant underwear	Pure mulberry silk	Set	2	
4	Safety belt of double insurance	Suspender	Nos.	2	
5	Safety helmet		Nos.	7	
6	Insulated transmission rope	φ12mm, with the length matching the lifting height	Nos.	1	
7	Insulated transmission rope	φ14 mm, with the length matching the lifting height	Nos.	2	
8	Insulated backup protection rope	φ16mm	Nos.	3	
9	Insulated noose	φ14mm	Nos.	3	
10	Insulated tackle	1T	Nos.	3	
11	Insulated pull-rod	8T	Set	1	
12	Tackle block	3-3	Set	1	
13	Mechanical leading screw	4T	Nos.	2	
14	Closed clamp	Tc4	Set	1	
15	End clamp	Tc4	Set	1	
16	Potential transfer rod		Nos.	2	
17	Pin puller		Nos.	1	
18	Insulation resistance tester	2,500V, with electrode width 2cm and interelectrode width 2cm	Set	1	
19	Multimeter		Set	1	
20	Wind speed, temperature and humidity tester	HT-8321	Nos.	1	
21	Safety fence		Set	Several	
22	Warning sign	"Work Here", "Access from Here" "Slow Down" and "Blocking"	Set	1	

Table (Cont'd)

S/N	Name	Specification	Unit	Qty.	Remarks
23	Red waistcoat	"Responsible Person" "Special Supervisor"	Piece	1	
24	Moisture-proof tarpaulin	3m×3m	Piece	2	
25	Personal tools	Wrench, vice	Set	1	
26	Falling protector	Corresponding to the type of iron tower anti-fall device	Nos.	2	
27	Towel	Cotton	Nos.	1	
28	Insulator	U550BP/240T	Piece	Several	

Table 2-6-4 Division of Operating Personnel Involved in Live Repair of 1000kV AC Transmission Line Conductor

S/N	Post	Qty. (person)	Responsibilities
1	Responsible Person	1	Be responsible for division of operating personnel of the task, work order reading, handling re-closing deactivation of the line and the formalities of work permit, getting work permits, holding pre-shift meeting, dealing with emergency situations in work, quality surveillance, and the summary after work
2	Special Supervisor	1	Be responsible for the safety monitoring of the work site
3	Intermediate potential operator	2	Be responsible for tool installation and insulator replacement
4	Equipotential electrician	2	Be responsible for installation of tools and instruments
5	Ground electrician	6	Be responsible for transferring tools and materials, and co-operating with the electricians on the tower to install tools and instruments and replace insulator

4. Work Procedure

The workflow of this task is shown in Table 2-6-5.

Table 2-6-5 Workflow Chart for Live Replacement of Arbitrary Section Insulator of Tensile Glass Insulator String of 1000kV AC Transmission Line

S/N	Work Content	Operation Steps and Standards	Safety Measures and Precautions	Responsible Person
1	Site re-survey	The Responsible Person shall complete the following work: (1) Check the line title, the number of the pole and tower and ensure the phases are correct; guarantee that the foundation and the pole and tower are intact and in normal condition; ensure that the cross and span distance meets the safety requirements; confirm the defect conditions and the specifications and models of earth wires. (2) Check that the site meteorological conditions such as wind speed and humidity should meet the operation requirements. (3) Check that the terrain and environment shall meet the operation requirements. (4) Check that the safety measures listed in the work order are in line with the actual situations on site, and the measures will be supplemented if necessary	(1) Correctly wear helmet, working clothes, work shoes and protective gloves. (2) Operation under meteorological conditions that may endanger the safety of operators is forbidden. (3) Non-operation personnel and vehicles are strictly prohibited from entering the working site.	
2	Work Permit	(1) The Responsible Person is responsible for contacting the on-duty control personnel and applying for stopping the line re-closing as per the contents of the work order. (2) Live working could be started only after being approved by the on-duty control personnel.	Live working shall not be started without the permission of the on-duty control personnel	
3	Site layout	Install the security fence and hang the signboards correctly: (1) The security fence should take full account of falling objects from the heights and the influence on road traffic. (2) The entrance and exit of the security fence shall be set reasonably. (3) Signs such as "Access from Here", "Work Here", "Access from Here" shall be properly and well arranged	When the influence on road traffic safety is uncontrollable, the traffic management department should be contacted in time to strengthen the on-site control of traffic safety.	

Table (Cont'd)

S/N	Work Content	Operation Steps and Standards	Safety Measures and Precautions	Responsible Person
4	Hold a pre-shift meeting	(1) All working personnel shall line up. (2) The Responsible Person shall wear red waistcoat and read out the work order and be clear with work task and division of personnel; explain safety measures and technical measures in work; check (inquire after) mental state of all working personnel; inform of hazards in work and precontrol measures. (3) All working personnel shall sign on the work order for confirmation.	(1) The work order shall be filled in, issued and approved in a standardized manner, and the signature shall be complete. (2) All working personnel shall be in good mental states. (3) All working personnel shall be clear with task division of works, safety measures and technical measures	
5	Inspect tools	(1) All necessary tools and instruments shall be prepared as per the operation requirements and placed on the moisture-proof tarpaulin regularly according to the category and location. The appearance and test certificate of tools and instruments shall be checked to ensure there is no omission. (2) The surface insulation resistance of insulating tools and insulating ropes shall be tested with insulation resistance tester in correct methods, and the value shall be not less than 700MΩ. (3) New insulator shall be tested and cleaned. (4) The internal resistance of full shielding clothes shall be tested with a multimeter in correct methods, and the value shall be not more than 20Ω. (5) The inspector shall report to the Responsible Person that all inspection results are in conformity with the operation requirements	(1) The waterproof tarpaulin shall be enough in quantity and reasonable in position, and be clean and dry. (2) Before using metal and insulating tools, they shall be carefully checked for damage, dampness, deformation and failure, and the certificate of conformity is within the validity period. (3) Insulating tools and insulating ropes are tested as acceptable. (4) The new insulator shall be provided with appearance inspection, and tested by an insulation resistance meter under dry and clean conditions. It cannot be used if its resistance is lower than 500MΩ.	

Part II Skill Module Training and Assessment Standards

Table (Cont'd)

S/N	Work Content	Operation Steps and Standards	Safety Measures and Precautions	Responsible Person
6	Climbing the tower	(1) The double name of line shall be checked, and the iron tower base shall be checked. (2) The electrician on the tower shall carry with the insulated transmission rope and climb the tower to the hanging point of cross arm strain insulator string, and properly tie the safety belt and protective rope.	(1) In case of no error after checking the double name of line and if the iron tower base is firm and reliable, it is allowed to climb the tower. (2) The falling protector shall be used in the process of tower climbing; when moving on the iron tower, the safety protection shall not be lost, and the operators must climb and grasp the components securely. (3) The intermediate potential electricians must wear the complete set of qualified shielding clothes which must be connected reliably	
7	Entering the intense electric field	(1) The electrician on the tower shall transfer the safety belt to the insulator set fitting, and carry the insulated tackle and insulated transmission rope. (2) After the equipotential electrician confirms that all parts of the shielding clothes are well connected, it shall be reported to the Responsible Person for approval. The electrician shall grasp one string with both hands and step on the other string with both feet, and enter the insulator string and move to the operation point along the insulator string in the operation mode of "two-span and three-short-circuit". (3) The intermediate potential electrician shall adopt the same ways to reach the operation point, and then tie the safety belt and place the insulated transmission rope in suitable place and make proper preparations of hoisting.	(1) The intermediate potential electrician must be approved by the Responsible Person before entering the insulator string. (2) After entering the operation point, the safety belt shall be tied at the side of insulator string which will not be replaced and in reasonable position convenient for operation. (3) Before entering the insulator string, the intermediate potential electrician must fasten the protection rope and adjust the insulated transmission rope. (4) The combined gap composed of the gaps between the intermediate potential electrician and the grounding body and the electrified body shall not be less than 6.9 m (medium phase) (6.7 m, side phase).	

Table (Cont'd)

S/N	Work Content	Operation Steps and Standards	Safety Measures and Precautions	Responsible Person
8	Install tools and transfer conductor tension	(1) The ground electrician shall respectively transfer the tackle block, closed clamps, etc. to the intermediate potential electrician, and transfer the hydraulic leading screw and end clamps to the equipotential electrician. The intermediate potential electrician and equipotential electrician shall respectively install the tools, and then the ground electrician shall transfer the insulated pull-rods to the intermediate potential electrician and equipotential electrician for proper installation through transmission rope. (2) The intermediate potential electrician shall properly install the closed clamps in correct positions before and after replacing the insulator string. (3) After checking the installation of all parts of the load-bearing tools correctly, the electrician on tower shall respectively report to the Responsible Person. After approval by the Responsible Person, the equipotential electrician shall tighten the leading screw. After the leading screw is suitably stressed, the impact inspection is required for the load-bearing tools. After the approval by the Responsible Person, the hydraulic stringing system shall be tightened to loosen the insulator requiring replacement	(1) The upper and lower working electricians shall cooperate closely and follow the command of the Responsible Person. (2) The binding rope buckles of the upper and lower transfer tools shall be correct and reliable, and the equipotential electrician shall not drop objects from high place. (3) Two hydraulic leading screws shall be stressed evenly. (4) After removing inferior insulators, insulators shorted by human body and tools, the number of good insulators (the structural height is 195 mm) cannot be less than 37	
9	Replacement of insulator	(1) After checking the load-bearing tools which are normally stressed, it is required to report to the Responsible Person. After the approval by the Responsible Person, the intermediate potential electrician shall take out the upper and lower fitting pins of the insulator to be replaced, and the equipotential electrician shall continue to tighten the hydraulic stringing system until taking out the used insulator.	(1) Before replacement of insulator, it is a must to check the stress of such parts as closed clamps and hydraulic leading screws for normality in details, and the insulator can be replaced after being reported to and approved by the Responsible Person and in case of no error in inspection.	

Part II Skill Module Training and Assessment Standards

Table (Cont'd)

S/N	Work Content	Operation Steps and Standards	Safety Measures and Precautions	Responsible Person
9	Replacement of insulator	(2) The intermediate potential electrician shall properly tie the used insulator by insulated transmission rope, and the ground electrician shall tie the insulator by another side of the insulated transmission rope as well. The new insulator shall be transferred to the ground potential electrician by means of the old lower and the new upper. (3) The intermediate potential electrician shall replace the new insulator and reset the upper and lower fitting pins	(2) The new insulator shall be measured by the insulation resistance meter of 5000V and above, and the insulation resistance cannot be less than 500MΩ. Appearance inspection is required, and it cannot be used with rust, damage, cracks, etc. (3) Operation shall be provided with uniform force, as small as possible. (4) Collision is not allowed when transferring the new and used insulators up and down. (5) After the clamps are stressed and the conductor load is completely transferred, the impact is required to judge its reliability.	
10	Remove tools	(1) After checking the reliable connection of the new insulator and being approved by the Responsible Person, the equipotential electrician loosens the hydraulic stringing system. (2) After re-checking the stress of new insulator without error and being approved by the Responsible Person, the intermediate potential electrician shall remove and replace tools and transfer them to the ground	(1) After the new insulator is replaced, it is a must to confirm the reliable installation, correct connection and all fitting pins rest. (2) The tools shall not collide with each other in the process of transfer, and the rope buckle shall be tied correctly and reliably.	

Table (Cont'd)

S/N	Work Content	Operation Steps and Standards	Safety Measures and Precautions	Responsible Person
11	Exiting the potential	(1) The electrician on tower shall fasten the insulated transmission rope after checking that there is no object left in the operation area to make preparation for exiting the potential. (2) The electrician on tower shall exit the insulator string by means of "two-span and three-short-circuit"	(1) The intermediate potential electrician must be approved by the Responsible Person before exiting the insulator string. (2) When the intermediate potential electrician exits the potential, the combined gap cannot be less than 6.9m (medium phase) (6.7 m, side phase).	
12	Return to the ground	After checking that there is no object left on the tower, the equipotential electricians on the tower shall report it to the Responsible Person and then climb down the tower with transmission rope after obtaining the consent of the Responsible Person.	The falling protector shall be used during climbing down. The safety protection is required while moving on the tower	
13	End of the work	(1) The Responsible Person shall organize all working members to put working apparatus and materials in order and put them in a special kit (bag) after cleaning; clean the site to ensure that "the materials are removed and the site is cleaned after construction". (2) In the post-shift meeting, Responsible Person will give work summaries and comments. Comments include the construction quality of this work and the implementation of safety measures from all working personnel. (3) Responsible Person shall report to the on-duty control personnel that the work is over, apply for the restoration of circuit re-closing and terminate the work order.	It is forbidden to restore the line re-closing in the appointed time	

Part II Skill Module Training and Assessment Standards

II. Assessment Standard

Detailed Rules for Assessment and Scoring of Operation and Inspection Skills of UHV AC Transmission Line

Fill-in Column of Examinee	No.:		Unit:		Name:		Position:		Date:		MM/DD/YYYY
Fill-in Column of Assessor	Grade:		Assessor:		Assessment Leader:		Team Starting time:		Closing time:		Operation Duration:
Assessment module	Live replacement of arbitrary section insulator of tensile glass insulator string of 1000kV AC transmission line				Assessee		Maintenance personnel of UHV AC transmission line		Assessment method	Operation	Assessment Time Limit 90min
Job Description	Live replacement of arbitrary section insulator of tensile glass insulator string of 1000kV AC transmission line										
Work Specifications and Requirements	1. Live working shall be carried out in good weather. In case of thunder, rain, snow or fog, no live working shall be carried out. When the wind force is greater than Level 5 and the humidity is greater than 80%, live working should not be carried out. 2. This operation requires 1 Responsible Person, 1 Special Supervisor, 2 ground electricians, 6 ground electricians and 2 equipotential electricians. They shall adopt the basket transfer method to enter the intense electric field for insulator replacement. 3. Responsibilities of Responsible Person: Be responsible for division of operating personnel of the task, work order reading, handling re-closing deactivation of the line and the formalities of work permit, getting work permits, holding pre-shift meeting, dealing with emergency situations in work, quality surveillance, and the summary after work. 4. Special Supervisor: Be responsible for safety control of the work site. 5. Responsibilities of ground electrician: Cooperate with ground electrician in installing wire lifting system (airfoil clamp, planar leading screw, special joint, insulating suspender, hydraulic screw rod), operating hydraulic leading screw to transfer wire load, and disassembling and assembling insulator string. 6. Responsibilities of ground electrician on tower: Be responsible for installation of basket, wire lifting system (plane leading screw, special joint, insulating suspender, hydraulic screw rod), insulating grinding rope and cooperating with equal potential electrician in entering and exiting the potential, disassembly and assembly of composite insulator string, etc. 7. Responsibilities of ground electrician: Be responsible for transferring tools and materials and cooperating with equipotential electrician in entering and exiting the equipotential. 8. During the live working, if thunder, rain, strong wind or any other circumstance threaten the safety of the staff, the Responsible Person or Supervisor may stop working temporarily according to the circumstances.										

· 451 ·

Table (Cont'd)

Work Specifications and Requirements	Given conditions: 1. Training base: insulator string at large size side of phase B of UHV AC 1000kV training line tension tower, insulator model: U550BP/240T. 2. Work orders have been handled, safety measures have been completed (re-closing has been deactivated), and oral application (dispatcher or assessor) shall be made at the beginning and end of the work. 3. The instrument shall be used safely and correctly to test the insulating tool. 4. The operation must be carried out according to the working procedures. The scores of the items to be carried out shall be deducted for the process error. In case of major hidden dangers of personal, equipment and operational safety, the assessor may order the termination of the operation (assessment)
Assessment scenario preparation	1. Line: Phase B of UHV AC 10000kV training line tension tower, work content: live replacement of 1000kV arbitrary section insulator, insulator model: U550BP/240T. 2. Required operation tools: 4 safety belts (including insulated backup protection rope), 4 sets of Type I shielding clothes, 2 sheets of moisture-proof tarpaulin, 1 multimeter, 1 insulation resistance detector, 1 anemometer, 1 2-in-one temperature and humidity detector, 2 hydraulic leading screws, 2 insulating ropes, 1 set of closed clamp, 1 tackle, 2 potential transfer rods, 1 pin puller and 1 set of hand-operated tools; 3. The work site shall be monitored, and the safety measures (fence, etc.) on the work site have been fully implemented; non-operation personnel are prohibited from entering the site, and the staff must wear safety helmets when entering the work site. 4. Examinees shall bring their own work clothes, flame retardant underwear, safety helmets, gloves, safety belts (including insulated back-up protection rope)
Remarks	1. The deduction shall be done until the scores of each item are deducted completely. In case of major hidden dangers of personal, equipment and operational safety, the assessor may order the termination of the operation. 2. When equipment, working environment, safety belt, safety helmet, tool, shielding clothes, etc., do not conform to the operation condition, the assessor may order the termination of the operation

Part II Skill Module Training and Assessment Standards

Table (Cont'd)

S/N	Project name	Quality requirements	Score	Deduction standard	Reasons for deduction	Deduction	Scoring
1	Site re-survey	1) The Responsible Person shall go to the work site to check the line name, iron tower number, on-site working conditions, defective parts and so on. 2) Check that the site meteorological conditions such as wind speed and humidity should meet the operation requirements. 3) Check whether the work order is complete and unmodified, check whether the safety measures listed are consistent with the actual situation on site, and supplement it if necessary	5	1) Deduct 1 point for failure to check the double title. 2) Deduct 1 point for failure to verify on-site working conditions (meteorology), defective parts. 3) Deduct 0.5 points/item for any alteration in the work order filling, and deduct 1 point for incorrect work order number. Deduct 1.5 points for each incomplete work order			
2	Work Permit	1) The Responsible Person is responsible for contacting the on-duty control personnel and applying for stopping the line re-closing as per the contents of the work order. 2) Reporting content is standardized and complete	2	1) Deduct 2 points for failure to contact the dispatching department (referee) for deactivating the re-closing. 2) Deduct 0.5 points for non-standard or incomplete terminology reporting respectively			

Table (Cont'd)

S/N	Project name	Quality requirements	Score	Deduction standard	Reasons for deduction	Deduction	Scoring
3	Site layout	Install the security fence and hang the signboards. 1) The security fence should take full account of falling objects from the heights and the influence on road traffic. 2) The entrance and exit of the security fence shall be set reasonably. 3) Signs such as "Access from Here", "Work Here", "Access from Here" shall be properly and well arranged	3	1) Deduct 0.5 points for failure to arrange the fence at the work site. 2) Deduct 0.5 points for failure to arrange the warning board. 3) Deduct 0.5 points for failure to hang the tower climbing operation sign			
4	Hold a pre-shift meeting	1) All staff and personnel shall wear safety helmets and work clothes correctly. 2) Responsible Person shall wear red vest and read out the work order and be clear with work task and division of personnel; explain safety measures and technical measures in work; check (inquire after) mental state of all working personnel; inform of hazards in work and precontrol measures. 3) All working personnel shall sign on the work order for confirmation.	3	1) Deduct 0.5 points/person for the staff not dressing uniformly. Deduct 0.5 points/person for the staff not dressing uniformly 2) Give no points to this item for no division of labor, and deduct 1 point for unclear division of labor. 3) Deduct 0.5 points for the on-site Responsible Person not wearing a safety monitoring vest. 4) Deduct 1 point for the work shift member failing to sign or signing incompletely on the work order.			

Part II Skill Module Training and Assessment Standards

Table (Cont'd)

S/N	Project name	Quality requirements	Score	Deduction standard	Reasons for deduction	Deduction	Scoring
5	Inspection of tools	1) The staff shall place the tools on the moisture-proof tarpaulin as required; the moisture-proof tarpaulin shall be clean and dry. 2) The tools shall be placed in category according to the requirements of the fixed management; the insulated tools shall not be mixed with metal tools and materials; and the appearance inspection shall be done on the tools. 3) The surface of insulated tools shall not be worn, deformed and damaged, and the operation shall be flexible. Carry out segment insulation detection with such insulated tools with insulation resistance meter of 2,500V or above and with the resistance no less than 700MΩ, and wipe it off with a clean dry towel. 4) The electrician on tower shall correctly wear a whole suit of qualified shield clothes and conductive shoes as required, with each part connected well, shall not wear chemical fiber clothes next to the skin in the shielding clothes and shall fasten safety belts; the Responsible Person shall carefully check whether they wears it correctly. 5) Tower climbing personnel shall check the double name, tower number and phase again, and report them.	7	1) Deduct 1 point for failure to use moisture-proof cloth and place tools to designed positions. 2) Deduct 0.5 points/item for failure to check qualified label of tool test and appearance inspection. 3) Deduct 1 point/item for failure to use testing instrument for testing the tools. 4) Deduct 2 points/person time for the operator failing to wear the shielding clothes correctly and each part connected well. 5) Deduct 1 point for the on-site Responsible Person failing to check the safety protective equipment of the tower climbing operators. 6) Deduct 2 points/person for the tower climbing personnel failing to check the double name of the line, pole number and phase. 7) Deduct 2 points/person for the tower climbing personnel failing to report the check results			

Table (Cont'd)

S/N	Project name	Quality requirements	Score	Deduction standard	Reasons for deduction	Deduction	Scoring
6	Climbing the tower	1) The ground electrician and the equipotential electrician on tower shall wear a whole suit of qualified shielding clothes, fasten the safety belt after performing the impulse test on the safety belt, and carry the insulated transmission rope to climb the tower one after another. 2) During the tower climbing process, they shall fasten the anti-fall protection device, climb the tower to an appropriate position, fasten the safety belt, arrange the insulated transmission rope, and then cooperate with the ground electrician to make lifting preparation of the insulated transmission rope separately. 3) During tower climbing, the electrician shall fasten the anti-fall protection device, climb the tower at a uniform speed, grasp the main material by hand, hang the safety belt on the shoulder and keep the safety distance of more than 6.8m away from the electrified body, and the Responsible Person shall strengthen the operation monitoring	5	1) Deduct 2 points for failure to fasten the safety belt or for failure to perform the impulse test on the safety belt and the backup protection rope. 2) Deduct 2 points for grasping the shackles with hands. 3) Deduct 1 point for inconvenient suspension position of tackle transmission rope for taking tools. 4) Deduct 2 points for metal tools that are difficult to ensure safe distance during transfer; deduct 2 points for tools that are not bound securely. 5) Deduct 2 points for falling object at high place. 6) Deduct 2 points for tools colliding with the tower body during the transfer process. 7) Deduct 1 point for knotting and twining of rope in tool transfer. 8) Deduct 2 points for the Responsible Person failing to monitor the operation in place. 9) Deduct 2 points for incorrect operation of electrician on tower			

Part II Skill Module Training and Assessment Standards

Table (Cont'd)

S/N	Project name	Quality requirements	Score	Deduction standard	Reasons for deduction	Deduction	Scoring
7	Detection of insulator	1) The ground electrician shall transfer the insulating bar and insulator detector to the tower. The equipotential electrician shall assist the ground potential electrician on tower in detecting the insulator. 2) The insulators must be detected string by string and piece by piece, and the contact must be reliable from the conductor side to the cross arm side successively. 3) When the number of any string of good insulators is less than 37, the testing shall be stopped immediately, and the live working shall be stopped.	2	1) Deduct 2 points if not testing the null value of insulator. 2) Deduct 2 points if the null value measurement method is wrong. 3) Deduct 2 points if not reporting the null value measurement result.			
8	Entering the intense electric field	1) Before entering the insulator string, the equipotential electrician shall fasten the protection rope and adjust the insulated transmission rope and potential transfer rod. 2) Adopt the "two-span and three-short-circuit" operation mode to enter the intense electric field along the insulator string. 3) When the equipotential electrician enters the intense electric field along the insulator string, the combined gap cannot be less than 6.9m (medium phase), and the potential must be transferred by potential transfer rod when entering the intense electric field	8	1) Deduct 2 points for insufficient distance of the exposed parts during potential transfer of equipotential electrician. 2) Deduct 2 points each time for the incorrect action of equipotential electrician entering the intense electric field and repeat discharging. 3) Deduct 1 point for unskillful potential transfer and 3 points for potential transfer without using the potential transfer rod. 4) Deduct 5 points for carrying out potential transfer without permission of the Responsible Person			

Table (Cont'd)

S/N	Project name	Quality requirements	Score	Deduction standard	Reasons for deduction	Deduction	Scoring
9	Installation tool	1) The electrician on tower shall arrange the transmission rope after arriving at the operation point. The ground electrician shall respectively transfer the tackle block, closed clamps, etc. to the intermediate potential electrician, and transfer the hydraulic leading screw and end clamps to the equipotential electrician. The intermediate potential electrician and equipotential electrician shall respectively install the tools, and then the ground electrician shall transfer the insulated pull-rods to the intermediate potential electrician and equipotential electrician for proper installation through transmission rope. 2) The intermediate potential electrician shall properly install the closed clamps in correct positions before and after replacing the insulator string. After checking the installation of all parts of the load-bearing tools correctly, the electrician on tower shall respectively report to the Responsible Person. After approval by the Responsible Person, the equipotential electrician shall pre-tighten the leading screw. After the leading screw is suitably stressed, the impact inspection is required for the load-bearing tools. After the approval by the Responsible Person, the hydraulic stringing system shall be tightened to loosen the insulator requiring replacement	17	1) Deduct 2 points for improper fixation of clamps; 2) Deduct point(s) for loosening the leading screw if not checking the installation of insulator string; deduct 1 point in case of no reporting after check; and deduct 1 point if loosening the leading screw after reporting without the approval by the Responsible Person; 3) Deduct 3 points/time for the number of shorted insulators exceeding 4 during the operation; 4) Deduct 2 points for collision and breakage of insulators in the clamp installation; 5) No point if the clamps are installation in other positions, and replacing wrong insulator			

· 458 ·

Part II Skill Module Training and Assessment Standards

Table (Cont'd)

S/N	Project name	Quality requirements	Score	Deduction standard	Reasons for deduction	Deduction	Scoring
10	Replacement of insulator	1) After checking the load-bearing tools which are normally stressed, the electrician on tower shall report to the Responsible Person. After the approval by the Responsible Person, the intermediate potential electrician shall take out the upper and lower fitting pins of the insulator to be replaced, and the equipotential electrician shall continue to tighten the hydraulic stringing system until taking out the used insulator. 2) The intermediate potential electrician shall properly tie the used insulator by insulated transmission rope, and the ground electrician shall tie the insulator by another side of the insulated transmission rope as well. The new insulator shall be transferred to the ground potential electrician by means of the old lower and the new upper. The intermediate potential electrician shall replace the new insulator and reset the upper and lower fitting pins	17	1) Deduct 3 points for failure to check the installation of load-bearing tools but starting to tighten the leading screw; deduct 1 point for failure to report after checking; deduct 2 points for starting tightening the leading screw without the consent of Responsible Person after reporting. 2) Deduct 3 points for failure to check the stress of load-bearing tools after pre-tightening the leading screw; deduct 1 point for failure to report after checking; deduct 2 points for continuously tightening the leading screw without the consent of Responsible Person after reporting. 3) Deduct 2 points for failure to tighten leading screw in a balanced way. 4) Deduct 1 point for wrong knot. 5) Deduct 1 point/times for falling object at high place. 6) Deduct 1 point for new and old insulators colliding with each other. 7) Deduct 1 point for the transmission insulator and tower body colliding with each other. 8) Deduct 0.5 points for winding insulating ropes			

Table (Cont'd)

S/N	Project name	Quality requirements	Score	Deduction standard	Reasons for deduction	Deduction	Scoring
11	Remove tools	1) After the intermediate potential electrician checks the reliable connection of the new insulator and is approved by the Responsible Person, the equipotential electrician loosens the hydraulic stringing system. 2) After re-checking the stress of new insulator without error and being approved by the Responsible Person, the intermediate potential electrician shall remove and replace tools and transfer them to the ground	6	1) Deduct 1 point for incorrect use of knot when binding the tools. 2) Deduct 1 point/times · pcs for falling object at high place. 3) Deduct 1 point/time for the tools colliding with each other; 4) deduct 1 point/time for the tool colliding with the electrified body or tower body; and deduct 0.5 points for winding insulating ropes. 5) Deduct 1 point/times for the ground personnel stepping on the moisture-proof tarpaulin; and deduct 1 point for the ground personnel right under the operation point			
12	Exit the intense electric field	1) After inspection, there is no residue at the operation point, the Responsible Person shall authorize the equipotential electrician to carry the insulated transmission rope and prepare for leaving potential.	10	1) Deduct 2 points for the potential transfer without applying to the Responsible Person;deduct 1 point for starting the work without the consent after applying. 2) Deduct 2 points for insufficient distance of the exposed parts during potential transfer of equipotential electrician.			

Part II Skill Module Training and Assessment Standards

Table (Cont'd)

S/N	Project name	Quality requirements	Score	Deduction standard	Reasons for deduction	Deduction	Scoring
12	Exit the intense electric field	2) The equipotential electrician shall hook the grading ring with the potential transfer rod and enter the third insulator of the grading ring. With one hand grasping the insulator tightly and the other holding the potential transfer rod, the equipotential electrician shall break himself away from the potential with the help of potential transfer rod. 3) The equipotential electrician exits the intense electric field by means of "two-span and three-short-circuit" operation mode	10	3) Deduct 2 points for the incorrect action of equipotential electrician exiting the intense electric field and repeat discharging. 4) Deduct 2 points for unskillful potential transfer and 3 points for potential transfer without using the potential transfer pole			
13	Return to the ground	After checking that there is no object left on the tower, the equipotential electrician on the tower shall report it to the Responsible Person and then climb down the tower with insulated transmission rope after obtaining the consent of the Responsible Person	5	1) Deduct 2 points for failure to use the falling protector when climbing down the tower. 2) Deduct 2 points for loosing the protection of the safety belt when moving on the tower. 3) Deduct 1 point for grasping the tower nail when climbing down the tower. 4) Deduct 2 points for any objects left on tower.			

Table (Cont'd)

S/N	Project name	Quality requirements	Score	Deduction standard	Reasons for deduction	Deduction	Scoring
14	End of the work	1) The Responsible Person shall organize all working members to put working apparatus and materials in order and put them in a special kit (bag) after cleaning; clean the site to ensure that "the materials are removed and the site is cleaned after construction". 2) In the post-shift meeting, Responsible Person will give work summaries and comments. Comments include the construction quality of this work and the implementation of safety measures from all working personnel. 3) Responsible Person shall report to the on-duty control personnel that the work is over, apply for the restoration of circuit re-closing and terminate the work order.	10	1) Deduct 2 points for failure to clean the tools. 2) Deduct 2 points for missing tools. 3) Deduct 2 points for failure to hold the post-shift meeting. 4) Deduct 2 points for failure to remove the fence. 5) Deduct 2 points for failure to report to dispatcher			
	Total		100				

Part II Skill Module Training and Assessment Standards

Module 7 Standards for Training and Assessment on Live Replacement of 1000kV AC Transmission Line Conductor Spacer

Ⅰ. Training Standard

(I) Training Requirements

Designation of module	Live replacement of 1000kV AC transmission line conductor spacer	Type of training	Operation
Training method	Practical operation training	Hours of training	21 training hours
Training objectives	1. Master the electrical significance of "two-span and three-short-circuit" operation mode in entering and exiting 1000kV intense electric field along strain insulator string. 2. Can complete the entry of 1000kV equipotential operation point along the strain insulator string. 3. Be able to independently complete the replacement of conductor spacer (equipotential operation method)		
Training venue	UHV AC training line		
Training content	The "two-span and three-short-circuit" operation mode is adopted to enter the intense electric field along the strain insulator string, and the equipotential operation method is adopted for live replacement the eight-bundle conductor spacer of 1000kV AC transmission line.		
Scope of application	Maintenance personnel of UHV AC transmission line		

(Ⅱ) Referenced Rules and Specifications

(1) Code for Design of 1000kV Overhead Transmission Line (GB50665-2011).

(2) Maintenance Code for 1000kV AC Transmission Line (DL/T209-2008).

(3) Operation Code for 1000kV AC Transmission Line (DL/T307-2010).

(4) Technical Guide for Live Working on 1000kV AC Transmission Line (DL/T392-2015).

(5) Calculation Method of Live Working Minimum Approach Distance on AC Transmission Line (GB/T 19185-2008).

(6)State Grid Corporation of China on the Management Regulations of Live Working (Trial Implementation) (SGCC [2007] No.751).

(7)State Grid Corporation of China Working Regulations of Power Safety (Transmission Line Section) (Q/GDW1799.2-2013).

(8)Electrotechnical Terminology - Overhead Line (GB/T 2900.51-1998).

(9)Electrotechnical Terminology- Live Working (GB/T2900.55-2002).

(10)Live Working - Terminology for Tools, Equipment and Devices (GB/T 14286-2002).

(11)Insulated Tackles for Live Working (GB/T13034-2008).

(12)Live Working-Insulating Ropes (GB 13035-2008).

(13)Minimum Requirements for Utilization of Tools, Devices and Equipment for Live Working (DL/T 877-2004).

(14)Preventive Test Code of Tools, Devices and Equipment for Live Working (DL/T 976-2005).

(15)Screen Clothes for Live Working on 1000kV AC (GB/T25726-2010).

(16)Technical Guide for Live Working on 1000kV AC Transmission Line (DL/T392-2010).

(Ⅲ) Teaching Design for Training

To complete the live replacement of 1000kV AC transmission line conductor spacer is the Work task during the design. Each training stage shall be designed according to the standard operation procedure for work task completion. Each stage includes specific training objectives, training content, hours of training, training methods (training resources), training environment, assessment and evaluation, etc, as shown in Table 2-7-1.

(Ⅳ) Operation Flow

1. Work Task

The "two-span and three-short-circuit" operation mode is adopted to enter the intense electric field and reach the operation point along the strain insulator string, and the equipotential operation method is adopted for live replacement of the eight-bundle conductor spacer of 1000kV AC transmission line.

(This operation task is applicable to the operation point of the first spacer for side phase conductor of resisting-tensile tower of 1000kV AC single circuit transmission line in 1,000m

Part II Skill Module Training and Assessment Standards

Table 2-7-1 Training Content Design for Live Replacement of Conductor Spacer of 1000kV AC Transmission Line

Training schedule	Training objectives	Training content	Hours of training	Training methods and resources	Preparation of training conditions	Assessment and evaluation
1. Theoretical teaching	1. Preliminarily master the basic method for entering and exiting 1000kV intense electric field along the insulator string. 2. Be familiar with the methods for potential transfer. 3. Be familiar with the method for replacing the damaged conductor spacer of power transmission line. 4. Be familiar with the safe distance, hazard identification and precontrol measures for live working on UHV AC line	1. The electrical significance of the "two-span and three-short-circuit" operation mode when entering and exiting the intense electric field along the insulator. 2. The use method for the potential transfer rod during entering and exiting the UHV intense electric field. 3. Method and quality standard for replacement of power transmission line conductor spacer. 4. Safe distance, hazard analysis and precontrol measures for live working on UHV AC line	2	Training methods: Lecture. Training resources: PPT, relevant regulations, specifications and technical guidelines	Multimedia classroom	Attendance, classroom questions and assignments
2. Preparations	Be able to complete the preparation before operation	1. Work site survey. 2. Preparation of the standardized operation card. 3. Filling of the work order. 4. Preparation of tools and materials for this operation	1	Training methods: 1. Site survey and preparation of tools, instruments and materials shall be practiced at site. 2. Preparation of operation card and the filling of work order shall adopt lecture method. Training resources: 1. 1000kV practical training line. 2. UHV tools warehouse. 3. Blank work order	1. 1000kV practical training line 2. Multi-Media classroom	

Table (Cont'd)

Training schedule	Training objectives	Training content	Hours of training	Training methods and resources	Preparation of training conditions	Assessment and evaluation
3. Work site preparation	Be able to complete the preparations of work site	1. Work site re-survey. 2. Job application. 3. Work site layout. 4. Pre-shift meeting. 5. Inspection of tools and materials. 6. The use method of special spanner for spacer	1	Training methods: demonstration and role play. Resources: 1000kV training line	1000kV training line	
4. Trainer's demonstration	The trainees can preliminarily understand the operation process of the task through inspecting and learning from each other's work	1. Detection of null-value insulator. 2. The equipotential electrician enters and exits the intense electric field and potential transfer along the strain insulator string. 3. The equipotential electrician reaches the replacement position of the spacer by means of walking along the line. 4. Equipotential electrician uses the special tools to complete the replacement of conductor spacer	2	Training methods: Demonstration. Resources: 1000kV training line	1000kV training line	

Part II Skill Module Training and Assessment Standards

Table (Cont'd)

Training schedule	Training objectives	Training content	Hours of training	Training methods and resources	Preparation of training conditions	Assessment and evaluation
5. Group training of trainees	1. Be able to complete the operations of entrance in and exit from 1000kV intense electric field along the insulator string and potential transfer. 2. Be able to complete the 1000kV transmission line conductor spacer replacement	1. The trainees are grouped (6 persons per group) to train the skill operation of entering and exiting 1000kV intense electric field, potential transfer and conductor spacer replacement. 2. Trainers guide the operation of trainees and conduct safety supervision	14	Training methods: Role play. Resources: 1000kV training line	1000kV training line	Score the operation of trainees according to the detailed rules for skill assessment and scoring
6. End of the work	1. Enable the trainees to further distinguish the shortcomings of the operation process and facilitate the promotion in the later stage. 2. Train the trainees in the working style of safe and civilized production	1. Cleaning up the work site. 2. Report to dispatcher. 3. Comment and summarize the work task this time at post-shift meeting	1	Training method: Lecture and inductive method	1000kV training line	

· 467 ·

and below altitude)

2. Requirements for Weather and Work Site

(1) The live replacement for eight-bundle conductor spacer of 1000kV AV transmission line shall be carried out in good weather.

In case of lightning (hearing thunder or seeing lightning), snow, hail, rain, fog and so on, live working is prohibited. When the wind force is greater than level 5, or the relative humidity of the air is greater than 80%, it is unsuitable for live working; when emergency live repair is required in bad weather, relevant personnel shall be organized to fully discuss and prepare necessary safety measures, which can be implemented after being approved by the unit.

(2) The operating personnel shall be in good spirits, and be familiar with the organizational and technical measures to ensure safety in work; they shall hold the live working qualification within the validity period.

(3) The Responsible Person shall organize relevant personnel to complete field investigation in advance, determine the operating methods, required working apparatus and necessary measures according to the results, and handle the work orders for live working.

(4) The work site shall be reasonably set up with fence and warning signs. Non-operating personnel is forbidden to enter.

(5) The line re-closing device shall be deactivated in the Project.

(6) Safety working distance and effective insulation length are shown in Table 2-7-2.

Table 2-7-2　Safe Distance for Live Replacement of Conductor Spacer of 1000kV AC Transmission Line

Voltage class	Safe distance between human body and electrified body /m	Minimum distance to adjacent phase conductor /m	Minimum effective insulation length/m		Minimum combination gap /m	Minimum distance between exposed part of human body and electrified body when potential is transferred /m
			Insulating bar	Insulated load-bearing tools and insulating ropes		
1000kV	6.0 (6.8)	6.9 (7.2)	6.8	6.8	6.7 (6.9)	0.5

Notes: ① In case of the altitude at 1000m and below, the data of 6.8 m in brackets, the data of 7.2 m in brackets and the data of 6.9 m in brackets shall be respectively adopted for the phase safe distance, the minimum distance and the combined gap during the live working of 1000kV AC single-loop transmission line.

② As the phase-to-phase spacing of 1000kV UHV line is large enough (generally speaking, the distance between the phase and the ground shall be controlled) and no important safety factors are taken into account, the "minimum distance to the adjacent phase conductor" is not given in the "Safety Regulation", and the actual work can refer to the data of 750kV.

(7) For operation on 1000kV transmission line, the number of insulators in good operation phase shall be no less than 37 pieces.

3. Preparations

3.1 Hazards and precontrol measures

(1) Hazard - precontrol measures for electric shock injury:

① Before work, Responsible Person should contact the control personnel on duty, deactivate the line re-closing, and perform the licensing procedures.

② Before climbing the tower, the equipotential operators on tower must carefully check the name of the line, the number of the pole and tower, and phase, and then the tower can be climbed after all have been confirmed correct.

③ If lines lose power suddenly during work, operators shall still deem it to be charged. The Responsible Person shall contact the control personnel as soon as possible, and no forced energization is allowed before the on-duty control personnel getting in touch with the Responsible Person.

④ The ground electrician shall wear clean and dry gloves when operating the insulating tools. Insulating tools and insulating ropes shall not be damaged, damped, deformed, and malfunctioned. It is not allowed to use non-insulating ropes (such as cotton rope, manila rope, and steel wire rope). The tools and instruments for live working at the work site shall be placed on the moisture-proof tarpaulin to prevent them from dirt and moisture in use.

⑤ The equipotential operator shall wear flame-retardant underwear and full set of shielding clothes for 1000kV live working shall be worn outside the clothes (including coverall, helmet, protective mask, gloves, conductive socks and shoes). All parts shall be in excellent connection conditions. The resistance between the farthest points of the whole shielding clothes shall not be greater than 20Ω.

⑥ Before potential transfer, equipotential operator shall obtain the approval of Responsible Person, and the minimum distance between exposed part of human body and electrified body shall not be less than 0.5m. During potential transfer, potential transfer rod shall be adopted. The operation shall be fast, and end shall not be used for power charging and discharging; During delivering tools and materials to ground potential operators, the effective length of insulating tools or insulating ropes shall not be lower than the requirements as specified in Table 2-7-2.

⑦ When transmitting large metal objects by using insulating ropes, ground potential operators shall not touch them before ground connection.

⑧ The special Supervisor shall continuously monitor the operators and correct their nonstandard operation or actions in violation at any time. Special attention shall be paid to operators work at heights to ensure that there is enough safety distance (meeting the requirements in Table 2-7-2). It is forbidden to contact two non-connected electrified bodies or make contact with electrified body and grounding body at the same time.

(2) Hazard - falling accident

Precontrol measures:

① Before climbing, operators work at heights must satisfy the requirements of this operation, such as physical condition, mental state and skill and quality.

② Before climbing the tower, operators working at heights shall conduct visual inspection and impulse test on safety belts and falling protector to ensure that their mechanical strength meets the requirements.

③ Operators who climb the tower shall first check whether the shackles is firm, shoe soles is clean, or the falling device is firm or is provided with the falling protector; grasp the main materials, tread on shackles and climb up (down) the tower at a uniform pace.

④ Supervisor shall correct irregularities and violations at any time. Special attention shall be paid to operators working at heights to prevent them from losing the protection of the safety belt or insulated backup protection rope during transposition. The safety belt shall be attached to a secure part and shall not be fastened lower than operating personnel.

⑤ When the equipotential operators move on the insulator string in parallel, they usually grasp one string with both hands and step on the other string with both feet to enter the intense electric field at a constant speed. During the movement, the backup protection rope holds two strings of insulators, and excessive hand waving and big step are not allowed.

⑥ Equipotential operators must fasten the safety belt and make the backup protection rope cover all the sub-conductors when they are walking along the conductor; they shall control the center of gravity during walking along the line to prevent the conductor from turning over.

Part II Skill Module Training and Assessment Standards

(3) Hazard - injury caused by objects falling from heights.

Precontrol measures:

① Operator working at heights should put personal tools and fragmentary materials into the tools bag. It is strictly forbidden to hang objects in heights or keep in the mouth.

② Ground operator should correctly wear a safety helmet and use the knots to deliver the tools, instruments and materials. The vertical distance from the operation point should not be less than the falling radius.

③ The work site shall be set up with fence and warning signs. The Supervisor shall maintain vigilance at all times and prohibit non-staff members and vehicles from entering the operation area.

3.2 Selection of tools and instruments and materials

Tools and materials required for live replacement of eight-bundle conductor spacer for 1000kV AC transmission line can be seen in Table 2-7-3. Before delivering tools and instruments out of warehouse, application voltage class and test period shall be carefully checked and they shall be inspected to ensure that appearance is intact, connection is firm, rotation is flexible and meet the working task requirements. After delivering tools and instruments out of warehouse, they shall be stored in tools bag or tool kit for transportation to avoid contamination and damp. Metal tools and insulated tools shall be separately loaded and transported to avoid deformation, damage or other defects caused by mixed loading and transportation.

Table 2-7-3 Tools, Instruments and Materials Required by Live Replacement of Conductor Spacer of 1000kV AC Transmission Line

S/N	Name	Specification	Unit	Qty.	Remarks
1	Insulated transmission rope	TJS-12, with the length matching the lifting height	Nos.	2	Insulating tool
2	Insulated backup protection rope	TJS-16, with buffer provided	Nos.	2	Insulating tool
3	Insulated tackle	JH10-0.5	Nos.	2	Insulating tool
4	Insulated noose	TJS-14	Nos.	2	Insulating tool
5	Potential transfer rod	0.4m	Nos.	1	Insulating tool
6	Insulation jacks		Nos.	4	Insulating tool
7	I-type shielding clothes (coverall, helmet, protective mask, gloves and conductive socks)	Shielding efficiency >= 60dB (shield efficiency of shielding mask >=20 dB)	Set	2	PPE

Table (Cont'd)

S/N	Name	Specification	Unit	Qty.	Remarks
8	Conductive shoes	The size depends on the wearer	Nos.	2	PPE
9	Flame retardant underwear	Pure mulberry silk	Set	2	PPE
10	Safety belt of double insurance	Full-body harness	Nos.	2	PPE
11	Falling protector	Corresponding to the type of pole and tower falling protector	Nos.	2	PPE
12	Safety helmet		Nos.	6	PPE
13	Special spanner for spacer	Used for eight-bundle spacer	Nos.	1	Special tools
14	Insulator detector		Set	1	Other tools
15	Insulation resistance tester	5,000V, with electrode width 2cm and interelectrode width 2cm	Set	1	Other tools
16	Wind speed, temperature and humidity tester	HT-8321	Set	1	Other tools
17	Multimeter		Nos.		Other tools
18	Interphone	As required by the work	Set	2	Other tools
19	Moisture-proof tarpaulin	2m×4m	Piece	2	Other tools
20	Security fence		Set	Several	Other tools
21	Warning sign	"Work Here", "Access from Here" "Access from Here"	Set	1	Other tools
22	Red waistcoat	"Responsible Person"	Piece	1	Other tools
23	Clean towel	Cotton	Nos.	1	Other tools
24	Shoe covers		Nos.	Several	Other tools
25	Work gloves		Nos.	Several	Other tools
26	Personal tools	Tool bag, flat tong, marker pen	Set	2	Other tools
27	Eight-bundle spacer	The same model with replaced spacer	Nos.	1	Material

Note: The electrical strength of the insulating tool shall meet the requirements of the "State Grid Corporation of China Working Regulations of Power Safety (Transmission Line Section)", the test is required to be qualified and within the validity period.

Part II Skill Module Training and Assessment Standards

3.3 Division of labor for operators

Division of labor for operators of the task is shown in Table 2-7-4.

Table 2-7-4 Division of Operating Personnel Involved in Live Replacement of 1000kV AC Transmission Line Conductor Spacer

S/N	Post	Qty. (person)	Responsibilities
1	Responsible Person	1	Be responsible for division of operating personnel of the task, work order reading, handling re-closing deactivation of the line and the formalities of work permit, getting work permits, holding pre-shift meeting, dealing with emergency situations in work, quality surveillance, and the summary after work
2	Special Supervisor	1	Be responsible for safety supervision and control during operation
3	Equipotential electrician	1	Be responsible for the replacement of eight-bundle conductor spacer after entering the equipotential
4	Ground potential electrician on tower	1	Be responsible for testing the null-value or low-value insulator and assisting equipotential in entering and exiting the intense electric field.
5	Ground electrician	2	Be responsible for taking site safety measures, arranging operation site, inspecting tools and instruments, delivering tools and materials, and cooperating with the equipotential electrician to enter and leave equipotential

4. Work Procedure

The workflow of this task is shown in Table 2-7-5.

Table 2-7-5 Workflow Chart for Live Replacement of 1000kV AC Transmission Line Conductor Spacer

S/N	Work Content	Operation Steps and Standards	Safety Measures and Precautions	Responsible Person
1	Site re-survey	The Responsible Person shall complete the following work: (1) Check the line title, the number of the pole and tower and ensure the phases are correct; guarantee that the foundation and the pole and tower are intact and in normal condition; ensure that the cross and span distance meets the safety requirements; confirm the defect conditions and the specifications and models of earth wires.	(1) Correctly wear helmet, working clothes, work shoes and protective gloves. (2) Operation under meteorological conditions that may endanger the safety of operators is forbidden. (3) Non-operation personnel and vehicles are strictly prohibited from entering the working site.	

Table (Cont'd)

S/N	Work Content	Operation Steps and Standards	Safety Measures and Precautions	Responsible Person
1	Site re-survey	(2) Check that the site meteorological conditions such as wind speed and humidity should meet the operation requirements. (3) Check that the terrain and environment shall meet the operation requirements. (4) Check that the safety measures listed in the work order are in line with the actual situations on site, and the measures will be supplemented if necessary		
2	Work Permit	(1) The Responsible Person is responsible for contacting the on-duty control personnel and applying for stopping the line re-closing as per the contents of the work order. (2) Live working could be started only after being approved by the on-duty control personnel.	Live working shall not be started without the permission of the on-duty control personnel	
3	Site layout	Install the security fence and hang the signboards correctly: (1) The security fence should take full account of falling objects from the heights and the influence on road traffic. (2) The entrance and exit of the security fence shall be set reasonably. (3) Complete signs such as "Access from Here", "Work Here", "Access from Here" shall be properly and well arranged	When the influence on road traffic safety is uncontrollable, the traffic management department should be contacted in time to strengthen the on-site control of traffic safety.	

Part II Skill Module Training and Assessment Standards

Table (Cont'd)

S/N	Work Content	Operation Steps and Standards	Safety Measures and Precautions	Responsible Person
4	Hold a pre-shift meeting	(1) All working personnel shall line up. (2) The Responsible Person shall wear red waistcoat and read out the work order and be clear with work task and division of personnel; explain safety measures and technical measures in work; check (inquire after) mental state of all working personnel; inform of hazards in work and precontrol measures. (3) All working personnel shall sign on the work order for confirmation.	(1) The work order shall be filled in, issued and approved in a standardized manner, and the signature shall be complete. (2) All working personnel shall be in good mental states. (3) All working personnel shall be clear with task division of works, safety measures and technical measures	
5	Inspect tools	(1) The ground electrician and equipotential electrician on tower shall wear the shielding clothes in a right way and pass the inspection, which shall be supervised and inspected by the Responsible Person. (2) Wear personal safety equipment correctly (proper size and easy lock), and the Responsible Person shall supervise and inspect it. (3) Measure the wind speed, wind direction and humidity, check the insulation performance of insulating tools, and make records	(1) Check carefully for damage, deformation and failure before using metal and insulating tools. Carry out segment insulation detection with such insulating tools as insulation resistance tester of 2500V or above and with the resistance no less than $700M\Omega$, and wipe them clean with the clean dry towel. (2) Use a multimeter to measure the resistance between the farthest ends of the shielding clothes and trousers, which shall not be greater than 20Ω. The Responsible Person shall check the connection of the electrician's shielding clothing. (3) Check the tool assembly and make sure the connection is reliable. (4) The live working tools used at site shall be placed on the moisture-proof tarpaulin	

Table (Cont'd)

S/N	Work Content	Operation Steps and Standards	Safety Measures and Precautions	Responsible Person
6	Climbing the tower	(1) After checking the line name and pole and tower number without error, the ground electrician and equipotential electrician on tower shall check the stress of the safety belt and the falling protector in the shock. (2) The ground electrician on tower carries the insulated transmission rope to climb the tower, and the equipotential electrician then climbs the tower. When they reach the operation point of the cross arm, they choose the appropriate position to fasten the safety belt, and the ground electrician on tower installs the insulated tackle and the insulated transmission rope in the appropriate position of the cross arm. Then he shall cooperate with the ground electrician to separate the insulated transmission rope for lifting preparation	(1) After checking the correct line name and pole and tower number, the tower can be climbed for operation. (2) The anti-falling device installed on tower shall be used in the process of climbing the tower; when moving and transposition on the pole and tower, the safety protection shall not be lost, and the operators must climb and grasp the components securely. (3) The working electrician must wear a whole suit of qualified shielding clothes which must be connected reliably. Before entering the equipotential at cross arm, the equipotential electrician shall re-check and confirm that each part of the shielding clothes are connected reliably before the next operation	
7	Detection of insulator	(1) The ground electrician shall transfer the insulating bar and insulator detector to the tower. The equipotential electrician shall assist the ground potential electrician on tower in detecting the insulator. (2) Testing order shall be from the conductor side to the cross arm side, and records shall be made properly. (3) After testing, the ground electrician shall cooperate with the electrician on tower to transfer the insulating bar to the ground, and transfer the potential transfer rod to the tower	(1) No testing is required if the glass insulator is confirmed in good status; and the following items (2) and (3) shall be conducted for porcelain insulator. (2) The insulators must be detected string by string and piece by piece, and the contact must be reliable. (3) When the number of any string of good insulators is less than 37, the testing shall be stopped immediately, and the live working shall be stopped. (4) During delivering the tools, the binding rope fastener shall be correct and reliable so as to prevent objects falling from heights.	

Part II Skill Module Training and Assessment Standards

Table (Cont'd)

S/N	Work Content	Operation Steps and Standards	Safety Measures and Precautions	Responsible Person
8	Entering the intense electric field	(1) The equipotential electrician shall transfer the safety belt to the insulator set fitting, and carry the potential transfer rod, insulated tackle and insulated transmission rope. (2) After the equipotential electrician confirms that all parts of the shielding clothes are well connected, it shall be reported to the Responsible Person for approval. The electrician shall grasp one string with both hands and step on the other string with both feet, and enter the intense electric field along the insulator string in the operation mode of "two-span and three-short-circuit". (3) When reaching 3 insulators outside the grading ring on the conductor side, the operator shall stop, and carry out the potential transfer with the potential transfer rod	(1) The equipotential electrician must obtain the permission of the Responsible Person before entering the potential. (2) When the equipotential electrician enters the insulator string, he shall use the safety belt and the backup protection rope alternately (holding two insulator strings with the backup protection rope, grasping one string with hands and stepping on the other string) without losing the protection of the safety belt, and adjust the insulated transmission rope and the potential transfer rod. (3) The equipotential electrician shall coordinate his hands and feet during entering the potential with uniform speed so as to avoid excessive hand waving, big step and other actions; The side phase combined gap composed of the grounding body and the electrified body shall be greater than 6.7 m (larger than 6.9 m for medium phase). (4) The minimum distance to adjacent conductors shall be greater than 6.9 m (7.2m for medium phase). (5) Before the potential transfer, the equipotential electrician shall check whether the electrical connection between the potential transfer rod and the shielding clothes is reliable and that the minimum distance between the exposed part of the human body and the electrified body is greater than 0.5m, and shall get the permission of the Responsible Person; The protection of safety belt shall not be lost during potential transfer, and the action shall be accurate, stable and rapid when entering the intense electric field	

Table (Cont'd)

S/N	Work Content	Operation Steps and Standards	Safety Measures and Precautions	Responsible Person
9	Replace the eight-bundle spacer	(1) After entering the equipotential, the equipotential electrician shall tie the safety belt to the upper sub-conductor, and install the insulated protection rope (all sub-conductors shall be covered). (2) The equipotential electrician shall carry the insulated transmission rope and walk to the operation point along the conductor for replacing the spacer, install the insulating noose at the appropriate place of the sub-conductor, and then connect the insulated tackle and insulated transmission rope and hook the insulated tackle into the insulating noose. (3) Equipotential electrician shall mark the installation points of the old spacer on conductor at 4 points symmetrically with the marker pen. (4) The equipotential electrician shall install 4 insulation jacks in the appropriate position next to the old spacer in the way of "two-two corresponding" to fix the sub-conductor reliably. (5) The equipotential electrician shall use the insulated transmission rope to fasten the old spacer by the way of slip knot, then use the special spanner for spacer to remove the old spacer, and put it on the ground with the insulated transmission rope in cooperation with the ground electrician. (6) The ground electrician shall lift the new spacer to the position of equipotential electrician. The equipotential electrician shall mark and install the new spacer correctly. After installation, the plane of the spacer shall be vertical to the sub-conductor. (7) The equipotential electrician shall remove the insulated jack, insulated tackle, insulated transmission rope and noose successively. (8) No tools, instruments or materials shall drop during the equipotential operation	(1) The equipotential electrician shall not lose the protection of safety belt. (2) Equipotential electrician shall cooperate closely with the ground electrician and follow the command of the Responsible Person. (3) The minimum distance to adjacent conductors shall be greater than 6.9 m (7.2m for medium phase). (4) In the process of delivering, the conductor spacer shall not be bumped, and the insulated transmission ropes shall not be intertwined. (5) During delivering the tools, the binding rope shall be correct and reliable so as to prevent objects falling from heights. (6) During delivering the tools, instruments and materials, the ground electrician is not allowed to stand directly below the operation point of equipotential electrician	

Part II Skill Module Training and Assessment Standards

Table (Cont'd)

S/N	Work Content	Operation Steps and Standards	Safety Measures and Precautions	Responsible Person
10	Exit the intense electric field	(1) After inspecting that the spacer is installed firmly and there is no object left at the operation point, the equipotential electrician, with the permission of the Responsible Person, can return to the grading ring with insulated transmission along the conductor for the preparation to leave the potential. (2) The equipotential electrician shall hook the grading ring with the potential transfer rod and enter the third insulator of the grading ring. With one hand grasping the insulator tightly and the other holding the potential transfer rod, the equipotential electrician shall break himself away from the equipotential with the help of potential transfer rod. (3) The equipotential electrician exits the intense electric field by means of "two-span and three-short-circuit" operation mode	(1) The equipotential electrician must obtain the permission of the Responsible Person before exiting the potential. (2) When the equipotential electrician returns to the insulator string, the safety belt and backup protection rope shall be used alternately. The protection of safety belt shall not be lost during potential transfer, and the instant action shall be accurate, stable and rapid when leaving the intense electric field. (3) The equipotential electrician shall coordinate his hands and feet during leaving the potential with uniform speed so as to avoid excessive hand waving, big step and other actions. The side phase of the combined gap composed of the grounding body and the electrified body shall be greater than 6.7m (larger than 6.9m for medium phase). The minimum distance between the exposed part of human body and the electrified body should not be less than 0.5m. (4) When moving along the insulator string, the equipotential electrician shall use the backup protection rope to hold two insulator strings with the hands grasping firmly and the feet stepping firmly. (5) The minimum distance to adjacent conductors shall be greater than 6.9 m (7.2m for medium phase). (6) The equipotential electrician shall not lose the protection of safety belt or backup protection rope when returning to cross arm	

Table (Cont'd)

S/N	Work Content	Operation Steps and Standards	Safety Measures and Precautions	Responsible Person
11	Return to the ground	After checking that there is no object left on the tower, the equipotential electrician on the tower shall report it to the Responsible Person and then climb down the tower with insulated transmission rope after obtaining the consent of the Responsible Person	The anti-falling device installed on the tower shall be used in the process of climbing down the tower; when moving and transposition on the pole and tower, the safety protection shall not be lost, and the operators must grasp the components securely	
12	End of the work	(1) The Responsible Person shall organize all working members to put working apparatus and materials in order and put them in a special kit (bag) after cleaning; clean the site to ensure that "the materials are removed and the site is cleaned after construction". (2) In the post-shift meeting, Responsible Person will give work summaries and comments. Comments include the construction quality of this work and the implementation of safety measures from all working personnel. (3) Responsible Person shall report to the on-duty control personnel that the work is over, apply for the restoration of circuit re-closing and terminate the work order.	It is forbidden to restore the line re-closing in the appointed time	

II. Assessment Standard

Part II Skill Module Training and Assessment Standards

Detailed Rules for Assessment and Scoring of Operation and Inspection Skills of UHV AC Transmission Line

Fill-in Column of Examinee	No.:	Name:	Position:	Date:	MM/DD/YYYY
Fill-in Column of Assessor	Grade:	Assessor:	Assessment Team Leader:	Starting time:	Closing time:	Operation Duration:

Assessment module	Live replacement of 1000kV AC transmission line conductor spacer	Assessee	Maintenance personnel of UHV AC transmission line	Assessment method	Operation	Assessment Time Limit	90min

Job Description	Enter the intense electric field along the strain insulator strings and carry out live replacement of the eight-bundle conductor spacer of 1000kV AC transmission line (equipotential operation method)

Work Specifications and Requirements	1. Live working shall be carried out in good weather. In case of thunder, rain, snow or fog, no live working shall be carried out. When the wind force is greater than Level 5 and the humidity is greater than 80%, live working should not be carried out. 2. Person required for this operation include 1 Responsible Person, 1 special Supervisor, 1 ground electrician on the tower, 1 equipotential electrician and 2 ground auxiliary electricians. Enter the intense electric field along the insulator strings and carry out the live replacement of the eight-bundle spacer of the damaged conductor of 1000kV AC transmission line. 3. Responsibilities of Responsible Person: Be responsible for division of operating personnel of the task, work order reading, handling re-closing deactivation of the line and the formalities of work permit, getting work permits, holding pre-shift meeting, dealing with emergency situations in work, quality surveillance, and the summary after work. 4. Special Supervisor: Be responsible for safety supervision and control during operation. 5. Equipotential electrician: Enter the intense electric field along the insulator strings and carry out the live replacement of the eight-bundle conductor spacer of the transmission line. 6. Responsibilities of electrician on tower: Be responsible for testing the null-value or low-value insulator and assisting equipotential in entering and exiting the intense electric field.

Table (Cont'd)

Work Specifications and Requirements	7. Responsibilities of ground electrician: Be responsible for implementing site safety measures, arranging work site, inspecting tools and instruments, delivering tools and materials, and cooperating with the equipotential electrician to enter and exit equipotential. 8. During the live working, if thunder, rain, strong wind or any other circumstance threaten the safety of the staff, the Responsible Person or Supervisor may stop working temporarily according to the circumstances.
Work Specifications and Requirements	Given conditions: 1. Training base: UHV AC 1000kV training line resisting-tensile tower large side phase A conductor first eight-bundle spacer, and model of the conductor is: 8×JL/G1A-630/45. 2. Work orders have been handled, safety measures have been completed (re-closing has been deactivated), and oral application (dispatcher or assessor) shall be made at the beginning and end of the work. 3. Safety fence shall be installed and signs such as "Work here" and "Access from Here" shall be hung at the work site. Safety measures have been completed. 4. The instruments and apparatus shall be used safely and correctly to test the insulating tool. 5. Fall-arrest device should be used in the process of up and down the tower to prevent falling from heights. 6. The operation must be carried out according to Standard Operation Procedures. The relevant item scores shall be deducted for the process error. In case of major hidden dangers of personal, equipment and operational safety, the assessor may order the termination of the operation (assessment) and the assessment of this module is recorded as "unqualified"
Assessment scenario preparation	1. Line: UHV AC 1000kV training line resisting-tensile tower large side phase A conductor, work content: live replacement of 1000kV UHV transmission line conductor spacer, and model of the conductor is: 8×JL/G1A-630/45. 2. Required operation tools and instruments: 2 insulated transmission ropes (TJS-12), 2 insulated backup protection ropes (TJS-16, with buffer), 2 insulated tackles (JH10-0.5), 2 insulated nooses (TJS-14), 4 insulation jacks, 1 potential transfer rod (0.4m), 2 Type-I shielding clothes (coverall, helmet, protective mask, gloves and conductive socks), 2 pairs of conductive shoes, 2 falling protectors, 2 falling protectors, 1 special spanner for vibration damper, 1 insulator detector, 1 set of insulation resistance tester (type 5000V), 1 set of wind speed and humidity tester (HT-8321), 1 multimeter, 2 moisture-proof tarpaulins (2m*4m), 1 red waistcoat (for Responsible Person), 2 clean towels, 2 sets of personal tools (workbasket, flat tongs, marker pen), and 1 eight-bundle spacer of the same model. 3. The work site shall be monitored, and the safety measures (fence, etc.) on the work site have been fully implemented; non-operation personnel are prohibited from entering the site, and the staff must wear safety helmets when entering the work site. 4. Examinees shall bring their own work clothes, flame retardant cotton underwear, safety helmets, gloves, safety belts (including double-protective ropes)

Part II Skill Module Training and Assessment Standards

Table (Cont'd)

Remarks	1. The total score of this module is 100 points, and score of each item is deducted until the corresponding points are finished. If the task is not completed within the specified time, the test shall be terminated immediately. The score of this module shall be calculated according to the actual score of the completed item, and the unfinished item shall not be scored.
	2. In the process of assessment, if the equipment, operating environment, safety measures, safety protection and safe distance do not meet the requirements of the operation, or human error occurs, which may endanger the safety of the operation, the assessor shall order the termination of the operation.
	3. Exam participants should be organized for site survey before the examination and work orders should be handled in advance

S/N	Project name	Quality requirements	Score	Deduction standard	Reasons for deduction	Deduction	Scoring
1	Site re-survey	1) The Responsible Person shall go to the work site to check the line name, pole and tower number, on-site working conditions, defective parts and so on. 2) Check that the site meteorological conditions such as wind speed and humidity should meet the operation requirements. 3) Check whether the work order is complete and unmodified, check whether the safety measures listed are consistent with the actual situation on site, and supplement it if necessary	5	1) No point shall be awarded without a work order. 2) Deduct 1 point for unchecked double title. 3) Deduct 1 point/item for failure to check on-site working conditions (meteorology) and defective parts. 4) Deduct 0.5 points/place for any alteration or frowziness on the work order, and deduct 1 point for incorrect work order number. Deduct 1 point/item for missing items in work order			

Table (Cont'd)

S/N	Project name	Quality requirements	Score	Deduction standard	Reasons for deduction	Deduction	Scoring
2	Work Permit	1) The Responsible Person is responsible for contacting the on-duty control personnel and applying for stopping the line re-closing as per the contents of the work order. 2) Report content is standardized and complete. The voice is loud and clear. 3) Relevant licensing procedures are completed in a timely manner	2	1) No point shall be awarded for items of work without the work permit of the Dispatching Department (Assessor). 2) Deduct 0.5 points/item for non-standard and incomplete reporting terminology or the voice is not loud or clear enough. 3) Deduct 1 point for failure to apply the re-closing lock. 4) Deduct 1 point for failure to repeat permission content. 5) Deduct 0.5 points / item for failure to repeat each permission content (name, time, task and re-closing lock status of the Licensor). 6) Deduct 1 point for failure to complete work order in time			
3	Site layout	Install the security fence and hang the signboards correctly: 1) The security fence should take full account of falling objects from the heights and the influence on road traffic. 2) The entrance and exit of the security fence shall be set reasonably. 3) Signs such as "Access from Here", "Work Here", "Access from Here" shall be properly and well arranged	3	1) Deduct 2 points for failure to arrange the safety fence at the work site. 2) Deduct 1 point/place for not setting the safety fence at the work site properly. 3) Deduct 1.5 points for failure to hang signboards. 4) Deduct 0.5 points per piece for incomplete signboards. 5) Deduct 0.5 points/person for non-operating personnel entering the fenced area			

Part II Skill Module Training and Assessment Standards

Table (Cont'd)

S/N	Project name	Quality requirements	Score	Deduction standard	Reasons for deduction	Deduction	Scoring
4	Hold a pre-shift meeting	1) All staff and personnel shall wear safety helmets and work clothes correctly. 2) The Responsible Person shall wear red waistcoat and read out the work order and be clear with work task and division of personnel; explain safety measures and technical measures in work; check (inquire after) mental state of all working personnel; inform of hazards in work and pre-control measures. 3) All working personnel shall sign on the work order for confirmation.	3	1) Deduct 0.5 points/person for the incorrect wear of safety helmet by working personnel. Deduct 0.5 points/person/time for improper dressing. 2) Deduct 0.5 points/person for the failure to wear safety red waistcoat by Responsible Person and Special Supervisor. 3) No point shall be awarded for unclear work task and division of work. 4) Deduct 1 point for unclear division of labor. 5) Deduct 1 point for incomplete explanation of safety measures or pre-control measures. 6) Deduct 1 point for failure to inform hazards at work. 7) Deduct 1 point for failure to confirm mental status of work shift members. 8) Deduct 1 point for the work shift member failing to sign or signing incompletely.			

Standard for Professional Training and Assessment for Operation Maintenance of UHV AC Transmission Line

Table (Cont'd)

S/N	Project name	Quality requirements	Score	Deduction standard	Reasons for deduction	Deduction	Scoring
5	Inspection of tools	1) The work shift members shall properly set the moisture-proof tarpaulin in the appropriate position. The tarpaulin shall be clean and dry. It is strictly prohibited to tread the tarpaulin. 2) The tools and instruments shall be classified and placed neatly on moisture-proof tarpaulin according to the requirements of the fixed location management; the insulated tools shall not be mixed with metal tools and materials; and the appearance inspection shall be done on tools, instruments and apparatus. 3) All kinds of tools shall be qualified and within the effective time by test. The surface of insulating tools shall not be worn, deformed and damaged, and the operation shall be flexible. Carry out segment insulation detection with such insulated tools with insulation resistance meter of 2, 500V or above and with the resistance no less than 700M Ω, and wipe it off with a clean dry towel. 4) The ground potential and equipotential personnel on tower shall correctly wear a whole suit of qualified shield clothes and conductive shoes as required, with each part connected well, shall not wear chemical fiber clothes next to the skin in the shielding clothes and shall fasten safety belts; the Responsible Person shall carefully check and confirm whether they wear them correctly and all parts are well connected.	7	1) Deduct 1 point for the inappropriate position of moisture-proof tarpaulin. 2) Deduct 0.5 points/time for stepping on moisture-proof tarpaulin. 3) Deduct 1 point for failure to classify or place tools and instruments at fixed position. 4) Deduct 1 point/item for failure to check qualified label and appearance of tools, instruments and apparatus. 5) Deduct 0.5 points per piece for missing items in inspection of tools, instruments, and apparatus. 6) Deduct 0.5 points per piece for incorrect inspection method of tools, instruments and apparatus. 7) Deduct 0.5 points per piece for failure to clean or wipe the hard insulating tools. 8) Deduct 0.5 points/time for holding insulating tools without wearing clean and dry cotton gloves. 9) Deduct 1 point/ item for the improper usage of testing instruments to test the tools, instruments and the full set of shielding clothes.			

Table (Cont'd)

S/N	Project name	Quality requirements	Score	Deduction standard	Reasons for deduction	Deduction	Scoring
5	Inspection of tools	5) The full set of shielding clothes shall be tested with a multimeter, and the resistance value between the farthest two points shall not be more than 20Ω, and the resistance value of a single set shall not be more than 15Ω. 6) The electrician climbing the tower shall conduct the visual inspection on the safety belt, backup protection rope and falling protector, which shall pass the impulse test.	7	10) Deduct 2 points/person for electrician climbing the tower failing to wear the shielding clothes correctly or to check the connecting portion. 11) Deduct 1 point/item for failing to conduct visual inspection or impulse test (or incorrect way) for safety belt, backup protection rope and falling protector 12) Deduct 1 point/item for the Responsible Person failing to check or missing to check the safety protection equipment of electrician climbing the tower			
6	Climbing the tower	1) Tower-climbing personnel shall check the double name, pole number and phase again, report to the Responsible Person and apply for climbing the tower. 2) The ground potential electrician on tower and equipotential electrician carry the insulated transmission rope to climb the tower one after another. 3) During tower climbing, the electrician must use anti-fall device, climb the tower at a uniform speed with the main materials in hand, hang the safety belt and backup protection rope on the shoulder and keep the safety distance of more than 6 m away from the electrified body, and the Responsible Person shall strengthen the operation monitoring.		1) Deduct 1 point/item for the tower electrician failing to check the double name, tower number and phase. 2) Deduct 1 point/item for the tower electrician failing to report the result or apply for climbing the tower. 3) The assessor shall order the termination of the operation (assessment) if the tower electrician fails to use anti-fall device during climbing down. 4) Deduct 0.5 points/time for grasping the shackles with hands during climbing down.			

Table (Cont'd)

S/N	Project name	Quality requirements	Score	Deduction standard	Reasons for deduction	Deduction	Scoring
6	Climbing the tower	4) When climbing to an appropriate position, ground potential electrician shall fasten the safety belt, arrange the insulated transmission rope, and then cooperate with the ground electrician on tower to make lifting preparation of the insulated transmission rope separately. 5) The Responsible Person shall carefully monitor and remind the whole process of tower climbing		5) Deduct 1 point/time for slipping or missing his step during climbing down. 6) Deduct 1 point/ person for main band of safety belt and backup protection rope failing to be hung on the shoulder. 7) Deduct 1 point/ time for safety belt and backup protection rope twining and getting caught. 8) Deduct 1 point/ time for fastening the safety belt and backup protection rope lower than operating personnel. 9) Deduct 1 point/ time for fastening the safety belt and backup protection rope on the same components. 10) No point shall be awarded for losing protection of safety belt during work at heights. 12) Deduct 1 point for installation of tackle without insulating noose. 13) Deduct 1 point for the suspension position of the tackle transmission rope inconvenient for getting tools. 14) Deduct 1 point for the Responsible Person failing to monitor the operation in place. 15) The Assessor shall order the termination of the operation (assessment) for insufficient safe distance			

Part II Skill Module Training and Assessment Standards

Table (Cont'd)

S/N	Project name	Quality requirements	Score	Deduction standard	Reasons for deduction	Deduction	Scoring
7	Detection of insulator	1) The ground electrician shall transfer the insulating bar and insulator detector to the tower. The ground potential electrician shall carry out commissioning on the spark gap detector, with the gap distance of 0.4mm. 2) The equipotential electrician shall assist the ground potential electrician on tower to test the insulators string by string and piece by piece, and the contact must be reliable from the conductor side to the cross arm side. 3) In case of null-value or low-value insulator during the detection, detection shall be carried out for 3 times. 4) When the number of any string of good insulators is less than 37, the testing shall be stopped immediately, and the live working shall be stopped. 5) The safe distance of at least 6m shall be maintained between the electrician on tower and electrified body, and the effective insulting length of insulating rope shall not be less than 6.8m. The minimum insulting length of insulated testing rod shall not be less than 6.8m. 6) After testing, the ground electrician shall cooperate with the electrician on tower to transfer the insulating bar to the ground, and transfer the potential transfer rod to the tower. 7) The Responsible Person shall carefully supervise and remind the operation process.	10	1) Deduct 0.5 points if the knot is tied wrongly while transferring the rope binding tools and instruments. 2) Deduct 0.5 points for each piece in case of collision while transferring tools and instruments. 3) Deduct 0.5 points for winding of transmission rope 4) Deduct 1 point for falling object from high place while transferring. 5) No point shall be awarded if not testing the null value of insulator. 6) Deduct 2 points for incorrect operation methods of electrician on tower. 7) Deduct 2 points for wrong insulator string testing sequence. 8) Deduct 2 points for not re-testing the null-value insulator. 9) The assessor shall order the termination of the operation for insufficient safe distance of operation. 10) Deduct 2 points if not reporting the null value measurement result. 11) Deduct 2 points if not transferring the potential transfer rod. 12) Deduct 2 points for the Responsible Person failing to monitor the operation in place.			

Table (Cont'd)

S/N	Project name	Quality requirements	Score	Deduction standard	Reasons for deduction	Deduction	Scoring
8	Entering the intense electric field	1) Check that all parts of the shielding clothes are well connected before entering the potential, and report to the Responsible Person for approval. 2) The equipotential electrician shall transfer the safety belt to the insulator set fitting, and carry the potential transfer rod, insulated tackle and insulated transmission rope. 3) Before entering the insulator string, the equipotential electrician shall fasten the protection rope (holding two insulator strings with the backup protection rope, grasping one string with hands and stepping on the other string). 4) When reaching 3 insulators outside the grading ring on the conductor side along the insulator string by the operating method of two-span and three-short-circuit, the operator shall stop, wait for the approval of the Responsible Person and then carry out the potential transfer with the potential transfer rod. 5) The side phase combined gap composed of the gaps between the equipotential electrician and the grounding body & the electrified body in the process of entering the potential shall not be less than 6.7 m (not less than 6.9 m for medium phase), and the potential transfer rod must be used for potential transfer when entering the intense electric field. The minimum distance between the exposed part of human body and the electrified body should not be less than 0.5m. 6) Protection of safety belt shall not be lost when entering the intense electric field	10	1) Deduct 2 points for the equipotential electrician failing to check the connecting parts of the shielding clothes. 2) Deduct 2 points for losing the protection of safety belt when equipotential electrician transfers to insulator set fitting. 3) Deduct 2 points for insufficient combined clearance between equipotential electrician and grounding body & electrified body. 4) Deduct 2 points for incorrect and unskilled action of equipotential electrician entering the intense electric field. 5) Deduct 2 points for repeated discharge due to insufficient distance of the exposed parts during potential transfer of equipotential electrician. 6) Deduct 1 point for being unskilled in potential transfer. 7) Deduct 5 points for potential transfer without using the potential transfer rod. 8) Deduct 3 points for carrying out potential transfer without permission of the Responsible Person.			

Part II Skill Module Training and Assessment Standards

Table (Cont'd)

S/N	Project name	Quality requirements	Score	Deduction standard	Reasons for deduction	Deduction	Scoring
8	Entering the intense electric field		10	9) Deduct 2 points for losing the protection of safety belt when equipotential electrician enters the intense electric field. 10) Deduct 2 points for the Responsible Person failing to monitor the operation in place			
9	Replace the eight-bundle spacer	1) After entering the equipotential, the equipotential electrician shall tie the pole belt to the upper sub-conductor, and install the insulated protection rope (all sub-conductors shall be covered). 2) The equipotential electrician shall carry the insulated transmission rope and walk to the operation point for replacing the spacer, and shall correctly install the insulating noose at the appropriate place of the sub-conductor and hang insulated tackle and insulated transmission rope. 3) The equipotential electrician shall mark the fixed point of old spacer symmetrically with marker pen. 4) The equipotential electrician shall install 4 insulator jacks in the appropriate position next to the old spacer in the way of "two-two corresponding" to fix the sub-conductor reliably.	30	1) Deduct 3 points for the incomplete covering of sub-conductor by insulated protection rope. 2) Deduct 2 points for failure to tie pole belt to upper sub-conductor during walking along the line. 3) Deduct 2 points for being unskilled in walking along the line. 4) Deduct 5 points for the insulated tackle hooked directly on the conductor. 5) Deduct 1 point/item for incorrect installation method or inappropriate position of insulating noose. 6) Deduct 1 point for failure to lock the insulated tackle after it is hooked on insulating noose. 7) Deduct 3 points for no marking. 8) Deduct 2 points per nos. for failure to fix the sub-conductor with 4 insulated jacks. 9) Deduct 5 points for removing the spacer without tying the knot.			

Table (Cont'd)

S/N	Project name	Quality requirements	Score	Deduction standard	Reasons for deduction	Deduction	Scoring
9	Replace the eight-bundle spacer	5) The equipotential electrician shall fasten the insulated transmission rope to the old spacer by the way of slip knot, then use the special spanner for spacer to remove the old spacer, and put it on the moisture-proof tarpaulin in cooperation with the ground electrician. 6) The ground electrician shall lift the new spacer to the position of equipotential electrician. 7) The equipotential electrician shall mark and install the new spacer with the special spanner for spacer correctly. The installation quality shall be that the plane of the spacer shall be vertical to the sub-conductor. 8) The equipotential electrician shall remove the insulated jack, insulated tackle, insulated transmission rope and noose successively. 9) No tools, instruments or materials shall drop during the equipotential operation	30	10) Deduct 2 points for removing tools and instruments in the hurry and confusion. 11) Deduct 3 points per piece for falling object from heights. 12) No point shall be awarded at this module for throwing old spacer at heights. 13) Deduct 2 points/person/time for the ground electrician standing directly below the work point of equipotential electrician. 14) Deduct 1 point for failure to put old spacer on moisture-proof tarpaulin. 15) Deduct 1 point for incorrect way the knot is tied when passing tools up and down. 16) Deduct 5 points for untying the knot before the new spacer is installed. 17) Deduct 3 points for new spacer deviating from original spacer position. 18) Deduct 3 points for the plane of new spacer failing to be perpendicular to the sub-conductor after installation.			

Part II Skill Module Training and Assessment Standards

Table (Cont'd)

S/N	Project name	Quality requirements	Score	Deduction standard	Reasons for deduction	Deduction	Scoring
9	Replace the eight-bundle spacer		30	19) Deduct 1 point per piece for not removing insulated jack, insulated tackle, insulated transmission rope and noose. 20) Deduct 2 points for not removing insulated jack, insulated tackle, insulated transmission rope and noose successively. 21) No point shall be awarded at this module for failure to replace the old spacer by the new spacer. 22) The assessor shall order the termination of the operation for insufficient safe distance between adjacent conductors			
10	Exit the intense electric field	1) After inspecting that the spacer is installed firmly and there is no object left at the operation point, the equipotential electrician, with the permission of the Responsible Person, can return to the grading ring with insulated transmission along the conductor for the preparation to leave the potential. 2) The equipotential electrician shall hook the grading ring with the potential transfer rod and enter the third insulator of the grading ring. With one hand grasping the insulator tightly and the other holding the potential transfer rod, the equipotential electrician shall break himself away from the equipotential with the help of potential transfer rod. 3) Protection of safety belt shall not be lost when leaving the intense electric field.	10	1) Deduct 2 points per item for objects left in operation point. 2) Deduct 3 points for failure to apply to the Responsible Person for potential transfer. 3) Deduct 2 points for carrying out potential transfer without permission of the Responsible Person. 4) Deduct 5 points for potential transfer without using the potential transfer rod. 5) Deduct 1 point for being unskilled in potential transfer. 6) Deduct 2 points for losing the protection of safety belt when equipotential electrician leaves the intense electric field.			

Table (Cont'd)

S/N	Project name	Quality requirements	Score	Deduction standard	Reasons for deduction	Deduction	Scoring
10	Exit the intense electric field	4) The equipotential electrician leaves the intense electric field along the insulator string in the "two-span and three-short-circuit" operation mode and transfers to the cross arm. 5) The side phase combined gap composed of the gaps between the equipotential electrician and the grounding body & the electrified body in the process of leaving the potential shall not be less than 6.7 m (not less than 6.9 m for medium phase), and the potential transfer rod must be used for potential transfer when leaving the intense electric field. The minimum distance between the exposed part of human body and the electrified body should not be less than 0.5m	10	7) Deduct 2 points for repeated discharge due to insufficient distance of the exposed parts during potential transfer of equipotential electrician. 8) Deduct 2 points for incorrect and unskilled action of equipotential electrician leaving the intense electric field. 9) Deduct 2 points for insufficient combined clearance between equipotential electrician and grounding body & electrified body. 10) Deduct 2 points for losing the protection of safety belt when equipotential electrician transfers to the cross arm. 11) Deduct 2 points for the Responsible Person failing to monitor the operation in place			
11	Return to the ground	1) After checking that there is no object left on the tower, the electrician on the tower shall report it to the Responsible Person and then climb down the tower one after another with insulated transmission rope after obtaining the consent of the Responsible Person.	5	1) Deduct 2 points per item for objects left on tower. 2) Deduct 1 point/item for the tower electrician failing to report the inspection results of the leftovers or apply for climbing the tower.			

Part II Skill Module Training and Assessment Standards

Table (Cont'd)

S/N	Project name	Quality requirements	Score	Deduction standard	Reasons for deduction	Deduction	Scoring
11	Return to the ground	2) When climbing downing the tower, the electrician must use anti-fall device, climb down the tower at a uniform speed with the main materials in hand, hang the safety belt and backup protection rope on the shoulder and keep the safe distance of more than 6 m away from the electrified body, and the Responsible Person shall strengthen the operation monitoring	5	3) The assessor shall order the termination of the operation (assessment) if the tower electrician fails to use anti-fall device during climbing down. 4) Deduct 0.5 points/time for grasping the shackles with hands during climbing down. 5) Deduct 1 point/time for slipping or missing his step during climbing down. 6) Deduct 1 point/ person for main band of safety belt and backup protection rope failing to be hung on the shoulder			
12	End of the work	1) The Responsible Person shall organize all working members to put working apparatus and materials in order and put them in a special kit (bag) after cleaning; clean the site to ensure that "the materials are removed and the site is cleaned after construction". 2) In the post-shift meeting, Responsible Person will give work summaries and comments. Comments include the construction quality of this work and the implementation of safety measures from all working personnel. 3) Responsible Person shall report to the on-duty control personnel that the work is over, apply for the restoration of circuit re-closing and terminate the work order.	10	1) Deduct 0.5 points per piece for failure to clean or wipe the insulating tools and instruments. 2) Deduct 1 point per piece for failure to classify or place tools and instruments at fixed position. 3) Deduct 0.5 points/time for throwing about tools and instruments or stepping on moisture-proof tarpaulin. 4) Deduct 1 point per piece for not removing fence and signs or leftovers. 5) Deduct 2 points for failure to hold the post-shift meeting.			

Table (Cont'd)

S/N	Project name	Quality requirements	Score	Deduction standard	Reasons for deduction	Deduction	Scoring
12	End of the work		10	6) Deduct 1 point when the team is not orderly or focused. 7) Deduct 1 point/person for members of the working group not attending post-shift meeting. 8) Deduct 1 point for unqualified comments. 9) Deduct 1 point for failure to report to the Dispatching Department (Assessor) the end of the work and apply for restoration of re-closing lock. 10) Deduct 0.5 points/item for non-standard and incomplete reporting terminology or the voice is not loud enough.			
12	End of the work		10	11) Deduct 1 point for failure to repeat permission content. 12) Deduct 0.5 points/item for missing items during repeating the contents (name of working unit, name of the Responsible Person, time, name of the line, work completion, device has restored to normal, the personnel have evacuated and re-closing lock can be restored). 13) Deduct 1 point/ item for failure to complete in time or error in the fill-in of work order termination procedure			
	Total		100				

Module 8 Standards for Training and Assessment on Live Replacement of 1000kV AC Transmission Line Conductor Vibration Damper

I. Training Standard

(I) Training Requirements

Designation of module	Live replacement of 1000kV AC transmission line conductor vibration damper	Type of training	Operation
Training method	Practical operation training	Hours of training	21 training hours
Training objectives	1. Master the electrical significance of "two-span and three-short-circuit" operation mode in entering and exiting 1000kV intense electric field along strain insulator string. 2. Can complete the entry of 1000kV equipotential operation point along the strain insulator string. 3. Be able to independently complete the replacement of conductor vibration damper (equipotential operation method).		
Training venue	UHV AC training line		
Training content	The "two-span and three-short-circuit" operation mode is adopted to enter the strong electric field along the strain insulator string, and the equipotential operation method is adopted for live replacement of the 1000kV AC transmission line conductor vibration damper		
Scope of application	Maintenance personnel of UHV AC transmission line		

(II) Referenced rules and specifications

(1) Code for Design of 1000kV Overhead Transmission Line (GB50665-2011).

(2) Maintenance Code for 1000kV AC Transmission Line (DL/T209-2008).

(3) Operation Code for 1000kV AC Transmission Line (DL/T307-2010).

(4) Technical Guide for Live Working on 1000kV AC Transmission Line (DL/T392-2015).

(5) Calculation Method of Live Working Minimum Approach Distance on AC Transmission Line (GB/T 19185-2008).

(6) Guidelines of Insulation Coordination for Live Working (DL/T867-2004).

(7)Live Working-Guidelines for the Installation of Power Transmission Line Conductors and Earthwires-stringing Equipment and Accessory Items (DL/T 1007-2006).

(8)State Grid Corporation of China on the Management Regulations of Live Working (Trial Implementation) (SGCC [2007] No.751).

(9)State Grid Corporation of China Working Regulations of Power Safety (Transmission Line Section) (Q/GDW1799.2-2013).

(10)Electrotechnical Terminology - Overhead Line (GB/T 2900.51-1998).

(11)Electrotechnical Terminology- Live Working (GB/T2900.55-2002).

(12)Live Working - Terminology for Tools, Equipment and Devices (GB/T 14286-2002).

(13)Minimum Requirements for Utilization of Tools, Devices and Equipment for Live Working (DL/T 877-2004).

(14)Preventive Test Code of Tools, Devices and Equipment for Live Working (DL/T 976-2005).

(15)Insulated Tackles for Live Working (GB/T13034-2008).

(16)Live Working-Insulating Ropes (GB 13035-2008).

(17)Screen Clothes for Live Working on 1000kV AC (GB/T25726-2010).

(Ⅲ) Teaching Design for Training

To complete the live replacement of 1000kV AC transmission line conductor vibration damper is the Work task during the design. Each training stage shall be designed according to the standard operation procedure for work task completion. Each stage includes specific training objectives, training content, hours of training, training methods (training resources), training environment, assessment and evaluation, etc, as shown in Table 2-8-1.

(Ⅳ) Operation Flow

1. Work Task

The "two-span and three-short-circuit" operation mode is adopted to enter the intense electric field and reach the operation point along the strain insulator string, and the equipotential operation method is adopted for live replacement of the conductor vibration damper of 1000kV AC transmission line.

(This operation task is applicable to the operation point of the first group of vibration dampers for side phase conductor of resisting-tensile tower of 1000kV AC single circuit transmission line in 1000m and below altitude)

2. Requirements for Weather and Work Site

(1) The live replacement for conductor vibration damper of 1000kV AV transmission

Part II Skill Module Training and Assessment Standards

Table 2-8-1 Training Content Design for Live Replacement of Conductor Vibration Damper of 1000kV AC Transmission Line

Training schedule	Training objectives	Training content	Hours of training	Training methods and resources	Preparation of training conditions	Assessment and evaluation
1. Theoretical teaching	1. Preliminarily master the basic method for entering and exiting 1000kV intense electric field along the insulator string. 2. Be familiar with the methods for potential transfer. 3. Be familiar with the method for replacing the damaged conductor vibration damper of power transmission line. 4. Be familiar with the safe distance, hazard identification and precontrol of live working on UHV AC line	1. The electrical significance of the "two-span and three-short-circuit" operation mode when entering and leaving the intense electric field along the insulator. 2. The use method for the potential transfer rod during entering and exiting the UHV intense electric field. 3. Method and quality standard for replacement of power transmission line conductor vibration damper. 4. Safe distance, hazard analysis and precontrol measures for live working on UHV AC line	2	Training methods: Lecture. Training resources: PPT, relevant regulations, specifications and technical guidelines	Multimedia classroom	Attendance, classroom questions and assignments
2. Preparations	Be able to complete the preparation before operation	1. Work site survey. 2. Preparation of the standardized operation card. 3. Filling of the work order. 4. Preparation of tools and materials for this operation	1	Training methods: 1. Site survey and preparation of tools, instruments and materials shall be practiced at site. 2. Preparation of operation card and the filling of work order shall adopt lecture method. Training resources: 1. 1000kV practical training line. 2. UHV tools warehouse. 3. Blank work order	1. 1000kV practical training line. 2. Multi-Media classroom	

Table (Cont'd)

Training schedule	Training objectives	Training content	Hours of training	Training methods and resources	Preparation of training conditions	Assessment and evaluation
3. Work site preparation	Be able to complete the preparations of work site	1. Work site re-survey. 2. Job application. 3. Work site layout. 4. Pre-shift meeting. 5. Inspection of tools and materials. 6. Usage of special spanner of vibration damper	1	Training methods: Demonstration and role play. Resources; 1000kV training line	1000kV training line	
4. Trainer's demonstration	The trainees can preliminarily understand the operation process of the task through inspecting and learning from each other's work	1. Testing of null-value insulator. 2. The equipotential electrician enters and exits the intense electric field and potential transfer along the strain insulator string. 3. The equipotential electrician reaches the replacement position of the vibration damper by means of walking along the line. 4. Equipotential electrician uses the special tools to complete the replacement of conductor vibration damper	2	Training methods: Demonstration method. Resources; 1000kV training line	1000kV training line	

Part II Skill Module Training and Assessment Standards

Table (Cont'd)

Training schedule	Training objectives	Training content	Hours of training	Training methods and resources	Preparation of training conditions	Assessment and evaluation
5. Group training of trainees	1. Be able to enter and leave the 1000kV intense electric field along the insulator string and operate the potential transfer. 2. Be able to complete the 1000kV transmission line conductor vibration damper replacement	1. The trainees are grouped (6 persons per group) to train the skill operation of entering and exiting 1000kV intense electric field, potential transfer and conductor vibration damper replacement. 2. Trainers guide the operation of trainees and conduct safety supervision	14	Training methods: Role play. Resources: 1000kV training line	1000kV training line	Score the operation of trainees according to the detailed rules for skill assessment and scoring
6. End of the work	1. Enable the trainees to further distinguish the shortcomings of the operation process and facilitate the promotion in the later stage. 2. Train the trainees in the working style of safe and civilized production	1. Cleaning up the work site. 2. Report to dispatcher. 3. Comment and summarize the work task this time at the post-shift meeting	1	Training methods: Lecture and inductive method	1000kV training line	

line shall be carried out in good weather.

In case of lightning (hearing thunder or seeing lightning), snow, hail, rain, fog and so on, live working is prohibited. When the wind force is greater than level 5, live working shall not be carried out; When the relative humidity is greater than 80%, insulating tool with moisture-proof performance shall be adopted if live working is required; When live rush repair must be carried out in bad weather, relevant personnel shall be organized to fully discuss and prepare necessary safety measures, which shall be carried out after approval of the Unit.

(2) The operating personnel shall be in good spirits, and be familiar with the organizational and technical measures to ensure safety in work; they shall hold the live working qualification within the validity period.

(3) The Responsible Person shall organize relevant personnel to complete field investigation in advance, determine the operating methods, required working apparatus and necessary measures according to the results, and handle the work orders for live working.

(4) The work site shall be reasonably set up with fence and warning signs. Non-operating personnel is forbidden to enter.

(5) The line re-closing device shall be deactivated in the Project.

(6) Safety working distance and effective insulation length are shown in Table 2-8-2.

Table 2-8-2　Safe Distance for Live Replacement of Conductor Vibration Damper of 1000kV AC Transmission Line

Voltage class/m	Safe distance between human body and electrified body /m	Minimum distance to adjacent phase conductor /m	Minimum effective insulation length/m		Minimum combined clearance /m	Minimum distance between exposed part of human body and electrified body when potential is transferred /m
			Insulating bar	Insulated load-bearing tools and insulating ropes		
1000kV	6.0 (6.8)	6.9 (7.2)	6.8	6.8	6.7 (6.9)	0.5

Notes: ① In case of the altitude at 1000m and below, the data of 6.8 m in brackets, the data of 7.2 m in brackets and the data of 6.9 m in brackets shall be respectively adopted for the phase safe distance, the minimum distance and the combined gap during the live working of 1000kV AC single-loop transmission line.

② As the phase-to-phase spacing of 1000kV UHV line is large enough (generally speaking, the distance between the phase and the ground shall be controlled) and no important safety factors are taken into account, the "minimum distance to the adjacent phase conductor" is not given in the "Safety Regulation", and the actual work can refer to the data of 750kV.

(7) For operation on 1000kV transmission line, the number of insulators in good operation phase shall be no less than 37 pieces.

3. Preparations

3.1 Hazards and precontrol measures

(1) Hazard - electric shock injury

Precontrol measures:

① Before work, Responsible Person should contact the control personnel on duty, deactivate the line re-closing, and perform the licensing procedures.

② Before climbing the tower, the equipotential operators on tower must carefully check the name of the line, the number of the pole and tower, and phase, and then the tower can be climbed after all have been confirmed correct.

③ If lines lose power suddenly during work, operators shall still deem it to be charged. The Responsible Person shall contact the control personnel as soon as possible, and no forced energization is allowed before the on-duty control personnel getting in touch with the Responsible Person.

④ The ground electrician shall wear clean and dry gloves when operating the insulating tools. Insulating tools and insulating ropes shall not be damaged, damped, deformed, and malfunctioned. It is not allowed to use non-insulating ropes (such as cotton rope, manila rope, and steel wire rope). The tools and instruments for live working at the work site shall be placed on the moisture-proof tarpaulin to prevent them from dirt and moisture in use.

⑤ The equipotential operator shall wear flame-retardant underwear and full set of shielding clothes for 1000kV live working shall be worn outside the clothes (including coverall, helmet, protective mask, gloves, conductive socks and shoes). All parts shall be in excellent connection conditions. The resistance between the farthest points of the whole shielding clothes shall not be greater than 20Ω.

⑥ Before potential transfer, equipotential operator shall obtain the approval of Responsible Person, and the minimum distance between exposed part of human body and electrified body shall not be less than 0.5m. During potential transfer, potential transfer rod shall be adopted. The operation shall be fast, and end shall not be used for power charging and discharging; During delivering tools and materials to ground potential operators, the effective length of insulating tools or insulating ropes shall not be lower than the requirements as specified in Table 2-8-2.

⑦ When transmitting large metal objects by using insulating rope, the ground potential operators shall ground the metal objects before contacting with the insulating rope.

⑧ The special Supervisor shall continuously monitor the operators and correct their nonstandard operation or actions in violation at any time. Special attention shall be paid to operators work at heights to ensure that there is enough safety distance (meeting the requirements in Table 2-8-2). It is forbidden to contact two non-connected electrified bodies or make contact with electrified body and grounding body at the same time.

(2) Hazard - falling accident

Precontrol measures:

① Before climbing, operators work at heights must satisfy the requirements of this operation, such as physical condition, mental state and skill and quality.

② Before climbing the tower, operators working at heights shall conduct visual inspection and impulse test on safety belts and falling protector to ensure that their mechanical strength meets the requirements.

③ Operators who climb the tower shall first check whether the shackles is firm, shoe soles is clean, or the falling device is firm or is provided with the falling protector; grasp the main materials, tread on shackles and climb up (down) the tower at a uniform pace.

④ Supervisor shall correct irregularities and violations at any time. Special attention shall be paid to operators working at heights to prevent them from losing the protection of the safety belt or insulated backup protection rope during transposition. The safety belt shall be attached to a secure part and shall not be fastened lower than operating personnel.

⑤ When the equipotential operators move on the insulator string in parallel, they usually grasp one string with both hands and step on the other string with both feet to enter the intense electric field at a constant speed. During the movement, the backup protection rope holds two strings of insulators, and excessive hand waving and big step are not allowed.

⑥ Equipotential operators must fasten the safety belt and make the backup protection rope cover all the sub-conductors when they are walking along the conductor; they shall control the center of gravity during walking along the line to prevent the conductor from turning over.

(3) Hazard - injury caused by objects falling from heights. Precontrol measures:

① Operator working at heights should put personal tools and fragmentary materials into the tools bag. It is strictly forbidden to hang objects in heights or keep in the mouth.

Part II Skill Module Training and Assessment Standards

② Ground operator should correctly wear a safety helmet and use the knots to deliver the tools, instruments and materials. The vertical distance from the operation point should not be less than the falling radius.

③ The work site shall be set up with fence and warning signs. The Supervisor shall maintain vigilance at all times and prohibit non-staff members and vehicles from entering the operation area.

3.2 Selection of tools and instruments and materials

Tools and materials required for live replacement of conductor vibration damper for 1000kV AC transmission line can be seen in Table 2-8-3. Before delivering tools and instruments out of warehouse, application voltage class and test period shall be carefully checked and they shall be inspected to ensure that appearance is intact, connection is firm, rotation is flexible and meet the working task requirements. After delivering tools and instruments out of warehouse, they shall be stored in tools bag or tool kit for transportation to avoid contamination and damp. Metal tools and insulated tools shall be separately loaded and transported to avoid deformation, damage or other defects caused by mixed loading and transportation.

Table 2-8-3 Tools, Instruments and Materials Required by Live Replacement of Conductor Vibration Damper of 1000kV AC Transmission Line

S/N	Name	Specification	Unit	Qty.	Remarks
1	Insulated transmission rope	TJS-12, with the length matching the lifting height	Nos.	2	Insulating tool
2	Insulated backup protection rope	TJS-16, with buffer provided	Nos.	2	Insulating tool
3	Insulated tackle	JH10-0.5	Nos.	2	Insulating tool
4	Insulated noose	TJS-14	Nos.	2	Insulating tool
5	Potential transfer rod	0.4m	Nos	1	Insulating tool
6	I-type shielding clothes (coverall, helmet, protective mask, gloves and conductive socks)	Shielding efficiency ≥ 60 dB (shield efficiency of shielding mask ≥ 20 dB)	Set	2	PPE
7	Conductive shoes	The size depends on the wearer	Nos.	2	PPE
8	Flame retardant underwear	Pure mulberry silk	Set	2	PPE
9	Safety belt of double insurance	Full-body harness	Nos.	2	PPE

Table (Cont'd)

S/N	Name	Specification	Unit	Qty.	Remarks
10	Falling protector	Corresponding to the type of pole and tower falling protector	Nos.	2	PPE
11	Safety helmet		Nos.	6	PPE
12	Special spanner for vibration damper		Nos.	1	Special tools
13	Insulator detector		Set	1	Other tools
14	Insulation resistance tester	5,000V, with electrode width 2cm and interelectrode width 2cm	Set	1	Other tools
15	Wind speed, temperature and humidity tester	HT-8321	Set	1	Other tools
16	Multimeter		Nos.		Other tools
17	Interphone	As required by the work	Set	2	Other tools
18	Moisture-proof tarpaulin	2m×4m	Piece	2	Other tools
19	Security fence		Set	Several	Other tools
20	Warning sign	"Work Here", "Access from Here" "Access from Here"	Set	1	Other tools
21	Red waistcoat	"Responsible Person"	Piece	1	Other tools
22	Clean towel	Cotton	Nos.	1	Other tools
23	Shoe covers		Nos.	Several	Other tools
24	Work gloves		Nos.	Several	Other tools
25	Personal tools	Tool bag, flat tong, marker pen	Set	2	Other tools
26	Vibration damper	The same model with replaced vibration damper	Nos.	1	Material
27	Aluminum armor tape		m	Proper	Material

Note: The electrical strength of the insulating tool shall meet the requirements of the "State Grid Corporation of China Working Regulations of Power Safety (Transmission Line Section)", the test is required to be qualified and within the validity period.

3.3 Division of labor for operators

Division of labor for operators of the task is shown in Table 2-8-4.

Part II Skill Module Training and Assessment Standards

Table 2-8-4 Division of Operating Personnel Involved in Live Replacement of 1000kV AC Transmission Line Conductor Vibration Damper

S/N	Post	Qty. (person)	Responsibilities
1	Responsible Person	1	Be responsible for division of operating personnel of the task, work order reading, handling re-closing deactivation of the line and the formalities of work permit, getting work permits, holding pre-shift meeting, dealing with emergency situations in work, quality surveillance, and the summary after work
2	Special Supervisor	1	Be responsible for safety supervision and control during operation
3	Equipotential electrician	1	Be responsible for entering the equipotential to replace the conductor vibration damper
4	Ground potential electrician on tower	1	Be responsible for testing the null-value or low-value insulator and assisting equipotential in entering and exiting the intense electric field.
5	Ground electrician	2	Be responsible for taking site safety measures, arranging operation site, inspecting tools and instruments, delivering tools and materials, and cooperating with the equipotential electrician to enter and leave equipotential

4. Work Procedure

The workflow of this task is shown in Table 2-8-5.

Table 2-8-5 Workflow Chart for Live Replacement of 1000kV AC Transmission Line Conductor Vibration Damper

S/N	Work Content	Operation Steps and Standards	Safety Measures and Precautions	Responsible Person
1	Site re-survey	The Responsible Person shall complete the following work: (1) Check the line title, the number of the pole and tower and ensure the phases are correct; guarantee that the foundation and the pole and tower are intact and in normal condition; ensure that the cross and span distance meets the safety requirements; confirm the defect conditions and the specifications and models of earth wires. (2) Check that the site meteorological conditions such as wind speed and humidity should meet the operation requirements.	(1) Correctly wear helmet, working clothes, work shoes and protective gloves. (2) Operation under meteorological conditions that may endanger the safety of operators is forbidden. (3) Non-operation personnel and vehicles are strictly prohibited from entering the working site.	

Table (Cont'd)

S/N	Work Content	Operation Steps and Standards	Safety Measures and Precautions	Responsible Person
1	Site re-survey	(3) Check that the terrain and environment shall meet the operation requirements. (4) Check that the safety measures listed in the work order are in line with the actual situations on site, and the measures will be supplemented if necessary		
2	Work Permit	(1) The Responsible Person is responsible for contacting the on-duty control personnel and applying for stopping the line re-closing as per the contents of the work order. (2) Live working could be started only after being approved by the on-duty control personnel.	Live working shall not be started without the permission of the on-duty control personnel.	
3	Site layout	Install the security fence and hang the signboards correctly: (1) The security fence should take full account of falling objects from the heights and the influence on road traffic. (2) The entrance and exit of the security fence shall be set reasonably. (3) Complete signs such as "Access from Here", "Work Here", "Access from Here" shall be properly and well arranged	When the influence on road traffic safety is uncontrollable, the traffic management department should be contacted in time to strengthen the on-site control of traffic safety.	
4	Hold a pre-shift meeting	(1) All working personnel shall line up. (2) The Responsible Person shall wear red waistcoat and read out the work order and be clear with work task and division of personnel; explain safety measures and technical measures in work; check (inquire after) mental state of all working personnel; inform of hazards in work and precontrol measures. (3) All working personnel shall sign on the work order for confirmation.	(1) The work order shall be filled in, issued and approved in a standardized manner, and the signature shall be complete. (2) All working personnel shall be in good mental states. (3) All working personnel shall be clear with task division of works, safety measures and technical measures	

Part II Skill Module Training and Assessment Standards

Table (Cont'd)

S/N	Work Content	Operation Steps and Standards	Safety Measures and Precautions	Responsible Person
5	Inspect tools	(1) The ground electrician and equipotential electrician on tower shall wear the shielding clothes in a right way and pass the inspection, which shall be supervised and inspected by the Responsible Person. (2) Wear personal safety equipment correctly (proper size and easy lock), and the Responsible Person shall supervise and inspect it. (3) Measure the wind speed, wind direction and humidity, check the insulation performance of insulating tools, and make records	(1) Check carefully for damage, deformation and failure before using metal and insulating tools. Carry out segment insulation detection with such insulating tools as insulation resistance tester of 2500V or above and with the resistance no less than 700 M Ω, and wipe them clean with the clean dry towel. (2) Use a multimeter to measure the resistance between the farthest ends of the shielding clothes and trousers, which shall not be greater than 20Ω. The Responsible Person shall check the connection of the electrician's shielding clothing. (3) Check the tool assembly and make sure the connection is reliable. (4) The live working tools used at site shall be placed on the moisture-proof tarpaulin	
6	Climbing the tower	(1) After checking the line name and pole and tower number without error, the ground electrician and equipotential electrician on tower shall check the stress of the safety belt and the falling protector in the shock. (2) The ground electrician on tower carries the insulated transmission rope to climb the tower, and the equipotential electrician then climbs the tower. When they reach the operation point of the cross arm, they choose the appropriate position to fasten the safety belt, and the ground electrician on tower installs the insulated tackle and the insulated transmission rope in the appropriate position of the cross arm. Then he shall cooperate with the ground electrician to separate the insulated transmission rope for lifting preparation	(1) After checking the correct line name and pole and tower number, the tower can be climbed for operation. (2) The anti-falling device installed on tower shall be used in the process of climbing the tower; when moving and transposition on the pole and tower, the safety protection shall not be lost, and the operators must climb and grasp the components securely. (3) The working electrician must wear a whole suit of qualified shielding clothes which must be connected reliably. Before entering the equipotential at cross arm, the equipotential electrician shall re-check and confirm that each part of the shielding clothes are connected reliably before the next operation	

Table (Cont'd)

S/N	Work Content	Operation Steps and Standards	Safety Measures and Precautions	Responsible Person
2	Work Permit	(1) The Responsible Person is responsible for contacting the on-duty control personnel and applying for stopping the line re-closing as per the contents of the work order. (2) Live working could be started only after being approved by the on-duty control personnel.	Live working shall not be started without the permission of the on-duty control personnel	
3	Site layout	Install the security fence and hang the signboards correctly: (1) The security fence should take full account of falling objects from the heights and the influence on road traffic. (2) The entrance and exit of the security fence shall be set reasonably. (3) Complete signs such as "Access from Here", "Work Here", "Access from Here" shall be properly and well arranged	When the influence on road traffic safety is uncontrollable, the traffic management department should be contacted in time to strengthen the on-site control of traffic safety.	
4	Hold a pre-shift meeting	(1) All working personnel shall line up. (2) The Responsible Person shall wear red waistcoat and read out the work order and be clear with work task and division of personnel; explain safety measures and technical measures in work; check (inquire after) mental state of all working personnel; inform of hazards in work and precontrol measures. (3) All working personnel shall sign on the work order for confirmation.	(1) The work order shall be filled in, issued and approved in a standardized manner, and the signature shall be complete. (2) All working personnel shall be in good mental states. (3) All working personnel shall be clear with task division of works, safety measures and technical measures	

Part II Skill Module Training and Assessment Standards

Table (Cont'd)

S/N	Work Content	Operation Steps and Standards	Safety Measures and Precautions	Responsible Person
5	Inspect tools	(1) The ground electrician and equipotential electrician on tower shall wear the shielding clothes in a right way and pass the inspection, which shall be supervised and inspected by the Responsible Person. (2) Wear personal safety equipment correctly (proper size and easy lock), and the Responsible Person shall supervise and inspect it. (3) Measure the wind speed, wind direction and humidity, check the insulation performance of insulating tools, and make records	(1) Check carefully for damage, deformation and failure before using metal and insulating tools. Carry out segment insulation detection with such insulating tools as insulation resistance tester of 2500V or above and with the resistance no less than 700M Ω, and wipe them clean with the clean dry towel. (2) Use a multimeter to measure the resistance between the farthest ends of the shielding clothes and trousers, which shall not be greater than 20Ω.The Responsible Person shall check the connection of the electrician's shielding clothing. (3) Check the tool assembly and make sure the connection is reliable. (4) The live working tools used at site shall be placed on the moisture-proof tarpaulin	
6	Climbing the tower	(1) After checking the line name and pole and tower number without error, the ground electrician and equipotential electrician on tower shall check the stress of the safety belt and the falling protector in the shock. (2) The ground electrician on tower carries the insulated transmission rope to climb the tower, and the equipotential electrician then climbs the tower. When they reach the operation point of the cross arm, they choose the appropriate position to fasten the safety belt, and the ground electrician on tower installs the insulated tackle and the insulated transmission rope in the appropriate position of the cross arm.Then he shall cooperate with the ground electrician to separate the insulated transmission rope for lifting preparation	(1) After checking the correct line name and pole and tower number, the tower can be climbed for operation. (2) The anti-falling device installed on tower shall be used in the process of climbing the tower; when moving and transposition on the pole and tower, the safety protection shall not be lost, and the operators must climb and grasp the components securely. (3) The working electrician must wear a whole suit of qualified shielding clothes which must be connected reliably.Before entering the equipotential at cross arm, the equipotential electrician shall re-check and confirm that each part of the shielding clothes are connected reliably before the next operation	

Table (Cont'd)

S/N	Work Content	Operation Steps and Standards	Safety Measures and Precautions	Responsible Person
9	Replacement of the sub-conduct or vibration damper	(3) Equipotential electrician shall mark the installation points of the old vibration damper on conductor at both ends with the marker pen. (4) The equipotential electrician shall use the insulated transmission rope to fasten the old vibration damper by the way of slip knot, then use the special spanner for vibration damper to remove the old vibration damper, and put it on ground with the insulated transmission rope in cooperation with the ground electrician. (5) The equipotential electrician shall remove the aluminum armor tape and place it in the tool kit; The new aluminum armor tape shall be wound according to the drawing mark. The winding shall be smooth and tight, and the twisting direction shall be the same as that of the outer conductor. (6) The ground electrician shall lift the new vibration damper to the position of equipotential electrician, and the equipotential electrician shall install it correctly. (7) Check the installation quality and make sure that the vibration damper is vertically downward with the hammer ball parallel to the main wire. The installation displacement should not exceed ±30mm; The both ends of the aluminum armor tape should be pressed back into the cord bracket of vibration damper, and both ends should be exposed 10mm. The spring washer at the bolt should be tight and flat, and the bolt direction should be consistent with other sub-conductor vibration damper. (8) The equipotential electrician shall remove the insulated tackle, insulated transmission rope and noose successively. (9) No tools, instruments or materials shall drop during the equipotential operation	(4) During delivering, the conductor vibration damper shall not be bumped, and the insulated transmission ropes on both sides shall not be intertwined. (5) During delivering the tools binding rope shall be correct and reliable so as to prevent objects falling from heights. (6) During delivering the tools, instruments and materials, the ground electrician is not allowed to stand directly below the operation point of equipotential electrician	

Part II Skill Module Training and Assessment Standards

Table (Cont'd)

S/N	Work Content	Operation Steps and Standards	Safety Measures and Precautions	Responsible Person
10	Exit the intense electric field	(1) After inspecting that the vibration damper is installed firmly and there is no object left at the operation point, the equipotential electrician, with the permission of the Responsible Person, can return to the grading ring with insulated transmission along the conductor for the preparation to leave the potential. (2) The equipotential electrician shall hook the grading ring with the potential transfer rod and enter the third insulator of the grading ring. With one hand grasping the insulator tightly and the other holding the potential transfer rod, the equipotential electrician shall break himself away from the equipotential with the help of potential transfer rod. (3) The equipotential electrician exits the intense electric field by means of "two-span and three-short-circuit" operation mode	(1) The equipotential electrician must obtain the permission of the Responsible Person before exiting the potential. (2) When the equipotential electrician returns to the insulator string, the safety belt and backup protection rope shall be used alternately. The protection of safety belt shall not be lost during potential transfer, and the instant action shall be accurate, stable and rapid when leaving the intense electric field. (3) The equipotential electrician shall coordinate his hands and feet during leaving the potential with uniform speed so as to avoid excessive hand waving, big step and other actions. The side phase of the combined gap composed of the grounding body and the electrified body shall be greater than 6.7m (larger than 6.9m for medium phase). The minimum distance between the exposed part of human body and the electrified body should not be less than 0.5m. (4) When moving along the insulator string, the equipotential electrician shall use the backup protection rope to hold two insulator strings with the hands grasping firmly and the feet stepping firmly. (5) The minimum distance to adjacent conductors shall be greater than 6.9 m (7.2m for medium phase). (6) The equipotential electrician shall not lose the protection of safety belt or backup protection rope when returning to cross arm	

· 513 ·

Table (Cont'd)

S/N	Work Content	Operation Steps and Standards	Safety Measures and Precautions	Responsible Person
11	Return to the ground	After checking that there is no object left on the tower, the equipotential electrician on the tower shall report it to the Responsible Person and then climb down the tower with insulated transmission rope after obtaining the consent of the Responsible Person	The anti-falling device installed on the tower shall be used in the process of climbing down the tower; when moving and transposition on the pole and tower, the safety protection shall not be lost, and the operators must grasp the components securely	
12	End of the work	(1) The Responsible Person shall organize all working members to put working apparatus and materials in order and put them in a special kit (bag) after cleaning; clean the site to ensure that "the materials are removed and the site is cleaned after construction". (2) In the post-shift meeting, Responsible Person will give work summaries and comments. Comments include the construction quality of this work and the implementation of safety measures from all working personnel. (3) Responsible Person shall report to the on-duty control personnel that the work is over, apply for the restoration of circuit re-closing and terminate the work order.	It is forbidden to restore the line re-closing in the appointed time	

II. Assessment Standard

Detailed Rules for Assessment and Scoring of Operation and Inspection Skills of UHV AC Transmission Line

Fill-in Column of Examinee	No.:		Name:		Position:		Date:	MM/DD/YYYY
Fill-in Column of Assessor	Grade:		Assessor:		Assessment Team Leader:		Closing time:	Operation Duration:
Assessment module	Live replacement of 1000kV AC transmission line conductor vibration damper		Assessee	Maintenance personnel of UHV AC transmission line		Assessment method	Operation	Assessment Time Limit
								90min
Job Description	Enter the intense electric field along the strain insulator strings and carry out live replacement of the conductor vibration damper of 1000kV AC transmission line (equipotential operation method)							
Work Specifications and Requirements	1. Live working shall be carried out in good weather. In case of thunder, rain, snow or fog, no live working shall be carried out. When the wind force is greater than Level 5 and the humidity is greater than 80%, live working should not be carried out. 2. Person required for this operation include 1 Responsible Person, 1 special Supervisor, 1 ground electrician on the tower, 1 equipotential electricians and 2 ground auxiliary electricians. Enter the intense electric field along the insulator strings and carry out the live replacement of the eight-bundle spacer of the damaged conductor vibration damper of 1000kV AC transmission line. 3. Responsibilities of Responsible Person: Be responsible for division of operating personnel of the task, work order reading, handling re-closing deactivation of the line and the formalities of work permit, getting work permits, holding pre-shift meeting, dealing with emergency situations in work, quality surveillance, and the summary after work. 4. Special Supervisor: Be responsible for safety supervision and control during operation. 5. Equipotential electrician: Enter the intense electric field along the insulator strings and replace the vibration damper. 6. Responsibilities of electrician on tower: Be responsible for testing the null-value or low-value insulator and assisting equipotential in entering and exiting the intense electric field. 7. Responsibilities of ground electrician: Be responsible for implementing site safety measures, arranging work site, inspecting tools and instruments, delivering tools and materials, and cooperating with the equipotential electrician to enter and exit equipotential. 8. During the live working, if thunder, rain, strong wind or any other circumstance threaten the safety of the staff, the Responsible Person or Supervisor may stop working temporarily according to the circumstances.							

Part II Skill Module Training and Assessment Standards

Table (Cont'd)

Work Specifications and Requirements	Given conditions: 1. Training base: UHV AC 1000kV Training line tensile tower large side phase A sub-conductor vibration damper, model of vibration damper: FRYJ-4/6. 2. Work orders have been handled, safety measures have been completed (re-closing has been deactivated), and oral application (dispatcher or assessor) shall be made at the beginning and end of the work. 3. Safety fence shall be installed and signs such as "Work here" and "Access from Here" shall be hung at the work site. Safety measures have been completed. 4. The instruments and apparatus shall be used safely and correctly to test the insulating tool. 5. Fall-arrest device should be used in the process of up and down the tower to prevent falling from heights. 6. The operation must be carried out according to Standard Operation Procedures. The relevant item scores shall be deducted for the process error. In case of major hidden dangers of personal, equipment and operational safety, the assessor may order the termination of the operation (assessment) and the assessment of this module is recorded as "unqualified"
Assessment scenario preparation	1. Line: UHV AC 1000kV training line resisting-tensile tower large side phase A conductor, work content: live replacement of 1000kV UHV transmission line conductor vibration damper, and model of the vibration damper is: FRYJ-4/6. 2. Required operation tools: 2 insulated transmission ropes (TJS-12), 2 insulated backup protection ropes (TJS-16, with buffer provided), 2 insulated tackles (JH10-0.5), 2 insulated nooses (TJS-14), 1 potential transfer rod (0.4m), 2 Type I shielding clothes (coverall, helmet, protective mask, gloves and conductive socks), 2 pairs of conductive shoes, 2 falling protectors, 1 special spanner for vibration damper, 1 insulator detector, 1 set of insulation resistance tester (type 5,000V), 1 set of wind speed and humidity tester (HT-8321), 1 multimeter, 2 Moisture-proof tarpaulins (2m*4m), 1 red waistcoat (for Responsible Person), 2 clean towels, 2 sets of personal tools (workbasket, flat tongs, marker pen), 1 vibration damper of same model and several aluminum armor tapes. 3. The work site shall be monitored, and the safety measures (fence, etc.) on the work site have been fully implemented; non-operation personnel are prohibited from entering the site, and the staff must wear safety helmets when entering the work site. 4. Examinees shall bring their own work clothes, flame retardant cotton underwear, safety helmets, gloves, safety belts (including double-protective ropes)
Remarks	1. The total score of this module is 100 points, and score of each item is deducted until the corresponding points are finished. If the task is not completed within the specified time, the test shall be terminated immediately. The score of this module shall be calculated according to the actual score of the completed item, and the unfinished item shall not be scored. 2. In the process of assessment, if the equipment, operating environment, safety measures, safety protection and safe distance do not meet the requirements of the operation, or human error occurs, which may endanger the safety of the operation, the assessor shall order the termination of the operation. 3. Exam participants should be organized for site survey before the examination and work orders should be handled in advance

Part II Skill Module Training and Assessment Standards

Table (Cont'd)

S/N	Project name	Quality requirements	Score	Deduction standard	Reasons for deduction	Deduction	Scoring
1	Site re-survey	1) The Responsible Person shall go to the work site to check the line name, pole and tower number, on-site working conditions, defective parts and so on. 2) Check that the site meteorological conditions such as wind speed and humidity should meet the operation requirements. 3) Check whether the work order is complete and unmodified, check whether the safety measures listed are consistent with the actual situation on site, and supplement it if necessary	5	1) No point shall be awarded without a work order. 2) Deduct 1 point for unchecked double title. 3) Deduct 1 point/item for failure to check on-site working conditions (meteorology) and defective parts. 4) Deduct 0.5 points/place for any alteration or frowziness on the work order, and deduct 1 point for incorrect work order number. Deduct 1 point/item for missing items in work order			
2	Work Permit	1) The Responsible Person is responsible for contacting the on-duty control personnel and applying for stopping the line re-closing as per the contents of the work order. 2) Report content is standardized and complete. The voice is loud and clear. 3) Relevant licensing procedures are completed in a timely manner	2	1) No point shall be awarded for items of work without the work permit of the Dispatching Department (Assessor). 2) Deduct 0.5 points/item for non-standard and incomplete reporting terminology or the voice is not loud or clear enough. 3) Deduct 1 point for failure to apply the re-closing lock. 4) Deduct 1 point for failure to repeat permission content. 5) Deduct 0.5 points / item for failure to repeat each permission content (name, time, task and re-closing lock status of the Licensor). 6) Deduct 1 point for failure to complete work order in time			

· 517 ·

Table (Cont'd)

S/N	Project name	Quality requirements	Score	Deduction standard	Reasons for deduction	Deduction	Scoring
3	Site layout	Install the security fence and hang the signboards correctly: 1) The security fence should take full account of falling objects from the heights and the influence on road traffic. 2) The entrance and exit of the security fence shall be set reasonably. 3) Signs such as "Access from Here", "Work Here", "Access from Here" shall be properly and well arranged	3	1) Deduct 2 points for failure to arrange the safety fence at the work site. 2) Deduct 1 point/place for not setting the safety fence at the work site properly. 3) Deduct 1.5 points for failure to hang signboards. 4) Deduct 0.5 points per piece for incomplete signboards. 5) Deduct 0.5 points/person for non-operating personnel entering the fenced area			
4	Hold a pre-shift meeting	1) All staff and personnel shall wear safety helmets and work clothes correctly. 2) Responsible Person shall wear red vest and read out the work order and be clear with work task and division of personnel; explain safety measures and technical measures in work; check (inquire after) mental state of all working personnel; inform of hazards in work and precontrol measures. 3) All working personnel shall sign on the work order for confirmation.	3	1) Deduct 0.5 points/person for the incorrect wear of safety helmet by working personnel. Deduct 0.5 points/person/time for improper dressing. 2) Deduct 0.5 points/person for the failure to wear safety red waistcoat by Responsible Person and Special Supervisor. 3) No point shall be awarded for unclear work task and division of work. 4) Deduct 1 point for unclear division of labor. 5) Deduct 1 point for incomplete explanation of safety measures or precontrol measures.			

Part II Skill Module Training and Assessment Standards

Table (Cont'd)

S/N	Project name	Quality requirements	Score	Deduction standard	Reasons for deduction	Deduction	Scoring
4	Hold a pre-shift meeting		3	6) Deduct 1 point for failure to inform hazards at work. 7) Deduct 1 point for failure to confirm mental status of work shift members. 8) Deduct 1 point for the work shift member failing to sign or signing incompletely			
5	Inspection of tools	1) The work shift members shall properly set the moisture-proof tarpaulin in the appropriate position. The tarpaulin shall be clean and dry. It is strictly prohibited to tread the tarpaulin. 2) The tools and instruments shall be classified and placed neatly on moisture-proof tarpaulin according to the requirements of the fixed location management; the insulated tools shall not be mixed with metal tools and materials; and the appearance inspection shall be done on tools, instruments and apparatus. 3) All kinds of tools shall be qualified and within the effective time by test. The surface of insulating tools shall not be worn, deformed and damaged, and the operation shall be flexible. Carry out segment insulation detection with such insulated tools with insulation resistance meter of 2,500V or above and with the resistance no less than 700 MΩ, and wipe it off with a clean dry towel.	7	1) Deduct 1 point for the inappropriate position of moisture-proof tarpaulin. 2) Deduct 0.5 points/time for stepping on moisture-proof tarpaulin. 3) Deduct 1 point for failure to classify or place tools and instruments at fixed position. 4) Deduct 1 point/item for failure to check qualified label and appearance of tools, instruments and apparatus. 5) Deduct 0.5 points per piece for missing items in inspection of tools, instruments, and apparatus. 6) Deduct 0.5 points per piece for incorrect inspection method of tools, instruments and apparatus. 7) Deduct 0.5 points per piece for failure to clean or wipe the hard insulating tools. 8) Deduct 0.5 points/time for holding insulating tools without wearing clean and dry cotton gloves.			

Table (Cont'd)

S/N	Project name	Quality requirements	Score	Deduction standard	Reasons for deduction	Deduction	Scoring
5	Inspection of tools	4) The ground potential and equipotential personnel on tower shall correctly wear a whole suit of qualified shield clothes and conductive shoes as required, with each part connected well, shall not wear chemical fiber clothes next to the skin in the shielding clothes and shall fasten safety belts; the Responsible Person shall carefully check and confirm whether they wear them correctly and all parts are well connected. 5) The full set of shielding clothes shall be tested with a multimeter, and the resistance value between the farthest two points shall not be more than 20Ω, and the resistance value of a single set shall not be more than 15Ω. 6) The electrician climbing the tower shall conduct the visual inspection on the safety belt, backup protection rope and falling protector, which shall pass the impulse test.		9) Deduct 1 point/item for the improper usage of testing instruments to test the tools, instruments and the full set of shielding clothes. 10) Deduct 2 points/person for electrician climbing the tower failing to wear the shielding clothes correctly or to check the connecting portion. 11) Deduct 1 point/item for failing to conduct visual inspection or impulse test (or incorrect way) for safety belt, backup protection rope and falling protector 12) Deduct 1 point/item for the Responsible Person failing to check or missing to check the safety protection equipment of electrician climbing the tower			

Part II Skill Module Training and Assessment Standards

Table (Cont'd)

S/N	Project name	Quality requirements	Score	Deduction standard	Reasons for deduction	Deduction	Scoring
6	Climbing the tower	1) Tower-climbing personnel shall check the double name, pole number and phase again, report to the Responsible Person and apply for climbing the tower. 2) The ground potential electrician on tower and equipotential electrician carry the insulated transmission rope to climb the tower one after another. 3) During tower climbing, the electrician must use anti-fall device, climb the tower at a uniform speed with the main materials in hand, hang the safety belt and back-up protection rope on the shoulder and keep the safety distance of more than 6 m away from the electrified body, and the Responsible Person shall strengthen the operation monitoring. 4) When climbing to an appropriate position, ground potential electrician shall fasten the safety belt, arrange the insulated transmission rope, and then cooperate with the ground electrician on tower to make lifting preparation of the insulated transmission rope separately. 5) The Responsible Person shall carefully monitor and remind the whole process of tower climbing	5	1) Deduct 1 point/item for the tower electrician failing to check the double name, tower number and phase. 2) Deduct 1 point/item for the tower electrician failing to report the result or apply for climbing the tower. 3) The assessor shall order the termination of the operation (assessment) if the tower electrician fails to use anti-fall device during climbing down. 4) Deduct 0.5 points/time for grasping the shackles with hands during climbing down. 5) Deduct 1 point/time for slipping or missing his step during climbing down. 6) Deduct 1 point/ person for main band of safety belt and backup protection rope failing to be hung on the shoulder. 7) Deduct 1 point/ time for safety belt and backup protection rope twining and getting caught. 8) Deduct 1 point/ time for fastening the safety belt and backup protection rope lower than operating personnel. 9) Deduct 1 point/ time for fastening the safety belt and backup protection rope on the same components.			

Table (Cont'd)

S/N	Project name	Quality requirements	Score	Deduction standard	Reasons for deduction	Deduction	Scoring
6	Climbing the tower		5	10) No point shall be awarded for losing protection of safety belt during work at heights. 12) Deduct 1 point for installation of tackle without insulating noose. 13) Deduct 1 point for the suspension position of the tackle transmission rope inconvenient for getting tools. 14) Deduct 1 point for the Responsible Person with improper supervision and reminding. 15) The Assessor shall order the termination of the operation (assessment) for insufficient safe distance			
7	Detection of insulator	1) The ground electrician shall transfer the insulating bar and insulator detector to the tower. The ground potential electrician shall carry out commissioning on the spark gap detector, with the gap distance of 0.4mm. 2) The equipotential electrician shall assist the ground potential electrician on tower to test the insulators string by string and piece by piece, and the contact must be reliable from the conductor side to the cross arm side.	10	1) Deduct 0.5 points if the knot is tied wrongly while transferring the tools and instruments. 2) Deduct 0.5 points for each piece in case of collision while transferring tools and instruments. 3) Deduct 0.5 points for winding of transmission rope 4) Deduct 1 point for falling object from high place while transferring. 5) No point shall be awarded if not testing the null value of insulator.			

Part II Skill Module Training and Assessment Standards

Table (Cont'd)

S/N	Project name	Quality requirements	Score	Deduction standard	Reasons for deduction	Deduction	Scoring
7	Detection of insulator	3) In case of null-value or low-value insulator during the detection, detection shall be carried out for 3 times. 4) When the number of any string of good insulators is less than 37, the testing shall be stopped immediately, and the live working shall be stopped. 5) The safe distance of at least 6m shall be maintained between the electrician on tower and electrified body, and the effective insulting length of insulating rope shall not be less than 6.8m. The minimum insulting length of insulated testing rod shall not be less than 6.8m. 6) After testing, the ground electrician shall cooperate with the electrician on tower to transfer the insulating bar to the ground, and transfer the potential transfer rod to the tower. 7) The Responsible Person shall carefully supervise and remind the operation process.	10	6) Deduct 2 points for incorrect operation methods of electrician on tower. 7) Deduct 2 points for wrong insulator string testing sequence. 8) Deduct 2 points for not re-testing the null-value insulator. 9) The assessor shall order the termination of the operation for insufficient safe distance of operation. 10) Deduct 2 points if not reporting the null value measurement result. 11) Deduct 2 points if not transferring the potential transfer rod. 12) Deduct 2 points for the Responsible Person failing to monitor the operation in place.			
8	Entering the intense electric field	1) Check that all parts of the shielding clothes are well connected before entering the potential, and report to the Responsible Person for approval. 2) The equipotential electrician shall transfer the safety belt to the insulator set fitting, and carry the potential transfer rod, insulated tackle and insulated transmission rope.		1) Deduct 2 points for the equipotential electrician failing to check the connecting parts of the shielding clothes. 2) Deduct 2 points for losing the protection of safety belt when equipotential electrician transfers to insulator set fitting.			

Standard for Professional Training and Assessment for Operation Maintenance of UHV AC Transmission Line

Table (Cont'd)

S/N	Project name	Quality requirements	Score	Deduction standard	Reasons for deduction	Deduction	Scoring
8	Entering the intense electric field	3) Before entering the insulator string, the equipotential electrician shall fasten the protection rope (holding two insulator strings with the backup protection rope, grasping one string with hands and stepping on the other string). 4) When reaching 3 insulators outside the grading ring on the conductor side along the insulator string by the operating method of two-span and three-short-circuit, the operator shall stop, wait for the approval of the Responsible Person and then carry out the potential transfer with the potential transfer rod. 5) The side phase combined gap composed of the gaps between the equipotential electrician and the grounding body & the electrified body in the process of entering the potential shall not be less than 6.7 m (not less than 6.9 m for medium phase), and the potential transfer rod must be used for potential transfer when entering the intense electric field. The minimum distance between the exposed part of human body and the electrified body should not be less than 0.5m. 6) Protection of safety belt shall not be lost when entering the intense electric field	10	3) Deduct 2 points for insufficient combined clearance between equipotential electrician and grounding body & electrified body. 4) Deduct 2 points for incorrect and unskilled action of equipotential electrician entering the intense electric field. 5) Deduct 2 points for repeated discharge due to insufficient distance of the exposed parts during potential transfer of equipotential electrician. 6) Deduct 1 point for being unskilled in potential transfer. 7) Deduct 5 points for potential transfer without using the potential transfer rod. 8) Deduct 3 points for carrying out potential transfer without permission of the Responsible Person. 9) Deduct 2 points for losing the protection of safety belt when equipotential electrician enters the intense electric field. 10) Deduct 2 points for the Responsible Person failing to monitor the operation in place.			

Part II Skill Module Training and Assessment Standards

Table (Cont'd)

S/N	Project name	Quality requirements	Score	Deduction standard	Reasons for deduction	Deduction	Scoring
9	Replacement of the sub-conductor vibration damper	1) After entering the equipotential, the equipotential electrician shall tie the pole belt to the upper sub-conductor, and install the insulated protection rope (all sub-conductors shall be covered). 2) The equipotential electrician shall carry the insulated transmission rope and walk to the operation point for replacing the vibration damper, and shall correctly install the insulating noose at the appropriate place of the sub-conductor and hang insulated tackle and insulated transmission rope. 3) The equipotential electrician shall mark on both ends of installation site of vibration damper with marker pen. 4) The equipotential electrician shall use the insulated transmission rope to fasten the old vibration damper by the way of slip knot, then use the special spanner for vibration damper to remove the old vibration damper, and puts it on the moisture-proof tarpaulin on ground with the insulated transmission rope in cooperation with the ground electrician.	30	1) Deduct 3 points for the incomplete covering of sub-conductor by insulated protection rope. 2) Deduct 1 point for failure to tie pole belt to upper sub-conductor during walking along the line. 3) Deduct 2 points for being unskilled in walking along the line. 4) Deduct 5 points for the insulated tackle hooked directly on the conductor. 5) Deduct 1 point/item for incorrect installation method or inappropriate position of insulating noose. 6) Deduct 1 point for failure to lock the insulated tackle after it is hooked on insulating noose. 7) Deduct 3 points for no marking. 8) Deduct 5 points for removing the vibration damper without tying the knot. 9) Deduct 2 points for removing tools and instruments in the hurry and confusion. 10) Deduct 3 points per piece for falling object from heights. 11) No point shall be awarded at this module for throwing old vibration damper at heights.			

· 525 ·

Table (Cont'd)

S/N	Project name	Quality requirements	Score	Deduction standard	Reasons for deduction	Deduction	Scoring
9	Replacement of the sub-conductor vibration damper	5) The equipotential electrician shall remove the aluminum armor tape and place it in the tool kit; The new aluminum armor tape shall be wound according to the drawing mark. The winding shall be smooth and tight, and the twisting direction shall be the same as that of the outer conductor. 6) The ground electrician shall lift the new vibration damper to the position of equipotential electrician. 7) The equipotential electrician should correctly install the new vibration damper and ensure that the vibration damper is vertically downward with the hammer ball parallel to the main wire. The installation displacement should not exceed ± 30mm; The both ends of the aluminum armor tape should be pressed back into the cord bracket of vibration damper, and both ends should be exposed 10mm. The spring washer at the bolt should be tight and flat, and the bolt direction should be consistent with other sub-conductor vibration damper.	30	12) Deduct 2 points/person/time for the ground electrician standing directly below the work point of equipotential electrician. 13) Deduct 1 point for failure to put old vibration damper on moisture-proof tarpaulin. 14) Deduct 1 point for incorrect way the knot is tied when passing tools up and down. 15) Deduct 5 points for untying the knot before the new vibration damper is installed. 16) Deduct 3 points when the new vibration damper deviates from the original position and exceeds the allowable value. 17) Deduct 3 points respectively when the vibration damper is not vertically downward or the hammer ball is not parallel to the main wire after installation. 18) Deduct 2 points when the both ends of the aluminum armor tape are not pressed back into the cord bracket of vibration damper after installation.			

Part II Skill Module Training and Assessment Standards

Table (Cont'd)

S/N	Project name	Quality requirements	Score	Deduction standard	Reasons for deduction	Deduction	Scoring
9	Replacement of the sub-conductor vibration damper	8) The equipotential electrician shall remove the insulated tackle, insulated transmission rope and noose successively. 9) No tools, instruments or materials shall drop during the equipotential operation	30	19) Deduct 2 points when the both ends of the aluminum armor tape expose less than 10mm after installation. 20) Deduct 1 point/place for wrong threading direction of vibration damper bolt. 21) Deduct 1 point/place for not removing insulated tackle, insulated transmission rope and noose. 22) Deduct 2 points for not removing insulated tackle, insulated transmission rope and noose in order. 23) No point shall be awarded at this module for not finishing the replacement of old vibration damper by new vibration damper. 24) The assessor shall order the termination of the operation for insufficient safe distance between adjacent conductors			
10	Exit the intense electric field	1) After inspecting that the vibration damper is installed firmly and there is no object left at the operation point, the equipotential electrician, with the permission of the Responsible Person, can return to the grading ring with insulated transmission along the conductor for the preparation to leave the potential.	10	1) Deduct 2 points per item for objects left in operation point. 2) Deduct 3 points for failure to apply to the Responsible Person for potential transfer.			

· 527 ·

Table (Cont'd)

S/N	Project name	Quality requirements	Score	Deduction standard	Reasons for deduction	Deduction	Scoring
10	Exit the intense electric field	2) The equipotential electrician shall hook the grading ring with the potential transfer rod and enter the third insulator of the grading ring. With one hand grasping the insulator tightly and the other holding the potential transfer rod, the equipotential electrician shall break himself away from the equipotential with the help of potential transfer rod. 3) Protection of safety belt shall not be lost when leaving the intense electric field. 4) The equipotential electrician leaves the intense electric field along the insulator string in the "two-span and three-short-circuit" operation mode and transfers to the cross arm. 5) The side phase combined gap composed of the gaps between the equipotential electrician and the grounding body & the electrified body in the process of leaving the potential shall not be less than 6.7m (not less than 6.9m for medium phase), and the potential transfer rod must be used for potential transfer when leaving the intense electric field. The minimum distance between the exposed part of human body and the electrified body should not be less than 0.5m	10	3) Deduct 2 points for carrying out potential transfer without permission of the Responsible Person. 4) Deduct 5 points for potential transfer without using the potential transfer rod. 5) Deduct 1 point for being unskilled in potential transfer. 6) Deduct 2 points for losing the protection of safety belt when equipotential electrician leaves the intense electric field. 7) Deduct 2 points for repeated discharge due to insufficient distance of the exposed parts during potential transfer of equipotential electrician. 8) Deduct 2 points for incorrect and unskilled action of equipotential electrician leaving the intense electric field. 9) Deduct 2 points for insufficient combined clearance between equipotential electrician and grounding body & electrified body. 10) Deduct 2 points for losing the protection of safety belt when equipotential electrician transfers to the cross arm. 11) Deduct 2 points for the Responsible Person failing to monitor the operation in place.			

Part II Skill Module Training and Assessment Standards

Table (Cont'd)

S/N	Project name	Quality requirements	Score	Deduction standard	Reasons for deduction	Deduction	Scoring
11	Return to the ground	1) After checking that there is no object left on the tower, the electrician on the tower shall report it to the Responsible Person and then climb down the tower one after another with insulated transmission rope after obtaining the consent of the Responsible Person. 2) When climbing downing the tower, the electrician must use anti-fall device, climb down the tower at a uniform speed with the main materials in hand, hang the safety belt and backup protection rope on the shoulder and keep the safe distance of more than 6 m away from the electrified body, and the Responsible Person shall strengthen the operation monitoring.	15	1) Deduct 2 points per item for objects left on tower. 2) Deduct 1 point/item for the tower electrician failing to report the inspection results of the leftovers or apply for climbing the tower. 3) The assessor shall order the termination of the operation (assessment) if the tower electrician fails to use anti-fall device during climbing down. 4) Deduct 0.5 points/time for grasping the shackles with hands during climbing down. 5) Deduct 1 point/time for slipping or missing his step during climbing down. 6) Deduct 1 point/ person for main band of safety belt and backup protection rope failing to be hung on the shoulder			
12	End of the work	1) The Responsible Person shall organize all working members to put working apparatus and materials in order and put them in a special kit (bag) after cleaning; clean the site to ensure that "the materials are removed and the site is cleaned after construction".	10	1) Deduct 0.5 points per piece for failure to clean or wipe the insulating tools and instruments. 2) Deduct 1 point per piece for failure to classify or place tools and instruments at fixed position. 3) Deduct 0.5 points/time for throwing about tools and instruments or stepping on moisture-proof tarpaulin.			

Table (Cont'd)

S/N	Project name	Quality requirements	Score	Deduction standard	Reasons for deduction	Deduction	Scoring
12	End of the work	2) In the post-shift meeting, Responsible Person will give work summaries and comments. Comments include the construction quality of this work and the implementation of safety measures from all working personnel. 3) Responsible Person shall report to the on-duty control personnel that the work is over, apply for the restoration of circuit re-closing and terminate the work order.	10	4) Deduct 1 point per piece for not removing fence and signs or leftovers. 5) Deduct 2 points for failure to hold the post-shift meeting. 6) Deduct 1 point when the team is not orderly or focused. 7) Deduct 1 point/person for members of the working group not attending post-shift meeting. 8) Deduct 1 point for unqualified comments. 9) Deduct 1 point for failure to report to the Dispatching Department (Assessor) the end of the work and apply for restoration of re-closing lock. 10) Deduct 0.5 points/item for non-standard and incomplete reporting terminology or the voice is not loud enough. 11) Deduct 1 point for failure to repeat permission content. 12) Deduct 0.5 points/item for missing items during repeating the contents (name of working unit, name of the Responsible Person, time, name of the line, work completion, device has restored to normal, the personnel have evacuated and re-closing lock can be restored). 13) Deduct 1 point/ item for failure to complete in time or error in the fill-in of work order termination procedure			
	Total		100				

Module 9 Standards for Training and Assessment on Live Repair of 1000kV AC Transmission Line Conductor

Ⅰ. Training Standard

(Ⅰ) Training Requirements

Designation of module	Live repair of 1000kV AC transmission line conductor	Type of training	Operation
Training method	Practical operation training	Hours of training	14 training hours
Training objectives	1. Master the electrical significance of "two-span and three-short-circuit" operation mode in entering and exiting 1000kV intense electric field along strain insulator string. 2. Can complete the entry of 1000kV equipotential operation point along the strain insulator string. 3. Be able to repair the conductor with pre-twisted armor rod independently (equipotential operation method)		
Training venue	UHV AC training line		
Training content	The "two-span and three-short-circuit" operation mode is adopted to enter the strong electric field along the strain insulator string, and the equipotential operation method is adopted for live repair of the 1000kV AC transmission line conductor		
Scope of application	Maintenance personnel of UHV AC transmission line		

(Ⅱ) Referenced Rules and Specifications

(1) Code for Design of 1000kV Overhead Transmission Line (GB50665-2011).

(2) Maintenance Code for 1000kV AC Transmission Line (DL/T209-2008).

(3) Operation Code for 1000kV AC Transmission Line (DL/T307-2010).

(4) Technical Guide for Live Working on 1000kV AC Transmission Line (DL/T392-2015).

(5) Calculation Method of Live Working Minimum Approach Distance on AC Transmission Line (GB/T 19185-2008).

(6)Guidelines of Insulation Coordination for Live Working (DL/T867-2004).

(7)Live Working-Guidelines for the Installation of Power Transmission Line Conductors and Earthwires-stringing Equipment and Accessory Items (DL/T 1007-2006).

(8)State Grid Corporation of China on the Management Regulations of Live Working (Trial Implementation) (SGCC [2007] No.751).

(9)State Grid Corporation of China Working Regulations of Power Safety (Transmission Line Section) (Q/GDW1799.2-2013).

(10)Electrotechnical Terminology - Overhead Line (GB/T 2900.51-1998).

(11)Electrotechnical Terminology- Live Working (GB/T2900.55-2002).

(12)Live Working - Terminology for Tools, Equipment and Devices (GB/T 14286-2002).

(13)Minimum Requirements for Utilization of Tools, Devices and Equipment for Live Working (DL/T 877-2004).

(14)Preventive Test Code of Tools, Devices and Equipment for Live Working (DL/T 976-2005).

(15)Insulated Tackles for Live Working (GB/T13034-2008).

(16)Live Working-Insulating Ropes (GB/T 13035-2008).

(17)Screen Clothes for Live Working on 1000kV AC (GB/T25726-2010).

(Ⅲ) Teaching Design for Training

To complete the live repair of 1000kV AC transmission line conductor is the Work task during the design. Each training stage shall be designed according to the standard operation procedure for work task completion. Each stage includes specific training objectives, training content, hours of training, training methods (training resources), training environment, assessment and evaluation, etc, as shown in Table 2-9-1.

(Ⅳ) Operation Flow

1. Work Task

The "two-span and three-short-circuit" operation mode is adopted to enter the intense electric field and reach the operation point along the strain insulator string, and the equipotential operation method is adopted for live repair of the conductor of 1000kV AC transmission line.

Part II Skill Module Training and Assessment Standards

Table 2-9-1 Training Content Design for Live Repair of Conductor of 1000kV AC Transmission Line

Training schedule	Training objectives	Training content	Hours of training	Training methods and resources	Preparation of training conditions	Assessment and evaluation
1. Theoretical teaching	1. Preliminarily master the basic method for entering and exiting 1000kV intense electric field along the insulator string; 2. Be familiar with the methods for potential transfer. 3. Be familiar with the method for repairing the damaged conductor of transmission line,	1. The electrical significance of the "two-span and three-short-circuit" operation mode when entering and leaving the intense electric field along the insulator. 2. The use method for the potential transfer rod during entering and exiting the UHV intense electric field. 3. Repair method and quality standard for transmission line conductor	2	Training methods: Lecture. Training resources: PPT, relevant regulations and specifications	Multimedia classroom	Attendance, classroom questions and assignments
2. Preparations	Be able to complete the preparation before operation	1. Work site survey. 2. Preparation of the standardized operation card. 3. Filling of the work order. 4. Preparation of tools and materials for this operation	1	Training methods: 1. Work site survey and cleaning of tools and materials shall be practiced at site; 2. Preparation of operation card and the filling of work order shall adopt lecture method. Training resources: 1. 1000kV practical training line. 2. UHV tools warehouse. 3. Blank work order	1. UHV training transmission line 2. Multi-Media classroom	

Table (Cont'd)

Training schedule	Training objectives	Training content	Hours of training	Training methods and resources	Preparation of training conditions	Assessment and evaluation
3. Work site preparation	Be able to complete the preparations of work site	1. Work site re-survey. 2. Job application. 3. Work site layout. 4. Pre-shift meeting. 5. Inspection of tools and materials	1	Training methods: demonstration and role play. Resources: 1000kV training line	1000kV training line	
4. Trainer's demonstration	The trainees can preliminarily understand the operation process of the task through inspecting and learning from each other's work	1. Detection of null-value insulator. 2. The equipotential electrician enters and exits the intense electric field along the strain insulator string. 3. The equipotential electrician reaches the conductor repair position by means of walking along the line. 4. Equipotential electrician uses the preformed armor rods to complete the repair of conductor	2	Training methods: Demonstration. Resources: 1000kV training line	1000kV training line	

Part II Skill Module Training and Assessment Standards

Table (Cont'd)

Training schedule	Training objectives	Training content	Hours of training	Training methods and resources	Preparation of training conditions	Assessment and evaluation
5. Group training of trainees	1. Can complete the operation of entering and exiting 1000kV intense electric field. 2. Be able to complete the 1000kV transmission line conductor repair	1. The trainees are grouped (6 persons per group) to train the skill operation of entering and exiting 1000kV intense electric field and conductor repair. 2. Trainers guide the operation of trainees and conduct safety supervision	7	Training methods: Role play. Resources: 1000kV training line	1000kV training line	Score the operation of trainees according to the detailed rules for skill assessment and scoring
6. End of the work	1. Enable the trainees to further distinguish the shortcomings of the operation process and facilitate the promotion in the later stage. 2. Train the trainees in the working style of safe and civilized production	1. Cleaning up the work site. 2. Report to dispatcher. 3. Comment and summarize the work task this time at post-shift meeting	1	Training method: Lecture and inductive method	1000kV training line	

2. Requirements for Weather and Work Site

(1) It should be carried out in good weather for live repair of 1000kV AC transmission line conductor.

In case of lightning (hearing thunder or seeing lightning), snow, hail, rain, fog and so on, live working is prohibited. When the wind force is greater than level 5, or the relative humidity of the air is greater than 80%, it is unsuitable for live working; when emergency live repair is required in bad weather, relevant personnel shall be organized to fully discuss and prepare necessary safety measures, which can be implemented after being approved by the unit.

(2) The operating personnel shall be in good spirits, and be familiar with the organizational and technical measures to ensure safety in work; they shall hold the live working qualification within the validity period.

(3) The Responsible Person shall organize relevant personnel to complete field investigation in advance, determine the operating methods, required working apparatus and necessary measures according to the results, and handle the work orders for live working.

(4) The work site shall be reasonably set up with fence and warning signs. Non-operating personnel is forbidden to enter.

(5) The line re-closing device shall be deactivated in the Project.

(6) Safety working distance and effective insulation length are shown in Table 2-9-2.

Table 2-9-2 Safe Distance for Live Repair of Conductor of 1000kV AC Transmission Line

Voltage class	Safe distance between human body and electrified body /m	Minimum effective insulation length/m		Minimum combined clearance /m	Minimum distance between exposed part of human body and electrified body when potential is transferred /m
		Insulating bar	Insulated load-bearing tools and insulating ropes		
1000kV	6.8	6.8	6.8	6.9	0.5

(7) For operation on 1000kV transmission line, the number of insulators in good operation phase shall be no less than 37 pieces.

3. Preparations

3.1 Hazards and precontrol measures

(1) Hazard - electric shock injury

Precontrol measures:

① Before work, Responsible Person should contact the control personnel on duty, deactivate the line re-closing, and perform the licensing procedures.

② Before climbing the tower, ground potential operators on tower must carefully check the name of the line, the number of the pole and tower, and phase, and then the tower can be climbed after all have been confirmed correct.

③ If lines lose power suddenly during work, operators shall still deem it to be charged. The Responsible Person shall contact the control personnel as soon as possible, and no forced energization is allowed before the on-duty control personnel getting in touch with the Responsible Person.

④ Insulating tool and insulating ropes shall be free of damage, moisture, deformation, and failure. It is not allowed to use non-insulating ropes (such as cotton rope, manila rope, and steel wire rope).

⑤ The equipotential operator shall wear flame-retardant underwear and full set of shielding clothes outside (including hat, dresses & trousers, gloves, socks and shoes). All parts shall be in excellent connection conditions.

⑥ Before potential transfer, equipotential operator shall obtain the approval of Responsible Person, and the minimum distance between exposed part of human body and electrified body shall not be lower than 0.5m. During potential transfer, the operation shall be fast, and end shall not be used for power charging and discharging; During delivering tools and materials to ground potential operators, the effective length of insulating tools or insulating ropes shall not be lower than the requirements as specified in Table 2-9-2.

⑦ When transmitting large metal objects by using insulating ropes, ground potential operators shall not touch them before ground connection.

⑧ The special Supervisor shall continuously monitor the operators and correct their nonstandard operation or actions in violation at any time. Special attention shall be paid to operators work at heights to ensure that there is enough safety distance (meeting the requirements in Table 2-9-2). It is forbidden to contact two non-connected electrified bodies or make contact with electrified body and grounding body at the same time.

(2) Hazard - precontrol measures for falling accident:

① Before climbing, operators work at heights must satisfy the requirements of this operation, such as physical condition, mental state and skill and quality.

② Supervisor shall correct irregularities and violations at any time. Special attention shall be paid to operators to prevent them from losing the protection of the safety belt or insulated backup protection rope during transposition. The safety belt or insulated backup protection rope shall not be fastened lower than operating personnel.

(3) Hazard - injury caused by objects falling from heights.

Precontrol measures:

① Operator working at heights should put personal tools and fragmentary materials into the tools bag. It is strictly forbidden to hang objects in heights or keep in the mouth.

② Ground operator should correctly wear a helmet and use the knots. The vertical distance from the work site should not be less than the falling radius.

③ The work site shall be set up with fence and warning signs. It shall be noted at any time that Supervisor shall prohibit irrelevant personnel and vehicles from entering operation area.

3.2 Selection of tools and instruments and materials

Tools and materials required for live repair of conductor for 1000kV AC transmission line can be seen in Table 2-9-3. Before delivering tools and instruments out of warehouse, application voltage class and test period shall be carefully checked and they shall be inspected to ensure that appearance is intact, connection is firm, rotation is flexible and meet the working task requirements. After delivering tools and instruments out of warehouse, they shall be stored in tools bag or tool kit for transportation to avoid contamination and damp. Metal tools and insulated tools shall be separately loaded and transported to avoid deformation, damage or other defects caused by mixed loading and transportation.

Table 2-9-3 Tools, Instruments and Materials Required by Live Repair of Conductor of 1000kV AC Transmission Line

S/N	Name	Specification	Unit	Qty.	Remarks
1	Insulated transmission rope	TJS-12	Nos.	2	
2	Insulated protection rope	TJS-16	Nos.	2	
3	Insulator detector		Set	1	
4	Insulated tackle	JH10-1	Nos.	2	
5	Safety helmet		Nos.	6	
6	Potential transfer rod		Nos.	1	
7	Insulation resistance meter	5000V	Piece	1	
8	Anemometer		Piece	1	
9	Temperature and humidity meter		Piece	1	
10	Multimeter		Piece	1	

Part II Skill Module Training and Assessment Standards

Table (Cont'd)

S/N	Name	Specification	Unit	Qty.	Remarks
11	Moisture-proof canvas	2m×4m	Piece	2	
12	Insulated noose		Nos.	4	
13	Shielding clothes	Shielding efficiency $\geqslant$ 60 dB (shield efficiency of shielding mask $\geqslant$ 20 dB)	Set	2	
14	Falling protector	Corresponding to the type of pole and tower falling protector	Nos.	2	
15	Safety belt		Nos.	2	
16	Security fence		Set	Several	
17	Warning sign	"Work Here", "Access from Here""Access from Here"	Set	1	
18	Red waistcoat	"Responsible Person"	Piece	1	
19	Preformed armor rod		Set	1	
20	Conductive paste		Box	1	
21	Sandpaper		Sheet	1	
22	Clean towel		Nos.	1	
23	Interphone		Set	4	
24	lever		Nos.	1	

3.3 Division of labor for operators

Division of labor for operators of the task is shown in Table 2-9-4.

Table 2-9-4 Division of Operating Personnel Involved in Live Repair of 1000kV AC Transmission Line Conductor

S/N	Post	Qty. (person)	Responsibilities
1	Responsible Person	1	Be responsible for division of operating personnel of the task, work order reading, handling re-closing deactivation of the line and the formalities of work permit, getting work permits, holding pre-shift meeting, dealing with emergency situations in work, quality surveillance, and the summary after work
2	Special Supervisor	1	Be responsible for the safety control of the work site
3	Equipotential electrician	1	Be responsible for entering the equipotential to repair the conductor
4	Ground potential electrician	1	Be responsible for testing the insulator and assisting equipotential in entering and exiting the intense electric field.
5	Ground electrician	2	Be responsible for transferring tools and materials and cooperating with equipotential electrician in entering and exiting the equipotential

4. Work Procedure

The workflow of this task is shown in Table 2-9-5.

Table 2-9-5 Workload Chart for Live Repair of 1000kV AC Transmission Line Conductor

S/N	Work Content	Operation Steps and Standards	Safety Measures and Precautions	Responsible Person
1	Site re-survey	The Responsible Person shall complete the following work: (1) Check the line title, the number of the pole and tower and ensure the phases are correct; guarantee that the foundation and the pole and tower are intact and in normal condition; ensure that the cross and span distance meets the safety requirements; confirm the defect conditions and the specifications and models of earth wires. (2) Check that the site meteorological conditions such as wind speed and humidity should meet the operation requirements. (3) Check that the terrain and environment shall meet the operation requirements. (4) Check that the safety measures listed in the work order are in line with the actual situations on site, and the measures will be supplemented if necessary	(1) Correctly wear helmet, working clothes, work shoes and protective gloves. (2) Operation under meteorological conditions that may endanger the safety of operators is forbidden. (3) Non-operation personnel and vehicles are strictly prohibited from entering the working site.	
2	Work Permit	(1) The Responsible Person is responsible for contacting the on-duty control personnel and applying for stopping the line re-closing as per the contents of the work order. (2) Live working could be started only after being approved by the on-duty control personnel.	Live working shall not be started without the permission of the on-duty control personnel	
3	Site layout	Install the security fence and hang the signboards correctly: (1) The security fence should take full account of falling objects from the heights and the influence on road traffic. (2) The entrance and exit of the security fence shall be set reasonably. (3) Signs such as "Access from Here", "Work Here", "Access from Here" shall be properly and well arranged	When the influence on road traffic safety is uncontrollable, the traffic management department should be contacted in time to strengthen the on-site control of traffic safety.	

Part II Skill Module Training and Assessment Standards

Table (Cont'd)

S/N	Work Content	Operation Steps and Standards	Safety Measures and Precautions	Responsible Person
4	Hold a pre-shift meeting	(1) All working personnel shall line up. (2) The Responsible Person shall wear red waistcoat and read out the work order and be clear with work task and division of personnel; explain safety measures and technical measures in work; check (inquire after) mental state of all working personnel; inform of hazards in work and precontrol measures. (3) All working personnel shall sign on the work order for confirmation.	(1) The work order shall be filled in, issued and approved in a standardized manner, and the signature shall be complete. (2) All working personnel shall be in good mental states. (3) All working personnel shall be clear with task division of works, safety measures and technical measures	
5	Inspection tool	(1) The ground electrician and equipotential electrician shall wear the shielding clothes in a right way and pass the inspection, which shall be supervised and inspected by the Responsible Person. (2) Wear personal safety equipment correctly (proper size and easy lock), and the Responsible Person shall supervise and inspect it. (3) Measure the wind speed, wind direction and humidity, check the insulation performance of insulating tools, and make records	(1) Check carefully for damage, deformation and failure before using metal and insulating tools. Carry out segment insulation detection with such insulated tools with insulation resistance meter of 2,500V or above and with the resistance no less than 700 MΩ, and wipe it off with a clean dry towel. (2) Use a multimeter to measure the resistance between the farthest ends of the shielding clothes and trousers, which shall not be greater than 20Ω. The Responsible Person shall check the connection of the electrician's shielding clothing. (3) Check the tool assembly and make sure the connection is reliable. (4) The live working tools used at site shall be placed on the moisture-proof canvas	

Table (Cont'd)

S/N	Work Content	Operation Steps and Standards	Safety Measures and Precautions	Responsible Person
6	Climbing the tower	(1) After checking the line name and pole and tower number without error, the ground electrician shall check the stress of the safety belt and the falling protector in the equipotential shock. (2) The ground electrician carries the insulated transmission rope to climb the tower, and the equipotential electrician then climbs the tower. When they reach the operation point of the cross arm, they choose the appropriate position to fasten the safety belt, and the ground electrician installs the insulated tackle and the insulated transmission rope in the appropriate position of the cross arm. Then he shall cooperate with the ground electrician to separate the insulated transmission rope for lifting preparation	(1) After checking the correct line name and pole and tower number, the tower can be climbed for operation. (2) The anti-falling device installed on tower shall be used in the process of climbing the tower; when moving and transposition on the pole and tower, the safety protection shall not be lost, and the operators must climb and grasp the components securely. (3) The working electrician must wear a whole suit of qualified shielding clothes which must be connected reliably. Before the cross arm enters the equipotential, the equipotential electrician shall check and confirm that each part of the shielding clothes are connected reliably before the next operation	
7	Detection of insulator	(1) The ground electrician shall transfer the insulating bar and insulator detector to the tower. The equipotential electrician shall assist the ground potential electrician in detecting the insulator. (2) Testing order shall be from the cross arm side to the electrified side, and records shall be made properly. (3) After testing, the ground electrician shall cooperate with the electrician on tower to transfer the insulating bar to the ground, and transfer the potential transfer rod to the tower	(1) No testing is required if the glass insulator is confirmed in good status; and the following items (2) and (3) shall be conducted for porcelain insulator. (2) The insulators must be detected string by string and piece by piece, and the contact must be reliable. (3) When the number of any string of good insulators is less than 37, the testing shall be stopped immediately, and the live working shall be stopped	

Part II Skill Module Training and Assessment Standards

Table (Cont'd)

S/N	Work Content	Operation Steps and Standards	Safety Measures and Precautions	Responsible Person
8	Entering the intense electric field	(1) The equipotential electrician shall transfer the safety belt to the insulator set fitting, and carry the potential transfer rod, insulated tackle and insulated transmission rope. (2) After the equipotential electrician confirms that all parts of the shielding clothes are well connected, it shall be reported to the Responsible Person for approval. The electrician shall grasp one string with both hands and step on the other string with both feet, and enter the equipotential along the insulator string in the operation mode of "two-span and three-short-circuit". (3) When reaching 3 insulators outside the grading ring on the conductor side, the operator shall stop, and carry out the potential transfer with the potential transfer rod	(1) The equipotential electrician must obtain the permission of the Responsible Person before entering the potential. (2) Before entering the insulator string, the equipotential electrician shall fasten the protection rope (holding insulator strings with the backup protection rope and stepping on a insulator string) and adjust the insulated transmission rope and the potential transfer rod. (3) The combined gap composed of the gaps between the equipotential electrician and the grounding body and the electrified body while entering the potential shall not be less than 6.9 m (medium phase) (6.7 m, side phase).	
9	Surface treatment of damaged conductor	(1) After entering the equipotential, the equipotential electrician shall tie the safety belt to the upper sub-conductor, and install the insulated protection rope (all sub-conductors shall be covered). (2) The equipotential electrician shall carry the insulated transmission rope and walk to the operation point along the conductor, and shall install the insulated tackle and the insulated transmission rope at the sub-conductor. (3) The electrician shall check the damage of the conductor and deal with the damage point with abrasive paper (0#) to grind the burr of the damaged part. (4) Equipotential electrician shall clean the ground conductor with a cleaning cloth, and apply the live paste evenly on the surface of the damaged part of the conductor	(1) Equipotential electrician shall not use much strength to grind the damaged point of the conductor in case of larger damage. (2) The surface should be thoroughly cleaned after the conductor is ground. (3) The conductive paste shall be applied evenly on the surface of the conductor	

Table (Cont'd)

S/N	Work Content	Operation Steps and Standards	Safety Measures and Precautions	Responsible Person
10	Conductor repair	(1) The ground electrician shall pass the preformed armor rods to the equipotential electrician by using an insulated transmission rope. (2) Equipotential electrician uses the preformed armor rods to repair and strengthen the damaged part of conductor	(1) The specification and model of preformed armor rod should match the conductor. (2) The center of the preformed armor rod should be located at the most serious damage. (3) The length of the preformed armor rod shall be enough to fully cover the damaged part. The unilateral length of the end of the preformed armor rod from the edge of damaged part shall not be less than 100 mm. (4) Bind preformed armor rod firmly to ensure no dumping, missing or loose strands	
11	Exit the intense electric field	(1) After inspection, the damaged strip line has been well strengthened and there is no residue at the operation point, the Responsible Person shall authorize the equipotential electrician to carry the insulated transmission rope back to the grading ring and prepare for leaving potential by means of walking along the line. (2) The equipotential electrician shall hook the grading ring with the potential transfer rod and enter the third insulator of the grading ring. With one hand grasping the insulator tightly and the other holding the potential transfer rod, the equipotential electrician shall break himself away from the potential with the help of potential transfer rod. (3) The equipotential electrician exits the equipotential according to the operation mode of "two-span and three-short-circuit".	(1) The equipotential electrician must obtain the permission of the Responsible Person before exiting the potential. (2) The combined gap composed of the gaps between the equipotential electrician and the grounding body and the electrified body while exiting the potential shall not be less than 6.9 m (medium phase) (6.7 m, side phase). (3) The equipotential electrician shall move along the insulator string with the hands grasping firmly and the feet stepping firmly	

Part II　Skill Module Training and Assessment Standards

Table (Cont'd)

S/N	Work Content	Operation Steps and Standards	Safety Measures and Precautions	Responsible Person
12	Return to the ground	After checking that there is no object left on the tower, the equipotential electrician on the tower shall report it to the Responsible Person and then climb down the tower with insulated transmission rope after obtaining the consent of the Responsible Person	The anti-falling device installed on the tower shall be used in the process of climbing down the tower; when moving and transposition on the pole and tower, the safety protection shall not be lost, and the operators must grasp the components securely	
13	End of the work	(1) The Responsible Person shall organize all working members to put working apparatus and materials in order and put them in a special kit (bag) after cleaning; clean the site to ensure that "the materials are removed and the site is cleaned after construction". (2) In the post-shift meeting, Responsible Person will give work summaries and comments. Comments include the construction quality of this work and the implementation of safety measures from all working personnel. (3) Responsible Person shall report to the on-duty control personnel that the work is over, apply for the restoration of circuit re-closing and terminate the work order.	It is forbidden to restore the line re-closing in the appointed time	

· 545 ·

II. Assessment Standard

Detailed Rules for Assessment and Scoring of Operation and Inspection Skills of UHV AC Transmission Line

Fill-in Column of Examinee	No.:		Name:		Position:		Date:	MM/DD/YYYY
Fill-in Column of Assessor	Grade:		Assessor:		Assessment Team Leader:	Starting time:	Closing time:	Operation Duration:
Assessment module	Live repair of 1000kV AC transmission line conductor		Assessee		Maintenance personnel of UHV AC transmission line	Assessment method	Operation	Assessment Time Limit
								60min
Job Description	Enter the intense electric field along the strain insulator strings and carry out live repair of the damaged conductor of 1000kV UHV transmission line							
Work Specifications and Requirements	1. Live working shall be carried out in good weather. In case of thunder, rain, snow or fog, no live working shall be carried out. When the wind force is greater than Level 5 and the humidity is greater than 80%, live working should not be carried out. 2. Workers required for this operation include 1 Responsible Person, 1 Special Supervisor, 1 equipotential electrician, 1 equipotential electrician and 2 ground auxiliary electricians. They should enter the intense electric field along the insulator strings and carry out the live repair of the damaged sub-conductor of transmission line. 3. Responsibilities of Responsible Person: Be responsible for division of operating personnel of the task, work order reading, handling re-closing deactivation of the line and the formalities of work permit, getting work permits, holding pre-shift meeting, dealing with emergency situations in work, quality surveillance, and the summary after work. 4. Special Supervisor: Be responsible for safety control of the work site. 5. Equipotential electrician: Enter the intense electric field along the insulator strings to replace the damaged conductor. 6. Responsibilities of ground potential electrician: Be responsible for testing the insulator and assisting equipotential in entering and exiting the intense electric field. 7. Responsibilities of ground electrician: Be responsible for transferring tools and materials and cooperating with equipotential electrician in entering and exiting the equipotential. 8. During the live working, if thunder, rain, strong wind or any other circumstance threaten the safety of the staff, the Responsible Person or Supervisor may stop working temporarily according to the circumstances.							

Table (Cont'd)

Work Specifications and Requirements	Given conditions: 1. Training line: some sub-conductor of UHV AC 1000kV practical training line phase A eight-bundle conductor, and the model of the conductor is: 8×JL/G1A-630/45. 2. Work orders have been handled, safety measures have been completed (re-closing has been deactivated), and oral application (dispatcher or assessor) shall be made at the beginning and end of the work. 3. The instrument shall be used safely and correctly to test the insulating tool. 4. The operation must be carried out according to the working procedures. The scores of the items to be carried out shall be deducted for the process error. In case of major hidden dangers of personal, equipment and operational safety, the assessor may order the termination of the operation (assessment)
Assessment scenario preparation	1. Line: UHV AC 1000kV training line phase A sub-conductor, work content: live repair of 1000kV UHV transmission line conductor, and model of the conductor is: 8×JL/G1A-630/45. 2. Required work tools: 1 insulated transmission rope (TJS-12), insulated protection ropes (TJS-16), 1 insulated tackle (JH10-1), insulation detector, 1 potential transfer rod, insulation resistance meter (5000V), 2 suits of shielding clothes (shielding efficiency ≥ 60dB), 1 multimeter, 1 tarpaulin, 1 insulation tester, 1 temperature and humidity meter, 1 anemometer, 2 cotton towels, 1 preformed armor rod, 1 # 0 abrasive paper, 1 wood hammer, and 1 operation rod. 3. The work site shall be monitored, and the safety measures (fence, etc.) on the work site have been fully implemented; non-operation personnel are prohibited from entering the site, and the staff must wear safety helmets when entering the work site. 4. Examinees shall bring their own work clothes, flame retardant cotton underwear, safety helmets, gloves, safety belts (including double-protective ropes)
Remarks	1. The deduction shall be done until the scores of each item are deducted completely. In case of major hidden dangers of personal, equipment and operational safety, the assessor may order the termination of the operation. 2. When equipment, working environment, safety belt, safety helmet, tool, shielding clothes, etc., do not conform to the operation condition, the assessor may order the termination of the operation

Table (Cont'd)

S/N	Project name	Quality requirements	Score	Deduction standard	Reasons for deduction	Deduction	Scoring
1	Site re-survey	1) The Responsible Person shall go to the work site to check the line name, pole and tower number, on-site working conditions, defective parts and so on. 2) Check that the site meteorological conditions such as wind speed and humidity should meet the operation requirements. 3) Check whether the work order is complete and unmodified, check whether the safety measures listed are consistent with the actual situation on site, and supplement it if necessary	5	1) Deduct 1 point/item for failure to check the line name, pole and tower number, on-site working conditions and defective parts. 2) Deduct 1 point/item for failure to detect wind speed, humidity and other on-site meteorological conditions. 3) Deduct 0.5 points/part for any alteration in the work order filling; deduct 1 point for incorrect work order number; deduct 1.5 points for incomplete work order filling			
2	Work Permit	1) The Responsible Person is responsible for contacting the on-duty control personnel and applying for stopping the line re-closing as per the contents of the work order. 2) Reporting content is standardized and complete	2	1) Deduct 2 points for failure to contact the dispatching department (referee) for deactivating the re-closing. 2) Deduct 1 point for non-standard or incomplete terminology reporting			

Part II Skill Module Training and Assessment Standards

Table (Cont'd)

S/N	Project name	Quality requirements	Score	Deduction standard	Reasons for deduction	Deduction	Scoring
3	Site layout	Install the security fence and hang the signboards correctly: 1) The security fence should take full account of falling objects from the heights and the influence on road traffic. 2) The entrance and exit of the security fence shall be set reasonably. 3) Signs such as "Access from Here", "Work Here", "Access from Here" shall be properly and well arranged	3	1) Deduct 1 point for failure to arrange the fence at the work site. 2) Deduct 1 point for failure to arrange the warning board. 3) Deduct 1 point for failure to hang the tower climbing operation sign			
4	Hold a pre-shift meeting	1) All staff and personnel shall wear safety helmets and work clothes correctly. 2) Responsible Person shall wear red vest and read out the work order and be clear with work task and division of personnel; explain safety measures and technical measures in work; check (inquire after) mental state of all working personnel; inform of hazards in work and precontrol measures. 3) All working personnel shall sign on the work order for confirmation.	3	1) Deduct 0.5 points/person for the staff not dressing uniformly. 2) Deduct 3 points for no division of labor; deduct 1 point for unclear division of labor. 3) Deduct 1 point for the on-site Responsible Person not wearing a safety monitoring vest. 4) Deduct 1 point for the work shift member failing to sign or signing incompletely on the work order.			

· 549 ·

Table (Cont'd)

S/N	Project name	Quality requirements	Score	Deduction standard	Reasons for deduction	Deduction	Scoring
5	Inspection of tools	1) The staff shall place the tools on the moisture-proof tarpaulin as required; the moisture-proof tarpaulin shall be clean and dry. 2) The tools shall be placed in category according to the requirements of the fixed management; the insulated tools shall not be mixed with metal tools and materials; and the appearance inspection shall be done on the tools. 3) The surface of insulated tools shall not be worn, deformed and damaged, and the operation shall be flexible. Carry out segment insulation detection with such insulated tools with insulation resistance meter of 2,500V or above and with the resistance no less than 700MΩ, and wipe it off with a clean dry towel. 4) The ground potential and equipotential personnel on tower shall correctly wear a whole suit of qualified shield clothes and conductive shoes as required, with each part connected well, shall not wear chemical fiber clothes next to the skin in the shielding clothes and shall fasten safety belts; the Responsible Person shall carefully check whether they wears it correctly. 5) Tower climbing personnel shall check the double name, pole number and phase again and report them	7	1) Deduct 1 point for failure to use moisture-proof tarpaulin and place tools to designed positions. 2) Deduct 0.5 points/item for failure to check the appearance of tools and pass the test certificate. 3) Deduct 1 point/item for failure to use testing instrument for testing the tools. 4) Deduct 2 points/person time for the operator failing to wear the shielding clothes correctly and each part connected well. 5) Deduct 1 point for the on-site Responsible Person failing to check the safety protective equipment of the tower climbing operators. 6) Deduct 2 points/person for the tower climbing personnel failing to check the double name of the line, pole number and phase. 7) Deduct 1 point for non-standard reporting of test results; deduct 0.5 points/item for incomplete reporting			

Part II Skill Module Training and Assessment Standards

Table (Cont'd)

S/N	Project name	Quality requirements	Score	Deduction standard	Reasons for deduction	Deduction	Scoring
6	Climbing the tower	1) The ground electrician and the equipotential electrician shall wear a whole suit of qualified shielding clothes, fasten the safety belt after performing the impulse test on the safety belt, and carry the insulated transmission rope to climb the tower one after another. 2) During the tower climbing process, they shall fasten the anti-fall protection device, climb the tower to an appropriate position, fasten the safety belt, arrange the insulated transmission rope, and then cooperate with the ground electrician to make lifting preparation of the insulated transmission rope separately. 3) During tower climbing, the electrician shall fasten the anti-fall protection device, climb the tower at a uniform speed, grasp the main material by hand, hang the safety belt on the shoulder and keep the safety distance of more than 6.8m away from the electrified body, and the Responsible Person shall strengthen the operation monitoring.	5	1) Deduct 2 points/item for failure to fasten the safety belt or for failure to perform the impulse test on the safety belt and the backup protection rope. 2) Deduct 0.5 points/time for hands grasping the shackles. 3) Deduct 1 point/item for the on-site Responsible Person failing to check the safety protection equipment of the ground potential electrician and the intermediate potential electrician. 4) Deduct 1 point/times for falling object at high place. 5) Deduct 1 point/item for tools colliding with the tower body during the transfer process. 6) Deduct 5 points for loosing the protection of the safety belt during transposition by the electrician on tower			

Table (Cont'd)

S/N	Project name	Quality requirements	Score	Deduction standard	Reasons for deduction	Deduction	Scoring
7	Detection of insulator	1) The ground electrician shall transfer the insulating bar and insulator detector to the tower. The equipotential electrician shall assist the ground potential electrician in detecting the insulator. 2) The insulators must be detected string by string and piece by piece, and the contact must be reliable from the cross arm side to the electrified side. 3) When the number of any string of good insulators is less than 37, the testing shall be stopped immediately, and the live working shall be stopped.	2	1) Deduct 2 points if not testing the null value of insulator. 2) Deduct 2 points if the null value measurement method is wrong. 3) Deduct 2 points if not reporting the null value measurement result.			
8	Entering the intense electric field	1) Before entering the insulator string, the equipotential electrician shall fasten the protection rope (holding insulator strings with the backup protection rope and stepping on a insulator string) and adjust the insulated transmission rope and the potential transfer rod. 2) After the equipotential electrician confirms that all parts of the shielding clothes are well connected, it shall be reported to the Responsible Person for approval. The electrician shall grasp one string with both hands and step on the other string with both feet, and enter the intense electric field along the insulator string in the operation mode of "two-span and three-short-circuit".	8	1) Deduct 2 points for insufficient distance of the exposed parts during potential transfer of equipotential electrician. 2) Deduct 2 points/time for the incorrect action of equipotential electrician entering the intense electric field and repeat discharging. 3) Deduct 1 point/times for unskillful potential transfer and 4 points/times for potential transfer without using the potential transfer rod. 4) Deduct 4 points/times for carrying out potential transfer without permission of the Responsible Person.			

Part II Skill Module Training and Assessment Standards

Table (Cont'd)

S/N	Project name	Quality requirements	Score	Deduction standard	Reasons for deduction	Deduction	Scoring
8	Entering the intense electric field	3) The combined gap composed of the gaps between the equipotential electrician and the grounding body and the electrified body in the process of entering the potential shall not be less than 6.9 m (medium phase) (6.7 m, side phase), and the potential transfer rod must be used for potential transfer when entering the intense electric field.	8				
9	Surface treatment of damaged conductor	1) The equipotential electrician shall carry the insulated transmission rope and walk to the operation point along the conductor, and shall install the insulated tackle and the insulated transmission rope at the sub-conductor and hoop all insulating ropes by sub-conductor. 2) The electrician shall check the damage of the conductor and deal with the damage point with abrasive paper (0#) to grind the burr of the damaged part. 3) Equipotential electrician shall clean the surface of the ground conductor with a cleaning cloth, and apply the live paste evenly on the surface of the damaged part of the conductor	20	1) Deduct 4 points for the incomplete covering of sub-conductor by insulating rope. 2) Deduct 2 points for failure to report the conductor damage situation to the Responsible Person. 3) Deduct 2 points for failure to smoothen the damaged conductor. 4) Deduct 2 points/item respectively for failure to remove surface oxide or apply conductive paste			

· 553 ·

Table (Cont'd)

S/N	Project name	Quality requirements	Score	Deduction standard	Reasons for deduction	Deduction	Scoring
10	Conductor repair	1) The ground electrician shall pass the preformed armor rods to the equipotential electrician by using an insulated transmission rope. Equipotential electrician uses the preformed armor rods to repair and strengthen the damaged part of conductor. 2) The specification and model of preformed armor rod should match the conductor and its center should be located at the most serious damage. 3) The length of the preformed armor rod shall be enough to fully cover the damaged part. The unilateral length of the end of the preformed armor rod from the edge of damaged part shall not be less than 100 mm. 4) Bind preformed armor rod firmly to ensure no dumping, missing or loose strands	30	1) Deduct 3 points for the deviation of repair center exceeding 5mm. 2) Deduct 2 points/positions for any gap at installation of preformed armor rods. 3) Deduct 10 points for deformation of preformed armor rods due to misoperation. 4) Deduct 1 point/Nos. for an uneven end. 5) Deduct 0.5 points/place when preformed armor rod is not bound firmly and there are dumping, missing or loose strands			
11	Exit the intense electric field	1) After inspection, the damaged strip line has been well strengthened and there is no residue at the operation point, the Responsible Person shall authorize the equipotential electrician to carry the insulated transmission rope back to the grading ring and prepare for leaving potential by means of walking along the line.	10	1) Deduct 2 points for the potential transfer without applying to the Responsible Person; deduct 1 point for starting the work without the consent after applying. 2) Deduct 2 points/times for insufficient distance of the exposed parts to the electrified body during potential transfer of equipotential electrician.			

Part II Skill Module Training and Assessment Standards

Table (Cont'd)

S/N	Project name	Quality requirements	Score	Deduction standard	Reasons for deduction	Deduction	Scoring
11	Exit the intense electric field	2) The equipotential electrician shall hook the grading ring with the potential transfer rod and enter the third insulator of the grading ring. With one hand grasping the insulator tightly and the other holding the potential transfer rod, the equipotential electrician shall break himself away from the potential with the help of potential transfer rod. 3) The equipotential electrician exits the equipotential according to the operation mode of "two-span and three-short-circuit".	10	3) Deduct 2 points/time for the incorrect action of equipotential electrician exiting the intense electric field and repeat discharging. 4) Deduct 2 points for unskillful potential transfer and 3 points for potential transfer without using the potential transfer pole			
12	Return to the ground	After checking that there is no object left on the tower, the equipotential electrician on the tower shall report it to the Responsible Person and then climb down the tower with insulated transmission rope after obtaining the consent of the Responsible Person	5	1) Deduct 5 points for failure to use the falling protector when climbing down the tower. 2) Deduct 5 points for loosing the protection of the safety belt when moving on the tower. 3) Deduct 1 point/time for hands grasping the shackles when climbing down the tower. 4) Deduct 2 points for any objects left on tower.			

· 555 ·

Table (Cont'd)

S/N	Project name	Quality requirements	Score	Deduction standard	Reasons for deduction	Deduction	Scoring
13	End of the work	1) The Responsible Person shall organize all working members to put working apparatus and materials in order and put them in a special kit (bag) after cleaning; clean the site to ensure that "the materials are removed and the site is cleaned after construction". 2) In the post-shift meeting, Responsible Person will give work summaries and comments.Comments include the construction quality of this work and the implementation of safety measures from all working personnel. 3) Responsible Person shall report to the on-duty control personnel that the work is over, apply for the restoration of circuit re-closing and terminate the work order.	10	1) Deduct 2 points for failure to clean the tools. 2) Deduct 2 points for missing tools. 3) Deduct 10 points for failure to hold the post-shift meeting. 4) Deduct 2 points for failure to remove the fence. 5) Deduct 2 points for failure to report to dispatcher			
	Total		100				

Module 10 Standards for Training and Assessment on Live Treatment of Heating Defects of Drainage Plate for Conductor of 1000kV AC Transmission Line

I. Training Standard

(I) Training Requirements

Designation of module	Live treatment of heating defects of 1000kV AC transmission line conductor drainage plate	Type of training	Operation
Training method	Practical operation training	Hours of training	14 hours
Training objectives	1. The "two-span and three-short-circuit" operation mode is adopted to enter and leave the 1000kV intense electric field along the strain insulator string. 2. Can complete the entry of 1000kV equipotential operation point along the strain insulator string. 3. Be able to independently complete the live treatment of Heating Defects of Drainage Plate for Conductor 1000kV AC transmission line (equipotential operation method)		
Training venue	UHV AC training line		
Training content	The "two-span and three-short-circuit" operation mode is adopted to enter the intense electric field along the strain insulator string, and the equipotential operation method is adopted for live treatment of Heating Defects of Drainage Plate for Conductor 1000kV AC transmission line		
Scope of application	Live Maintenance Personnel of UHV AC Transmission Line		

(II) Referenced Rules and Specifications

(1) Code for Design of 1000kV Overhead Transmission Line (GB50665-2011).

(2) Maintenance Code for 1000kV AC Transmission Line (DL/T209-2008).

(3) Operation Code for 1000kV AC Transmission Line (DL/T307-2010).

(4) Technical Guide for Live Working on 1000kV AC Transmission Line (DL/

T392-2015).

(5) Calculation Method of Live Working Minimum Approach Distance on AC Transmission Line (GB/T 19185-2008).

(6) Guidelines of Insulation Coordination for Live Working (DL/T867-2004).

(7) Live Working-Guidelines for the Installation of Power Transmission Line Conductors and Earthwires-stringing Equipment and Accessory Items (DL/T 1007-2006).

(8) State Grid Corporation of China on the Management Regulations of Live Working (Trial Implementation) (SGCC [2007] No.751).

(9) State Grid Corporation of China Working Regulations of Power Safety (Transmission Line Section) (Q/GDW1799.2-2013).

(10) Electrotechnical Terminology - Overhead Line (GB/T 2900.51-1998).

(11) Electrotechnical Terminology- Live Working (GB/T2900.55-2002).

(12) Live Working - Terminology for Tools, Equipment and Devices (GB/T 14286-2002).

(13) Minimum Requirements for Utilization of Tools, Devices and Equipment for Live Working (DL/T 877-2004).

(14) Preventive Test Code of Tools, Devices and Equipment for Live Working (DL/T 976-2005).

(15) Insulated Tackles for Live Working (GB/T13034-2008).

(16) Live Working-Insulating Ropes (GB/T 13035-2008).

(17) Screen Clothes for Live Working on 1000kV AC (GB/T25726-2010).

(Ⅲ) Teaching Design for Training

To complete the work task of "live treatment of heating defects of drainage plate for conductor of 1000kV AC transmission line", each training stage shall be designed according to the standard operation procedure for work task completion. Each stage includes specific training objectives, training content, hours of training, training methods (training resources), training environment, assessment and evaluation, etc, as shown in the Table 2-10-1.

(Ⅳ) Operation Flow

1. Work Task

The "two-span and three-short-circuit" mode shall be adopted to enter the intense electric field along the insulator string to arrive at the operation point, and the equipotential oper-

Part II Skill Module Training and Assessment Standards

Table 2-10-1 Training Content Design for Live Treatment of Heating Defects of 1000kV AC Transmission Line Conductor Drainage Plate

Training schedule	Training objectives	Training content	Hours of training	Training methods and resources	Preparation of training conditions	Assessment and evaluation
1. Theoretical teaching	1. Preliminarily master the basic method for entering and exiting 1000kV AC intense electric field along the insulator string. 2. Be familiar with the methods for potential transfer. 3. Be familiar with the handling method of heating defects of transmission line conductor drainage plate	1. Adopt the "two-span and three-short-circuit" operation mode to enter and leave the intense electric field along the insulator string. 2. The use method for the potential transfer rod during entering and exiting the UHV intense electric field. 3. The handling method and quality standard of heating defects of transmission line conductor drainage plate	2	Training methods: Lecture. Training resources: PPT, relevant regulations and specifications	Multimedia classroom	Attendance, classroom questions and assignments
2. Preparations	Be able to complete the preparation before operation	1. Work site survey. 2. Preparation of the standardized operation card. 3. Filling of the work order. 4. Preparation of tools and materials for this operation. 5. The on-duty control personnel shall contact and apply to stop work line re-closing lock device	1	Training methods: 1. Site survey and cleaning of tools and materials shall be practiced at site. 2. Preparation of operation card and the filling of work order shall adopt lecture method. Training resources: 1. 1000kV practical training line. 2. UHV tools warehouse. 3. Blank work order	1. UHV training transmission line 2.Multi-Media classroom	

Table (Cont'd)

Training schedule	Training objectives	Training content	Hours of training	Training methods and resources	Preparation of training conditions	Assessment and evaluation
3. Work site preparation	Be able to complete the preparations of work site	1. Re-closing lock has been de-activated and the dispatching permission has been obtained. 2. Work site re-survey. 3. Work site layout. 4. Pre-shift meeting. 5. Inspection of tools and materials	1	Training methods: Demonstration and role play. Resources; 1000kV training line	1000kV training line	
4. Trainer's demonstration	The trainees can preliminarily understand the operation process of the task through inspecting and learning from each other's work	1. Detection of null-value insulator. 2. The equipotential electrician enters and exits the intense electric field along the strain insulator string and reach the connection of jumper wire. 3. Equipotential electrician fastens the bolt of conductor jumper wire with socket spanner	2	Training methods: Demonstration. Resources; 1000kV training line	1000kV training line	

Part II Skill Module Training and Assessment Standards

Table (Cont'd)

Training schedule	Training objectives	Training content	Hours of training	Training methods and resources	Preparation of training conditions	Assessment and evaluation
5. Group training of trainees	1. Can complete the operation of entering and exiting 1000kV intense electric field. 2. Be able to complete the live treatment of heating of conductor jumper wire of 1000kV transmission line	1. The trainees are grouped (6 persons per group) to train the skill operation of entering and exiting 1000kV intense electric field and treatment of heating of conductor jumper wire. 2. Trainers guide the operation of trainees and conduct safety supervision	7	Training methods: Role play. Resources: 1000kV training line	1000kV training line	Score the operation of trainees according to the detailed rules for skill assessment and scoring
6. End of the work	1. Enable the trainees to further distinguish the shortcomings of the operation process and facilitate the promotion in the later stage. 2. Train the trainees in the working style of safe and civilized production	1. Cleaning up the work site. 2. Report the completion of work to the dispatch and apply for restoration of re-closing lock device. 3. Comment and summarize the work task this time at post-shift meeting	1	Training method: Lecture and inductive method	1000kV training line	

· 561 ·

ation method shall be adopted for live treatment Heating defects of 1000kV AC transmission line conductor drainage plate.

2. Requirements for Weather and Work Site

(1) The live treatment of heating defects of 1000kV AC transmission line conductor drainage plate shall be carried out in good weather. In case of lightning (hearing thunder or seeing lightning), snow, hail, rain, fog and so on, live working is prohibited. When the wind force is greater than level 5, or the relative humidity of the air is greater than 80%, it is unsuitable for live working; when emergency live repair is required in bad weather, relevant personnel shall be organized to fully discuss and prepare necessary safety measures, which can be implemented after being approved by the unit.

(2) The operating personnel shall be in good spirits, and be familiar with the organizational and technical measures to ensure safety in work; they shall hold the UHV AC live working qualification within the validity period.

(3) The Responsible Person shall organize relevant personnel to complete field investigation in advance, determine the operating methods, required working apparatus and necessary measures according to the results, and handle the work orders for live working.

(4) The work site shall be reasonably set up with fence and warning signs. Non-operating personnel is forbidden to enter.

(5) The line re-closing device shall be deactivated in the Project.

(6) Safety working distance and effective insulation length are shown in Table 2-10-2.

(7) For operation on 1000kV transmission line, the number of insulators in good operation phase shall be no less than 37 pieces.

Table 2-10-2 Safe Distance for Live Repair of Conductor of 1000kV AC Transmission Line

Altitude/m	Minimum safe distance/m		Minimum combined clearance /m		Minimum effective insulation length of insulating tools /m	Minimum distance between exposed part of human body and electrified body when potential is transferred /m
	Middle phase	Side phase	Middle phase	Side phase		
H≤500	6.5	5.8	6.7	6.4	6.8	0.5
500<H≤1000	6.8	6.0	6.9	6.7	6.8	0.5
1000<H≤1500	7.0	6.3	7.2	7.0	7.2	0.5
1500<H≤2000	7.4	6.6	7.6	7.3	7.2	0.5

Part II Skill Module Training and Assessment Standards

3. Preparations

3.1 Hazards and precontrol measures

(1) Hazard - precontrol measures for electric shock injury:

① Before work, Responsible Person should contact the control personnel on duty, deactivate the line re-closing, and perform the licensing procedures.

② Before climbing the tower, ground potential operators must carefully check the name of the line and then can climb the tower after confirmation.

③ If lines lose power suddenly during work, operators shall still deem it to be charged. The Responsible Person shall contact the control personnel as soon as possible, and no forced energization is allowed before the on-duty control personnel getting in touch with the Responsible Person.

④ Insulating tool and insulating ropes shall be free of damage, moisture, deformation, and failure. It is not allowed to use non-insulating ropes (such as cotton rope, manila rope, and steel wire rope).

⑤ The equipotential operator shall wear flame-retardant underwear and qualified full set of shielding clothes over the underwear (including hat, mask, dresses & trousers, gloves, socks and shoes). All parts shall be in excellent connection conditions.

⑥ During the operation, the equipotential operator shall ensure the minimum safe distance, namely 6.8 m for medium phase and 6.0 m for side phase; the minimum combined gap, namely 6.9 m for medium phase and 6.7 m for side phase; and the minimum effective insulating length of insulating tools of 6.8 m. Before potential transfer, equipotential operator shall obtain the approval of Responsible Person, and the minimum distance between exposed part of human body and electrified body shall not be lower than 0.5m. During potential transfer, the operation shall be fast, and end shall not be used for power charging and discharging; During delivering tools and materials to ground potential operators, the effective length of insulating tools or insulating ropes shall not be lower than the requirements as specified in Table 2-10-2.

⑦ The Supervisor and Responsible Person shall continuously monitor the operators and correct their nonstandard operation or actions in violation of rules and regulations at any time. The operators working at heights shall be monitored to ensure that there is enough safety distance (meeting the requirements in Table 2-10-2). It is forbidden to contact two non-connected electrified bodies or make contact with electrified body and grounding body at the same time.

(2) Hazard - falling accident.

① Before climbing, operators work at heights must satisfy the requirements of this operation, such as physical condition, mental state and skill and quality.

The equipotential electrician shall steadily coordinate his hands and feet during entering the intense electric field with uniform speed; backup protection rope of human body shall be used in the whole process of movement.

③Supervisors shall correct the nonstandard or illegal actions and behaviors at any time. Special monitoring shall be conducted to operators to prevent them from losing the protection of safety belt or insulated backup protection rope during transposition, and it is forbidden to fasten the safety belt or insulated backup protection rope in a position lower than the operating personnel.

(3) Hazard - injury caused by objects falling from heights. Precontrol measures:

① Operator working at heights should put personal tools and fragmentary materials into the tools bag. It is strictly forbidden to hang objects in heights or keep in the mouth.

② Ground operator should correctly wear a helmet and use the knots. The vertical distance from the work site should not be less than the falling radius.

③ The work site shall be set up with fence and warning signs. It shall be noted at any time that Supervisor shall prohibit irrelevant personnel and vehicles from entering operation area.

3.2 Selection of tools and instruments and materials

Tools and materials required for live treatment of heating defects of 1000kV AC transmission line conductor drainage plate can be seen in Table 2-10-3. Before delivering tools and instruments out of warehouse, application voltage class and test period shall be carefully checked and they shall be inspected to ensure that appearance is intact, connection is firm, rotation is flexible and meet the working task requirements. After delivering tools and instruments out of warehouse, they shall be stored in tools bag or tool kit for transportation to avoid contamination and damp. Metal tools and insulated tools shall be separately loaded and transported to avoid deformation, damage or other defects caused by mixed loading and transportation.

Part II Skill Module Training and Assessment Standards

Table 2-10-3 Tools, Instruments and Materials Required by live treatment of Heating Defects of 1000kV AC Transmission Line Conductor Drainage Plate

S/N	Name	Specification	Unit	Qty.	Remarks
1	Insulated transmission rope	φ12	Nos.	2	
2	Insulated backup protection rope	φ16	Nos.	2	
3	Insulator detector		Set	1	
4	Insulated tackle	1T	Nos.	2	
5	Socket wrench	Consistent with the model of bolt of drainage plate	Set	1	
6	Potential transfer rod		Nos.	1	
7	Insulation resistance tester	5000V	Set	1	
8	Anemometer		Set	1	
9	Temperature and humidity meter		Set	1	
10	Multimeter		Set	1	
11	Moisture-proof canvas	2m×4m	Sheet	1	
12	Shielding clothes	Shielding efficiency ≥ 60 dB (shield efficiency of shielding mask ≥ 20 dB)	Set	2	
13	Falling protector	Corresponding to the type of falling protector of iron tower	Nos.	2	
14	Safety fence		Set	Several	
15	Warning sign	"Work Here", "Access from Here""Access from Here"	Set	1	
16	Red waistcoat	"Responsible Person" "Special Supervisor"	Piece	2	
17	Bolt	Consistent with the model of bolt of drainage plate	Nos.	Several	Standby
18					
19	Backpack safety belt	With backup protection rope	Set	2	
20	Clean towel		Nos.	1	
21	Insulated noose	φ12	Nos.	2	
22	Safety helmet		Nos.	6	
23	Testing rod		Nos.	1	
24	Tool kit		Nos.	2	
25	Interphone		Nos.	4	

3.3 Division of labor for operators

Division of labor for operators of the task is shown in Table 2-10-4.

Table 2-10-4 Division of Labor for Operators for Live Treatment of Heating Defects of 1000kV AC Transmission Line Conductor Drainage Plate

S/N	Post	Qty. (person)	Responsibilities
1	Responsible Person	1	Be responsible for organizing various work on the work site
2	Special Supervisor	1	Be responsible for the safety monitoring of the work site
3	Equipotential electrician	1	Be responsible for entering equipotential and treatment of heating defects of drainage plate of conductor
4	Ground potential electrician	1	Be responsible for transferring tools and materials and co-operating with equipotential electrician in entering and exiting the equipotential
4	Ground electrician	2	Be responsible for transferring tools and materials and co-operating with equipotential electrician in entering and exiting the equipotential

4. Work Procedure

The workflow of this task is shown in Table 2-10-5.

Table 2-10-5 Workflow Chart for Live Treatment of Heating Defects of 1000kV AC Transmission Line Conductor Drainage Plate

S/N	Work Content	Operation Steps and Standards	Safety Measures and Precautions	Responsible Person
1	Work Permit	(1) The Responsible Person is in charge of contacting the on-duty control personnel and applies to stop line re-closing lock as per the content of work order. (2) Live working could be started only after being approved by the on-duty control personnel	Live working shall not be started without the permission of the on-duty control personnel	

Part II Skill Module Training and Assessment Standards

Table (Cont'd)

S/N	Work Content	Operation Steps and Standards	Safety Measures and Precautions	Responsible Person
2	Site re-survey	The Responsible Person shall complete the following work: (1) Check the line name on spot to ensure it is correct; guarantee that the foundation and the iron tower are intact and in normal condition, and ensure that the cross and span distance meets the safety requirements; confirm the defect conditions and the specifications and models of conductor or ground wires. (2) Check that the terrain and environment shall meet the operation requirements. (3) Check that the safety measures listed in the work order are in line with the actual situations on site, and the measures will be supplemented if necessary	(1) Correctly wear helmet, working clothes, work shoes and protective gloves. (2) Operation under meteorological conditions that may endanger the safety of operators is forbidden. (3) Non-operation personnel and vehicles are strictly prohibited from entering the working site.	
3	Site layout	Install the security fence and hang the signboards correctly: (1) The security fence should take full account of falling objects from the heights and the influence on road traffic. (2) The entrance and exit of the security fence shall be set reasonably. (3) Signs such as "Access from Here", "Work Here", "Access from Here" shall be properly and well arranged	When the influence on road traffic safety is uncontrollable, the traffic management department should be contacted in time to strengthen the on-site control of traffic safety.	
4	Hold a pre-shift meeting	(1) All working personnel shall line up. (2) The Responsible Person shall wear red waistcoat and read out the work order and be clear with work task and division of personnel; explain safety measures and technical measures in work; check (inquire after) mental state of all working personnel; inform of hazards in work and precontrol measures. (3) All working personnel shall sign on the work order for confirmation.	(1) The work order shall be filled in, issued and approved in a standardized manner, and the signature shall be complete. (2) All working personnel shall be in good mental states. (3) All working personnel shall be clear with task division of works, safety measures and technical measures	

Table (Cont'd)

S/N	Work Content	Operation Steps and Standards	Safety Measures and Precautions	Responsible Person
5	Inspection tool	(1) The equipotential electrician and ground potential electrician shall wear the shielding clothes in a right way and pass the inspection, which shall be supervised and inspected by the Responsible Person. (2) Wear personal safety equipment correctly (proper size and easy lock), and the Responsible Person shall supervise and inspect it. (3) Measure the wind speed and humidity, check the insulation performance of insulating tools, and make records	(1) Check carefully for damage, deformation and failure before using metal and insulating tools. Carry out segment insulation detection with such insulating tools as insulation resistance meter of 5000V and with the resistance no less than 700M Ω, and wipe them clean with the clean dry towel. (2) Use a multimeter to measure the resistance between the farthest ends of the shielding clothes and trousers, which shall not be greater than 20Ω. The Responsible Person shall check the connection of the electrician's shielding clothing. (3) Check the tool assembly and make sure the connection is reliable. (4) The live working tools and instruments used at site shall be placed on the moisture-proof canvas	
6	Climbing the tower	(1) After checking the line name without error, the electrician on tower shall make impulse test on the safety belt and falling protector. (2) The electrician on tower carries the insulated transmission rope to climb the tower. When reaching the operation point of the cross arm, he shall choose the appropriate position to fasten the safety belt, and install the insulated tackle and the insulated transmission rope in the appropriate position of the cross arm. Then he shall cooperate with the ground electrician to separate the insulated transmission rope for lifting preparation	(1) After checking the correct line name, it is allowed to climb the tower for operation. (2) The anti-falling device installed on tower shall be used in the process of climbing the tower; when moving and transposition on the tower, the safety protection shall not be lost, and the operators must climb and grasp the components securely. (3) The working electrician must wear a whole suit of qualified shielding clothes which must be connected reliably. Before entering the equipotential, the equipotential electrician shall check and confirm that each part of the shielding clothes are connected reliably before the next operation	

Part II Skill Module Training and Assessment Standards

Table (Cont'd)

S/N	Work Content	Operation Steps and Standards	Safety Measures and Precautions	Responsible Person
7	Detection of insulator	(1) The ground electrician shall transfer the insulating bar and insulator detector to the tower. The equipotential electrician shall test the insulator. (2) The ground potential electrician shall cooperate with the equipotential electrician and hold the insulating bar with hand for testing from the conductor side to the cross arm side successively. The testing shall be repeated for 2-3 times for low-value or null-value insulator, and records shall be made properly; (3) The ground potential electrician shall transfer the insulating bar and insulator detector to the place below the tower	(1) The insulators must be detected piece by piece, and the contact must be reliable. (2) After removing the human body short circuit, if the number of good insulators is less than 37, the operation shall be stopped immediately. (3) The operating electrician shall clearly know the way for judgment (no sound for null-value insulator and small sound for low-value insulator), and distinguish the sounds of corona and electric discharge	
8	Entering the intense electric field	(1) The ground electrician transmits the potential transfer rod to the tower. (2) The equipotential electrician shall move to the insulator set fitting with potential transfer rod, and carry the insulated tackle and insulated transmission rope. (3) After the equipotential electrician confirms that all parts of the shielding clothes are well connected, it shall be reported to the Responsible Person for approval. The electrician shall grasp one string with both hands and step on the other string with both feet, and enter the intense electric field along the insulator string in the mode of "two-span and three-short-circuit". (4) When reaching 3 insulator outside the grading ring on the conductor side, the equipotential electrician shall stop, and carry out the potential transfer with the potential transfer rod	(1) The equipotential electrician must obtain the permission of the Responsible Person before entering the potential. (2) Before entering the insulator string, the equipotential electrician shall fasten the backup protection rope and adjust the insulated transmission rope. (3) The combined gap composed of the gaps between the equipotential electrician and the grounding body and the electrified body while entering the potential shall not be less than 6.9 m (medium phase) (6.7 m, side phase).	

Table (Cont'd)

S/N	Work Content	Operation Steps and Standards	Safety Measures and Precautions	Responsible Person
9	Treatment of heating defects of drainage plate of conductor	(1) After entering the equipotential, the equipotential electrician shall tie the safety belt to the sub-conductor, and ground electrician pass the socket spanner to the equipotential electrician with transmission rope. (2) Equipotential electrician fastens the connecting bolt of conductor drainage plate with socket spanner. (3) The equipotential electrician passes the socket spanner down the tower with transmission rope	(1) The equipotential electrician should pay attention to avoid excessive action range, and avoid stretching the limbs to the insulator. (2) The equipotential electrician should pay attention to avoid scalding	
10	Exiting the potential	(1) After inspection, there is no residue at the operation point, the Responsible Person shall authorize the equipotential electrician to carry the insulated transmission rope and prepare for leaving potential. (2) The equipotential electrician shall hook the grading ring with the potential transfer rod and enter the third insulator of the grading ring. With one hand grasping the insulator tightly and the other holding the potential transfer rod, the equipotential electrician shall break himself away from the potential with the help of potential transfer rod. (3) The equipotential electrician exits the equipotential according to the mode of "two-span and three-short-circuit".	(1) The equipotential electrician must obtain the permission of the Responsible Person before exiting the potential. (2) The combined gap composed of the gaps between the equipotential electrician and the grounding body and the electrified body while exiting the potential shall not be less than 6.9 m (medium phase) (6.7 m, side phase). (3) The equipotential electrician shall move along the insulator string with the hands grasping firmly and the feet stepping firmly	

Part II Skill Module Training and Assessment Standards

Table (Cont'd)

S/N	Work Content	Operation Steps and Standards	Safety Measures and Precautions	Responsible Person
11	Return to the ground	The electrician on tower shall pass the tools on tower down the tower with transmission rope.After checking that there is no object left on the tower, it is required to report it to the Responsible Person and then climb down the tower with insulated transmission rope after obtaining the consent of the Responsible Person	The anti-falling device installed on the tower shall be used in the process of climbing down the tower; when moving and transposition on the tower, the safety protection shall not be lost, and the operators must grasp the components securely	
12	End of the work	(1) The Responsible Person shall organize all working members to put working apparatus and materials in order and put them in a special kit (bag) after cleaning; clean the site to ensure that "the materials are removed and the site is cleaned after construction". (2) In the post-shift meeting, Responsible Person will give work summaries and comments.Comments include the construction quality of this work and the implementation of safety measures from all working personnel. (3) Responsible Person shall report to the on-duty control personnel that the work is over, apply for the restoration of circuit re-closing and terminate the work order.	It is forbidden to restore the line re-closing in the appointed time	

II. Assessment Standard

Detailed Rules for Assessment and Scoring of Operation and Inspection Skills of UHV AC Transmission Line

Fill-in Column of Examinee	No.:		Name:		Position:		Date:	MM/DD/YYYY
Fill-in Column of Assessor	Grade:		Assessor:		Assessment Team Leader:	Starting time:	Closing time:	Operation Duration:
Assessment module	Live treatment of heating defects of 1000kV AC transmission line conductor drainage plate		Assessee	Maintenance personnel of UHV AC transmission line		Assessment method	Operation	Assessment Time Limit
								60min
Job Description	Live Treatment of Heating Defects of 1000kV AC Transmission Line Conductor Drainage Plate by the Method of Entering Intense Electric Field Along Insulator String with "Two-span and three-short-circuit"							
Work Specifications and Requirements	1. Live working shall be carried out in good weather. In case of thunder, rain, snow or fog, no live working shall be carried out. When the wind force is greater than Level 5 and the humidity is greater than 80%, live working should not be carried out. 2. This operation requires 6 people, including one Responsible Person, one Special Supervisor, one equipotential electrician, one ground potential electrician, and two ground electricians. Heating defects of the drainage plate for the conductor of 1000kV AC transmission line are treated by the method of entering the intense electric field along the insulator string with strain insulator string. 3. Responsibilities of Responsible Person (Supervisor): Be responsible for division of operating personnel of the task, work order reading, handling re-closing deactivation of the line and the formalities of work permit, getting work permits, holding pre-shift meeting, safety supervision in the operation process, dealing with emergency situations in work, quality surveillance, and the summary after work. 4. Responsibilities of equipotential electricians: Be responsible for the main work in this operation process, installing and dismantling the work tools according to the position of the work, and entering the equipotential for the treatment of the heating defects of the drainage plate for the conductor of the transmission line. 5. Responsibilities of ground potential electrician: Be responsible for transferring tools and materials and cooperating with equipotential electrician in entering and exiting the equipotential. 6. Responsibilities of ground electrician: Be responsible for transferring tools and materials. 7. During the live working, if thunder, rain, strong wind or any other circumstance threaten the safety of the staff, the Responsible Person or Supervisor may stop working temporarily according to the circumstances.							

Part II Skill Module Training and Assessment Standards

Table (Cont'd)

Work Specifications and Requirements	Given conditions: 1. Side phase conductor of resisting-tensile tower of 1000kV practical training line. 2. Work orders have been handled, safety measures have been completed (re-closing has been deactivated), and oral application (dispatcher or assessor) shall be made at the beginning and end of the work. 3. The instrument shall be used safely and correctly to test the insulating tool. 4. The operation must be carried out according to the working procedures. The scores of the items to be carried out shall be deducted for the process error. In case of major hidden dangers of personal, equipment and operational safety, the assessor may order the termination of the operation (assessment)
Assessment scenario preparation	1. Tower shape: Side phase conductor of resisting-tensile tower of 1000kV practical training line. Work content: live treatment of heating defects of the drainage plate for the conductor of 1000kV AC transmission line. 2. Required working tools: 2 × Φ12 insulated transmission ropes, 2 × Φ16 insulated protection ropes, 1 set of insulator detector, 2 × 1T insulated pulley tackles, 1 set of socket spanners of different types, 1 potential transfer rod, 1 5000V insulation resistance tester, 1 anemorumbometer, 1 thermohygrometer, 1 multimeter, 1 2m×4m moisture-proof canvas, 2 suits of type II shielding clothes, 2 falling protectors, 1 set of safety fence, 1 set of warning signs, 2 suits of red waistcoats, several bolts of different types, several pins of different types, 2 sets of dual fail-safe safety belts, 1 cleaning towel, 2 Φ12 insulating rope sleeves, 6 safety helmets, 1 operating rod, 2 tool kits, and 4 walkie-talkies. 3. The work site shall be monitored, and the safety measures (fence, etc.) on the work site have been fully implemented; non-operation personnel are prohibited from entering the site, and the staff must wear safety helmets when entering the work site. 4. Examinees shall bring their own work clothes, flame retardant cotton underwear, safety helmets, gloves, safety belts (including double-protective ropes).
Remarks	1. The deduction shall be done until the scores of each item are deducted completely. In case of major hidden dangers of personal, equipment and operational safety, the assessor may order the termination of the operation. 2. When equipment, working environment, safety belt, safety helmet, tool, shielding clothes, etc., do not conform to the operation condition, the assessor may order the termination of the operation

Table (Cont'd)

S/N	Project name	Quality requirements	Score	Deduction standard	Reasons for deduction	Deduction	Scoring
1	Site re-survey	1) The Responsible Person shall go to the work site to check the line name, pole and tower number, on-site working conditions, defective parts and so on. 2) Check that the site meteorological conditions such as wind speed and humidity should meet the operation requirements. 3) Check whether the work order is complete and unmodified, check whether the safety measures listed are consistent with the actual situation on site, and supplement it if necessary	5	1) Deduct 1 point/item for failure to check the line name, pole and tower number, on-site working conditions and defective parts. 2) Deduct 1 point/item for failure to detect wind speed, humidity and other on-site meteorological conditions. 3) Deduct 0.5 points/part for any alteration in the work order filling; deduct 1 point for incorrect work order number; deduct 1.5 points for incomplete work order filling			
2	Work Permit	1) The Responsible Person is responsible for contacting the on-duty control personnel and applying for stopping the line re-closing as per the contents of the work order. 2) Reporting content is standardized and complete	2	1) Deduct 2 points for failure to contact the dispatching department (referee) for deactivating the re-closing. 2) Deduct 1 point for non-standard or incomplete terminology reporting			

Part II Skill Module Training and Assessment Standards

Table (Cont'd)

S/N	Project name	Quality requirements	Score	Deduction standard	Reasons for deduction	Deduction	Scoring
3	Site layout	Install the security fence and hang the signboards correctly: 1) The security fence should take full account of falling objects from the heights and the influence on road traffic. 2) The entrance and exit of the security fence shall be set reasonably. 3) Signs such as "Access from Here", "Work Here", "Access from Here" shall be properly and well arranged	3	1) Deduct 1 point for failure to arrange the fence at the work site. 2) Deduct 1 point for failure to arrange the warning board. 3) Deduct 1 point for failure to hang the tower climbing operation sign			
4	Hold a pre-shift meeting	1) All staff and personnel shall wear safety helmets and work clothes correctly. 2) Responsible Person shall wear red vest and read out the work order and be clear with work task and division of personnel; explain safety measures and technical measures in work; check (inquire after) mental state of all working personnel; inform of hazards in work and precontrol measures. 3) All working personnel shall sign on the work order for confirmation.	3	1) Deduct 0.5 points/person for the staff not dressing uniformly. 2) Deduct 3 points for no division of labor; deduct 1 point for unclear division of labor. 3) Deduct 1 point for the on-site Responsible Person not wearing a safety monitoring vest. 4) Deduct 1 point for the work shift member failing to sign or signing incompletely on the work order.			

Table (Cont'd)

S/N	Project name	Quality requirements	Score	Deduction standard	Reasons for deduction	Deduction	Scoring
5	Inspection of tools	1) The staff shall place the tools on the moisture-proof tarpaulin as required; the moisture-proof tarpaulin shall be clean and dry. 2) The tools shall be placed in category according to the requirements of the fixed management; the insulated tools shall not be mixed with metal tools and materials; and the appearance inspection shall be done on the tools. 3) The surface of insulated tools shall not be worn, deformed and damaged, and the operation shall be flexible. Carry out segment insulation detection with such insulating tools as insulation resistance meter of 5,000M Ω, and with the resistance no less than 700M Ω, and wipe them clean with the clean dry towel. 4) The ground potential and equipotential personnel on tower shall correctly wear a whole suit of qualified shield clothes and conductive shoes as required, with each part connected well, shall not wear chemical fiber clothes next to the skin in the shielding clothes and shall fasten safety belts; the Responsible Person shall carefully check whether they wears it correctly. 5) Tower climbing personnel shall check the double name, pole number and phase again and report them	7	1) Deduct 1 point for failure to use moisture-proof tarpaulin and place tools to designed positions. 2) Deduct 0.5 points/item for failure to check the appearance of tools and pass the test certificate. 3) Deduct 1 point/item for failure to use testing instrument for testing the tools. 4) Deduct 2 points/person time for the operator failing to wear the shielding clothes correctly and each part connected well. 5) Deduct 1 point for the on-site Responsible Person failing to check the safety protective equipment of the tower climbing operators. 6) Deduct 2 points/person for the tower climbing personnel failing to check the double name of the line, pole number and phase. 7) Deduct 1 point for non-standard reporting of test results; deduct 0.5 points/item for incomplete reporting			

Part II Skill Module Training and Assessment Standards

Table (Cont'd)

S/N	Project name	Quality requirements	Score	Deduction standard	Reasons for deduction	Deduction	Scoring
6	Climbing the tower	1) The ground electrician and the equipotential electrician shall wear a whole suit of qualified shielding clothes, fasten the safety belt after performing the impulse test on the safety belt, and carry the insulated transmission rope to climb the tower one after another. 2) During the tower climbing process, they shall fasten the anti-fall protection device, climb the tower to an appropriate position, fasten the safety belt, arrange the insulated transmission rope, and then cooperate with the ground electrician to make lifting preparation of the insulated transmission rope separately. 3) During tower climbing, the electrician shall fasten the anti-fall protection device, climb the tower at a uniform speed, grasp the main material by hand, hang the safety belt on the shoulder and keep the safety distance of more than 6.8m away from the electrified body, and the Responsible Person shall strengthen the operation monitoring.	5	1) Deduct 2 points/item for failure to fasten the safety belt or for failure to perform the impulse test on the safety belt and the backup protection rope. 2) Deduct 0.5 points/time for hands grasping the shackles. 3) Deduct 1 point/item for the on-site Responsible Person failing to check the safety protection equipment of the ground potential electrician and the intermediate potential electrician. 4) Deduct 1 point/times for falling object at high place. 5) Deduct 1 point/item for tools colliding with the tower body during the transfer process. 6) Deduct 5 points for loosing the protection of the safety belt during transposition by the electrician on tower			

Table (Cont'd)

S/N	Project name	Quality requirements	Score	Deduction standard	Reasons for deduction	Deduction	Scoring
7	Detection of insulator	1) The ground electrician shall transfer the insulating bar and insulator detector to the tower. The equipotential electrician shall assist the ground potential electrician in detecting the insulator. 2) The insulators must be detected string by string and piece by piece, and the contact must be reliable from the cross arm side to the electrified side. 3) When the number of any string of good insulators is less than 37, the testing shall be stopped immediately, and the live working shall be stopped.	2	1) Deduct 2 points if not testing the null value of insulator. 2) Deduct 2 points if the null value measurement method is wrong. 3) Deduct 2 points if not reporting the null value measurement result.			
8	Entering the intense electric field	1) Before entering the insulator string, the equipotential electrician shall fasten the protection rope (holding insulator strings with the backup protection rope and stepping on a insulator string) and adjust the insulated transmission rope and the potential transfer rod.	15	1) Deduct 2 points for insufficient distance of the exposed parts during potential transfer of equipotential electrician. 2) Deduct 2 points/time for the incorrect action of equipotential electrician entering the intense electric field and repeat discharging. 3) Deduct 1 point/times for unskillful potential transfer and 4 points/times for potential transfer without using the potential transfer rod.			

Part II Skill Module Training and Assessment Standards

Table (Cont'd)

S/N	Project name	Quality requirements	Score	Deduction standard	Reasons for deduction	Deduction	Scoring
8	Entering the intense electric field	2) After the equipotential electrician confirms that all parts of the shielding clothes are well connected, it shall be reported to the Responsible Person for approval. The electrician shall grasp one string with both hands and step on the other string with both feet, and enter the intense electric field along the insulator string in the operation mode of "two-span and three-short-circuit". 3) The combined gap composed of the gaps between the equipotential electrician and the grounding body and the electrified body in the process of entering the potential shall not be less than 6.9 m (medium phase) (6.7 m, side phase), and the potential transfer rod must be used for potential transfer when entering the intense electric field.	15	4) Deduct 4 points/times for carrying out potential transfer without permission of the Responsible Person.			
9	Surface treatment of damaged conductor	1) After entering the equipotential, the equipotential electrician shall tie the safety belt to the sub-conductor, and ground electrician pass the socket spanner to the equipotential electrician with transmission rope. 2) Equipotential electrician fastens the connecting bolt of conductor drainage plate with socket spanner.	30	1) The equipotential electrician should pay attention to avoid excessive action range, and avoid stretching the limbs to the insulator, otherwise, 3 points/times shall be deducted.			

Table (Cont'd)

S/N	Project name	Quality requirements	Score	Deduction standard	Reasons for deduction	Deduction	Scoring
9	Surface treatment of damaged conductor	3) The equipotential electrician passes the socket spanner down the tower with transmission rope	30	2) Deduct 2 points/items if the tools and materials are not transferred by insulating tools and insulating rope. 3) Deduct 3 points/times for tools, instruments or materials dropped during the operation. 4) Deduct 3 points/item during operation transposition if the protection by safety belt is lost. 5) Deduct 2 points/item for the distance between exposed part of human body and the electrified body less than 0.5 m during potential transfer. 6) Deduct 3 points/times for the leftover tools, instruments and materials			
10	Exit the intense electric field	1) After inspection, the heating positions have disappeared and there is no residue at the operation point, the Responsible Person shall authorize the equipotential electrician to carry the insulated transmission rope back to the grading ring and prepare for leaving potential by means of walking along the line.	13	1) Deduct 2 points for the potential transfer without applying to the Responsible Person; deduct 1 point for starting the work without the consent after applying. 2) Deduct 2 points/times for insufficient distance of the exposed parts to the electrified body during potential transfer of equipotential electrician.			

Part II Skill Module Training and Assessment Standards

Table (Cont'd)

S/N	Project name	Quality requirements	Score	Deduction standard	Reasons for deduction	Deduction	Scoring
10	Exit the intense electric field	2) The equipotential electrician shall hook the grading ring with the potential transfer rod and enter the third insulator of the grading ring. With one hand grasping the insulator tightly and the other holding the potential transfer rod, the equipotential electrician shall break himself away from the potential with the help of potential transfer rod. 3) The equipotential electrician exits the ecuipotential according to the operation mode of "two-span and three-short-circuit".	13	3) Deduct 2 points/time for the incorrect action of equipotential electrician exiting the intense electric field and repeat discharging. 4) Deduct 2 points for unskillful potential transfer and 3 points for potential transfer without using the potential transfer pole			
11	Return to the ground	After checking that there is no object left on the tower, the equipotential electrician on the tower shall report it to the Responsible Person and then climb down the tower with insulated transmission rope after obtaining the consent of the Responsible Person	5	1) Deduct 5 points for failure to use the falling protector when climbing down the tower. 2) Deduct 5 points for loosing the protection of the safety belt when moving on the tower. 3) Deduct 1 point/time for hands grasping the shackles when climbing down the tower. 4) Deduct 2 points for any objects left on tower.			

Table (Cont'd)

S/N	Project name	Quality requirements	Score	Deduction standard	Reasons for deduction	Deduction	Scoring
12	End of the work	1) The Responsible Person shall organize all working members to put working apparatus and materials in order and put them in a special kit (bag) after cleaning; clean the site to ensure that "the materials are removed and the site is cleaned after construction". 2) In the post-shift meeting, Responsible Person will give work summaries and comments.Comments include the construction quality of this work and the implementation of safety measures from all working personnel. 3) Responsible Person shall report to the on-duty control personnel that the work is over, apply for the restoration of circuit re-closing and terminate the work order.	10	1) Deduct 2 points for failure to clean the tools. 2) Deduct 2 points for missing tools. 3) Deduct 10 points for failure to hold the post-shift meeting. 4) Deduct 2 points for failure to remove the fence. 5) Deduct 2 points for failure to report to dispatcher			
	Total		100				

Part II Skill Module Training and Assessment Standards

Module 11 Training and Assessment Standards for Live Replacement of Ground Wire Connecting Fitting for 1000kV AC Transmission Line Tangent Tower

I. Training Standard

(I) Training Requirements

Designation of module	Live replacement of ground wire connecting fitting for 1000kV AC transmission line tangent tower	Type of training	Operation
Training method	Practical operation training	Hours of training	14 hours
Training objectives	1. Master the operation flow for live replacement of ground wire connecting fitting for 1000kV AC transmission line tangent tower. 2. Be able to complete the operation of transferring the load of ground wire connecting fitting for 1000kV AC transmission line tangent tower. 3. Be able to independently complete the operation of replacing the ground wire connecting fitting for 1000kV AC transmission line tangent tower (ground potential operation method).		
Training venue	UHV AC training line		
Training content	Adopt the ground potential operation method for live replacement of ground wire connecting fitting for 1000kV AC transmission line tangent tower		
Scope of application	Maintenance personnel of UHV AC transmission line		

(II) Referenced Rules and Specifications

(1) Electrotechnical Terminology (GB/T2900.55-2002).

(2) Insulated Tackles for Live Working (GB/T13034-2008).

(3) Live Working-Insulating Ropes (GB/T 13035-2008).

(4) Calculation Method of Live Working Minimum Approach Distance on AC Transmission Line (GB/T 18037-2000)

(5) Screen Clothes for Live Working on 1000kV AC (GB/T25726-2010).

(6)Code for Design of 1000kV Overhead Transmission Line (GB50665-2011).

(7)Maintenance Code for 1000kV AC Transmission Line (DL/T209-2008).

(8)Operation Code for 1000kV AC Transmission Line (DL/T307-2010).

(9)Technical Guide for Live Working on 1000kV AC Transmission Line (DL/T392-2015).

(10)Minimum Requirements for Utilization of Tools, Devices and Equipment for Live Working (DL/T 877-2004)

(11)State Grid Corporation of China Working Regulations of Power Safety (Transmission Line Section) (Q/GDW1799.2-2013).

(Ⅲ) **Teaching Design for Training**

To complete the work task of "live replacement of ground wire connecting fitting for 1000kV AC transmission line tangent tower", each training stage shall be designed according to the standard operation procedure for work task completion. Each stage includes specific training objectives, training content, hours of training, training methods (training resources), training environment, assessment and evaluation, etc, as shown in the Table 2-11-1.

(Ⅳ) **Operation Flow**

1. Work Task

The ground potential electrician shall carry with the insulated transmission rope and climb the tower to the overhead ground wire operation point, and adopt the ground potential operation method to replace the ground wire connecting fitting for 1000kV AC transmission line tangent tower.

2. Requirements for Weather and Work Site

(1) The live replacement of ground wire connecting fitting for 1000kV AC transmission line tangent tower shall be carried out in good weather. In case of lightning (hearing thunder or seeing lightning), snow, hail, rain, fog and so on, live working is prohibited. When the wind force is greater than level 5, it is unsuitable for live working; when the relative humidity of the air is greater than 80%, insulating tools with moisture-proof properties shall be used if live working needs to be carried out. When emergency live repair is required in bad weather, relevant personnel shall be organized to fully discuss and prepare necessary safety measures, which can be implemented after being approved by the production Supervisor of the unit.

Part II Skill Module Training and Assessment Standards

Table 2-11-1 Training Content Design for Operators for Live Replacement of Ground Wire Connecting Fitting for 1000kV AC Transmission Line Tangent Tower

Training schedule	Training objectives	Training content	Hours of training	Training methods and resources	Preparation of training conditions	Assessment and evaluation
1. Theoretical teaching	1. Grasp the ultimate principle of ground potential operation method. 2. Be familiar with the methods of transferring the load of ground wire connecting fitting for 1000kV AC transmission line tangent tower. 3. Be familiar with the methods of replacing the ground wire connecting fitting for 1000kV AC transmission line tangent tower	1. Basic principle and equivalent circuit diagram of ground potential operation method. 2. Usage of ground wire lifter during load transfer. 3. Replacement method and quality standard of ground wire connecting armor clamp	2	Training methods: Lecture. Training resources: PPT, relevant regulations and specifications	Multimedia classroom	Attendance, classroom questions and assignments
2. Preparations	Be able to complete the preparation before operation	1. Work site survey. 2. Preparation of the standardized operation card. 3. Filling of the work order. 4. Preparation of tools and materials for this operation. 5. Work application	1	Training methods: 1. Site survey and cleaning of tools and materials shall be practiced at site. 2. Preparation of operation card and the filling of work order shall adopt lecture method. Training resources: 1. 1000kV training line; 2. UHV tools warehouse; 3. Blank work order	1. UHV training transmission line 2. Multi-Media classroom	

· 585 ·

Table (Cont'd)

Training schedule	Training objectives	Training content	Hours of training	Training methods and resources	Preparation of training conditions	Assessment and evaluation
3. Work site preparation	Be able to complete the preparations of work site	1. Work site layout. 2. Pre-shift meeting. 3. Inspection of tools and materials	1	Training methods: demonstration and role play. Resources: 1000kV training line	1000kV training line	
4. Trainer's demonstration	The trainees can preliminarily understand the operation process of the task through inspecting and learning from each other's work	1. Climb the tower, install the insulated tackle and prepare for hoisting. 2. The ground potential electrician shall reliably ground both ends of the operation point by ground wire. 3. Install the wire lifting tools and transfer the ground wire load. 4. Remove the former ground wire armor clamp. 5. Install new ground wire armor clamp. 6. Removal tools returned to the ground	2	Training methods: Demonstration. Resources: 1000kV training line	1000kV training line	

Part II Skill Module Training and Assessment Standards

Table (Cont'd)

Training schedule	Training objectives	Training content	Hours of training	Training methods and resources	Preparation of training conditions	Assessment and evaluation
5. Group training of trainees	1. Be able to complete the operation of transferring the load of ground wire connecting fitting for replace the ground wire connecting fitting of 1000kV AC transmission line tangent tower. 2. Be able to complete the operation of replacing the linear ground wire armor clamp	1. The trainees are divided into groups (5 persons for a group) to be trained about the skills to replace the ground wire connecting fitting of 1000kV transmission line tangent tower. 2. Trainers guide the operation of trainees and conduct safety supervision	7	Training methods: Role play. Resources: 1000kV training line	1000kV training line	Score the operation of trainees according to the detailed rules for skill assessment and scoring
6. End of the work	1. Enable the trainees to further distinguish the shortcomings of the operation process and facilitate the promotion in the later stage. 2. Train the trainees in the working style of safe and civilized production	1. Cleaning up the work site. 2. Report to dispatcher. 3. Comment and summarize the work task this time at the post-shift meeting	1	Training methods: Lecture and inductive method	1000kV training line	

· 587 ·

(2) The operating personnel shall be in good spirits, and be familiar with the organizational and technical measures to ensure safety in work; they shall hold the UHV AC live working qualification within the validity period.

(3) The Responsible Person shall organize the personnel with UHV AC live working experience to arrive at the field for investigation in advance, determine the operating methods, required working apparatus and necessary measures according to the results, handle the work orders for live working, and prepare the standard operation instructions.

(4) The work site shall be reasonably set up with fence and warning signs. Non-operating personnel is forbidden to enter.

(5) The deactivation of line re-closing device shall not be applied in the Project. If equipment loses power suddenly during live working, the operating electrician shall consider the equipment as still charged.

(6) The minimum safe distance of the ground potential electrician and the electrified body shall not be less than 6.8m (medium phase)/6.0m (side phase). The effective insulating length of insulating tool cannot be less than 6.8m.

3. Preparations

3.1 Hazards and precontrol measures

(1) Hazard - electric shock injury

Precontrol measures:

① Before work, the Responsible Person shall contact the control personnel on duty, and perform the work licensing procedures.

② Before climbing the tower, ground potential operators must carefully check the double name of the line, the number of the pole and tower, and ground wire location, and then the tower can be climbed after all have been confirmed correct.

③ If lines lose power suddenly during work, operators shall still deem it to be charged. The Responsible Person shall contact the control personnel as soon as possible, and no forced energization is allowed before the on-duty control personnel getting in touch with the Responsible Person.

④ Insulating tool and insulating ropes shall be free of damage, moisture, deformation, and failure. It is not allowed to use non-insulating ropes (such as cotton rope, manila rope, and steel wire rope).

⑤ The ground electrician shall wear flame-retardant underwear and full set of shielding clothes outside (including hat, dresses & trousers, gloves, socks, shoes and mask). All parts shall be in excellent connection conditions.

⑥ Before contacting the overhead ground wire, the ground potential electrician shall reliably ground both ends of the overhead ground wire at the operation point.

⑦ During the moving or transposition of the ground potential electrician on the pole and tower, the safe distance with the electrified body cannot be less than 6.8 m (medium phase)/6.0 m (side phase). The effective length of insulating tool or insulating rope when the ground operator transfers the tools and materials cannot be less than 6.8 m.

⑧ When transmitting large metal objects by using insulating ropes, ground electrician shall not touch them before ground connection.

(2) Hazard - falling accident

Precontrol measures:

① Before climbing the tower, the ground potential electrician must satisfy the requirements of this operation, such as physical condition, mental state, and skill and quality.

② After the ground potential electrician climbs the tower to the operation point, the safety belt and protection rope shall be respectively tied on the firm component.

③ Supervisors shall correct the nonstandard or illegal actions and behaviors at any time. Special monitoring shall be conducted to operators to prevent them from losing the protection of safety belt or insulated backup protection rope during transposition, and it is forbidden to fasten the safety belt or insulated backup protection rope in a position lower than the operating personnel.

(3) Hazard - injury caused by objects falling from heights.

Precontrol measures:

① The ground potential electrician shall put personal tools and fragmentary materials into the tools bag. It is strictly forbidden to hang objects in high place or keep in the mouth.

② Ground electrician should correctly wear a helmet and use the knots. The underface distance from the work site should not be less than the falling radius.

③ The work site shall be set up with fence and warning signs. It shall be noted at any time that Supervisor shall prohibit irrelevant personnel and vehicles from entering operation area.

3.2 Selection of tools and instruments and materials

Tools and materials required for live replacement of ground wire connecting fitting for 1000kV AC transmission line tangent tower can be seen in Table 2-11-2. Before delivering tools and instruments out of warehouse, application voltage class and test period shall be carefully checked and they shall be inspected to ensure that appearance is intact, connection

is firm, rotation is flexible and meet the working task requirements. After delivering tools and instruments out of warehouse, they shall be stored in tools bag or tool kit for transportation to avoid contamination and damp. Metal tools and insulated tools shall be separately loaded and transported to avoid deformation, damage or other defects caused by mixed loading and transportation.

Table 2-11-2　Tools and Materials Required for Live Replacement of Ground Wire Connecting Fitting for 1000kV AC Transmission Line Tangent Tower

S/N	Name	Specification	Unit	Qty.	Remarks
1	Shielding clothes	Shielding efficiency ≥ 60 dB	Set	1	PPE
2	Conductive shoes	The size depends on the wearer	Nos.	1	PPE
3	Flame retardant underwear	Pure mulberry silk	Set	1	PPE
4	Safety belt of double insurance	Backpack	Nos.	1	PPE
5	Safety helmet		Nos.	5	PPE
6	Wire elevator used for ground wire		Set	1	Metal tool
7	Special grounding rod for overhead ground wire		Set	2	Metal tool
8	Insulated transmission rope	φ12mm, with the length matching the lifting height	Nos.	1	Insulating tool
9	Insulated backup protection rope	φ16mm, with the length matching the work site	Nos.	1	Insulating tool
10	Insulated noose	φ14mm	Nos.	1	Insulating tool
11	Insulated tackle	0.5T	Nos.	1	Insulating tool
12	Portable phone system		Set	1	Other tools
13	Insulation resistance tester	5,000V, with electrode width 2cm and interelectrode width 2cm	Set	1	Other tools
14	Ground wire backup protection device		Set	1	Other tools
15	Multimeter		Set	1	Other tools
16	Wind speed, temperature and humidity tester	HT-8321	Nos.	1	Other tools
17	Security fence		Set	Several	Other tools
18	Warning sign	"Work Here", "Access from Here" "Slow Down" and "Blocking"	Set	1	Other tools

Part II Skill Module Training and Assessment Standards

Table (Cont'd)

S/N	Name	Specification	Unit	Qty.	Remarks
19	Red waistcoat	"Responsible Person"	Piece	1	Other tools
20	Moisture-proof tarpaulin	2m×4m	Piece	1	Other tools
21	Personal tools	Flat tong, special wrench, and marking pen	Set	1	Other tools
22	Towel	Cotton	Nos.	2	Other tools
23	Twisted clevis-clevis	ZH-10	Piece	1	Material
24	Hanging-point armor clamp	GD-12S	Piece	1	Material
25	U-shaped hanging panel	U-10	Piece	1	Material
26	Hanging panel	ZS-10	Piece	1	Material
27	Suspension clamp	XGU-2F	Set	1	Material

3.3 Division of labor for operators

Division of labor for operators of the task is shown in Table 2-11-3.

Table 2-11-3 Division of Labor for Operators for Live Replacement of Ground Wire Connecting Fitting for 1000kV AC Transmission Line Tangent Tower

S/N	Post	Qty. (person)	Responsibilities
1	Responsible Person (Supervisor)	1	Be responsible for the labor division of operating personnel of the work task, field investigation before work, preparation of the operation plan, work order filling, field re-investigation, performing work permit procedures, holding pre-shift meeting, implementation of the site safety measures, safety supervision in the operation process, dealing with emergency situations in work, quality surveillance of the work, and the summary after the work
2	Special Supervisor	1	Explain the safety measures within the scope of supervision to the supervised personnel, and inform the hazards and safety precautions before work. The Supervisor shall monitor the supervised personnel to comply with this procedures and implement the on-site safety measures, and timely correct the unsafe acts and behaviors of the supervised personnel.
3	Ground potential electrician	1	Be responsible for replacing the overhead ground wire armor clamp of ground potential
4	Ground electrician	2	Be responsible for transferring the materials, tools and instruments used in operation

4. Work Procedure

The workflow of this task is shown in Table 2-11-4.

Table 2-11-4 Workflow Chart for Live Replacement of Ground Wire Connecting Fitting for 1000kV AC Transmission Line Tangent Tower

S/N	Work Content	Operation Steps and Standards	Safety Measures and Precautions	Responsible Person
1	Site re-survey	The Responsible Person shall complete the following work: (1) Check the line title and the number of the pole and tower correct; guarantee that the foundation and the pole and tower are intact and in normal condition; ensure that the cross and span distance meets the safety requirements; confirm the defect conditions and the specifications and models of earth wires. (2) Check that the terrain and environment shall meet the operation requirements. (3) Check that the safety measures listed in the work order are in line with the actual situations on site, and the measures will be supplemented if necessary	(1) Correctly wear helmet, working clothes, work shoes and protective gloves. (2) Operation under meteorological conditions that may endanger the safety of operators is forbidden. (3) Non-operation personnel and vehicles are strictly prohibited from entering the working site.	
2	Work Permit	The Responsible Person shall be responsible for contacting the on-duty control personnel to start the live working after the approval by the on-duty control personnel	Live working shall not be started without the permission of the on-duty control personnel	
3	Site layout	Install the security fence and hang the signboards correctly: (1) The security fence should take full account of falling objects from the heights and the influence on road traffic. (2) The entrance and exit of the security fence shall be set reasonably. (3) Signs such as "Access from Here", "Work Here", "Access from Here" shall be properly and well arranged	When the influence on road traffic safety is uncontrollable, the traffic management department should be contacted in time to strengthen the on-site control of traffic safety.	

Part II Skill Module Training and Assessment Standards

Table (Cont'd)

S/N	Work Content	Operation Steps and Standards	Safety Measures and Precautions	Responsible Person
4	Hold a pre-shift meeting	(1) All working personnel shall line up. (2) The Responsible Person shall wear red waistcoat and read out the work order and be clear with work task and division of personnel; explain safety measures and technical measures in work; check (inquire after) mental state of all working personnel; inform of hazards in work and precontrol measures. (3) All working personnel shall sign on the work order for confirmation.	(1) The work order shall be filled in, issued and approved in a standardized manner, and the signature shall be complete. (2) All working personnel shall be in good mental states. (3) All working personnel shall be clear with task division of works, safety measures and technical measures	
5	Inspect tools	(1) All necessary tools and instruments shall be prepared as per the operation requirements and placed on the moisture-proof tarpaulin regularly according to the category and location. The appearance and test certificate of tools and instruments shall be checked to ensure there is no omission. (2) The surface insulation resistance of insulating tools shall be tested with insulation resistance tester in correct methods, and the resistance value shall be not less than 700MΩ. (3) The internal resistance of full shielding clothes shall be tested with a multimeter in correct methods, and the resistance value shall be not more than 20Ω. (4) Measure the wind speed and humidity, check the insulation performance of insulating tools, and make records. (5) The inspector shall report to the Responsible Person that all inspection results are in conformity with the operation requirements. (6) The ground potential electrician shall wear the shielding clothes in a right way and pass the inspection, which shall be supervised and inspected by the Responsible Person. (7) Wear personal safety equipment correctly (proper size and easy lock), and the Responsible Person shall supervise and inspect it	(1) The waterproof tarpaulin shall be reasonable in position, and be clean and dry. (2) The appearance of tools and instruments is acceptable after inspection, there is no damage, damp, deformation and malfunction, and the certificate is valid. (3) Insulating tools and insulating ropes are tested as acceptable. (4) Check the connection and assembly of tools.	

S/N	Work Content	Operation Steps and Standards	Safety Measures and Precautions	Responsible Person
6	Climbing the tower	(1) After checking the double name of line without error, the ground potential electrician shall make impulse test on the safety belt and falling protector. (2) The ground potential electrician who carry an insulated transmission rope climbs the tower to the operation point of ground wire support, selects a suitable position, fastens the safety belt and backup protection rope, installs the insulated transmission rope of insulated tackle at a suitable position on the ground wire support. (3) The ground electrician hoists the ground rod of the ground wire, the ground potential electrician makes reliable grounding for both ground wires at two ends of the working point	(1) After checking the correct double name of line, it is allowed to climb the tower for operation. (2) While climbing the tower, it is required to grasp main materials by hands. During the moving or transposition on tower, the protection by the safety belt is required, and operators must grasp the firm components. (3) The ground potential electrician must wear a whole suit of qualified shielding clothes which must be connected reliably, and then start the next operation. (4) The ground terminal shall be connected first during grounding, following the connection of ground wire terminal to ensure excellent insulation of insulating rod of the ground wire.	
7	Install tools and transfer ground wire load	(1) The ground electrician conveys the backup protective device and ground wire lifter to the ground potential electrician who is responsible for installing wire lifting tools correctly. (2) After tools are installed and confirmed to be correct by inspection, the ground potential electrician tightens the screw rod of the ground wire lifter to transfer the vertical load on the armor clamp of the ground wire to the ground wire lifter	(1) The operating electricians up and down shall closely cooperate, and the ground electrician shall listen to the command of the ground potential electrician. (2) The minimum safe distance of the ground potential electrician with the electrified body shall not be less than 6.8m. The effective insulation length of the insulating rope shall not be less than 6.8m. (3) When transferring the tools and instruments up and down on the pole and tower, firm binding is required, and the objects falling from heights are forbidden by the ground potential electrician during operation. (4) The ground wire lifter shall be evenly stressed	

Part II Skill Module Training and Assessment Standards

Table (Cont'd)

S/N	Work Content	Operation Steps and Standards	Safety Measures and Precautions	Responsible Person
8	Remove the former ground wire armor clamp.	Check the impulse test on the ground wire lifter, and remove the original ground wire armor clamp if there are no defects	Before replacement of ground wire armor clamp, it is required to check the impulse test on the ground wire lifter, and remove the original ground wire armor clamp if there are no defects after the consent of the Responsible Person	
9	Install the armor clamp of ground wire	The ground electrician shall hoist the new ground wire armor clamp to the place around the ground wire support, and the ground potential electrician shall install the new ground wire armor clamp	(1) Attention shall be paid to that the ground wire lifter cannot be collided during operation. (2) The tools shall not collide with each other in the process of transfer, and the rope buckling method shall be correct	
10	Removal tools returned to the ground	(1) Release the ground wire lifter to transfer all loads on it to the new ground wire armor clamp, and carry out impulse test inspection for the ground wire and the armor clamp, then remove the ground wire lifter and the backup protective device of the ground wire to transfer them to the ground after the inspection is free from defects and the permission is gotten from the Responsible Person. (2) Remove the ground wire. (3) The ground potential electrician confirms there are no defects on the tower by thorough inspection and carries the insulated transmission rope to climb down after permission of the Responsible Person	(1) When removing the grounding wire, the ground wire terminal and ground terminal shall be removed successively in the process. (2) Firmly grasp the tower while climbing down the tower to prevent missing steps and slip	

Table (Cont'd)

S/N	Work Content	Operation Steps and Standards	Safety Measures and Precautions	Responsible Person
11	End of the work	(1) The site and tools shall be cleaned up, and any left objects on the site is not allowed. The Responsible Person shall comprehensively inspect the completion of the work, and count the number of people. (2) In the post-shift meeting, Responsible Person will give work summaries and comments. Comments include the construction quality of this work and the implementation of safety measures from all working personnel. (3) Responsible Person shall report to the on-duty control personnel that the work is over, terminate the work order, and gather the operators to leave the work site		

II. Assessment Standard

Detailed Rules for Assessment and Scoring of Operation and Inspection Skills of UHV AC Transmission Line

Fill-in Column of Examinee	No.:		Name:		Position:		Date:	MM/DD/YYYY	
Fill-in Column of Assessor	Grade:		Assessor:		Assessment Team Leader:		Starting time:	Closing time:	Operation Duration:
Assessment module	Live replacement of ground wire connecting fitting for 1000kV AC transmission line tangent tower		Assessee		Maintenance personnel of UHV AC transmission line		Assessment method	Operation	Assessment Time Limit 60min
Job Description	Live replacement of ground wire connecting fitting for 1000kV AC transmission line tangent tower								
Work Specifications and Requirements	1. Live working shall be carried out in good weather. In case of thunder, rain, snow or fog, no live working shall be carried out. When the wind force is greater than Level 5 and the humidity is greater than 80%, live working should not be carried out. 2. Five persons are required for this operation, including 1 Responsible Person, 1 special Supervisor, 1 ground electrician and 2 ground electricians. 3. The minimum safe distance of the ground potential electrician and the electrified body shall not be less than 6.8m (medium phase)/ 6.0m (side phase). The effective length of insulating tool cannot be less than 6.8m. 4. Responsibilities of Responsible Person (Supervisor): Be responsible for division of operating personnel of the task, work order reading, handling the formalities of work permit, getting work permits, holding pre-shift meeting, safety supervision in the operation process, dealing with emergency situations in work, quality surveillance, and the summary after work. 5. During the live working, if thunder, rain, strong wind or any other circumstance threaten the safety of the staff, the Responsible Person or Supervisor may stop working temporarily according to the circumstances Given conditions: 1. Training line: 1000kV AC practical training line tangent tower, model of ground wire: JLB20A-150. 2. Work orders have been handled, safety measures have been completed, and oral application (dispatcher or assessor) shall be made at the beginning and end of the work. 3. The instrument shall be used safely and correctly to test the insulating tool. 4. The operation must be carried out according to the working procedures. The scores of the items to be carried out shall be deducted for the process error. In case of major hidden dangers of personal, equipment and operational safety, the assessor may order the termination of the operation (assessment)								

Part II Skill Module Training and Assessment Standards

Standard for Professional Training and Assessment for Operation Maintenance of UHV AC Transmission Line

Table (Cont'd)

Assessment scenario preparation	1. Line: 1000kV AC practical training line tangent tower.Work content: live replacement of ground wire connecting fitting for 1000kV AC transmission line tangent tower.Ground wire model: JLB20A-150. 2. Required operation tools and instruments: 1 set of shielding clothes, 1 pair of conductive shoes, a set of ground wire lifter, 2 sets of special grounding rods of overhead ground wire, 1 insulated transmission rope, 1 insulated backup protection rope, 1 set of ground wire backup protection device, 1 insulating rope sleeve, 1 insulated tackle, 1 set of portable communication system, 1 insulation resistance tester, 1 multimeter, 1 wind speed, temperature and humidity tester, several sets of security fences, 1 set of warning signboard, 1 red waistcoat, 1 moisture-proof tarpaulin, 1 set of personal tools, and 2 towels. 3. The work site shall be monitored, and the safety measures (fence, etc.) on the work site have been fully implemented; non-operation personnel are prohibited from entering the site, and the staff must wear safety helmets when entering the work site. 4. Examinees shall bring their own work clothes, flame retardant cotton underwear, safety helmets, gloves, safety belts (including backup protection rope)
Remarks	1. The deduction shall be done until the scores of each item are deducted completely. In case of major hidden dangers of personal, equipment and operational safety, the assessor may order the termination of the operation. 2. When equipment, working environment, safety belt, safety helmet, tool, shielding clothes, etc., do not conform to the operation condition, the assessor may order the termination of the operation

S/N	Project name	Quality requirements	Score	Deduction standard	Reasons for deduction	Deduction	Scoring
1	Site survey and material investigation	Line name and pole and tower number, on-site working conditions, safe distance and conductor model	2	1) No point shall be awarded in case of no field survey. 2) Deduct 0.5 points for each item not detailed and omitted in the field survey.			
2	Fill in the live working work order	1) Fill in the live working work order with correct and complete content and tidy order. 2) The standard operation instructions shall be correct, complete and tidy	3	1) No point shall be awarded without a work order. 2) Deduct 0.5 points for any error filled in the work order. 3) Deduct 1 point for untidy writing			

Part II Skill Module Training and Assessment Standards

Table (Cont'd)

S/N	Project name	Quality requirements	Score	Deduction standard	Reasons for deduction	Deduction	Scoring
3	Work Permit	Obtain the work permit after contacting the on-duty control personnel	2	1) Deduct 0.5 points for any omitted item and incomplete application and permission			
4	Hold a pre-shift meeting	1) All staff shall correctly wear safety helmets and work clothes, line up and read out the work orders, explain the safety measures, technical measures, hazards and precontrol measures of the work, check (inquire) the mental state of work team members and sign for confirmation after the work task is clear. 2) The personnel allocation shall be accurate and comprehensive, and the Responsible Person shall check the personnel working for clearness	4	1) No point shall be awarded in case of no division of labor. 2) Deduct 1 point for wearing not as per procedures. 3) Deduct 1 point for unclear division of labor. 4) Deduct 1 point for no wearing armband (red waistcoat). 5) Deduct 0.5 points for omitted signing on the work order on site for confirmation			
5	Layout of work site	1) Safety railings and operation signs shall be set properly around the positions placed with operating tools and instruments, and the influence on the objects falling from heights and the road traffic during operation shall be considered for the scope of security fence. 2) The entrance and exit of the fence shall be set reasonably. 3) The warning sign shall include "Access from Here" and "Work Here", and there shall be signs or roadblocks, such as "Slow Down" or "Blocking" on both sides of the road	2	No point shall be awarded in case of no layout of site			

Table (Cont'd)

S/N	Project name	Quality requirements	Score	Deduction standard	Reasons for deduction	Deduction	Scoring
6	Place and check the tools and instruments	1) Place the tools and instruments on moisture-proof tarpaulin as required. 2) The moisture-proof tarpaulin shall be clean and dry. 3) The tools and instruments shall be placed in category according to the requirements of the fixed management. 4) The insulated tools and instruments cannot be mixed with metal tools and materials. 5) The appearance inspection is required for tools and instruments. The inspectors shall wear clean and dry cotton gloves. The surface of insulating tool cannot be worn, deformed and damaged. The operation shall be flexible. 6) In the qualified test period, the safety belt shall be provided with appearance inspection and impulse test. The insulation resistance value on the surface of insulating tool shall be tested section by section by the insulation resistance detector. The measuring electrode shall conform to the procedures (electrode width of 2cm and space between the electrodes of 2cm).	6	1) Deduct 1 point for no placement. 2) Deduct 0.5 points for no checking the qualified test label. 3) Deduct 1 point for wrong usage of testing tools. 4) Deduct 2 points for no testing the insulating tool. 5) Deduct 1 point if the operator hand or hold the insulating tool. 6) Deduct 0.5 points for not using the towel to clean the rigid insulating tool. 7) Deduct 1 point for no shaking measurement of shielding clothes resistance. 8) Deduct 1 point for no impact inspection for safety belt and falling protector. 9) Deduct 0.5 points for no reporting of testing results. 10) Deduct 1 point/item for omitted inspection item			

Part II Skill Module Training and Assessment Standards

Table (Cont'd)

S/N	Project name	Quality requirements	Score	Deduction standard	Reasons for deduction	Deduction	Scoring
6	Place and check the tools and instruments	7) The insulation resistance detector shall be correctly used, and the insulation resistance value cannot be lower than 700MΩ. The internal resistance of full set of qualified shielding clothes cannot be larger than 20Ω. 8) After the insulated tools and instruments are checked, the inspection results shall be reported to the Responsible Person	6				
7	Preparations before climbing the tower	1) It is required to correctly wear a whole suit of qualified shield clothes and conductive shoes as required, with each part connected well, wear flame retardant underwear in the shielding clothes and fasten safety belts. 2) The Responsible Person shall carefully check the correct wearing	4	1) No point shall be awarded if not wearing the shielding clothes. 2) Deduct 1 point for not connecting the shielding clothes properly. 3) Deduct 1 point for no inspection. 4) Deduct 1 point for non-standard wearing of shielding clothes			
8	Check the name, pole number and phase group of the line	Check and report the name, tower number and ground wire location of the line	2	1) No score for unchecked name, tower number and ground wire position of the line. 2) Deduct 1 point for missing items when check. 3) Deduct 1 point for not reporting check results			

· 601 ·

Table (Cont'd)

S/N	Project name	Quality requirements	Score	Deduction standard	Reasons for deduction	Deduction	Scoring
9	Climbing the tower	1) The ground potential electrician who carry an insulated transmission rope climbs the tower to the operation point of ground wire support, selects a suitable position, fastens the safety belt and backup protection rope, installs the insulated tackle at a suitable position on the ground wire support. 2) The ground electrician hoists the ground rod of the ground wire, the ground potential electrician makes reliable grounding for both ground wires at two ends of the working point	10	1) Deduct 2 points for the tower electrician begins to climb the tower without applying to the Responsible Person; deduct 1 point for starting work without agreement after applying. 2) Deduct 2 points for not grasping the main material during the operation of climbing the tower. 3) Deduct 1 point for not putting up the safety belt on shoulder, deduct 2 points per time for insulating rope or safety belt twining or hooking. 4) Deduct 1 point per time for grasping the shackles, hands or feet skidding or missing the step when climbing the tower. 5) Deduct 2 points for not fastening the safety belt. 6) Deduct 1 point for not fastening off the buckle of the safety belt. 7) During the process of grounding, it shall connect ground terminal prior to that of ground wire terminal. For those not meeting the requirements, deduct 2 points. 8) When making displacement on tower, it shall not loss safety protection. For those not meeting the requirements, deduct 1 point			

Table (Cont'd)

S/N	Project name	Quality requirements	Score	Deduction standard	Reasons for deduction	Deduction	Scoring
10	Install tools and transfer ground wire load	1) The ground electrician conveys the backup protective device and ground wire lifter to the ground potential electrician who is responsible for installing all tools correctly. 2) After tools are installed and confirmed to be correct by inspection, the ground potential electrician tightens the screw rod of the ground wire lifter to transfer the vertical load on the armor clamp of the ground wire to the ground wire lifter	15	1) Deduct 1 point for inconsistent cooperation between the ground electrician and the ground potential electrician, deduct 1 point for not reporting to the Responsible Person. 2) Deduct 2 points for not checking after completion of tool installation. 3) Deduct 2 points for the ground potential electrician's failing to tighten the screw rod of the ground wire lifter. 4) Deduct 2 points for the ground potential electrician's objects falling at heights during the process of operation. 5) Deduct 4 points for not installing ground wire backup protective device			
11	Remove the former ground wire armor clamp.	Check the impulse test on the ground wire lifter, remove the original wire lifter armor clamp if there are no defects	15	1) Deduct 2 points for not carrying out impulse test on the ground wire lifter. 2) Deduct 2 points for removing the armor clamp of the ground wire without reporting to the Responsible Person. 3) Deduct 1 point for removing the armor clamp of the ground wire without agreement after reporting to the Responsible Person			

Table (Cont'd)

S/N	Project name	Quality requirements	Score	Deduction standard	Reasons for deduction	Deduction	Scoring
12	Install the new armor clamp of the ground wire	The ground electrician shall hoist the new ground wire armor clamp to the place around the ground wire support, and the ground potential electrician shall install the new ground wire armor clamp	15	1) Deduct 2 points for impacting the ground wire lifter when operation. 2) Deduct 2 points for collision of tools during the transfer process. 3) Deduct 1 point for incorrectness of rope buckling. 4) Deduct 1 point for insecurity of rope buckling			
13	Remove the tools and return to the ground	1) Release the ground wire lifter to transfer all loads on it to the new ground wire armor clamp, and carry out impulse test inspection for the ground wire and the armor clamp, then remove the ground wire lifter and the backup protective device of the ground wire to transfer them to the ground after the inspection is free from defects and the permission is gotten from the Responsible Person. 2) Remove the ground wire. 3) The ground potential electrician confirms there are no defects on the tower by thorough inspection and carries the insulated transmission rope to climb down after permission of the Responsible Person	5	1) Deduct 1 point for not carrying out the impulse test to the ground wire and the armor clamp. 2) Deduct 2 points for removing the ground wire lifter and the backup protection of the ground wire without reporting to the Responsible Person. 3) Deduct 1 point for removing the ground wire lifter and the backup protection of the ground wire without the permission from the Responsible Person after reporting.			

Part II Skill Module Training and Assessment Standards

Table (Cont'd)

S/N	Project name	Quality requirements	Score	Deduction standard	Reasons for deduction	Deduction	Scoring
13	Remove the tools and return to the ground		5	4) It shall remove the ground wire terminal prior to the ground terminal during process of removing grounding wire. For the removing not meeting the requirement, deduct 2 points. 5) Deduct 2 points for starting climb down without application to the Responsible Person. 6) Deduct 0.5 points per time for hands or feet skidding, missing the step, hands grasping the shackles. 7) Deduct 0.5 points per time for insulating rope or safety belt twining or hooking. 8) Deduct 1 point per piece for leaving tools on tower			
14	Clear the tools and the site	1) Clear and classify all operating tools and instrument and put them away. 2) Confirm there are no omissions after inspection (including materials, tools, instrument, etc.)	6	1) Deduct x points and add no points for not clearing. 2) Deduct 2 points for omissions			

· 605 ·

Table (Cont'd)

S/N	Project name	Quality requirements	Score	Deduction standard	Reasons for deduction	Deduction	Scoring
15	Post-shift meeting	1) The Responsible Person makes work summary report. 2) Remove the fence, the staff leave the site	6	1) Add no points for failure to hold the post-shift meeting. 2) Deduct 2 points for not removing the fence			
16	Report the end of the work	The Responsible Person reports the end of the work to the dispatcher	3	No scoring for not reporting to the dispatcher			
	Total		100				

Module 12　Standards for Training and Assessment on Power-cut Replacement of Type I Composite Insulator for 1000kV AC Transmission Line Tangent Tower

I. Training Standard

(I) Training Requirements

Designation of module	Power-cut replacement of Type I composite insulator for 1000kV AC transmission line tangent tower	Type of training	Operation
Training method	Practical operation training	Hours of training	21 training hours
Training objectives	1. Master the operational versions and stress structures of all kinds of tools and instruments, machines and tools, as well as the technical key points of replacing the composite insulator. 2. Acquire proficiency in the operation procedures, technical method and construction hazard of power-cut replacement of type I composite insulator for 1000kV AC transmission line tangent tower. 3. The main operators shall be able to proficiently complete the replacement of the whole tangent type I composite insulator for 1000kV AC transmission line tangent tower.		
Training venue	UHV AC training line		
Training content	Correctly use the operation methods of all kinds of stressed tools and instruments as well as install all kinds of tools and instruments, and use the method of power-cut replacement of type I composite insulator for 1000kV AC transmission line tangent tower.		
Scope of application	Maintenance personnel of UHV AC transmission line		

(II) Referenced Rules and Specifications

(1) Operating Code for Overhead Transmission Line (DL/T741-2010).

(2) Technical Code for Designing 110~500kV Overhead Transmission Line (DL/T5092-1999)

(3) State Grid Corporation of China Working Regulations of Power Safety (Transmis-

sion Line Section) (Q/GDW1799.2-2013).

(4) Code for Design of 1000kV Overhead Transmission Line (GB 50665-2011).

(5) Maintenance Code for 1000kV AC Transmission Line (DL/T 209-2008).

(6) Operation Code for 1000kV AC Transmission Line (DL/T 307-2010).

(7) Maintenance Specification for 110(66)kV ~ 500kV Overhead Transmission Line (State Grid Corporation of China).

(8) Guideline of Maintenance of Overhead Transmission Line State (DLT 1248-2013)

(9) Maintenance Management Regulations of Electric Transmission and Transformation Equipment State (State Grid Corporation of China).

(10) Maintenance and Test Specification of Electric Transmission and Transformation Equipment State (State Grid Corporation of China).

(Ⅱ) **Teaching Design for Training**

To complete the work task of "power-cut replacement of Type I composite insulator for 1000kV AC transmission line tangent tower", each training stage shall be designed according to the standard operation procedure for work task completion. Each stage includes specific training objectives, training content, hours of training, training methods (training resources), training environment, assessment and evaluation, etc, as shown in the Table 2-12-1.

(Ⅳ) **Operation Flow**

1. Work Task

Complete the power-cut replacement of type I composite insulator for 1000kV AC transmission line tangent tower.

2. Requirements for weather and work site

(1) The power-cut replacement of type I composite insulator for 1000kV AC transmission line tangent tower shall be carried out in good weather.

High-place operation in the open air shall be stopped in severe weather, such as heavy wind of 5 degree ad above, rainstorm, thunder, hail, heavy fog and sand storm. In particular cases, if emergency maintenance has to be carried out in severe whether, it shall fully discuss necessary safety measures by related personnel, which can only be implemented after being approved by the Unit.

(2) The operators shall be in good mental state, the members of the working group shall carefully study the work order and safety technical measures, and all the personnel shall be

Part II Skill Module Training and Assessment Standards

Table 2-12-1 Training Content Design for Power-cut Replacement of Type I Composite Insulator for 1000kV AC Transmission Line Tangent Tower

Training schedule	Training objectives	Training content	Hours of training	Training methods and resources	Preparation of training conditions	Assessment and evaluation
1. Theoretical teaching	1. Master the operational versions and stress structures of all kinds of tools and instruments, machines and tools, as well as the technical key points of replacing the composite insulator. 2. Acquire proficiency in the operation procedures, technical method and construction hazard of power-cut replacement of type I composite insulator for 1000kV AC transmission line tangent tower	1. Operating methods of correctly using all kinds of stressed tools and instruments. 2. Correctly install all kinds of tools and instruments 3. Replace whole composite insulator for 1000kV AC transmission line tangent tower by using the power-cut operation method	2	Training methods: Lecture. Training resources: PPT, relevant regulations and specifications	Multimedia classroom	Attendance, classroom questions and assignments
2. Preparations	Be able to complete the preparation before operation	1. Work site survey. 2. Preparation of the standardized operation card. 3. Filling of the work order. 4. Preparation of tools and materials for this operation	1	Training methods: 1. Site survey and cleaning of tools and materials shall be practiced at site. 2. Preparation of operation card and the filling of work order shall adopt lecture method. Training resources: 1. 1000kV practical training line. 2. UHV tools warehouse. 3. Blank work order	1. UHV training transmission line 2. Multi-Media classroom	

· 609 ·

Table (Cont'd)

Training schedule	Training objectives	Training content	Hours of training	Training methods and resources	Preparation of training conditions	Assessment and evaluation
3. Work site preparation	Be able to complete the preparations of work site	1. Work site re-survey. 2. Job application. 3. Work site layout. 4. Pre-shift meeting. 5. Inspection of tools and materials	1	Training methods: Demonstration and role play. Resources; 1000kV training line	1000kV training line	
4. Trainer's demonstration	The trainees can preliminarily understand the operation process of the task through inspecting and learning from each other's work	1. Explain the usage of all kinds of tools and instruments. 2. Demonstrate the connection mode of the tools and instruments on the tower for replacing the composite insulator. 3. The operators working at heights cooperate to demonstrate the operation process of replacing the composite insulator. 4. Replace the composite insulator for 1000KV AC transmission line tangent tower by taking advantages of cooperation of grounding personnel.	2	Training methods: Demonstration. Resources; 1000kV training line	1000kV training line	

Table (Cont'd)

Training schedule	Training objectives	Training content	Hours of training	Training methods and resources	Preparation of training conditions	Assessment and evaluation
5. Group training of trainees	1. Be able to master the usage methods and precautions of all kinds of stressed tools and instruments. 2. Master all operation procedures of replacing the composite whole insulator string. 3. Be able to complete the replacement of type I composite insulator for 1000kV transmission line tangent tower	1. The trainees will be divided into groups (3 persons for working at high altitude and 5 persons for ground coordination) to perform the training of the operation methods of tools, instruments and machines, and the actual operation of replacing the type I composite insulator on site. 2. Trainers guide the operation of trainees and conduct safety supervision	14	Training methods: Role play. Resources; 1000kV training line	1000kV practical training line	Score the operation of trainees according to the detailed rules for skill assessment and scoring
6. End of the work	1. Enable the trainees to further distinguish the shortcomings of the operation process and facilitate the promotion in the later stage. 2. Train the trainees in the working style of safe and civilized production	1. Cleaning up the work site. 2. Report to dispatcher. 3. Comment and summarize the work task this time at post-shift meeting	1	Training methods: Lecture and inductive method	1000kV training line	

conform to the "four clear" (i.e., clear to the operation tasks, the hazards, the operation procedures and the safety measures).

(3) The contact responsible for power-cut and transmission must fulfill the working licensing procedures with the dispatcher and it is strictly prohibited to make an appointment for stop or transfer electricity. Responsible Person must obtain the license work order of the licensor before checking the electricity on the line to be overhauled, setting up the ground wire and carrying out the maintenance work.

(4) After power-cut, the Responsible Person shall make records carefully.

(5) It is strictly prohibit to climb the tower if a honeycomb is found on tower by checking before climbing the pole.

(6) The personnel who works on the tower must use the double-safety belt and wears the safety goggles.

3. Preparations

3.1 Hazards and precontrol measures

(1) Hazard - Climb the live line.

Precontrol measures:

① Before climbing the pole and tower for operation, the Responsible Person and the members of the working team shall check whether the double-name and identification marks carefully (color code, distinguishing mark, etc.) are matched the name of the power-cut line.

② It shall check the root, base, etc. of the iron tower to keep them solid and reliable before climbing the tower.

③ It shall check the climbing tools, instruments and facilities such as safety belt, shackles, tower materials to keep them complete and firm before climbing the pole and tower.

④ Before climbing the pole, the intermediate operators who do not involved the installation of the grounding wire shall carefully verify that the phase sequence, color code, name, and number of the line are consistent with the power-cut line, and confirm that the line name is not error or in opposite order.

(2) Hazard - Operation in violation of safety operation procedures when climbing the tower and working on the tower may cause falling from height.

Precontrol measures:

① During climbing, in order to prevent the operator climbing the pole from falling one after another, the space between them shall not be less than 1.6m.

② Before climbing the iron tower, the dirt on the shoe soles shall be removed, and it shall check whether toolkit is complete to make sure that persons will not be hurt by the falling objects during climbing.

③ Operator shall wear safety helmet and soft soles, keep in narrow movement range with uniform steps when climbing pole and tower.

④ During climbing, the safety belt shall be properly packed, the long tail rope shall be placed in the toolkit, and the main belt shall be hung on the shoulder to prevent the safety belt from hooking shackles and tower materials during climbing, which can lead to falling of the operator from high-altitude.

⑤ Operators shall not lose the protection from the safety belt, and make firm step and clench when moving on the pole and tower.

⑥ When reaching the position of operation point, fasten the safety belt (rope), it shall be firm and reliable, and shall not be hung low for high use.

⑦ The safe distance between the human body, rope, etc. and the conductor must not be less than 9.5m before the verification of live part, and special person shall be assigned to perform supervision during the work.

(3) Hazard - injury caused by objects falling from heights.

Precontrol measures:

① The ground personnel shall not stand under the position vertical to the operation point. The personnel on tower shall prevent falling objects from injury, and the tools and materials used shall be transmitted by ropes.

② Tool bags shall be used when working at heights, relatively large tools shall be fixed on solid components and not allowed to be randomly placed.

③ In the process of lifting with a winching, special personnel shall be assigned to command and cooperate in a unified way. Impulse inspection shall be carried out immediately after the insulator string are lifted off the ground.

(4) Hazard - Prevent injury caused by induced electricity.

① In order to prevent injury from induced electricity, operators on tower shall wear full set of shielding clothes.

② The overhead ground wire shall be reliably grounded before necessary contact.

(5) Hazard - Site operating safety supervision.

① From the beginning to the end of the operation, the safety Supervisor must always carry out uninterrupted safety supervision on the operators on the site.

② The Responsible Person or the Supervisor must wear a safety vest.

(6) Hazard - Transportation safety. It is necessary to pay attention to the driving safety of the vehicle, drive carefully when driving, and illegal driving is not allowed.

3.2 Selection of tools and instruments and materials

See Table 2-12-2 for tools, instruments and materials required for power-cut replacement of whole strain insulator string for 1000kV AC transmission line. Before delivering the tools and instruments outbound the warehouse, application voltage class and test cycle of the tools and instruments shall be carefully checked and they shall be inspected to ensure that appearance is intact, connection is firm, rotation is flexible and meeting the requirements of the work task. After delivering the tools and instruments outbound the warehouse, they shall be kept from contamination and damp. Metal tools and insulated tools shall be separately loaded and transported to avoid deformation and damage caused by mixed loading and transportation.

Table 2-12-2　Tools, Instruments and Materials Required for Power-cut Replacement of Whole Strain Insulator String for 1000kV AC Transmission Line

S/N	Name	Specification	Unit	Qty.	Remarks
1	Grounding wire	Only for 1000kV	Group	2	Insulating tool
2	Insulating gloves	10kV	Nos.	2	Insulating tool
3	Electricity tester	Only for 1000kV	pcs	2	Other tools
4	Full-covered safety belt	Including 20m rear safety rope with buffer bag	Set	3	PPE
5	Personal security wire	(The diameter shall be not less than 16mm²)	Nos.	1	Other tools
6	Steel wire sleeve	Φ22	Nos.	4	Other tools
7	Eight-hook clamp	Applicable to conductor s of eight-splitting 900mm² or above	Set	2	Metal tool
8	Shackle	10T	Nos.	4	Metal tool
9	Rope ladder	The length shall be not less than 15M	Nos.	1	Other tools
10	Hand-operated hoist	9T	Nos.	2	Metal tool
11	Individual hand-operated tool		Set	3	Insulating tool
12	Interphone		Set	5	Other tools
13	Lifting rope tackle	1T	Nos.	1	Other tools

Part II Skill Module Training and Assessment Standards

Table (Cont'd)

S/N	Name	Specification	Unit	Qty.	Remarks
14	Transmission rope	φ16	Set	1	Other tools
15	Pin puller		Nos.	2	Other tools
16	Safety vest		Piece	2	Other tools
17	Security fence		Volume	4	Other tools
18	Moisture-proof tarpaulin		Sheet	1	Other tools
19	Sole timber		Piece	Several	Other tools
20	Composite insulators	FXBW-1000/420	String	1	Material

3.3 Division of labor for operators

Division of labor for operators of the task is shown in Table 2-12-3.

Table 2-12-3 Labor Division of Operators Involving in Power-cut Replacement of Type I Composite Insulator for 1000kV Transmission Line

S/N	Post	Qty. (person)	Responsibilities
1	Responsible Person	1	Be responsible for the labor division of operating personnel of the work task, field investigation before work, preparation of the operation plan, work order filling, field re-investigation, performing work permit procedures, holding pre-shift meeting, implementation of the site safety measures, safety supervision in the operation process, dealing with emergency situations in work, quality surveillance of the work, and the summary after the work
2	Safety Supervisor	2	Be responsible for the safety monitoring work during this operation
3	Workers working at height	3	Be responsible for the operation of the power-cut replacement of insulator string for 1000kV
4	Ground auxiliary personnel	5	Be responsible for ground auxiliary works during the operation.

4. Work Procedure

The workflow of this task is shown in Table 2-12-4.

Standard for Professional Training and Assessment for Operation Maintenance of UHV AC Transmission Line

Table 2-12-4   Working Flow Table of Operators Involving in Power-cut Replacement of Type I Composite Insulator String for 1000kV AC Transmission Line

S/N	Work Content	Operation Standard	Safety Precautions	Responsible Person
1	Work Permit	The contact person for power off/on before the operation must contact the dispatcher to complete the work permit procedure	(1) No work shall be started without the work permit of the working approver. (2) It is strictly prohibit to make appointment for power transmission or outage	
2	Site layout	Install the security fence and hang the signboards correctly: (1) The security fence should take full account of falling objects from the heights and the influence on road traffic. (2) The entrance and exit of the security fence shall be set reasonably. (3) Signs such as "Access from Here", "Work Here", "Slow Down" or "Blocking" shall be properly arranged.	When the influence on road traffic safety is uncontrollable, the traffic management department shall be contacted in time to strengthen the on-site control of traffic safety	
3	Hold a pre-shift meeting	(1) All working personnel shall line up. (2) The Responsible Person shall wear red waistcoat and read out the work order and be clear with work task and division of personnel; explain safety measures and technical measures in work; check (inquire after) mental state of all working personnel; inform of hazards in work and precontrol measures. (3) All working personnel shall sign on the work order for confirmation.	(1) The work order shall be filled in, issued and approved in a standardized manner, and the signature shall be complete. (2) All working personnel shall be in good mental states. (3) All working personnel shall be clear with task division of works, safety measures and technical measures	
4	Inspect tools	(1) All necessary tools and instruments shall be prepared as per the operation requirements and placed on the moisture-proof tarpaulin regularly according to the category and location. The appearance and test certificate of tools and instruments shall be checked to ensure there is no omission. (2) The inspector shall report to the Responsible Person that all inspection results are in conformity with the operation requirements	(1) The waterproof tarpaulin shall be enough in quantity and reasonable in position, and be clean and dry. (2) The appearance of tools and instruments is acceptable after inspection, there is no damage, damp, deformation and malfunction, and the certificate is valid.	

Part II Skill Module Training and Assessment Standards

Table (Cont'd)

S/N	Work Content	Operation Standard	Safety Precautions	Responsible Person
5	Climbing the tower	(1) Check the name and serial number of the line before climbing the pole and tower for operation. For multi-circuit lines on the same tower, the Responsible Person and the members of the working team shall check the double name and identification mark carefully (color code, distinguishing mark, etc.). (2) It shall check the root, base, etc. of the pole and tower to keep them solid and reliable before climbing the tower. (3) During climbing, in order to prevent the pole climbing operator from falling, the distance between the operators shall not be less than 1.6m, the safety belt shall be properly packed, the backup protective rope shall be placed in the kit, and the main belt shall be hung on the shoulder to prevent safety during climbing to prevent the safety belt from hooking shackles and tower materials during climbing, resulting in falling from of tower. (4) When the pole and tower is close to the cross arm, the Supervisor and the operator shall check the identification mark and dual tags of the power-cut line again, and can climb to the operation point only after they are confirmed to be correct.	(1) Operator shall wear safety helmet and soft soles, and keep in narrow movement range with uniform steps when climbing pole and tower. (2) During climbing, the safety belt shall be properly packed, the long tail rope shall be placed in the toolkit, and the main belt shall be hung on the shoulder to prevent the safety belt from hooking shackles and tower materials during climbing, which can lead to falling of the operator from high-altitude. (3) Operators shall not lose the protection from the safety belt, and make firm step and clench when moving on the pole and tower. (4) When reaching the position of operation point, fasten the safety belt (rope), it shall be firm and reliable, and shall not be hung low for high use. (5) The safe distance between the human body, transmission rope, etc. and the conductor must not be less than 9.5m before the verification of live part, and special person shall be assigned to perform supervision during the work.	

Table (Cont'd)

S/N	Work Content	Operation Standard	Safety Precautions	Responsible Person
6	Inspection power and install ground wire	(1) After the pole for climbing is in place, fasten the safety belt to a firm and reliable component or pole, and check whether the buckle is in place correctly. Appliances such as test pen (electroscope) must be transferred by rope. (2) Check whether the ground wire is intact, and install the ground wire according to the procedures (connect the ground terminal prior to the conductor terminal). (3) When ground wire is installed, it must be operated with insulating rope or insulated handle, and it is forbidden to install the ground wire by directly operating the metal part of the ground wire by hand. Make sure the collection of the ground wire is tightly connected to the conductor	(1) Check whether the electroscope is normal during receiving and before using. (2) It is forbidden to install the ground wire by the way of winding the conductor.	
7	Replace composite insulator string I	(1) After reaching the operation point, the operator working at heights shall build a rope ladder to enter the conductor, fix two sets of eight-hook clamps and 9T lever hoist at both sides of insulator string vertical to the direction of the line and reserve sufficient personnel operating space. (2) After confirmation of fixation, the operators working at heights tighten the lever hoists at the same time. When the lever hoists bear the completely bear the tensile force of composite insulator, it shall again check the cross arm for deformation, the connection part for abnormality. Only after confirming everything is normal, can the next operation be carried out. (3) Use rope to bind the composite insulator firmly. After taking off the connection parts of the insulator set fitting at the ends of the composite insulator and the conductor, the operators working at heights on the iron tower cooperate to take off the connection parts of the insulator set fitting at the cross arm end of the composite insulator.	(1) In the process of replacing insulator string with the lever hoist and the eight-hook clamp, it shall check the connection and stress of the lever hoist, wire rope noose (lifting belt) and shackle after the lever hoist begins to bear the conductor load. And the impulse test shall be conducted to confirm that it is completely reliable before continuing to tighten the lever hoist. (2) It shall prevent damaging the composite insulator from colliding with the insulator during using the lever hoist and the rope ladder.	

Table (Cont'd)

S/N	Work Content	Operation Standard	Safety Precautions	Responsible Person
		(4) Slowly place the old composite insulator to the ground, the personnel on the ground takes off the old composite insulator and replaces it with the one in good condition. Only after binding the composite insulator firmly, can it be transferred to the operators on tower. After the insulator is lifted to the operation point, the operators on tower shall first connect the connection parts of the insulator set fitting at the cross arm end and ensure that the pin is in place before continuing the connection of the composite insulator conductor end. (5) It shall check whether the ball head, the socket and the pin are installed in place after the composite insulator is installed, and the impulse test shall be conducted to confirm that the installation is correct and the connection is reliable before slowly releasing the two lever hoists	(3) When using the transmission rope to hoist the insulator string, the ground staff and the staff on the tower must cooperate closely to prevent the lifting rope from winding or the insulator string damaging the conductor due to colliding. When hoisting the insulator string, the ground personnel shall not stand under the vertical position of it.	
8	Removal of tools and instruments	Remove the eight-hook clamp, lever hoist, steel wire sleeves and other tools and instruments		
9	End of the work	(1) Responsible Person shall organize the staff to put working apparatus and materials in order, and clean the site to ensure "materials are removed and the site is cleaned after completion of construction". (2) In the post-shift meeting, Responsible Person will give work summaries and comments. Comments include the construction quality of this work and the implementation of safety measures from all working personnel. (3) Responsible Person shall report the end of the work to the work approver, resume the power transmission of the power-cut line and terminate the work order		

II. Assessment Standard

Detailed Rules for Assessment and Scoring of Operation and Inspection Skills of UHV AC Transmission Line

Fill-in Column of Examinee	No.:		Name:		Position:		Date:	MM/DD/YYYY
Fill-in Column of Assessor	Grade:		Assessor:		Assessment Team Leader:	Starting time:	Closing time:	Operation Duration:
Assessment module	Power-cut replacement of type I composite insulator for 1000kV AC transmission line tangent tower		Assessee		Maintenance personnel of UHV AC transmission line		Assessment method	Operation	Assessment Time Limit	150min
Job Description	Power-cut replacement of phase C type I composite insulator for 1000kV transmission line								
Work Specifications and Requirements	1. Given conditions: C-phase type I composite insulator string for 1000kV AC practical training line is aging, which needs to be replaced. The line is power off, test electricity and connection of the ground wire, the used composite insulators have been tested, the work order has been handled, and the safety measures have been taken. 2. The main operation procedures of the whole process are completed by cooperation of 1 Responsible Person, 1 special Supervisor, 3 electricians on the tower while 5 auxiliary workers on the ground assist the exam participants to finish the up and down transfer of tools, materials and other non-technical work. 3. The exam participants shall make necessary safety check before operation. 4. An oral application shall be made at the beginning of work, and an oral report also shall be made at the end of the work
Assessment scenario preparation	1. Tools and instruments: 2 sets of eight-hook clamps, 1 set of rope ladder, 1 personal security wire, 4 steel wire sleeves with Φ22, 4 10T shackles, 2 9T lever hoists, 1 transmission rope, 1 1T tackle, 1 noose, 5 interphones, 3 sets of double-insurance safety belts and 1 moisture-proof tarpaulin. 2. Material: one composite insulator string with the same model. 3. Operated on the training line
Remarks	1. Personal tools and instruments shall be provided by the exam participants themselves. 2. The deduction shall be done until the scores of each item are deducted completely

Part II Skill Module Training and Assessment Standards

Table (Cont'd)

S/N	Project name	Quality requirements	Score	Deduction standard	Reasons for deduction	Deduction	Scoring
1	Preparation for tools and materials						
1.1	Inspection of personal tools	Adjustable wrench, flat pliers, pin-pulling pliers and tool kits shall meet the quality requirements	2	Deduct 1 point for each error and missing item			
1.2	Inspection of stressed tools	Check whether the eight-hook clamp, lever hoist and rope ladder are within the test qualified period	3	Deduct 1 point for each error and missing item			
1.3	Inspection of safety tools and instruments	Double safety belts, personal safety wires and insulated gloves shall meet the quality requirements and shall be within the test qualified period	3	Deduct 1 point for each error and missing item			
1.4	Material inspection	Check whether the model number of composite insulator string, the appearance are in line with the requirements	2	Deduct 1 point for each error and missing item			
2	Venue layout						
2.1	Site fence	Layout of site fence	2	Deduct 2 points for failure to lay out			
3	Climbing the tower and operation on cross arm						

Table (Cont'd)

S/N	Project name	Quality requirements	Score	Deduction standard	Reasons for deduction	Deduction	Scoring
3.1	Climbing the tower	(1) Check whether there is no abnormality at pole and tower base. (2) Correctly carry the transmission rope (The lifting rope head is double folded, fast knot, and hung across over the shoulder). (3) Correctly climb the tower along the main material of the shackles side	6	(1) Deduct 1 point for every unchecked item. (2) Deduct 2 points for not carrying the lifting rope, and deduct 1 point for nonstandard carrying method of lifting rope. (3) Deduct 1 point for grasping the shackles per time. (4) Deduct 2 points for not climbing the tower along the main material of the shackles side			
3.2	Access the working point on the cross arm	The safety belt protection shall not be lost from the tower body to the working point on the cross arm.	3	Deduct 3 points for failure to use safety belt properly			
3.3	Install tackle and personal security wire	The transfer tackle shall be installed at the correct position for easy operation and is equipped with personal security wire.	3	(1) Deduct 1 point for nonstandard installation of tackle. (2) Deduct 2 points for not hanging the personal security wire			
4	Operation on insulator string						

Part II Skill Module Training and Assessment Standards

Table (Cont'd)

S/N	Project name	Quality requirements	Score	Deduction standard	Reasons for deduction	Deduction	Scoring
4.1	Enter into the working point	(1) Fasten the backup protective rope of the double safety belt to the cross arm in the proper position. (2) Install the rope ladder and enter the operation point along it, fasten the pole belt on the conductor. (3) Check the fitting pin of composite insulator string	5	(1) Deduct 5 points for failure to use double safety belt. (2) Deduct 5 points for incorrect usage of rope ladder. (3) Deduct 3 points per time for failure to use double safety belt properly. (4) Deduct 3 points for no check			
4.2	Installation of tools	(1) Correctly install the noose of the wire rope and the 9T lever hoist. (2) Correctly install the eight-hook clamp. (3) The selected models of noose of wire rope and shackle are correct and the installation is reliable. (4) The installed lever hoist, personal security wire and rope ladder have no effect on the operation of the personnel	9	(1) Deduct 1 point for not taking protection measures for the tower materials (2) Deduct 1 point for incorrect installation of lever hoist, eight-hook clamp (3) Deduct 1 point for every collision or winding of lever hoist after being stressed. (4) Deduct 1 point for every incorrect model selection of noose of wire rope and shackle. (5) Deduct 4 points for every leaving out of the lever hoist			

· 623 ·

Table (Cont'd)

S/N	Project name	Quality requirements	Score	Deduction standard	Reasons for deduction	Deduction	Scoring
4.3	Tighten the composite insulator string	(1) Tighten the 2 9T lever hoists at the same time to ensure that the lever hoists bear balanced stress (2) After the 9T lever hoists bear the stress, check the connection parts of the tackle, shackle, wire rope, etc. to confirm their connection reliability. (3) After confirming that the stress is correct, continue to tighten the lever hoist until the composite insulator string is released, carry out the impulse test. (4) During the use of lever hoist, it shall prevent the lever hoist from damage the composite insulator	7	(1) Deduct 1 point per time for touching the insulator. (2) Deduct 3 points for failure to carry out the impulse test. (3) Deduct 2 points for incorrect use of the lever hoist. (4) Deduct 2 point for every insulator damage due to hit			
4.4	Replace the composite type I insulator string	(1) Bind the transmission rope to the proper position of the composite insulator, which the operators working at heights take off the composite insulator sub-conductor end first. (2) The ground auxiliary personnel tightens the transmission rope to cooperate with the operators working at heights to take off the cross arm of the composite insulator and transfer it to the ground. (3) Transfer the new composite insulator to the pole and tower with the transmission rope, the operators working at heights first connect the cross arm end and install the pin in place.	15	(1) Deduct 2 points for failure to carry out the impulse test. (2) Deduct 2 points for the winding of the transmission rope. (3) Deduct 2 points per time for bump of transmission objects. (4) Deduct 5 points for dropping object. (5) Deduct 1 point per piece for not installing the fitting pin.			

Part II Skill Module Training and Assessment Standards

Table (Cont'd)

S/N	Project name	Quality requirements	Score	Deduction standard	Reasons for deduction	Deduction	Scoring
		(4) The operators working at heights install the sub-conductor end of the composite insulator, R pin of the armor clamp and check whether they have been installed in place. (5) After confirming that the connection is correct, release the lever hoist to make the composite insulator bear the stress, and carry out the impulse test		(6) Deduct 4 points for wrong sequence of mounting and dismounting the conductor end and the cross arm end			
4.5	Remove tools and instruments	(1) Check whether the fitting pin and ball head are in complete and in place, and the orientation of the bowl rim is correct, (2) Remove the lever hoist, eight-hook clamp and noose of wire rope, and transfer them to the ground	7	(1) Deduct 2 points for rope winding. (2) Deduct 2 points per time for the bump of the transmission objects. (3) Deduct 1 point for the incorrect position of the fitting pin. (4) Deduct 1 point for incorrect orientation of the bowl rim			
4.6	Clean the tools and instruments on the tower	Confirm that there are no leftovers	4	Deduct 4 points for any object left behind			
4.7	From conductor to tower body	(1) Enter the cross arm of the iron tower along the rope ladder (2) Safety belt protection shall not be lost during climbing the rope ladder	8	(1) Deduct 4 points for loosing the protection of the safety belt. (2) Deduct 4 points for incorrect use of rope ladder			

· 625 ·

Standard for Professional Training and Assessment for Operation Maintenance of UHV AC Transmission Line

Table (Cont'd)

S/N	Project name	Quality requirements	Score	Deduction standard	Reasons for deduction	Deduction	Scoring
5	Step down the tower	(1) It must correctly climb down along the main material at the shackles side. (2) Correctly carry the transmission rope (lifting rope head shall be double folded, fast knot, and hung over the shoulder)	8	(1) Deduct 4 points for not carrying the transmission rope, deduct 2 points for nonstandard carrying method of transmission rope. (2) Deduct 1 point per time for grasping the shackles. (3) Deduct 4 points for not climbing the tower along the main material of the shackles side			
6	Other requirements						
6.1	Tower operation	(1) No falling objects at heights. (2) Coordinate with both hands to operate during operation. (3) No floating objects. (4) No holding object in mouth	5	(1) Deduct 5 points for falling objects at heights. (2) Deduct 2 points for inconsistent actions. (3) Deduct 2 points for floating objects. (4) Deduct 2 points for holding object in mouth			
6.2	Dress	Correctly wear working clothes, rubber work shoes, safety helmet and protective gloves.	2	Deduct 2 points for missing one item			
6.3	Clean site	Clean up work site after completion and meet the requirements of civilized production	2	Deduct 2 points for failure to clean up work site			
6.4	Completion time	Complete operation as required within the specified time	3	Deduct 1 point if the time exceeds 10 minutes, terminate the operation when the time reaches 180 minutes, and only record the score of completed part			
	Total		100				

Module 13 Standards for Training and Assessment on Power-cut Replacement of Type Single V Composite Insulator for 1000kV AC Transmission Line Tangent Tower

Ⅰ. Training Standard

(Ⅰ) Training Requirements

Designation of module	Power-cut replacement of single-V composite insulator for 1000kV AC transmission line tangent tower	Type of training	Operation
Training method	Practical operation training	Hours of training	21 training hours
Training objectives	1. Master the operational versions and stress structures of all kinds of tools and instruments, machines and tools, as well as the technical key points of replacing the whole string insulator. 2. Acquire proficiency in the operation procedures, technical method and construction hazard of power-cut replacement of type single V composite insulator string for 1000kV AC transmission line tangent tower. 3. The main operators shall be able to proficiently complete the replacement of the whole tangent type single V composite insulator for 1000kV AC transmission line tangent tower		
Training venue	UHV AC training line		
Training content	Correctly use the operation methods of all kinds of stressed tools and instruments as well as install all kinds of tools and instruments, and use the method of power-cut replacement of type single V composite insulator for 1000kV AC transmission line tangent tower		
Scope of application	Maintenance personnel of UHV AC transmission line		

(Ⅱ) Referenced Rules and Specifications

(1) Operating Code for Overhead Transmission Line (DL/T741-2010).

(2) Technical Code for Designing 110~500kV Overhead Transmission Line (DL/T5092-1999)

(3) State Grid Corporation of China Working Regulations of Power Safety (Transmission Line Section) (Q/GDW1799.2-2013).

(4) Code for Design of 1000kV Overhead Transmission Line (GB50665-2011).

(5) Maintenance Code for 1000kV AC Transmission Line (DL/T209-2008).

(6) Operation Code for 1000kV AC Transmission Line (DL/T307-2010).

(7) Maintenance Specification for 110(66)kV ~ 500kV Overhead Transmission Line (State Grid Corporation of China).

(8) Guideline of Maintenance of Overhead Transmission Line State (DLT1248-2013).

(9) Maintenance Management Regulations of Electric Transmission and Transformation Equipment State (State Grid Corporation of China).

(10) Maintenance and Test Specification of Electric Transmission and Transformation Equipment State (State Grid Corporation of China).

(III) Teaching Design for Training

To complete the work task of "power-cut replacement of single V composite insulator for 1000kV AC transmission line tangent tower", each training stage shall be designed according to the standard operation procedure for work task completion. Each stage includes specific training objectives, training content, hours of training, training methods (training resources), training environment, assessment and evaluation, etc, as shown in the Table 2-13-1.

(IV) Operation Flow

1. Work Task

Complete the power-cut replacement of type single V composite insulator for 1000kV AC transmission line tangent tower.

2. Requirements for Weather and Work Site

(1) The power-cut replacement of type V composite insulator string for 1000kV shall be carried out in good weather.

High-place operation in the open air shall be stopped in severe weather, such as heavy wind of 5 degree ad above, rainstorm, thunder, hail, heavy fog and sand storm. In special cases, if emergency maintenance needs to be performed in severe weather, arrange related personnel to fully discuss necessary safety measures, which shall be taken after approved by the company.

(2) The operators shall be in good mental state, the members of the working group shall carefully study the work order and safety technical measures, and all the personnel shall be conform to the "four clear" (i.e., clear to the operation tasks, the hazards, the operation proce-

Part II Skill Module Training and Assessment Standards

Table 2-13-1 Training Content Design for Replacement of Type I Composite Insulator for 1000kV AC Transmission Line Tangent Tower

Training schedule	Training objectives	Training content	Hours of training	Training methods and resources	Preparation of training conditions	Assessment and evaluation
1. Theoretical teaching	1. Master the operational versions and stress structures of all kinds of tools and instruments, machines and tools, as well as the technical key points of replacing the composite insulator. 2. Acquire proficiency in the operation procedures, technical method and construction hazard of power-cut replacement of type single V composite insulator string for 1000kV AC transmission line tangent tower.	1. Correctly use all kinds of stressed tools and equipment, and be familiar with the operation methods of winching machines and tools. 2. Correctly install all kinds of tools and instruments 3. Replace whole composite insulator for 1000kV AC transmission line tangent tower by using the power-cut operation method	2	Training methods: Lecture. Training resources: PPT, relevant regulations and specifications	Multimedia classroom	Attendance, classroom questions and assignments
2. Preparations	Be able to complete the preparation before operation	1. Work site survey. 2. Preparation of the standardized operation card. 3. Filling of the work order. 4. Preparation of tools and materials for this operation	1	Training methods: 1. Site survey and cleaning of tools and materials shall be practiced at site. 2. Preparation of operation card and the filling of work order shall adopt lecture method. Training resources: 1. 1000kV practical training line. 2. UHV tools warehouse. 3. Blank work order	1. UHV training transmission line 2.Multi-Media classroom	

· 629 ·

Table (Cont'd)

Training schedule	Training objectives	Training content	Hours of training	Training methods and resources	Preparation of training conditions	Assessment and evaluation
3. Work site preparation	Be able to complete the preparations of work site	1. Work site re-survey. 2. Job application. 3. Work site layout. 4. Pre-shift meeting. 5. Inspection of tools and materials	1	Training methods: demonstration and role play. Resources; 1000kV training line	1000kV training line	
4. Trainer's demonstration	The trainees can preliminarily understand the operation process of the task through inspecting and learning from each other's work	1. Explain the usage of all kinds of tools and instruments. 2. Demonstrate the connection mode of the tools and instruments on the tower for replacing the composite insulator. 3. The operators working at heights cooperate to demonstrate the operation process of replacing the composite insulator. 4. Replace the composite insulator for 1000kV AC transmission line tangent tower by taking advantages of cooperation of grounding personnel.	2	Training methods: Demonstration method. Resources; 1000kV training line	1000kV training line	

Part II Skill Module Training and Assessment Standards

Table (Cont'd)

Training schedule	Training objectives	Training content	Hours of training	Training methods and resources	Preparation of training conditions	Assessment and evaluation
5. Group training of trainees	1. Be able to master the usage methods and precautions of all kinds of stressed tools and instruments. 2. Master all operation procedures of replacing the composite insulator. 3. Be able to complete the replacement of type I composite insulator for 1000kV transmission line tangent tower	1. The trainees will be divided into groups (3 persons for working at high altitude and 5 persons for ground coordination) to perform the training of the operation methods of tools, instruments and machines, and the actual operation of replacing the type V composite insulator on site. 2. Trainers guide the operation of trainees and conduct safety supervision	14	Training methods: Role play. Resources: 1000kV training line	1000kV training line	Score the operation of trainees according to the detailed rules for skill assessment and scoring
6. End of the work	1. Enable the trainees to further distinguish the shortcomings of the operation process and facilitate the promotion in the later stage. 2. Train the trainees in the working style of safe and civilized production	1. Cleaning up the work site. 2. Report to dispatcher. 3. Comment and summarize the work task this time at post-shift meeting	1	Training method: Lecture and inductive method	1000kV training line	

· 631 ·

dures and the safety measures).

(3) The contact responsible for power-cut and transmission must fulfill the working licensing procedures with the dispatcher and it is strictly prohibited to make an appointment for stop or transfer electricity. Responsible Person must obtain the license work order of the licensor before checking the electricity on the line to be overhauled, setting up the ground wire and carrying out the maintenance work.

(4) After power-cut, the Responsible Person shall make records carefully.

(5) It is strictly prohibit to climb the tower if a honeycomb is found on tower by checking before climbing the pole.

(6) The personnel who works on the tower must use the double-safety belt and wears the safety goggles.

3. Preparations

3.1 Hazards and precontrol measures

(1) Hazard - Climb the live line.

Precontrol measures:

① Before climbing the pole and tower for operation, the Responsible Person and the members of the working team shall check whether the double-name and identification marks carefully (color code, distinguishing mark, etc.) are matched the name of the power-cut line.

② It shall check the root, base, etc. of the iron tower to keep them solid and reliable before climbing the tower.

③ It shall check the climbing tools, instruments and facilities such as safety belt, shackles, tower materials to keep them complete and firm before climbing the pole and tower.

④ Before climbing the pole, the intermediate operators who do not involved the installation of the grounding wire shall carefully verify that the phase sequence, color code, name, and number of the line are consistent with the power-cut line, and confirm that the line name is not error or in opposite order.

(2) Hazard - Operation in violation of safety operation procedures when climbing the tower and working on the tower may cause falling from height.

Precontrol measures:

① During climbing, in order to prevent the operator climbing the pole from falling one after another, the space between them shall not be less than 1.6m.

② Before climbing the iron tower, the dirt on the shoe soles shall be removed, and it shall check whether toolkit is complete to make sure that persons will not be hurt by the falling objects during climbing.

③ Operator shall wear safety helmet and soft soles, keep in narrow movement range with uniform steps when climbing pole and tower.

④ During climbing, the safety belt shall be properly packed, the long tail rope shall be placed in the toolkit, and the main belt shall be hung on the shoulder to prevent the safety belt from hooking shackles and tower materials during climbing, which can lead to falling of the operator from high-altitude.

⑤ Operators shall not lose the protection from the safety belt, and make firm step and clench when moving on the pole and tower.

⑥ When reaching the position of operation point, fasten the safety belt (rope), it shall be firm and reliable, and shall not be hung low for high use.

⑦ The safe distance between the human body, rope, etc. and the conductor must not be less than 9.5m before the verification of live part, and special person shall be assigned to perform supervision during the work.

(3) Hazard - injury caused by objects falling from heights.

Precontrol measures:

① The ground personnel shall not stand under the position vertical to the operation point. The personnel on tower shall prevent falling objects from injury, and the tools and materials used shall be transmitted by ropes.

② Tool bags shall be used when working at heights, relatively large tools shall be fixed on solid components and not allowed to be randomly placed.

③ In the process of lifting with a winching, special personnel shall be assigned to command and cooperate in a unified way. Impulse inspection shall be carried out immediately after the composite insulator string are lifted off the ground.

(4) Hazard - Prevent injury caused by induced electricity.

① In order to prevent injury from induced electricity, operators on tower shall wear full set of shielding clothes.

② The overhead ground wire shall be reliably grounded before necessary contact.

(5) Hazard - Site operating safety supervision.

① From the beginning to the end of the operation, the safety Supervisor must always carry out uninterrupted safety supervision on the operators on the site.

② The Responsible Person or the Supervisor must wear a safety vest.

(6) Hazard - Transportation safety.

It is necessary to pay attention to the driving safety of the vehicle, drive carefully when driving, and illegal driving is not allowed.

3.2 Selection of tools and instruments and materials

See Table 2-13-2 for Tools, Instruments and Materials Required for Power-cut Replacement of Single V Composite Insulator for 1000kV Power Transmission Line Tangent Tower. Before delivering the tools and instruments out of the warehouse, application voltage class and test cycle of the tools and instruments shall be carefully checked and they shall be inspected to ensure that appearance is intact, connection is firm, rotation is flexible and meeting the requirements of the work task. After delivering the tools and instruments out of the warehouse, they shall be kept from contamination and damp. Metal tools and insulated tools shall be separately loaded and transported to avoid deformation and damage caused by mixed loading and transportation.

Table 2-13-2　Tools, Instruments and Materials Required for Power-cut Replacement of Type Single V Composite Insulator String for 1000kV AC Transmission Line

S/N	Name	Specification	Unit	Qty.	Remarks
1	Grounding wire	1000kV	Group	2	Insulating tool
2	Insulating gloves	10kV	Nos.	2	Insulating tool
3	Electricity tester	Only for 1000kV	pcs	2	Other tools
4	Full-covered safety belt	Including 20m rear safety rope with buffer bag	Set	3	PPE
5	Personal security wire	(The diameter shall be not less than 16mm²)	Nos.	1	Other tools
6	Steel wire sleeve	φ22	Nos.	4	Other tools
7	Eight-hook clamp	Applicable to conductors of eight-splitting 900mm² or above	Set	2	Metal tool
8	Shackle	10T	Nos.	4	Metal tool
9	Rope ladder	The length shall be not less than 15M	Nos.	1	Other tools
10	Hand-operated hoist	9T	Nos.	2	Metal tool
11	Individual hand-operated tool		Set	3	Insulating tool
12	Interphone		Set	5	Other tools

Part II Skill Module Training and Assessment Standards

Table (Cont'd)

S/N	Name	Specification	Unit	Qty.	Remarks
13	Lifting rope tackle	1T	Nos.	1	Other tools
14	Transmission rope	φ16	Set	1	Other tools
15	Pin puller		Nos.	2	Other tools
16	Safety vest		Piece	2	Other tools
17	Security fence		Volume	4	Other tools
18	Sole timber		Piece	Several	Other tools
19	Moisture-proof tarpaulin		Sheet	1	Other tools
20	Composite insulators	FXBW-1000/420	String	1	Material

3.3 Division of labor for operators

Division of labor for operators of the task is shown in Table 2-13-3.

Table 2-13-3 Labor Division of Operators Involving in Power-cut Replacement of Type V Composite Insulator for 1000kV Transmission Line

S/N	Post	Qty. (person)	Responsibilities
1	Responsible Person	1	Be responsible for the labor division of operating personnel of the work task, field investigation before work, preparation of the operation plan, work order filling, field re-investigation, performing work permit procedures, holding pre-shift meeting, implementation of the site safety measures, safety supervision in the operation process, dealing with emergency situations in work, quality surveillance of the work, and the summary after the work
2	Safety Supervisor	2	Be responsible for the safety monitoring work during this operation
3	Workers working at height	3	Be responsible for the operation of power-cut replacement of type V composite insulator for 1000kV
4	Ground auxiliary personnel	5	Be responsible for ground auxiliary works during the operation.

4. Work Procedure

The workflow of this task is shown in Table 2-13-4.

Table 2-13-4 Working Procedures for Power-cut Replacement of Whole Strain Insulator String for 1000kV AC Transmission Line

S/N	Work Content	Operation Standard	Safety Precautions	Responsible Person
1	Work Permit	The contact person for power off/on before the operation must contact the dispatcher to complete the work permit procedure	(1) No work shall be started without the work permit of the working approver. (2) It is strictly prohibit to make appointment for power transmission or outage	
2	Site layout	Install the security fence and hang the signboards correctly: (1) The security fence should take full account of falling objects from the heights and the influence on road traffic. (2) The entrance and exit of the security fence shall be set reasonably. (3) Signs such as "Access from Here", "Work Here", "Slow Down" or "Blocking" shall be properly arranged.	When the influence on road traffic safety is uncontrollable, the traffic management department shall be contacted in time to strengthen the on-site control of traffic safety	
3	Hold a pre-shift meeting	(1) All working personnel shall line up. (2) The Responsible Person shall wear red waistcoat and read out the work order and be clear with work task and division of personnel; explain safety measures and technical measures in work; check (inquire after) mental state of all working personnel; inform of hazards in work and precontrol measures. (3) All working personnel shall sign on the work order for confirmation.	(1) The work order shall be filled in, issued and approved in a standardized manner, and the signature shall be complete. (2) All working personnel shall be in good mental states. (3) All working personnel shall be clear with task division of works, safety measures and technical measures	
4	Inspect tools	(1) All necessary tools and instruments shall be prepared as per the operation requirements and placed on the moisture-proof tarpaulin regularly according to the category and location. The appearance and test certificate of tools and instruments shall be checked to ensure there is no omission. (2) The inspector shall report to the Responsible Person that all inspection results are in conformity with the operation requirements	(1) The waterproof tarpaulin shall be enough in quantity and reasonable in position, and be clean and dry. (2) The appearance of tools and instruments is acceptable after inspection, there is no damage, damp, deformation and malfunction, and the certificate is in its validity	

Part II Skill Module Training and Assessment Standards

Table (Cont'd)

S/N	Work Content	Operation Standard	Safety Precautions	Responsible Person
5	Climbing the tower	(1) Check the name and serial number of the line before climbing the pole and tower for operation. For multi-circuit lines on the same tower, the Responsible Person and the members of the working team shall check the double name and identification mark carefully (color code, distinguishing mark, etc.). (2) It shall check the root, base, etc. of the pole and tower to keep them solid and reliable before climbing the tower. (3) During climbing, in order to prevent the pole climbing operator from falling, the distance between the operators shall not be less than 1.6m, the safety belt shall be properly packed, the backup protective rope shall be placed in the kit, and the main belt shall be hung on the shoulder to prevent safety during climbing to prevent the safety belt from hooking shackles and tower materials during climbing, resulting in falling from of tower. (4) When the pole and tower is close to the cross arm, the Supervisor and the operator shall check the identification mark and dual tags of the power-cut line again, and can climb to the operation point only after they are confirmed to be correct	(1) Operator shall wear safety helmet and soft soles, and keep in narrow movement range with uniform steps when climbing pole and tower. (2) During climbing, the safety belt shall be properly packed, the long tail rope shall be placed in the toolkit, and the main belt shall be hung on the shoulder to prevent the safety belt from hooking shackles and tower materials during climbing, which can lead to falling of the operator from high-altitude. (3) Operators shall not lose the protection from the safety belt, and make firm step and clench when moving on the pole and tower. (4) When reaching the position of operation point, fasten the safety belt (rope), it shall be firm and reliable, and shall not be hung low for high use. (5) The safe distance between the human body, headless rope, etc. and the conductor must not be less than 9.5m before the verification of live part, and special person shall be assigned to perform supervision during the work	
6	· Inspection power and install ground wire	After the pole is in place, fasten the safety belt to a firm and reliable component or pole, and check whether the buckle is in place correctly. Appliances such as electroscope must be transferred by transmission rope. (2) Check whether the ground wire is intact, and install the ground wire according to the procedures (connect the ground terminal prior to the conductor terminal).	(1) Check whether the electroscope is normal during receiving and before using. (2) It is forbidden to install the ground wire by the way of winding the conductor	

Table (Cont'd)

S/N	Work Content	Operation Standard	Safety Precautions	Responsible Person
6	Inspection power and install ground wire	(3) When ground wire is installed, it must be operated with insulating rope or insulated handle, and it is forbidden to install the ground wire by directly operating the metal part of the ground wire by hand. Make sure the collection of the ground wire is tightly connected to the conductor		
7	Replace type V composite insulator	(1) After reaching the operation point, the operator working at heights shall build a rope ladder at the proper position on the tower for 1 operator working at heights to enter the conductor, fix two sets of eight-hook clamps and 9T lever hoist at both sides of the armor clamps of the composite insulator vertical to the direction of the line and reserve sufficient personnel operating space. (2) After confirming that all connecting parts have been fixed, 2 operators working at heights tighten the 2 9T lever hoists at the same time. After the 9T lever hoists fully bear the stress of the insulator string, it shall check whether there is deformation on the cross arm and abnormality at the connection part. After confirmation of all normality, use the transmission rope to firmly bind the composite insulator, then take off the connection part of the insulator set fitting between the composite insulator and the conductor end, the operators at heights on the iron tower cooperate to take off the connection part of the insulator set fitting between the composite insulator and the iron tower end. (3) Only after the connection part of the composite insulator has been taken off, can the composite insulator be slowly lowered. The personnel on the ground takes off the composite insulator and replaces it with the one in good condition. Only after binding the composite insulator firmly, can it be transferred to the operators on tower.	(1) In the process of replacing insulator string with the lever hoist and the eight-hook clamp, it shall check the connection and stress of the lever hoist, wire rope noose (lifting belt) and shackle after the lever hoist begins to bear the conductor load. And the impulse test shall be conducted to confirm that it is completely reliable before continuing to tighten the lever hoist. (2) It shall prevent damaging the composite insulator from colliding with the insulator during using the lever hoist and the rope ladder. (3) When using the transmission rope to hoist the insulator string, the ground staff and the staff on the tower must cooperate closely to prevent the transmission rope from winding or the insulator string damaging the conductor due to colliding. When hoisting the insulator string, the ground personnel shall not stand under the vertical position of it.	

Part II Skill Module Training and Assessment Standards

Table (Cont'd)

S/N	Work Content	Operation Standard	Safety Precautions	Responsible Person
7	Replace type V composite insulator	(4) After the insulator is lifted to the operation point, the operators on tower shall first connect the connection parts of the insulator set fitting at the iron tower end and ensure that the pin is in place before continuing the connection of the composite insulator conductor end. (5) Check whether the insulator set fitting and the pin are installed in place after the composite insulator string is installed, and the impulse test shall be conducted to confirm that the installation is correct and the connection is reliable before slowly releasing two lever hoists		
8	Removal of tools and instruments	Remove the eight-hook clamp, lever hoist, steel wire sleeves and other tools and instruments		
9	End of the work	(1) Responsible Person shall organize the staff to put working apparatus and materials in order, and clean the site to ensure "materials are removed and the site is cleaned after completion of construction". (2) In the post-shift meeting, Responsible Person will give work summaries and comments. Comments include the construction quality of this work and the implementation of safety measures from all working personnel. (3) Responsible Person shall report the end of the work to the work approver, resume the power transmission of the power-cut line and terminate the work order		

Standard for Professional Training and Assessment for Operation Maintenance of UHV AC Transmission Line

II. Assessment Standard

Detailed Rules for Assessment and Scoring of Operation and Inspection Skills of UHV AC Transmission Line

Fill-in Column of Examinee	No.:		Unit:		Name:		Position:		Date:		MM/DD/YYYY
Fill-in Column of Assessor	Grade:		Assessor:		Assessment Team Leader:		Starting time:		Closing time:		Operation Duration:
Assessment module	Power-cut replacement of type single V composite insulator for 1000kV AC transmission line tangent tower			Assessee		Maintenance personnel of UHV AC transmission line		Assessment method		Operation	Assessment Time Limit	150min
Job Description	Power-cut replacement of phase C type V composite insulator for 1000kV AC transmission line										
Work Specifications and Requirements	1. Given conditions: C-phase type V composite insulator for 1000kV AC practical training line is aging, which needs to be replaced. The line is power off, test electricity and connection of the ground wire, the used insulators have been tested, the work order has been handled, and the safety measures have been taken. 2. The main operation procedures of the whole process are completed by cooperation of 1 Responsible Person, 1 special Supervisor, 3 electricians on the tower while 5 auxiliary workers on the ground assist the exam participants to finish the up and down transfer of tools, materials and other non-technical work. 3. The exam participants shall make necessary safety check before operation. 4. An oral application shall be made at the beginning of work, and an oral report also shall be made at the end of the work										
Assessment scenario preparation	1. Tools and instruments: 2 sets of eight-hook clamps, 1 set of rope ladder, 1 personal security wire, 4 steel wire sleeves with Φ22, 4 10T shackles, 2 9T lever hoists, 1 1.5T lever hoist, 1 transmission rope, 1 1T tackle, 1 noose, 5 interphones, 3 sets of double-insurance safety belts and 1 moisture-proof tarpaulin. 2. Material: one composite insulator string with the same model. 3. Operated on the training line										
Remarks	1. Personal tools and instruments shall be provided by the exam participants themselves. 2. The deduction shall be done until the scores of each item are deducted completely										

Part II Skill Module Training and Assessment Standards

Table (Cont'd)

S/N	Project name	Quality requirements	Score	Deduction standard	Reasons for deduction	Deduction	Scoring
1	Preparation for tools and materials						
1.1	Inspection of personal tools	Adjustable wrench, flat pliers, pin-pulling pliers and tool kits shall meet the quality requirements	2	Deduct 1 point for each error and missing item			
1.2	Inspection of stressed tools	Check whether the eight-hook clamp, lever hoist and rope ladder are within the test qualified period	3	Deduct 1 point for each error and missing item			
1.3	Inspection of safety tools and instruments	Double safety belts, personal safety wires and insulated gloves shall meet the quality requirements and shall be within the test qualified period	3	Deduct 1 point for each error and missing item			
1.4	Material inspection	Check whether the model number of composite insulator string, the appearance are in line with the requirements	2	Deduct 1 point for each error and missing item			
2	Venue layout						
2.1	Site fence	Layout of site fence	2	Deduct 2 points for failure to lay out			
3	Climbing the tower and operation on cross arm						

Standard for Professional Training and Assessment for Operation Maintenance of UHV AC Transmission Line

Table (Cont'd)

S/N	Project name	Quality requirements	Score	Deduction standard	Reasons for deduction	Deduction	Scoring
3.1	Climbing the tower	1) Check whether there is no abnormality at pole and tower base. 2) Correctly carry the transmission rope (The lifting rope head is double folded, fast knot, and hung across over the shoulder). 3) Correctly climb the tower along the main material of the shackles side	6	1) Deduct 1 point per item for no check. 2) Deduct 2 points for not carrying the transmission rope, and deduct 1 point for nonstandard carrying method of lifting rope. 3) Deduct 1 point per time for grasping the shackles. 4) Deduct 2 points for not climbing the tower along the main material of the shackles side			
3.2	Access the working point on the cross arm	The safety belt protection shall not be lost from the tower body to the working point on the cross arm	3	Deduct 3 points for failure to use safety belt properly			
3.3	Install tackle and personal security wire	The transfer tackle shall be installed at the correct position for easy operation and is equipped with personal security wire.	3	Deduct 1 point for non-standard pulley installation; deduct 2 points for not hanging the personal security wire			
4	Operation on insulator string						

Part II Skill Module Training and Assessment Standards

Table (Cont'd)

S/N	Project name	Quality requirements	Score	Deduction standard	Reasons for deduction	Deduction	Scoring
4.1	Enter into the working point	1) Fasten the safety rope of the double safety belt to the cross arm in the proper position. 2) Enter the operation point along the insulator string, and tie the pole belt to the insulator string. 3) Check the insulator string fitting pin and insulator set fitting	6	1) Deduct 4 points for failure to use double safety belt. 2) Deduct 2 points per time for failure to use double safety belt properly. 3) Deduct 2 points for no inspection			
4.2	Installation of tools	1) Correctly install the noose of the wire rope and the 9T lever hoist. 2) Correctly install the eight-hook clamp. 3) The selected models of noose of wire rope and shackle are correct and the installation is reliable 4) The installed lever hoist, personal security wire and rope ladder have no effect on the operation of the personnel	9	1) Deduct 1 point for not taking protection measures for the tower materials 2) Deduct 1 point for incorrect installation of lever hoist, eight-hook clamp. 3) Deduct 1 point for every collision or winding of lever hoist after being stressed. 4) Deduct 1 point for every incorrect model selection of noose of wire rope and shackle. 5) Deduct 4 points for every leaving out of the lever hoist			

· 643 ·

Table (Cont'd)

S/N	Project name	Quality requirements	Score	Deduction standard	Reasons for deduction	Deduction	Scoring
4.3	Tighten the composite insulator string	1) Tighten the two 9T lever hoists at the same time to ensure that the lever hoists bear balanced stress 2) After the lever hoist is stressed, check the tackle, shackle, wire rope and other connection parts to confirm the connection is reliable, and impact test shall be taken for the insulator string. 3) After confirming that the stress is correct, continue to tighten the lever hoist until the composite insulator string is loosed.	6	1) Deduct 1 point per time for touching the insulator. 2) Deduct 3 points for failure to carry out the impulse test. 3) Deduct 2 points for incorrect use of the lever hoist			
4.4	Replace the composite type I insulator string	1) Bind the transmission rope to the proper position of the composite insulator, which the operators working at heights take off the composite insulator sub-conductor end first. 2) The ground auxiliary personnel tightens the insulator to cooperate with the operators working at heights to take off the iron tower end of the insulator and transfer it to the ground. 3) Transfer the new composite insulator to the pole and tower with the transmission rope, the operators on tower first connect the iron tower end and install the pin in place.	16	1) Deduct 2 points for failure to carry out the impulse test. 2) Deduct 2 points for the winding of the transmission rope. 3) Deduct 2 points per time for bump of transmission objects. 4) Deduct 5 points for dropping objects. 5) Deduct 1 point per Nos. for not installing the fitting pin. 6) Deduct 4 points for wrong sequence of mounting and dismounting the conductor end and the iron tower end			

Part II Skill Module Training and Assessment Standards

Table (Cont'd)

S/N	Project name	Quality requirements	Score	Deduction standard	Reasons for deduction	Deduction	Scoring
4.4	Replace the composite type I insulator string	4) Continue to install the sub-conductor end of the composite insulator, R pin of the armor clamp and check whether they have been installed in place. 5) After confirming that the connection is correct, release the lever hoist to make the composite insulator bear the stress, and carry out the impulse test	16				
4.5	Remove tools and instruments	1) Check whether the fitting pin and ball head are in position, the bowl is oriented correctly, and remove the dirty on the surface of the insulator string. 2) Remove the lever hoist, eight-hook clamp and noose of wire rope, and transfer them to the ground	7	1) Deduct 2 points for rope winding. 2) Deduct 2 points per time for the bump of the transmission objects. 3) Deduct 1 point for the incorrect position of the fitting pin. 4) Deduct 1 point for incorrect orientation of the bowl rim 5) Deduct 1 point for not cleaning the insulator string			
4.6	Clean the tools and instruments on the tower	Confirm that there are no objects left behind	4	Deduct 4 points for any object left behind			
4.7	From conductor to tower body	1) Enter the cross arm of the iron tower along the rope ladder. 2) Safety belt protection shall not be lost during climbing the rope ladder	8	1) Deduct 4 points for loosing the protection of the safety belt 2) Deduct 4 points for incorrect use of rope ladder			

Table (Cont'd)

S/N	Project name	Quality requirements	Score	Deduction standard	Reasons for deduction	Deduction	Scoring
5	Step down the tower	1) It must correctly climb down along the main material at the shackles side. 2) Correctly carry the transmission rope (lifting rope head shall be double folded, fast knot, and hung over the shoulder)	8	1) Deduct 4 points for not carrying the transmission rope, deduct 2 points for nonstandard carrying method of transmission rope 2) Deduct 1 point per time for grasping the shackles 3) Deduct 4 points for not climbing the tower along the main material of the shackles side			
6	Other requirements						
6.1	Tower operation	1) No falling objects at heights. 2) Coordinate with both hands to operate during operation. 3) No floating objects. 4) No holding object in mouth	5	1) Deduct 5 points for falling objects at heights. 2) Deduct 2 points for inconsistent actions. 3) Deduct 2 points for floating objects. 4) Deduct 2 points for holding object in mouth			
6.2	Dress	Correctly wear working clothes, rubber work shoes, safety helmet and protective gloves	2	Deduct 2 points for missing one item			
6.3	Clean site	Clean up work site after completion and meet the requirements of civilized production	2	Deduct 2 points for failure to clean up work site			
6.4	Completion time	Complete operation as required within the specified time	3	Deduct 1 point if the time exceeds 10 minutes, terminate the operation when the time reaches 480 minutes, and only record the score of completed part			
	Total		100				

Module 14 Training and Assessment Standard of the Power-cut Replacement of the Whole Strain Insulator String for 1000kV AC Transmission Line

I. Training Standard

(I) Training Requirements

Designation of module	Power-cut replacement of the whole strain insulator string on 1000kV AC transmission line	Type of training	Operation
Training method	Practical operation training	Hours of training	28 hours
Training objectives	1. Master the operational versions and stress structures of all kinds of tools and instruments, machines and tools, as well as the technical key points of replacing the whole string insulator. 2. Acquire proficiency in the operation procedures, technical method and construction work hazard of power-cut replacement of whole strain insulator string on 1000kV AC transmission line. 3. The personnel, as the principal operator work at height, should skillfully complete the replacement of whole strain insulator string on 1000kV AC transmission line.		
Training venue	UHV AC training line		
Training content	Correctly use all kinds of stressed tools and instruments; be familiar with the operation methods of tools and machines, such as the winching, etc; correctly install all kinds of tools and instruments by using the "pulley block" operation method; and replace whole strain insulator string on 1000kV AC transmission line by using the power-cut operation method.		
Scope of application	Maintenance personnel of UHV AC transmission line		

(II) Referenced Rules and Specifications

(1) Operating Code for Overhead Transmission Line (DL/T741-2010).

(2) Technical Code for Designing 110~500kV Overhead Transmission Line (DL/T5092-1999).

(3) State Grid Corporation of China Working Regulations of Power Safety (Transmission Line Section) (Q/GDW1799.2-2013).

(4) Code for Design of 1000kV Overhead Transmission Line (GB50665-2011).

(5) Maintenance Code for 1000kV AC Transmission Line (DL/T209-2008).

(6) Operation Code for 1000kV AC Transmission Line (DL/T307-2010).

(7) Maintenance Specification for 110(66)kV ~ 500kV Overhead Transmission Line (State Grid Corporation of China).

(8) Guideline of Maintenance of Overhead Transmission Line State (DLT1248-2013).

(9) Maintenance Management Regulations of Electric Transmission and Transformation Equipment State (State Grid Corporation of China).

(10) Maintenance and Test Specification of Electric Transmission and Transformation Equipment State (State Grid Corporation of China).

(III) Teaching Design for Training

To complete the power-cut replacement of whole strain insulator string on 1000kV AC transmission line is the Work task during the design. Each training stage shall be designed according to the standard operation procedure for work task completion. Each stage includes specific training objectives, training content, hours of training, training methods (training resources), training environment, assessment and evaluation, etc, as shown in Table 2-14-1.

(IV) Operation Flow

1. Work Task

Complete the power-cut replacement of the whole strain insulator string on 1000kV AC transmission line.

2. Requirements for Weather and Work Site

(1) The power-cut replacement of strain insulator string on 1000kV shall be carried out in good weather.

High-place operation in the open air shall be stopped in severe weather, such as heavy wind of 5 degree ad above, rainstorm, thunder, hail, heavy fog and sand storm. In special cases, if emergency maintenance needs to be performed in severe weather, arrange related personnel to fully discuss necessary safety measures, which shall be taken after approved by the company.

(2) The operators shall be in good mental state, the members of the working group shall carefully study the work order and safety technical measures, and all the personnel shall be conform to the "four clear" (i.e., clear to the operation tasks, the hazards, the operation proce-

Part II Skill Module Training and Assessment Standards

Table 2-14-1 Training Content Design for Power-cut Replacement of Whole Strain Insulator String on 1000kV AC Transmission Line

Training schedule	Training objectives	Training content	Hours of training	Training methods and resources	Preparation of training conditions	Assessment and evaluation
1. Theoretical teaching	1. Master the operational versions and stress structures of all kinds of tools and instruments, machines and tools, as well as the technical key points of replacing the whole string insulator. 2. Acquire proficiency in the operation procedures, technical method and construction work hazard of power-cut replacement of whole strain insulator string on 1000kV AC transmission line	1. Correctly use all kinds of stressed tools and equipment, and be familiar with the operation methods of winching machines and tools. 2. Correctly install all kinds of tools and instruments by using the "pulley block" operation method. 3. Replace whole strain insulator string on 1000kV AC transmission line by using the power-cut operation method	2	Training methods: Lecture. Training resources: PPT, relevant regulations and specifications	Multimedia classroom	Attendance, classroom questions and assignments
2. Preparations	Be able to complete the preparation before operation	1. Work site survey. 2. Preparation of the standardized operation card. 3. Filling of the work order. 4. Preparation of tools and materials for this operation	1	Training methods: 1. Site survey and cleaning of tools and materials shall be practiced at site. 2. Preparation of operation card and the filling of work order shall adopt lecture method. Training resources: 1. 1000kV practical training line. 2. UHV tools warehouse. 3. Blank work order	1. UHV training transmission line 2. Multi-Media classroom	

· 649 ·

Table (Cont'd)

Training schedule	Training objectives	Training content	Hours of training	Training methods and resources	Preparation of training conditions	Assessment and evaluation
3. Work site preparation	Be able to complete the preparations of work site	1. Work site re-survey. 2. Job application. 3. Work site layout. 4. Pre-shift meeting. 5. Inspection of tools and materials	3	Training methods: demonstration and role play. Resources: 1000kV training line	1000kV practical training line	
4. Trainer's demonstration	The trainees can preliminarily understand the operation process of the task through inspecting and learning from each other's work	1. Explain the usage of all kinds of tools and instruments. 2. Demonstrate the connection mode of tools and instruments on the tower for replacing the whole insulator string. 3. The personnel, who work in high altitude, shall cooperate to demonstrate the operation procedures of replacing the whole insulator string. 4. Replace the whole strain insulator string on 1000kV line by winching	7	Training methods: Demonstration. Resources: 1000kV training line	1000kV training line practical training line	

Part II Skill Module Training and Assessment Standards

Table (Cont'd)

Training schedule	Training objectives	Training content	Hours of training	Training methods and resources	Preparation of training conditions	Assessment and evaluation
5. Group training of trainees	1. Be able to master the usage methods and precautions of all kinds of stressed tools and instruments, winching machines, etc. 2. Master the whole operation procedures of replacing the whole insulator string. 3. Be able to complete the replacement of the whole strain insulator string on 1000kV transmission line	1. The trainees will be divided into groups (A group is with 6 persons for working at high altitude and 9 persons for ground coordination) to train the operation methods of tools, instruments and machines, and the actual operation of replacing the whole insulator string on site. 2. Trainers guide the operation of trainees and conduct safety supervision	14	Training methods: Role play. Resources: 1000kV training line	1000kV training line	Score the operation of trainees according to the detailed rules for skill assessment and scoring
6. End of the work	1. Enable the trainees to further distinguish the shortcomings of the operation process and facilitate the promotion in the later stage. 2. Train the trainees in the working style of safe and civilized production	1. Cleaning up the work site. 2. Report to dispatcher. 3. Comment and summarize the work task this time at post-shift meeting	1	Training method: Lecture and inductive method	1000kV training line	

dures and the safety measures).

(3) The contact responsible for power-cut and transmission must fulfill the working licensing procedures with the dispatcher and it is strictly prohibited to make an appointment for stop or transfer electricity. Responsible Person must obtain the license work order of the licensor before checking the electricity on the line to be overhauled, setting up the ground wire and carrying out the maintenance work.

(4) After power-cut, the Responsible Person shall make records carefully.

(5) It is strictly prohibit to climb the tower if a honeycomb is found on tower by checking before climbing the pole.

(6) The personnel who works on the tower must use the double-safety belt and wears the safety goggles.

3. Preparations

3.1 Hazards and precontrol measures

(1) Hazard - Climb the live line.

Precontrol measures:

① Before climbing the pole and tower for operation, the Responsible Person and the members of the working team shall check whether the double-name and identification marks carefully (color code, distinguishing mark, etc.) are matched the name of the power-cut line.

② It shall check the root, base, etc. of the iron tower to keep them solid and reliable before climbing the tower.

③ It shall check the climbing tools, instruments and facilities such as safety belt, shackles, tower materials to keep them complete and firm before climbing the pole and tower.

④ Before climbing the pole, the intermediate operators who do not involved the installation of the grounding wire shall carefully verify that the phase sequence, color code, name, and number of the line are consistent with the power-cut line, and confirm that the line name is not error or in opposite order.

(2) Hazard - Operation in violation of safety operation procedures when climbing the tower and working on the tower may cause falling from height.

Precontrol measures:

① During climbing, in order to prevent the operator climbing the pole from falling one after another, the space between them shall not be less than 1.6m.

② Before climbing the iron tower, the dirt on the shoe soles shall be removed, and it

Part II Skill Module Training and Assessment Standards

shall check whether toolkit is complete to make sure that persons will not be hurt by the falling objects during climbing.

③ Operator shall wear safety helmet and soft soles, keep in narrow movement range with uniform steps when climbing pole and tower.

④ During climbing, the safety belt shall be properly packed, the long tail rope shall be placed in the toolkit, and the main belt shall be hung on the shoulder to prevent the safety belt from hooking shackles and tower materials during climbing, which can lead to falling of the operator from high-altitude.

⑤ Operators shall not lose the protection from the safety belt, and make firm step and clench when moving on the pole and tower.

⑥ When reaching the position of operation point, fasten the safety belt (rope), it shall be firm and reliable, and shall not be hung low for high use.

⑦ The safe distance between the human body, headless rope, etc. and the conductor must not be less than 9.5m before the power test, and special person shall be assigned to supervise the work.

(3) Hazard - injury caused by objects falling from heights.

Precontrol measures:

① The ground personnel shall not stand under the position vertical to the operation point. The personnel on tower shall prevent falling objects from injury, and the tools and materials used shall be transmitted by ropes.

② Tool bags shall be used when working at heights, relatively large tools shall be fixed on solid components and not allowed to be randomly placed.

③ In the process of lifting with a winching, special personnel shall be assigned to command and cooperate in a unified way. Impulse inspection shall be carried out immediately after the insulator string are lifted off the ground.

(4) Hazard - Prevent injury caused by induced electricity.

① In order to prevent injury from induced electricity, operators on tower shall wear full set of shielding clothes.

② The overhead ground wire shall be reliably grounded before necessary contact.

(5) Hazard - Site operating safety supervision.

① From the beginning to the end of the operation, the safety Supervisor must always carry out uninterrupted safety supervision on the operators on the site.

② The Responsible Person or the Supervisor must wear a safety vest.

(6) Hazard - Transportation safety.

It is necessary to pay attention to the driving safety of the vehicle, drive carefully when driving, and illegal driving is not allowed.

3.2 Selection of tools and instruments and materials

See Table 2-14-2 for tools, instruments and materials required for power-cut replacement of whole strain insulator string for 1000kV AC transmission line. Before delivering the tools and instruments outbound the warehouse, application voltage class and test cycle of the tools and instruments shall be carefully checked and they shall be inspected to ensure that appearance is intact, connection is firm, rotation is flexible and meeting the requirements of the work task. After delivering the tools and instruments outbound the warehouse, they shall be kept from contamination and damp. Metal tools and insulated tools shall be separately loaded and transported to avoid deformation and damage caused by mixed loading and transportation.

Table 2-14-2 Tools, Instruments and Materials Required for Power-cut Replacement of Whole Strain Insulator String for 1000kV AC Transmission Line

S/N	Name	Specification	Unit	Qty.	Remarks
1	Grounding wire	Only for 1000kV	Group	2	Insulating tool
2	Insulating gloves	10kV	Nos.	2	Insulating tool
3	Electricity tester	Only for 1000kV	pcs	2	Other tools
4	Full-covered safety belt	Including 24m rear safety rope with buffer bag	Set	6	PPE
5	Personal security wire	(The diameter shall be not less than 16mm²)	Nos.	1	Other tools
6	Steel wire sleeve	Φ24	Nos.	20	Other tools
7	Iron tackle	15T	Nos.	12	Metal tool
8	Shackle	18T	Nos.	12	Metal tool
9	Iron tackle	5T	Nos.	8	Metal tool
10	Wire rope	Φ24	m	150	Other tools
11	Grinding rope	Φ17.5	m	600	Other tools
12	Winching	5T	Set	2	Motor tool
13	Hand-operated hoist	9T	Nos.	2	Metal tool

Part II Skill Module Training and Assessment Standards

Table (Cont'd)

S/N	Name	Specification	Unit	Qty.	Remarks
14	Individual hand-operated tool		Set	6	Other tools
15	Interphone		Set	8	Other tools
16	Lifting rope tackle	1T	Nos.	2	Metal tool
17	Transmission rope	φ16	Set	2	Other tools
18	Pin puller		Nos.	3	Metal tool
19	Safety vest		Piece	3	Other tools
20	Goggles		Nos.	6	Other tools
21	Security fence		Volume	5	Other tools
22	Sole timber		Piece	Several	Other tools
23	Moisture-proof tarpaulin		Sheet	1	Other tools
24	Porcelain insulator	U550BP/240T	Piece	56	Material

3.3 Division of labor for operators

Division of labor for operators of the task is shown in Table 2-14-3.

Table 2-14-3 Division of Operating Personnel Involving in Power-cut Replacement of Whole Strain Insulator String on 1000kV AC Transmission Line

S/N	Post	Qty. (person)	Responsibilities
1	Responsible Person	1	Be responsible for the labor division of operating personnel of the work task, field investigation before work, preparation of the operation plan, work order filling, field re-investigation, performing work permit procedures, holding pre-shift meeting, implementation of the site safety measures, safety supervision in the operation process, dealing with emergency situations in work, quality surveillance of the work, and the summary after the work
2	Safety Supervisor	2	Be responsible for the safety monitoring work during this operation
3	Workers working at height	6	Be responsible for the operation of power-cut replacement of whole strain insulator string on 1000kV
4	Ground auxiliary personnel	5	Be responsible for ground auxiliary works during the operation.
5	Winching operator	2	Be responsible for the winching work during this operation
6	Signal commander	2	Be responsible for commanding the start and stop of 2 sets of winches

4. Work Procedure

The work flow of this task is shown in Table 2-14-4.

Table 2-14-4 Working Procedures for Power-cut Replacement of Whole Strain Insulator String for 1000kV AC Transmission Line

S/N	Work Content	Operation Standard	Safety Precautions	Responsible Person
1	Work Permit	The contact person for power off/on before the operation must contact the dispatcher to complete the work permit procedure.	(1) No work shall be started without the Work Permit of approver. (2) It is strictly prohibit to make appointment for power transmission or outage	
2	Site layout	Install the security fence and hang the signboards correctly: (1) The security fence should take full account of falling objects from the heights and the influence on road traffic. (2) The entrance and exit of the security fence shall be set reasonably. (3) Signs such as "Access from Here", "Work Here", "Slow Down" or "Blocking" shall be properly arranged.	When the influence on road traffic safety is uncontrollable, the traffic management department should be contacted in time to strengthen the on-site control of traffic safety.	
3	Hold a pre-shift meeting	(1) All working personnel shall line up. (2) The Responsible Person shall wear red waistcoat and read out the work order and be clear with work task and division of personnel; explain safety measures and technical measures in work; check (inquire after) mental state of all working personnel; inform of hazards in work and precontrol measures. (3) All working personnel shall sign on the work order for confirmation.	(1) The work order shall be filled in, issued and approved in a standardized manner, and the signature shall be complete. (2) All working personnel shall be in good mental states. (3) All working personnel shall be clear with task division of works, safety measures and technical measures	

Part II Skill Module Training and Assessment Standards

Table (Cont'd)

S/N	Work Content	Operation Standard	Safety Precautions	Responsible Person
4	Inspect tools	(1) All necessary tools and instruments shall be prepared as per the operation requirements and placed on the moisture-proof tarpaulin regularly according to the category and location. The appearance and test certificate of tools and instruments shall be checked to ensure there is no omission. (2) The inspector shall report to the Responsible Person that all inspection results are in conformity with the operation requirements	(1) The moisture-proof tarpaulin shall be enough in quantity and reasonable in position, and be clean and dry. (2) The appearance of tools and instruments is acceptable after inspection, there is no damage, damp, deformation and malfunction, and the certificate is valid.	
5	Climbing the tower	(1) Check the name and serial number of the line before climbing the pole and tower for operation. For multi-circuit lines on the same tower, the Responsible Person and the members of the working team shall check the double name and identification mark carefully (color code, distinguishing mark, etc.). (2) It shall check the root, base, etc. of the pole and tower to keep them solid and reliable before climbing the tower. (3) During climbing, in order to prevent the pole climbing operator from falling, the distance between the operators shall not be less than 1.6m, the safety belt shall be properly packed, the long tail rope shall be placed in the kit, and the main belt shall be hung on the shoulder to prevent safety during climbing to prevent the safety belt from hooking shackles and tower materials during climbing, resulting in falling from of tower. (4) When the pole and tower is close to the cross arm, the Supervisor and the operator shall check the identification mark and dual tags of the power-cut line again, and can enter the cross arm at the power-cut line side only after they are confirmed to be correct	(1) Operator shall wear safety helmet and soft soles, and keep in narrow movement range with uniform steps when climbing pole and tower. (2) During climbing, the safety belt shall be properly packed, the long tail rope shall be placed in the toolkit, and the main belt shall be hung on the shoulder to prevent the safety belt from hooking shackles and tower materials during climbing, which can lead to falling of the operator from high-altitude. (3) Operators shall not lose the protection from the safety belt, and make firm step and clench when moving on the pole and tower. (4) When reaching the position of operation point, fasten the safety belt (rope), it shall be firm and reliable, and shall not be hung low for high use. (5) The safe distance between the human body, headless rope, etc. and the conductor must not be less than 9.5m before the verification of live part, and special person shall be assigned to perform supervision during the work	

Table (Cont'd)

S/N	Work Content	Operation Standard	Safety Precautions	Responsible Person
6	Inspection power and install ground wire	(1) After the pole is in place, fasten the safety belt to a firm and reliable component or pole, and check whether the buckle is in place correctly. Appliances such as test pen (electroscope) must be transferred by rope. (2) Check whether the ground wire is in good condition, and install the ground wire according to the procedures (first connect the ground terminal, then the conductor terminal). (3) When setting the ground wire, the insulating rope or insulating handle must be used for operation, and it is forbidden to install the ground wire with directly operating the metal part of the ground wire by hand. Make sure the collection of the ground wire is tightly connected to the conductor	(1) Check whether the test pole is normal during receiving and before using. (2) It is forbidden to install the ground wire by the way of winding the conductor	
7	Replacement of tension whole strain insulator string	(1) After the high-altitude workers arrive at the operation point, three high-altitude workers install three 15T iron tackles on the firm tower material between the two strain insulator strings at the cross arm end. Three high-altitude workers install three 15T iron tackles on the corresponding live side link plate, and fix the steel wire sleeve and the lever hoist pothook on one side of the three iron tackles at the cross arm side. Then put the steel wire rope connected by the chain of the lever hoist into the 6 tackles on both sides in turn to form a set of pulley block, and reserve enough operation space for personnel. Tighten the 9T lever hoist after confirming that the connection part between the lever hoist and the pulley block is firmly connected. (2) Three (3) high-altitude personnel shall pass the grinding rope on No. 1 winching through the 5T pulley on the cross arm, and then fix it on the third piece of cross arm end of insulator string. After it is fixed firmly, the grinding rope on No. 2 winching shall be passed through the two 5T pulleys on the insulator set fitting between the cross arm end and the live end in turn, and then fix the grinding rope on No. 2 winching at the third piece of insulator on the live end. After all connecting parts have been connected firmly, the winching can be started.	(1) In the process of replacing insulator string with lever hoist, it is necessary to check the connection and stress of lever hoist, wire rope noose (lifting belt) and shackle after the lever hoist begins to bear the conductor load. And the impulse test shall be conducted to confirm that it is completely reliable before continuing to tighten the lever hoist. (2) To avoid damaging the insulator, it shall be prevented from colliding with the insulator when using the lever hoist.	

Part II Skill Module Training and Assessment Standards

Table (Cont'd)

S/N	Work Content	Operation Standard	Safety Precautions	Responsible Person
7	Replacement of tension whole strain insulator string	(3) The ground grinding operator shall start the No. 2 winching. Firstly, tighten the live end of the insulator string. and then slowly release the insulator string at the live end after the high-altitude personnel taking down the connecting part of the insulator string at the live end. Secondly, start the No. 1 winching after the insulator string is vertical to the ground. After the insulator string is tensioned and lifted up at one end, the high-altitude personnel take down the connecting part of the insulator string at the cross arm end, and then slowly loosen the two ends at the same time to the ground, and the ground personnel cooperate to take down the insulator. After the ground personnel take off the insulator and replace the good insulator, start the No. 1 winching firstly and then start the No. 2 winching. The installation sequence is opposite to the removal sequence. Install the cross arm end and then fasten the live end. (4) After the insulator is in place, the operators on the tower shall first connect the insulator set fitting at the tower end and ensure that the pin is in place before continuing the connection between the metal fitting and the iron tower and conductor. (5) Check whether the insulator set fitting and the pin are installed in place after the insulator string is installed, and the impulse test shall be conducted to confirm that the installation is correct and the connection is reliable before slowly releasing the 9T lever hoist	(3) When using the transmission rope to hang the whole insulator string, the ground operators and the operators on the tower must cooperate closely to prevent the lifting rope from winding the insulator string and damaging the conductor. When hanging the insulator string, the ground operators shall not stand under the vertical. (4) When the whole insulator string is lifted by winching, there shall be a special person for command. The two winches shall cooperate closely. Upon the insulator strings leaving the ground, check again if the insulator strings are bound firmly, and carry out the impulse test. The lifting can be continued only after confirmation	
8	Removal of tools and instruments	Remove the ropes, lever hoist, steel wire sleeves, tackles and other tools and instruments in order		

Table (Cont'd)

S/N	Work Content	Operation Standard	Safety Precautions	Responsible Person
9	End of the work	(1) Responsible Person shall organize the staff to put working apparatus and materials in order, and clean the site to ensure "materials are removed and the site is cleaned after completion of construction". (2) In the post-shift meeting, Responsible Person will give work summaries and comments. Comments include the construction quality of this work and the implementation of safety measures from all working personnel. (3) Responsible Person shall report the end of the work to the work approver, resume the power transmission of the power-cut line and terminate the work order		

II. Assessment Standard

Part II Skill Module Training and Assessment Standards

Detailed Rules for Assessment and Scoring of Operation and Inspection Skills of UHV AC Transmission Line

Fill-in Column of Examinee	No.:		Name:		Position:	Date:	MM/DD/YYYY
Fill-in Column of Assessor	Grade:	Assessor:		Assessment Team Leader:	Starting time:	Closing time :	Operation Duration:
Assessment module	Power-cut replacement of the whole strain insulator string on 1000kV AC transmission line		Assessee	Maintenance personnel of UHV AC transmission line		Assessment method	Operation	Assessment Time Limit	360min
Job Description	Power-cut replacement of C-phase whole right porcelain insulator string on 1000kV AC transmission line							
Work Specifications and Requirements	1. Given conditions: C-phase whole right porcelain insulator string on 1000kV AC training line is aging, which needs to be replaced. The line is power off, test electricity and connect the ground wire, the used insulator has been tested, the work order has been handled, and the safety measures have been taken. 2. The main operation procedures of the whole process are completed by 1 Responsible Person, 1 special Supervisor, 6 electricians on the tower, 7 auxiliary workers on the ground, 2 winching operators to assist the exam participant to finish the up and down transfer of tools, materials and other non-technical work. 3. The exam participants shall make necessary safety check before operation. 4. The tools used to replace the whole insulator string shall meet the stress requirements. 5. An oral application shall be made at the beginning of work, and an oral report also shall be made at the end of the work							
Assessment scenario preparation	1. Tools and instruments: 150m of φ24 wire thread insert, 12 nos. of 15T iron tackle, 8 nos. of 5T iron tackle, 12 nos. of 18T shackle, 150m of φ24 wire rope, a bundle of φ16 grinding rope,2 bundle of 9T lever hoist , 2 sets of 5T winching, 2 sets of lifting rope tackle, 2 nos. of ncose, several intercoms, 6 sets of double safety belt. 2. Material: a string of porcelain insulator, which is with the same model.							
Remarks	1. Personal tools and instruments shall be provided by the exam participants themselves. 2. The deduction shall be done until the scores of each item are deducted completely							

Table (Cont'd)

S/N	Project name	Quality requirements	Score	Deduction standard	Reasons for deduction	Deduction	Scoring
1	Preparation for tools and materials						
1.1	Inspection of personal tools	Adjustable wrench, flat pliers, pin-pulling pliers and tool kits shall meet the quality requirements	2	Deduct 1 point for each error and missing item			
1.2	Inspection of tools and instruments	Test the machine with winching, check whether the gear is normal and check the lever hoist	2	Deduct 1 point for each error and missing item			
1.3	Inspection of wire rope and tackle	Check the wire rope and tackle, confirm the connection is reliable and the force meets the requirements.	2	Deduct 1 point for each error and missing item			
1.4	Safety belt	The double safety belt meets the quality requirements within the test period	1	Deduct 1 point for no check			
1.5	Inspection of special tools	The appearance inspection of personal security wire and insulating gloves shall meet the requirements within the test cycle	1	Deduct 0.5 point for each error and missing item			
1.6	Material inspection	Clean the insulators and check the number of insulator strings, the appearance are in line with the requirements	2	Deduct 1 point for each error and missing item			

Part II Skill Module Training and Assessment Standards

Table (Cont'd)

S/N	Project name	Quality requirements	Score	Deduction standard	Reasons for deduction	Deduction	Scoring
2	Venue layout						
2.1	Arrangement of winching site	Arrangement of winching site and angle tower pulley	2	Deduct 1 point for each error and missing item			
2.2	Site fence	Layout of site fence	1	Deduct 1 point for failure to lay out			
3	Climbing the tower and operation on cross arm						
3.1	Climbing the tower	1) Check whether there is no abnormality at pole and tower base. 2) Correctly carry the transmission rope (The lifting rope head is double folded, fast knot, and hung across over the shoulder). 3) Correctly climb the tower along the main material of the shackles side	6	1) Deduct 1 point for no check of one item. 2) Deduct 2 points for not carrying the transmission rope, and deduct 1 point for nonstandard carrying method of lifting rope. 3) Deduct 1 point for one time hands grasping the shackles. 4) Deduct 2 points for not climbing the tower along the main material of the shackles side			
3.2	Access the working point on the cross arm	The safety belt protection shall not be lost from the tower body to the working point on the cross arm.	3	Deduct 3 points for failure to use safety belt properly.			

Table (Cont'd)

S/N	Project name	Quality requirements	Score	Deduction standard	Reasons for deduction	Deduction	Scoring
3.3	Installation of transfer tackle	The transfer tackle shall be installed at the correct position for easy operation.	1	Deduct 1 point if the installation is not standardized			
4	Operation on insulator string						
4.1	Enter into the working point	1) Fasten the safety rope of the double safety belt to the cross arm in the proper position. 2) Enter the operation point along the insulator string, and tie the pole belt to the insulator string. 3) Check the locking pin of insulator string	5	1) Deduct 5 points for failure to use double safety belt warning board. 2) Deduct 3 points every one time for failure to use double safety belt properly. 3) Deduct 3 points for no check			
4.2	Installation of 3-3 tackle block	1) Make sure that the 3-3 tackle blocks at both the conductor end and tower end are installed correctly. 2) The wire rope shall be threaded in the correct direction without winding. 3) The selected models of noose of wire rope and shackle are correct and the installation is reliable	6	1. Deduct 1 point for every one wrong installation of tackle block. 2. Deduct 1 point for every one incorrect direction of the wire rope. 3. Deduct 1 point for every one collision or winding after the steel wire rope is stressed. 4. Deduct 1 point for every one incorrect selection of the model of wire rope or shackle			

Part II Skill Module Training and Assessment Standards

Table (Cont'd)

S/N	Project name	Quality requirements	Score	Deduction standard	Reasons for deduction	Deduction	Scoring
4.3	Installation of lever hoist	1) Transfer the lever hoist to the tower and install it correctly. 2) Connect the lever hoist with the 3-3 tackle block to make it slightly stressed	5	1) Deduct 1 point for rope winding 2) Deduct 2 points for incorrect position 3) Deduct 1 point for one touch of insulator			
4.4	Tighten the insulator string	1) Tighten the lever hoist. After the chain block is stressed, check the tackle, shackle, wire rope and other connection parts to confirm the connection is reliable, and impact test shall be taken for the insulator string. 2) After confirming that the force is correct, continue to tighten the lever hoist until the insulator string is loose.	5	1) Deduct 3 points for no impulse test 2) Deduct 2 points for incorrect use of lever hoist			
4.5	Replacement of insulator string	1) Correctly install a 6T auxiliary lever hoist on the loose insulator string. 2) Connect the wear-proof lifting rope at a proper position on the insulator string. 3) Tighten up the 9T lever hoist, conduct impulse test after being forced, and take out R pin when there is no abnormality. 4) Remove the insulator set fitting at both ends of the insulator string and tighten up the wear-proof lifting rope. 5) Check the connection part of the rope and conduct impulse test after the rope is stressed.	18	1) Deduct 2 points each time for wrong installation position or operation of the lever hoist 2) Deduct 2 points for incorrect position of lifting tackle by wear-proof rope 3) Deduct 3 points for no impulse test 4) Deduct 2 points for winding of transmission rope 5) Deduct 2 points each time for impact phenomenon of conveying objects			

Table (Cont'd)

S/N	Project name	Quality requirements	Score	Deduction standard	Reasons for deduction	Deduction	Scoring
4.5	Replacement of insulator string	6). Remove 9T lever hoist and transfer insulator string to the ground by wear-proof lifting rope. 7). Transfer the new insulator string to the pole and tower, and install the 9T lever hoist. 8) Tighten the lever hoist, install the R pins at both ends, and check whether they are installed in place.	18	6) Deduct 5 points for falling object 7) Deduct 2 points for each non-installed locking pin.			
4.6	Remove tools and instruments	1) Check whether the locking pin and ball head are in position, the bowl is oriented correctly, and clean the insulator string. 2) After the insulator string of 3-3 tackle block is released and the insulator is stressed, the impulse test shall be conducted on the insulator string. 3) Remove the lever hoist, tackle block and wire rope, and transfer them to the ground.	10	1) Deduct 2 points for rope winding. 2) Deduct 2 points per time for the bump of the transmission objects. 3) Deduct 5 points for not transferring pole belt to the insulator string. 4) Deduct 2 points for the incorrect position of the fitting pin. 5) Deduct 2 points for incorrect orientation of insulator big mouth warning board. 6) Deduct 3 points for no impulse test. 7) Deduct 1 point for not cleaning the insulator string			

Part II Skill Module Training and Assessment Standards

Table (Cont'd)

S/N	Project name	Quality requirements	Score	Deduction standard	Reasons for deduction	Deduction	Scoring
4.7	Clean the tools and instruments on the tower	Confirm that there are no leftovers	4	Deduct 4 points for any object left behind			
4.8	From the insulator string to the tower	The safety belt protection shall not be lost from tower to the insulator string	5	Deduct 5 points for loosing the protection of the safety belt			
5	Step down the tower	1) It must correctly climb down along the main material at the shackles side. 2) Correctly carry the transmission rope (The lifting rope head is double folded, fast knot, and hung over the shoulder).	8	1) Deduct 4 points for not carrying the transmission rope, and deduct 2 points for nonstandard carrying method of lifting rope. 2) Deduct 1 point for one time hands grasping the shackles. 3) Deduct 4 points for not climbing the tower along the main material of the shackles side.			
6	Other requirements						
6.1	Tower operation	1) No falling objects at heights. 2) Coordinate with both hands to operate during operation. 3) No floating objects. 4) None holding object in mouth	5	1) Deduct 5 points for falling objects at heights. 2) Deduct 2 points for inconsistent actions. 3) Deduct 2 points for floating objects. 4) Deduct 2 points for holding object in mouth			

Table (Cont'd)

S/N	Project name	Quality requirements	Score	Deduction standard	Reasons for deduction	Deduction	Scoring
6.2	Dress	Correctly wear working clothes, rubber work shoes, safety helmet and protective gloves	2	Deduct 2 points for missing one item			
6.3	Clean site	Clean up work site after completion and meet the requirements of civilized production	2	Deduct 2 points for failure to clean up work site			
6.4	Completion time	Complete operation as required within the specified time	2	Deduct 1 point if the time exceeds 10 minutes, terminate the operation when the time reaches 480 minutes, and only record the score of completed part			
	Total		100				

Part II Skill Module Training and Assessment Standards

Module 15   Training and Assessment Standard for Power-cut Repair of 1000kV AC Transmission Line Overhead Ground Wire

I. Training Standard

(I) Training Requirements

Designation of module	Power-cut repair of 1000kV AC transmission line overhead ground wire	Type of training	Operation
Training method	Practical operation training	Hours of training	14 training hours
Training objectives	1. Be able to use the hanging wheel to reach the specified work position along the overhead ground wire of 1000kV AC transmission line; 2. Be able to independently complete the operation of repairing overhead ground wire with preformed armor rods repair strip		
Training venue	UHV AC training line		
Training content	Use the hanging wheel to reach the specified work position along the overhead ground wire of 1000kV transmission line, and repair the overhead ground wire of 1000kV AC transmission line with preformed armor rods repair strip		
Scope of application	Maintenance personnel of UHV AC transmission line		

(II) Referenced Rules and Specifications

(1) Electrotechnical Terminology - Overhead Line (GB/T 2900.51-1998).

(2) Code for Design of 1000kV Overhead Transmission Line (GB50665-2011).

(3) Maintenance Code for 1000kV AC Transmission Line (DL/T209-2008).

(4) Operation Code for 1000kV AC Transmission Line (DL/T307-2010).

(5) State Grid Corporation of China Working Regulations of Power Safety (Transmission Line Section) (Q/GDW1799.2-2013).

(III) Teaching Design for Training

To complete the power-cut repair of 1000kV AC transmission line overhead ground wire is the Work task during the design. Each training stage shall be designed according to the standard operation procedure for work task completion. Each stage includes specific training objectives, training content, hours of training, training methods (training resources), training environment, assessment and evaluation, etc, as shown in Table 2-15-1.

Standard for Professional Training and Assessment for Operation Maintenance of UHV AC Transmission Line

Table 2-15-1   Training Content Design for Power-cut Repair of 1000kV AC Transmission Line Overhead Ground Wire

Training schedule	Training objectives	Training content	Hours of training	Training methods and resources	Preparation of training conditions	Assessment and evaluation
1. Theoretical teaching	1. Master the basic method of using hanging wheel to reach the specified work position along the overhead ground wire of 1000kV AC transmission line 2. Be familiar with the repair methods for the damage of overhead ground wire of transmission line	1. Classification, structure and precautions of the hanging wheel. 2. The method of using the hanging wheel to reach the specified work position along the overhead ground wire of 1000kV AC transmission line. 3. Repair method and quality standard for transmission line overhead ground wire	2	Training methods: Lecture. Training resources: PPT, relevant regulations and specifications	Multimedia classroom	Attendance, classroom questions and assignments
2. Preparations	Be able to complete the preparation before operation	1. Work site survey. 2. Preparation of the standardized operation card. 3. Fill in the first work order of transmission line. 4. Preparation of tools and materials for this operation	1	Training methods: 1. Site survey and cleaning of tools and materials shall be practiced at site. 2. Preparation of operation card and the filling of work order shall adopt lecture method. Training resources: 1. 1000kV training line; 2. UHV tools warehouse; 3. Blank work order	1. UHV transmission practical training line 2. Multi-Media classroom	

Part II Skill Module Training and Assessment Standards

Table (Cont'd)

Training schedule	Training objectives	Training content	Hours of training	Training methods and resources	Preparation of training conditions	Assessment and evaluation
3. Work site preparation	Be able to complete the preparations of work site	1. Work application. 2. Work site re-survey. 3. Work site layout. 4. Pre-shift meeting. 5. Inspection of tools and materials	1	Training methods: demonstration and role play. Resources: 1000kV training line	1000kV training line	
4. Trainer's demonstration	The trainees can preliminarily understand the operation process of the task through inspecting and learning from each other's work	1. Install the ground wire. 2. Installation the hanging wheel. 3. Use the hanging wheel to reach the specified work position along the overhead ground wire of 1000kV AC transmission line. 4. Repair the overhead ground wire with preformed armor rods repair strip	2	Training methods: Demonstration. Resources: 1000kV training line	1000kV training line	

Table (Cont'd)

Training schedule	Training objectives	Training content	Hours of training	Training methods and resources	Preparation of training conditions	Assessment and evaluation
5. Group training of trainees	1. Be able to complete the correct installation of the hanging wheel. 2. Be able to use the hanging wheel to reach the specified operation position along the overhead ground wire of 1000kV AC transmission line. 3. Be able to complete the repair of 1000kV transmission line overhead ground wire	1. The trainees are grouped (6 persons per group) to train the skills of using hanging wheel and repairing overhead ground wire. 2. Trainers guide the operation of trainees and conduct safety supervision	7	Training methods: Role play. Resources: 1000kV training line	1000kV training line	Score the operation of trainees according to the detailed rules for skill assessment and scoring
6. End of the work	1. Enable the trainees to further distinguish the shortcomings of the operation process and facilitate the promotion in the later stage. 2. Train the trainees in the work awareness of safe and civilized production	1. Cleaning up the work site. 2. Report to dispatcher. 3. Comment and summarize the work task this time at post-shift meeting	1	Training method: Lecture and inductive method	1000kV training line	

(Ⅳ) Operation Flow

1. Work Task

Finish the task of power-cut repair of 1000kV AC transmission line overhead ground wire.

2. Requirements for Weather and Work Site

(1) It should be carried out in good weather for power-cut repair of 1000kV AC transmission line overhead ground wire.

In case of thunder (hear thunder, see lightning), snow, hail, rain, fog, etc., and the wind force is greater than level 5, any working shall not be conducted. When the working must be carried out in bad weather, relevant personnel shall be organized to fully discuss and prepare necessary safety measures, which can only be carried out after the approval of the production leader in charge of the unit.

(2) The operators shall be in good mental state, the members of the working group shall carefully study the work order and safety technical measures, and all the personnel shall be conform to the "four clear" (i.e., clear to the operation tasks, the hazards, the operation procedures and the safety measures).

(3) The Responsible Person must contact with the dispatcher to fulfill the Work permit procedures, and it is strictly prohibited to stop and transfer electricity at the appointed time. Responsible Person must obtain the license work order of the licensor before checking the electricity on the line to be overhauled, setting up the ground wire and carrying out the maintenance work.

(4) After power-cut, the Responsible Person shall make records carefully.

(5) Check whether there is a honeycomb on the tower before climbing the tower. It is forbidden to climb the tower if a honeycomb is found.

(6) Safety belt shall be used by the operators on the tower.

3. Preparations

3.1 Hazards and precontrol measures

(1) Hazard - Climb the live line.

Precontrol measures:

① Before climbing the tower, the Responsible Person and the members of the work team shall carefully check that the double name and identification mark (color code, identification mark, etc.) are consistent with the name of the power-cut line.

② Check the root and base, etc. of iron tower before climbing the tower to keep it solid and reliable.

③ Before climbing the tower, the shackles and tower materials shall be checked to confirm completely reliable.

(2) Hazard - Operation in violation of safety operation procedures when climbing the tower and working on the tower may cause falling from height.

Precontrol measures:

① In order to prevent tower climbing personnel from colliding with each other, the distance between tower climbing operators shall not be less than 1.6m during climbing the tower.

② Before climbing the tower, the soil on the sole shall be removed, the tool kit shall be completed, and do not carry a heavy load during climbing.

③ Operators shall wear safety helmet and soft work shoes, and climb evenly when climbing the tower.

④ During climbing the tower, the safety belt shall be properly packed, the safety rope at the back shall be placed in the kit, and the main belt shall be hung on the shoulder to prevent safety during climbing to prevent the safety belt from hooking shackles and tower materials during climbing, resulting in falling from the tower.

⑤ Operators shall not lose the protection from the safety belt and step and hold it firmly when moving on the tower.

⑥ When reaching the position of operation point, fasten the safety belt. And the safety belt and backup protection rope shall be firmly fastened on the component respectively, and shall not be hung low for high use.

⑦ The safe distance between the human body, transmission rope, etc. and the conductor shall not be less than 9.5m before the power test, and special person shall be assigned to supervise the work

(3) Hazard - injury caused by objects falling from heights.

Precontrol measures:

① The ground staff shall not stand under the position vertical to the operation point. The personnel on tower shall prevent falling objects from injury, and the tools and materials used shall be transmitted by ropes.

② Tool bags should be used when working at heights. Larger tools should be fixed on solid components and are not allowed to be randomly placed.

(4) Hazard - Prevent injury caused by induced electricity.

① Before connecting the overhead ground wire, the ground terminal of the ground wire shall be reliably grounded.

② When the ground wire is placed, the operator shall wear insulating gloves and hold the insulating part.

(5) Hazard - Site operating safety supervision.

① The safety Supervisor shall continuously monitor the operators during the operation.

② The Responsible Person and the Supervisor must have an clear identification which is suitable for their identity.

(6) Hazard - Transportation safety.

It is necessary to pay attention to the driving safety of the vehicle, drive carefully when driving, and illegal driving is not allowed.

3.2 Selection of tools and instruments and materials

See Table 2-15-2 for Tools, Instruments and Materials Required for the Power-cut Repair of 1000kV AC Transmission Line Overhead Ground Wire. Before delivering the tools and instruments out of the warehouse, application voltage class and test period of the tools and instruments shall be carefully checked and they shall be inspected to ensure that appearance is intact, connection is firm, rotation is flexible and meeting the requirements of the work task. After delivering the tools and instruments out of the warehouse, they shall be kept from contamination and damp. Metal tools and insulated tools shall be separately loaded and transported to avoid deformation and damage caused by mixed loading and transportation.

Table 2-15-2 Tools, Instruments and Materials Required for the Power-cut Repair of 1000kV AC Transmission Line Overhead Ground Wire

S/N	Name	Specification	Unit	Qty.	Remarks
1	Operation hanging wheel		Nos.	1	Metal tool
2	Safety belt	Including 20m rear safety rope with buffer bag	Nos.	2	PPE
3	Insulating gloves		Nos.	1	Insulating tool
4	Transmission rope	φ16	Nos.	1	Other tools
5	Iron tackle	0.5T	Nos.	1	Metal tool
6	Steel wire sleeve	φ8	Nos.	1	Other tools

Table (Cont'd)

S/N	Name	Specification	Unit	Qty.	Remarks
7	Personal tools		Set	2	Other tools
8	Sandpaper		Sheet	1	Other tools
9	Steel tap		Nos.	1	Other tools
10	Marker pen		Nos.	1	Other tools
11	Wood hammer		Nos.	1	Other tools
12	Interphone		Set	3	Other tools
13	Safety vest		Piece	2	Other tools
14	Security fence		Volume	4	Other tools
15	Moisture-proof tarpaulin		Sheet	1	Other tools
16	Hanging ground wire	1000kV	Group	3	Other tools
17	Ground wire	25mm2	Group	1	Other tools
18	Electricity tester	1000kV	Nos.	1	Other tools
19	Falling protector	T-Shape	Nos.	1	Other tools
20	Preformed armor rods	Corresponding ground wire type	Group	1	Material

3.3 Division of labor for operators

Division of labor for operators of the task is shown in Table 2-15-3.

Table 2-15-3 Division of Operating Personnel Involving in Power-cut Repair of 1000kV AC Transmission Line Overhead Ground Wire

S/N	Post	Qty. (person)	Responsibilities
1	Responsible Person	1	Be responsible for the labor division of operating personnel of the work task, field investigation before work, preparation of the operation plan, work order filling, field re-investigation, performing work permit procedures, holding pre-shift meeting, implementation of the site safety measures, safety supervision in the operation process, dealing with emergency situations in work, quality surveillance of the work, and the summary after the work
2	Safety Supervisor	1	Explain the safety measures within the scope of supervision to the supervised personnel, and inform the hazards and safety precautions before work. The Supervisor shall monitor the supervised personnel to comply with this procedures and strictly implement the on-site safety measures, and timely correct the unsafe acts and behaviors of the supervised personnel.
3	Workers working at heights	2	Be responsible for the operation of power-cut repair of 1000kV AC transmission line overhead ground wire
4	Ground auxiliary personnel	2	Be responsible for ground auxiliary works during the operation.

Part II Skill Module Training and Assessment Standards

4. Work Procedure

The work flow of this task is shown in Table 2-15-4.

Table 2-15-4 Work Procedures for Power-cut Repair of 1000kV AC Transmission Line Overhead Ground Wire

S/N	Work Content	Operation Steps and Standards	Safety Measures and Precautions	Responsible Person
1	Work Permit	(1) The Responsible Person must contact with the dispatcher to fulfill the Work licensing procedures	(1) No work shall be started without the Work Permit of approver. (2) It is strictly prohibit to make appointment for power transmission or outage	
2	Site re-survey	The Responsible Person shall complete the following work: (1) Check the line name, the number of the iron tower number on spot to ensure they are correct; guarantee that the foundation and the tower body are intact and in normal condition, and ensure that the cross and span distance meets the safety requirements; confirm the defect conditions and the specification and model of ground wire; (2) Check that the terrain and environment shall meet the operation requirements; (3) Check that the safety measures listed in the work order are in line with the actual situations on site, and the measures will be supplemented if necessary	(1) Correctly wear helmet, working clothes, work shoes and protective gloves. (2) Non-operation personnel and vehicles are strictly prohibited from entering the work site	
3	Site layout	Arrange the security fence and hang the signboards correctly: (1) The security fence should take full account of falling objects from the heights and the influence on road traffic; (2) The entrance and exit of the security fence shall be set reasonably; (3) Signs such as "Access from Here", "Work Here", "Slow Down" or "Blocking" shall be properly arranged.	When the influence on road traffic safety is uncontrollable, the traffic management department should be contacted in time to strengthen the on-site control of traffic safety.	

Table (Cont'd)

S/N	Work Content	Operation Steps and Standards	Safety Measures and Precautions	Responsible Person
4	Hold a pre-shift meeting	(1) All working personnel shall line up; (2) The Responsible Person will read out the work order and be clear with the work task and division of personnel; explain safety measures and technical measures in work; check and inquire the mental state of all working personnel; and inform of hazards in work and precontrol measures; (3) All working personnel shall sign on the work order for confirmation.	(1) The work order shall be filled in, issued and approved in a standardized manner, and the signature shall be complete. (2) All working personnel shall be in good mental states. (3) All working personnel shall be clear with task division of works, safety measures and technical measures	
5	Inspect tools	(1) All necessary tools and instruments shall be prepared as per the operation requirements and placed on the moisture-proof tarpaulin regularly according to the category and location. The appearance and test certificate of tools and instruments shall be checked to ensure there is no omission; (2) The inspector shall report to the Responsible Person that all inspection results are in conformity with the operation requirements	(1) The waterproof tarpaulin shall be reasonable in position, and be clean and dry. (2) The appearance of tools and instruments is acceptable after inspection, there is no damage, damp, deformation and malfunction, and the certificate is in its validity	
6	Climbing the tower	(1) Check the double name and serial number of the line before climbing the tower. For multi-circuit lines on the same tower, the Responsible Person and the members of the working team shall check the double name and identification mark carefully (color code, distinguishing mark, etc.); (2) Check the body and base, etc. of iron tower before climbing the tower to keep it solid and reliable;	(1) Operators shall wear safety helmet and soft work shoes, and climb evenly when climbing the tower. (2) During climbing the tower, the safety belt shall be properly packed, the safety rope at the back shall be placed in the kit, and the main belt shall be hung on the shoulder to prevent safety during climbing to prevent the safety belt from hooking shackles and tower materials during climbing, resulting in falling from heights.	

Table (Cont'd)

S/N	Work Content	Operation Steps and Standards	Safety Measures and Precautions	Responsible Person
6	Climbing the tower	(3) During climbing the tower, in order to prevent the operators from colliding with each other, the distance between the operators shall not be less than 1.6m, the safety belt shall be properly packed, the backup protection rope shall be placed in the kit, and the main belt shall be hung on the shoulder. (4) When climbing the tower to the cross arm, see the walking passage clearly, and the safe distance between the walking passage and the conductor shall not be less than 9.5m	(3) Operators shall not lose the protection from the safety belt and step and hold it firmly when moving on the tower. (4) When reaching the position of operation point, the safety belt and backup protection rope shall be firmly fastened on the component respectively, and shall not be hung low for high use. (5) The safe distance between the human body, transmission rope, etc. and the conductor shall not be less than 9.5m before the power test, and special person shall be assigned to supervise the work	
7	Electricity testing and installing ground wire	(1) After climbing the tower to the specified position, fasten the safety belt to a firm component and check whether the buckle is fastened. The installation position of the tackle is correct and convenient for operation. Tools and instruments must be transferred by ropes. (2) Before using the electroscope to test electricity, wear insulating gloves for self inspection and the signal shall be normal. When telescopic electroscope is used, all the insulation rods of each section should be pulled out in place to ensure the effective insulation length of the insulation rods. When testing electricity, the operator should hold the insulation handle of electroscope to ensure enough safe distance between human body and conductor. Lower layer first then upper layer, near side first then far side. The electrical inspection of power line shall be performed phase by phase.	(1) Check whether the electroscope is normal during receiving and before using. (2) The ground wire shall be in good contact with the conductor and tower material. It is forbidden to install grounding wires by winding wires. (3) An enough safe distance shall be kept between human body and the conductor	

Table (Cont'd)

S/N	Work Content	Operation Steps and Standards	Safety Measures and Precautions	Responsible Person
7	Electricity testing and installing ground wire	(3) The conductor ground wire shall be installed immediately after there is no voltage. The installation position of the ground terminal shall be polished with sandpaper. The ground terminal shall be installed first, then the conductor end. (4) When installing the conductor ground wire, the insulating gloves with throw and hang methods must be used for operation, and it is forbidden to install the ground wire with directly operating the metal part of the conductor ground wire by hand. Make sure the hook of the ground wire is tightly connected to the conductor		
8	Installing ground wire	(1) After climbing the tower to the specified position, fasten the safety belt to a firm component and check whether the buckle is fastened. Use ropes to transfer tools and instruments. (2) Check whether the ground wire is in good condition, polish the tower material ground terminal with sandpaper, and install the conductor terminal after installing the ground terminal. (3) When installing the ground wire, the insulating gloves must be used for operation, and it is forbidden to install the ground wire with directly operating the metal part of the ground wire by hand. Make sure the hook of the ground wire is tightly connected to the ground wire	(1) It is forbidden to install the ground wire by winding the conductor. (2) When installing the ground wire, the operator shall wear insulating gloves and hold the insulating rod	

Part II Skill Module Training and Assessment Standards

Table (Cont'd)

S/N	Work Content	Operation Steps and Standards	Safety Measures and Precautions	Responsible Person
9	Installation the hanging wheel	(1) The operator shall check the corrosion of insulator set fitting before entering the ground wire. (2) Check whether the ground wire insulator is in good condition and whether the pin is complete. (3) Conduct impulse test on the ground wire. (4) Auxiliary personnel on the tower shall cooperate with installing the hanging wheel on the ground wire	(1) The operator shall check the corrosion of insulator set fitting before entering the ground wire. (2) The ground wire shall be conducted the impulse test before the installation of hanging wheel	
10	Access to operation point	Correctly use the work hanging wheel with uniform moving speed and no dangerous action; when moving on the ground wire, the safety belt protection shall not be lost.	(1) Safety belt protection shall not be lost during the whole process. (2) The hanging wheel shall not move too fast	
11	Repair the broken ground wire	(1) Fix the hanging wheel after arriving at the operation point; (2) The operator shall polish the damaged part of the ground wire. (3) Measure the length of the preformed armor rods, and use a steel tape to measure the position of 1/2 length of the preformed armor rods at one end of the conductor damage, then paint and print. (4) The center of the preformed armor rods shall be installed at the seriously damaged part of the conductor, and without any gap. (5) Tap the end of the preformed armor rods with a wooden hammer, and the end shall be flat	(1) Correctly use the work hanging wheel with uniform moving speed and no dangerous action; when moving on the ground wire, the safety belt protection shall not be lost. (2) Before repairing the broken ground wire, the damaged point shall be polished. (3) The preformed armor rods shall be in the same direction as the broken ground wire. The winding shall be smooth and tight. The damaged part shall be in the middle of the preformed armor rods. (4) When winding the preformed armor rods, it is not allowed to pry it with force. To prevent it from deforming, both ends shall be kept flat during winding	

Table (Cont'd)

S/N	Work Content	Operation Steps and Standards	Safety Measures and Precautions	Responsible Person
12	Access to cross arm	(1) The operator moves along the ground wire by the hanging wheel to return to the cross arm. (2) Protection of safety belt shall not be lost during entering the cross arm. Use the transmission rope to transfer the hanging wheel to the ground	(1) Safety belt protection shall not be lost during the whole process. (2) The hanging wheel shall not move too fast	
13	Clean the tools and instruments on the iron tower	(1) Remove the tools and instruments in turn, such as the hanging wheel, ground wire, conductor ground wire, steel wire sleeves and tackle. (2) Remove the ground wire on the ground and conductor. Remove the lead (ground) wire terminal firstly, and then the ground terminal. Confirm that there are no objects left behind	(1) Remove the ground wire on the ground and conductor. The operator shall wear insulating gloves and hold the insulating rod. (2) An enough safe distance shall be kept between human body and the conductor. (3) Objects falling from high place shall be avoided	
14	Step down the tower	(1) Correctly climb down the tower along the main material of the shackles side. (2) Correctly carry the transmission rope (The transmission rope head is double folded, fast knot, and hung over the shoulder)	to Prevent falling from heights	
15	End of the work	(1) Responsible Person shall organize the staff to put working apparatus and materials in order, and clean the site to ensure "materials are removed and the site is cleaned after completion of construction". (2) In the post-shift meeting, Responsible Person will give work summaries and comments. Comments include the construction quality of this work and the implementation of safety measures from all working personnel. (3) Responsible Person shall report the end of the work to the work approver, resume the power transmission of the power-cut line and terminate the work order	There must be no object left on site.	

II. Assessment Standard

Detailed Rules for Assessment and Scoring of Operation and Inspection Skills of UHV AC Transmission Line

Fill-in Column of Examinee	No.:		Name:		Position:		Date:	MM/DD/YYYY
Fill-in Column of Assessor	Grade:		Assessor:		Assessment Team Leader:	Starting time:	Closing time:	Operation Duration:
Assessment module	Power-cut repair of 1000kV AC transmission line overhead ground wire		Assessee	Maintenance personnel of UHV AC transmission line		Assessment method	Operation	Assessment Time Limit
								100min
Job Description	Power-cut repair of 1000kV AC transmission line left overhead ground wire							
Work Specifications and Requirements	1. Given conditions: 1000kV AC practical training line left overhead ground wire needs to be repaired. The line is power off, test electricity and install the ground wire, the used tools and instruments have been tested, the work order has been handled, and the safety measures have been taken. 2. The exam participants shall make necessary safety check before operation. 3. The operation hanging wheel shall meet the stress requirements. 4. The main operation procedures of the whole process are completed by 1 Responsible Person, 1 special Supervisor, 1 electrician on the tower, 1 auxiliary workers on the tower and 2 auxiliary workers on the ground. Responsibilities of Responsible Person: Be responsible for the labor division of operating personnel of the work task, field investigation before work, preparation of the operation plan, work order filling, field re-investigation, performing work permit procedures, holding pre-shift meeting, implementation of the site safety measures, safety supervision in the operation process, dealing with emergency situations in work, quality surveillance of the work, and the summary after the work. Responsibility of Safety Supervisor: Explain the safety measures within the scope of supervision to the supervised personnel, and inform the hazards and safety precautions before work. Supervise the supervised personnel to comply with this procedures and implement the on-site safety measures, and timely correct the unsafe acts and behaviors of the supervised personnel. Responsibilities of operators on the tower: Be responsible for power-cut repair of 1000kV AC transmission line overhead ground wire. Auxiliary worker on the tower: be responsible for transferring tools and materials to match the installation of hanging wheel. Responsibilities of ground electricians: Assist the exam participant to complete the up and down transfer of tools and materials, and other auxiliary work.							

Standard for Professional Training and Assessment for Operation Maintenance of UHV AC Transmission Line

Table (Cont'd)

Work Specifications and Requirements	Given conditions: 1. Training line: UHV 1000kV AC practical training line left overhead ground wire, model of overhead ground wire: JLB20A-170. 2. The operation must be carried out according to the working procedure. The relevant item scores shall be deducted for the process error. In case of major hidden dangers of personal, equipment and operational safety, the assessor may order the termination of the assessment					
Assessment scenario preparation	1. Line: UHV 1000kV AC training line left overhead ground wire. Working content: power-cut repair of 1000kV AC transmission line left overhead ground wire. Model of overhead ground wire: JLB20A-170. 2. Required work tools and instruments: 1 set of work hanging wheel, 2 sets of safety belt, 3 nos. of walkie-talkies, 1 nos. of conductor ground wire, 1 nos. of ground wire, 1 pair of insulating gloves, 1 nos. of transmission rope, 1 nos. of 0.5T tackle, 1 nos. of steel wire sleeve, 1 sheet of sandpaper, 2 sets of personal tools and 1 nos. of wooden hammer. 3. Material: a set of preformed armor rods corresponding to the model of ground wire					
Remarks	1. Personal tools and instruments shall be provided by the exam participants themselves. 2. The deduction shall be done until the scores of each item are deducted completely					
S/N	Project name	Quality requirements	Score	Deduction standard	Reasons for deduction	Deduction	Scoring
1	Dress	Correctly wear working clothes, work shoes, safety helmet and protective gloves	4	Deduct 1 point for missing one item			
2	Inspection of personal tools	Adjustable wrench, flat pliers, and tool kits shall meet the quality requirements	3	Deduct 1 point for failure of inspection			
3	Inspection of safety belts	The safety belt meets the quality requirements within the test period	2	(1) Deduct 1 point for no visual inspection; (2) Deduct 1 point for not checking the factory certificate and test certificate			

Part II Skill Module Training and Assessment Standards

Table (Cont'd)

S/N	Project name	Quality requirements	Score	Deduction standard	Reasons for deduction	Deduction	Scoring
4	Inspection of operation tools	The operation hanging wheel shall meet the requirements within the test cycle	2	(1) Deduct 1 point for no visual inspection; (2) Deduct 1 point for not checking the factory certificate and test certificate			
5	Material inspection	Confirm that the preformed armor rods are corresponding to the ground wire, and there shall be no damage in the appearance inspection	2	(1) Deduct 1 point for wrong selection of the model of preformed armor rods; (2) Deduct 1 point for no inspection			
6	Climbing the tower	(1) Check the line name, the number of the iron tower number on spot to ensure they are correct; guarantee that the foundation and the tower body are intact and in normal condition, and ensure that the cross and span distance meets the safety requirements; confirm the defect conditions and the specification and model of ground wire; (2) Correctly carry the transmission rope (The end of the transmission rope is double folded, fast knot, and hung over the shoulder); (3) Correctly climb the tower along the main material of the shackles side; (4) After climbing the tower to the specified position, fasten the safety belt to a firm component and check whether the buckle is fastened	9	(1) Deduct 2 points for not confirming the double title of the line. Deduct 1 point for not checking the base, tower body and crossing. Deduct 1 point for not confirming the ground wire defect; (2) Deduct 2 points for not carrying transmission rope Deduct 1 point for nonstandard carry way; (3) Deduct 1 point for one time hands grasping the shackles; (4) Deduct 2 points for not climbing the tower along the main material of the shackles side; (5) Deduct 2 points in case of nonstandard use of safety belt; (6) Deduct 1 point each time when stepping on air or sliding			

· 685 ·

Table (Cont'd)

S/N	Project name	Quality requirements	Score	Deduction standard	Reasons for deduction	Deduction	Scoring
7	Power test and installation of conductor and ground wire	(1) After climbing the tower to the specified position, fasten the safety belt to a firm component and check whether the buckle is fastened. The installation position of the tackle is correct and convenient for operation. Tools and instruments must be transferred by ropes; (2) Before using the electroscope, check the sound and light signal again and the signal shall be normal. When telescopic electroscope is used, all the insulation rods of each section should be pulled out in place to ensure the effective insulation length of the insulation rods. When testing electricity, the operator should hold the insulated handle of electroscope to ensure enough safe distance between human body and conductor. Lower layer first then upper layer, near side first then far side. The electrical inspection of power line shall be performed phase by phase; (3) The conductor ground wire shall be installed immediately after there is no voltage. The tower material position of the ground terminal shall be polished with sandpaper. The ground terminal shall be installed first, then the conductor end. The installation sequence is middle phase first, then two side phase;	8	(1) Deduct 2 points for non standard installation of tackle (2) Deduct 2 points in case of nonstandard use of safety belt. (3) Deduct 2 points if no insulating gloves are worn for testing electricity and installation of ground wire. (4) Deduct 2 points for nonstandard use of electroscope, and 2 points for incorrect sequence for electricity testing. (5) Deduct 2 points for wrong installation sequence of the ground wire two sides. Deduct 1 point for unstable connection. Deduct 1 point for not checking of installed ends; (6) Deduct 2 points for winding ground wire			

· 686 ·

Table (Cont'd)

S/N	Project name	Quality requirements	Score	Deduction standard	Reasons for deduction	Deduction	Scoring
7	Power test and installation of conductor and ground wire	(4) When installing the conductor ground wire, the insulating gloves must be used for operation, and it is forbidden to install the ground wire with directly operating the metal part of the conductor ground wire by hand.Make sure the hook of the ground wire is tightly connected to the conductor	8				
8	Hanging ground wire	(1) After climbing the tower to the specified position, fasten the safety belt to a firm position and check whether the buckle is fastened.Use ropes to transfer tools and instruments; (2) Check whether the ground wire is in good condition, polish the tower material ground terminal with sandpaper, and install the conductor terminal after installing the ground terminal; (3) When installing the ground wire, the insulating gloves must be used for operation, and it is forbidden to install the ground wire with directly operating the metal part of the ground wire by hand.Make sure the hook of the ground wire is tightly connected to the ground wire	6	(1) Deduct 2 points in case of non-standard use of safety belt; (2) Deduct 2 points for installing the ground wire without wearing insulating gloves; (3) Deduct 2 points for wrong installation sequence of the ground wire two sides. Deduct 1 point for unstable connection. Deduct 1 point for not checking of installed ends; (4) Deduct 2 points for winding ground wire			

Table (Cont'd)

S/N	Project name	Quality requirements	Score	Deduction standard	Reasons for deduction	Deduction	Scoring
9	Installation the hanging wheel	(1) The operator shall check the corrosion of insulator set fitting before entering the ground wire; (2) Check whether the ground wire insulator is in good condition and whether the pin is complete; (3) Conduct impulse test on the ground wire; (4) Auxiliary personnel on the tower shall cooperate with the installation work of hanging wheel	9	(1) Deduct 2 points if the insulator set fitting is not inspected for corrosion; (2) Deduct 1 point for each item for not checking if the ground wire insulator is not in good condition and pin is complete; (3) Deduct 2 points for not conducting impulse test on the ground wire. (4) Deduct 2 points for the installation of working hanging wheel is not standard			
10	Access to operation point	Correctly use the work hanging wheel with uniform moving speed and no dangerous action; when moving on the ground wire, the safety belt protection shall not be lost.	9	(1) Deduct 3 points each time for the safety belt protection is lost; (2) Deduct 2 points for the hanging wheel speed moves too fast; 3) Deduct 3 points each time for there is dangerous action in the process of movement			

Part II Skill Module Training and Assessment Standards

Table (Cont'd)

S/N	Project name	Quality requirements	Score	Deduction standard	Reasons for deduction	Deduction	Scoring
11	Repair the broken ground wire	(1) Fix the hanging wheel after arriving at the operation point; (2) Repair and polish and level the damaged ground wire; (3) Measure the length of the preformed armor rods, and use a steel tape to measure the position of 1/2 length of the preformed armor rods at one end of the conductor damage, then paint and print; (4) The center of the preformed armor rods shall be installed at the seriously damaged part of the conductor, and without any gap; (5) Tap the end of the preformed armor rods with a wooden hammer, and the end shall be flat	16	(1) Deduct 3 points for not fixing the hanging wheel in the proper position; (2) Deduct 3 points if the ground wire is not polished and flat; (3) Deduct 2 points if not correctly painted and printed; (4) Deduct 4 points if the installation sequence of preformed armor rods is not standard; (5) Deduct 2 points if the end of preformed armor rods is not smooth; (6) Deduct 1 point for each installation of preformed armor rods with gap			
12	Access to cross arm	(1) The operator moves along the ground wire by the hanging wheel to return to the cross arm; (2) Protection of safety belt shall not be lost during entering the cross arm; (3) Use the transmission rope to transfer the hanging wheel to the ground	8	(1) Deduct 2 points for improper control of the hanging wheel moving speed; (2) Deduct 2 point each time for losing the protection of safety belt during entering the cross arm; (3) Deduct 2 points for nonstandard transmission of the hanging wheel			

· 689 ·

Table (Cont'd)

S/N	Project name	Quality requirements	Score	Deduction standard	Reasons for deduction	Deduction	Scoring
13	Clean the tools and instruments on the iron tower	(1) Remove the tools and instruments in turn, such as the hanging wheel, ground wire, conductor ground wire, steel wire sleeves and tackle; (2) Remove the ground wire on the ground and conductor. Remove the lead (ground) wire terminal firstly, and then the ground terminal; (3) Confirm that there is no leftover	7	(1) Deduct 3 points for the wrong removal sequence; (2) Deduct 4 points for any leftovers			
14	Step down the tower	(1) Carry the transmission rope and climb down the tower evenly along the shackles; (2) Correctly carry the transmission rope (The end of the transmission rope is double folded, fast knot, and hung over the shoulder)	8	(1) Deduct 4 points for not carrying transmission rope, and deduct 2 points for nonstandard carrying method of transmission rope; (2) Deduct 1 point for one time hands grasping the shackles; (3) Deduct 4 points for not climbing the tower along the main material of the shackles side			
15	Place tools and instruments	Clean up work site after completion and arrange the tools neatly as required	2	Deduct 2 points for not placing tools and instruments as required			

Part II Skill Module Training and Assessment Standards

Table (Cont'd)

S/N	Project name	Quality requirements	Score	Deduction standard	Reasons for deduction	Deduction	Scoring
16	Housekeeping	(1) No falling objects at heights; (2) No floating objects; (3) No holding object in mouth	5	(1) Deduct 5 points for falling objects; (2) Deduct 2 points for floating articles; (3) Deduct 2 points for holding object in mouth			
17	Completion time	Complete operation as required within the specified time.		Complete as required within the specified time, terminate the operation after 10 minutes, and only record the score of the completed part			
18	Total		100				

Module 16 Training and Assessment Standard for Power-cut Replacement of 1000kV AC Transmission Line Sub-conductor

I. Training Standard

(I) Training Requirements

Designation of module	Power-cut replacement of 1000kV AC transmission line sub-conductor	Type of training	Operation
Training method	Practical operation training	Hours of training	21 training hours
Training objectives	1. Be familiar with the structure and installation method of 1000kV AC transmission line sub-conductor 2. Be able to test electricity of 1000kV AC transmission line conductor in proper position and hang the ground wire 3. Be able to complete the work of power-cut replacement of 1000kV AC transmission line eight-bundle sub-conductor		
Training venue	UHV AC training line		
Training content	Power-cut replacement of 1000kV AC transmission line sub-conductor		
Scope of application	Maintenance personnel of UHV AC transmission line		

(II) Referenced Rules and Specifications

(1) Electrotechnical Terminology - Overhead Line (GB/T 2900.51-1998).

(2) Code for Design of 1000kV Overhead Transmission Line (GB50665-2011).

(3) Maintenance Code for 1000kV AC Transmission Line (DL/T209-2008).

(4) Operation Code for 1000kV AC Transmission Line (DL/T307-2010).

(5) State Grid Corporation of China Working Regulations of Power Safety (Transmission Line Section) (Q/GDW1799.2-2013).

Part II Skill Module Training and Assessment Standards

(Ⅲ) Teaching Design for Training

To complete the power-cut replacement of 1000kV AC transmission line sub-conductor is the Work task during the design. Each training stage shall be designed according to the standard operation procedure for work task completion. Each stage includes specific training objectives, training content, hours of training, training methods (training resources), training environment, assessment and evaluation, etc, as shown in Table 2-16-1.

(Ⅳ) Operation Flow

1. Work Task

Finish the power-cut replacement of 1000kV AC transmission line sub-conductor.

2. Requirements for Weather and Work Site

(1) It should be carried out in good weather for power-cut replacement of 1000kV AC transmission line sub-conductor.

In case of thunder (hear thunder, see lightning), snow, hail, rain, fog, etc., and the wind force is greater than level 5, any working shall not be conducted. When emergency live repair is required in bad weather, relevant personnel shall be organized to fully discuss and prepare necessary safety measures, which can be implemented after being approved by the unit.

(2) The operators shall be in good mental state, the members of the working group shall carefully study the work order and safety technical measures, and all the personnel shall be conform to the "four clear" (i.e., clear to the operation tasks, the hazards, the operation procedures and the safety measures).

(3) The contact responsible for power-cut and transmission must fulfill the working licensing procedures with the dispatcher and it is strictly prohibited to make an appointment for stop or transfer electricity. Responsible Person must obtain the license work order of the licensor before checking the electricity on the line to be overhauled, setting up the ground wire and carrying out the maintenance work.

(4) After power-cut, the Responsible Person shall make records carefully.

(5) It is strictly prohibit to climb the tower if a honeycomb is found on tower by checking before climbing the pole.

(6) Double safety belt shall be used by the personnel who work on the tower.

3. Preparations

3.1 Hazards and precontrol measures

(1) Hazard - Climb the live line.

Standard for Professional Training and Assessment for Operation Maintenance of UHV AC Transmission Line

Table 2-16-1 Training Content Design for Power-cut Replacement of 1000kV AC Transmission Line Sub-conductor

Training schedule	Training objectives	Training content	Hours of training	Training methods and resources	Preparation of training conditions	Assessment and evaluation
1. Theoretical teaching	1. Be familiar with the structure and installation method of 1000kV AC transmission line sub-conductor. 2. Be familiar with the operation process for the work of power-cut replacement of 1000kV AC transmission line sub-conductor	Teach the operation process for the work of power-cut replacement of 1000kV AC transmission line sub-conductor	2	Training methods: Lecture. Training resources: PPT, relevant regulations and specifications	Multimedia classroom	Attendance, classroom questions and assignments
2. Preparations	Be able to complete the preparation before operation	1. Work site survey. 2. Preparation of the standardized operation card. 3. Filling of the work order. 4. Preparation of tools and materials for this operation	1	Training methods: 1. Site survey and cleaning of tools and materials shall be practiced at site. 2. Preparation of operation card and the filling of work order shall adopt lecture method. Training resources: 1. 1000kV practical training line. 2. UHV tools warehouse. 3. Blank work order	1. UHV training transmission line 2. Multi-Media classroom	
3. Work site preparation	Be able to complete the preparations of work site	1. Work site re-survey. 2. Job application. 3. Work site layout. 4. Pre-shift meeting. 5. Inspection of tools and materials	2	Training methods: demonstration and role play. Resources: 1000kV training line	1000kV training line	

Part II Skill Module Training and Assessment Standards

Table (Cont'd)

Training schedule	Training objectives	Training content	Hours of training	Training methods and resources	Preparation of training conditions	Assessment and evaluation
4. Trainer's demonstration	The trainees can preliminarily understand the operation process of the task through inspecting and learning from each other's work	1. Climbing the tower. 2. Test electricity and hang ground wire. 3. Operation on conductor. 4. Installation of crossover and accessories. 5. Remove tools and instruments	8	Training methods: Demonstration. Resources: 1000kV training line	1000kV training line	
5. Group training of trainees.	Be able to finish the work of power-cut replacement of 1000kV AC transmission line sub-conductor	1. The trainees are divided into groups (22 persons per group) to train the power-cut replacement of 1000kV AC transmission line sub-conductor. 2. Trainers guide the operation of trainees and conduct safety supervision	7	Training method: Role play. Resources; 1000kV training line	1000kV training line	Score the operation of trainees according to the detailed rules for skill assessment and scoring
6. End of the work	1. Enable the trainees to further distinguish the shortcomings of the operation process and facilitate the promotion in the later stage. 2. Train the trainees in the working style of safe and civilized production	1. Cleaning up the work site. 2. Report to dispatcher. 3. Comment and summarize the work task this time at post-shift meeting	1	Training method: Lecture and inductive method	1000kV training line	

Precontrol measures:

① Before climbing the pole and tower for operation, the Responsible Person and the members of the working team shall check whether the double-name and identification marks carefully (color code, distinguishing mark, etc.) are matched the name of the power-cut line.

② It shall check the root, base, etc. of the iron tower to keep them solid and reliable before climbing the tower.

③ It shall check the climbing tools, instruments and facilities such as safety belt, shackles, tower materials to keep them complete and firm before climbing the pole and tower.

④ Before climbing the pole, the intermediate operators who do not involved the installation of the grounding wire shall carefully verify that the phase sequence, color code, name, and number of the line are consistent with the power-cut line, and confirm that the line name is not error or in opposite order.

(2) Hazard - Operation in violation of safety operation procedures when climbing the tower and working on the tower may cause falling from height.

Precontrol measures:

① During climbing, in order to prevent the operator climbing the pole from falling one after another, the space between them shall not be less than 1.6m.

② Before climbing the iron tower, the dirt on the shoe soles shall be removed, and it shall check whether toolkit is complete to make sure that persons will not be hurt by the falling objects during climbing.

③ Operator shall wear safety helmet and soft soles, keep in narrow movement range with uniform steps when climbing pole and tower.

④ During climbing, the safety belt shall be properly packed, the long tail rope shall be placed in the toolkit, and the main belt shall be hung on the shoulder to prevent the safety belt from hooking shackles and tower materials during climbing, which can lead to falling of the operator from high-altitude.

⑤ Operators shall not lose the protection from the safety belt, and make firm step and clench when moving on the pole and tower.

⑥ When reaching the position of operation point, fasten the safety belt (rope), it shall be firm and reliable, and shall not be hung low for high use.

⑦ The safe distance between the human body, headless rope, etc. and the conductor must not be less than 9.5m before the power test, and special person shall be assigned to supervise the work.

(3) Hazard - injury caused by objects falling from heights

Precontrol measures:

① The ground personnel shall not stand under the position vertical to the operation point. The personnel on tower shall prevent falling objects from injury, and the tools and materials used shall be transmitted by ropes.

② Tool bags shall be used when working at heights, relatively large tools shall be fixed on solid components and not allowed to be randomly placed.

③ In the process of lifting with a winching, special personnel shall be assigned to command and cooperate in a unified way. Impulse inspection shall be carried out immediately after the insulator string are lifted off the ground.

(4) Hazard - Prevent injury caused by induced electricity

① In order to prevent injury caused by induced electricity, insulating gloves should be used in electricity testing and hanging ground wire

② The overhead ground wire shall be reliably grounded before necessary contact.

(5) Hazard - Site operating safety supervision.

① From the beginning to the end of the operation, the safety Supervisor must always carry out uninterrupted safety supervision on the operators on the site.

② The Responsible Person or the Supervisor must wear a safety vest.

(6) Hazard - Transportation safety.

It is necessary to pay attention to the driving safety of the vehicle, drive carefully when driving, and illegal driving is not allowed.

3.2 Selection of tools and instruments and materials

See Table 2-16-2 for tools, instruments and materials required for power-cut replacement of 1000kV AC transmission line sub-conductor. Before the tools and instruments are delivered out of the warehouse, the service voltage class and test period of the tools and instruments shall be carefully checked to ensure that appearance is intact, connection is firm, rotation is flexible and meeting the requirements of the work task. After the tools and instruments are delivered out of the warehouse, they shall be kept from contamination and damp. Metal tools and insulated tools shall be separately loaded and transported to avoid deformation and damage caused by mixed loading and transportation.

Table 2-16-2 Tools, Instruments and Materials Required for Power-cut Replacement of 1000kV AC Transmission Line Sub-conductor

S/N	Name	Specification	Unit	Qty.	Remarks
1	Transmission rope		Nos.	2	
2	Tackle		Nos.	2	
3	Safety helmet		Nos.	22	
4	Anemometer		Piece	1	
5	Moisture-proof canvas	2m×4m	Piece	2	
6	Noose		Nos.	4	
7	Falling protector	Corresponding to the type of pole and tower falling protector	Nos.	7	
8	Safety belt		Nos.	7	
9	Security fence		Set	Several	
10	Warning sign	"Work Here", "Access from Here" "Access from Here"	Set	1	
11	Red waistcoat	Responsible Person and Safety Supervisor	Piece	2	
12	Conductive paste		Box	1	
13	Sandpaper		Sheet	1	
14	Clean towel		Nos.	1	
15	Interphone		Set	4	
16	lever		Nos.	1	
17	Electricity tester		Nos.	1	
18	Hanging ground wire		Group	1	
19	Rope ladder		Nos.	1	

3.3 Division of labor for operators

Division of labor for operators of the task is shown in Table 2-16-3.

Table 2-16-3   Division of Operating Personnel Involving in Power-cut Replacement of 1000kV AC Transmission Line Sub-conductor

S/N	Post	Qty. (person)	Responsibilities
1	Responsible Person	1	Be responsible for the labor division of operating personnel of the work task, field investigation before work, preparation of the operation plan, work order filling, field re-investigation, performing work permit procedures, holding pre-shift meeting, implementation of the site safety measures, safety supervision in the operation process, dealing with emergency situations in work, quality surveillance of the work, and the summary after the work
2	Safety Supervisor	1	Be responsible for the safety monitoring work during this operation
3	Operators on tower	7	Be responsible for the operation of the power-cut replacement of 1000kV AC transmission line sub-conductor
4	Ground auxiliary personnel	10	Be responsible for ground auxiliary works during the operation.
5	Winching operator	2	Be responsible for the winching work during this operation
6	Signal commander	2	Be responsible for commanding the start and stop of 2 sets of winches

4. Work Procedure

The workflow of this task is shown in Table 2-16-4.

Table 2-16-4   Working Procedures for Power-cut Replacement of 1000kV AC Transmission Line Sub-conductor

S/N	Work Content	Operation Standard	Safety Precautions	Responsible Person
1	Work Permit	(1) The contact person for power off/on before the operation must contact the dispatcher to complete the work permit procedure.	(1) No work shall be started without the work permit of the working approver. (2) It is strictly prohibit to make appointment for power transmission or outage	

Table (Cont'd)

S/N	Work Content	Operation Standard	Safety Precautions	Responsible Person
2	Site layout	Install the security fence and hang the signboards correctly: (1) The security fence should take full account of falling objects from the heights and the influence on road traffic. (2) The entrance and exit of the security fence shall be set reasonably. (3) Signs such as "Access from Here", "Work Here", "Slow Down" or "Blocking" shall be properly arranged.	When the influence on road traffic safety is uncontrollable, the traffic management department should be contacted in time to strengthen the on-site control of traffic safety.	
3	Hold a pre-shift meeting	(1) All working personnel shall line up. (2) The Responsible Person shall wear red waistcoat and read out the work order and be clear with work task and division of personnel; explain safety measures and technical measures in work; check (inquire after) mental state of all working personnel; inform of hazards in work and precontrol measures. (3) All working personnel shall sign on the work order for confirmation.	(1) The work order shall be filled in, issued and approved in a standardized manner, and the signature shall be complete. (2) All working personnel shall be in good mental states. (3) All working personnel shall be clear with task division of works, safety measures and technical measures	
4	Inspect tools	(1) All necessary tools and instruments shall be prepared as per the operation requirements and placed on the moisture-proof tarpaulin regularly according to the category and location. The appearance and test certificate of tools and instruments shall be checked to ensure there is no omission. (2) The inspector shall report to the Responsible Person that all inspection results are in conformity with the operation requirements	(1) The moisture-proof tarpaulin shall be enough in quantity and reasonable in position, and be clean and dry. (2) The appearance of tools and instruments is acceptable after inspection, there is no damage, damp, deformation and malfunction, and the certificate is in its validity	

Part II Skill Module Training and Assessment Standards

Table (Cont'd)

S/N	Work Content	Operation Standard	Safety Precautions	Responsible Person
5	Climbing the tower (Electricity testing, hanging ground wire tower)	(1) Before climbing the tower for operation, operators must check the line designation and number first. The tower to be climbed shall be two base towers in the working area. The Responsible Person and the members of the working team shall check the double-designation and identification mark carefully (color code, distinguishing mark, etc.). (2) It shall check the root, base, etc. of the pole and tower to keep them solid and reliable before climbing the tower. (3) During climbing, in order to prevent the pole climbing operator from falling, the distance between the operators shall not be less than 1.6m, the safety belt shall be properly packed, the long tail rope shall be placed in the kit, and the main belt shall be hung on the shoulder to prevent safety during climbing to prevent the safety belt from hooking shackles and tower materials during climbing, resulting in falling from of tower. (4) When climbing the tower to the cross arm, the Supervisor and the operator shall check the identification mark and double-designation of the line outage again	(1) Operator shall wear safety helmet and soft soles, and keep in narrow movement range with uniform steps when climbing pole and tower. (2) During climbing, the safety belt shall be properly packed, the long tail rope shall be placed in the toolkit, and the main belt shall be hung on the shoulder to prevent the safety belt from hooking shackles and tower materials during climbing, which can lead to falling of the operator from high-altitude. (3) Operators shall not lose the protection from the safety belt, and make firm step and clench when moving on the pole and tower. (4) When reaching the position of operation point, fasten the safety belt (rope), it shall be firm and reliable, and shall not be hung low for high use. (5) The safe distance between the human body, headless rope, etc. and the conductor must not be less than 9.5m before the verification of live part, and special person shall be assigned to perform supervision during the work	
6	Inspection power and install ground wire	(1) After climbing the tower in place, fasten the safety belt to a firm and reliable component, and check whether the buckle is in place correctly. Appliances such as test pen (electroscope) must be transferred by rope.	(1) Check whether the test pole is normal during receiving and before using. (2) It is forbidden to install the ground wire by the way of winding the conductor. (3) Operator must wear Insulating gloves testing electricity and installing ground wires	

· 701 ·

Table (Cont'd)

S/N	Work Content	Operation Standard	Safety Precautions	Responsible Person
6	Inspection power and install ground wire	(2) Electricity testing should be carried out for the front and back towers in the working section by the specially-assigned person wearing qualified insulating gloves with qualified electroscope. (3) After verifying that there is no voltage in the line, start to install the ground wire, check that the ground wire is in good condition, and install the ground wire according to the procedure (connect the ground terminal first, and connect the conductor end later). (4) When ground wire is installed, it must be operated with insulating rope or insulated handle, and it is forbidden to install the ground wire by directly operating the metal part of the ground wire by hand. Make sure the collection of the ground wire is tightly connected to the conductor		
7	Climbing the tower (Operation point)	(1) Before climbing the tower for operation, operators must check the line designation and number first. Responsible Person and the members of the working team shall check the double designation and identification mark carefully (color code, distinguishing mark, etc.). (2) It shall check the root, base, etc. of the pole and tower to keep them solid and reliable before climbing the tower. (3) During climbing, in order to prevent the pole climbing operator from falling, the distance between the operators shall not be less than 1.6m, the safety belt shall be properly packed, the long tail rope shall be placed in the kit, and the main belt shall be hung on the shoulder to prevent safety during climbing to prevent the safety belt from hooking shackles and tower materials during climbing, resulting in falling from of tower.	(1) Operator shall wear safety helmet and soft soles, and keep in narrow movement range with uniform steps when climbing pole and tower. (2) During climbing, the safety belt shall be properly packed, the long tail rope shall be placed in the toolkit, and the main belt shall be hung on the shoulder to prevent the safety belt from hooking shackles and tower materials during climbing, which can lead to falling of the operator from high-altitude. (3) Operators shall not lose the protection from the safety belt, and make firm step and clench when moving on the pole and tower.	

Part II Skill Module Training and Assessment Standards

Table (Cont'd)

S/N	Work Content	Operation Standard	Safety Precautions	Responsible Person
7	Climbing the tower (Operation point)	(4) When climbing the tower to the cross arm, the Supervisor and the operator shall check the identification mark and double-designation of the line outage again and can enter the cross arm on the line outage side only after confirming no error.	(4) When reaching the position of operation point, fasten the safety belt (rope), it shall be firm and reliable, and shall not be hung low for high use. (5) The safe distance between the human body, headless rope, etc. and the conductor must not be less than 9.5m before the verification of live part, and special person shall be assigned to perform supervision during the work	
8	Operation on the conductor	(1) The operator should enter the conductor, remove the spacer from the conductor and mark the removal position. (2) The operator should enter the conductor at the tangent tower along the rope ladder and install the single-sheave tackle, then remove the vibration damper, lift the sub-conductor with the wire lifter and 6T lever hoist, take off the armor clamp, and turn the sub-conductor into the single-sheave tackle. (3) After removing part armor clamps for connection, turn the ground wire into the single-sheave tackle fixed on the ground wire cross arm. (4) The operator should connect the 5T winching to the sub-conductor with wire clamp, and then tighten the sub-conductor with 9T lever hoist to take off the conductor. Loose the conductors at both ends to the ground for the resisting-tensile towers on both sides. (5) The operator shall remove the strain clamp at the conductor end, install the net cover on the old and new conductors, bind them firmly, and connect the old and new conductors with the rotary joint		

Table (Cont'd)

S/N	Work Content	Operation Standard	Safety Precautions	Responsible Person
9	Installation of crossover and accessories	(1) Start 2 winchings on the traction field and tension field at the same time under unified command by specially-assigned person. (2) After the new conductor arrives at the traction field, the tension field shall be crimped with the strain clamp. After crimping, the crimped conductor shall be connected to the insulator armor clamp by winding and winching. (3) Tighten the ground wire at the traction field, and mark when the conductor sag is consistent with the original sag. (4) Loosen the conductor to the ground, crimp the strain clamp at the other end, then tighten the grinding rope. Connect the ground wire to the insulator string armor clamp and the new conductor to the cable clamp by the tangent tower, fasten the bolts, and install the vibration damper according to the previously marked position.	(1) During the paying-off construction, special personnel shall be assigned to watch over temporary stay wire, cross span, tangent tower and stressed tools, and check that the stress conditions are in good condition. (2) When paying-off, communication should be kept smooth with unified signals and commands. (3) During construction, it is forbidden to cross or linger under the ground wire. (4) In the process of crossover, special personnel shall check the stress of ground anchor, steering tackle and grinding rope at any time and make timely adjustment. (5) During paying-off, the winding and winching operation should be stable. Maintain the balance of grounding traction, and prevent the ground wire from slipping.	
10	Removal of tools and instruments	Remove tools in sequence, such as grinding rope, lever hoist, tackle, wire clamp and backup protection rope, etc.		
11	Remove of ground wire	Remove the ground wire at the position where the ground wire is installed. The removal of the ground wire must be operated with an insulating rope or an insulating handle. It is forbidden to install the ground wire by directly operating the metal part of the ground wire by hand	Wear insulating gloves when removing ground wires	

Table (Cont'd)

S/N	Work Content	Operation Standard	Safety Precautions	Responsible Person
12	Step down the tower	After checking that there is no object left on the tower, the equipotential electricians on the tower shall report it to the Responsible Person and then climb down the tower with transmission rope after obtaining the consent of the Responsible Person.	(1) Use anti-falling device when climbing down the tower. (2) Do not grasp tower nails when climbing down the tower	
13	End of the work	(1) Responsible Person shall organize the staff to put working apparatus and materials in order, and clean the site to ensure "materials are removed and the site is cleaned after completion of construction". (2) In the post-shift meeting, Responsible Person will give work summaries and comments.Comments include the construction quality of this work and the implementation of safety measures from all working personnel. (3) Responsible Person shall report the end of the work to the work approver, resume the power transmission of the power-cut line and terminate the work order		

Standard for Professional Training and Assessment for Operation Maintenance of UHV AC Transmission Line

II. Assessment Standard

Detailed Rules for Assessment and Scoring of Operation and Inspection Skills of UHV AC Transmission Line

Fill-in Column of Examinee	No.:		Name:		Position:		Date:	MM/DD/YYYY
Fill-in Column of Assessor	Grade:		Assessor:		Assessment Team Leader:		Starting time:	Closing time:	Operation Duration:
Assessment module	Power-cut replacement of 1000kV AC transmission line sub-conductor			Assessee	Maintenance personnel for UHV 1000kV AC transmission line	Assessment method	Operation	Assessment Time Limit	360min
Job Description	Power-cut replacement of phase X sub-conductor X# of strain section of 1000kV AC transmission line								
Work Specifications and Requirements	1. The whole process is completed by the team members including 7 electricians on tower, 10 auxiliary workers on the ground, 2 grinding operators to assist the exam participant to finish the up and down transfer of tools, materials and other non-technical work. 2. The exam participants shall make necessary safety check before operation. 3. The tools used to replace the sub-conductor should meet the stress requirements. 4. An oral application shall be made at the beginning of work, and an oral report also shall be made at the end of the work. Given conditions: 1. Training base: UHVAC 1000kV AC practical training line strain section phase A 7# sub-conductor, and the model of the conductor is: 8×JL/G1A-630/45. 2. Work orders have been handled, safety measures have been completed, and oral application (dispatcher or assessor) shall be made at the beginning and end of the work. 3. The operation must be carried out according to the working procedures. The scores of the items to be carried out shall be deducted for the process error. In case of major hidden dangers of personal, equipment and operational safety, the assessor may order the termination of the operation (assessment)								

Part II Skill Module Training and Assessment Standards

Table (Cont'd)

Assessment scenario preparation	1. Tools and instruments: a bundle of Φ16 grinding rope, 2 lever hoists (9T), 1 lever hoist (6T), 3 winchings (5T), 3 sets of lifting rope tackles, 3 nooses, 2 sets of rotating joints, 2 pairs of net covers, 2 wire clamps, 1 wire lifter, 1 wire tumble tool, 2 sets of special tools for spacers, 3 single-sheave tackles, 2 sets of pay-off racks, 2 sets of pay-off reels, 8 steel wire sleeves (Φ20), 8 iron tackles (15T), 8 shackles (18T), 6 shackles (10T), 2 sets of hydraulic presses (including hydraulic pliers), some walkie-talkies, 1 rope ladder, 7 double-insurance safety belts, and several wood blocks. 2. Materials: 3 reels of conductors of the same model, and 2 sets of strain clamps.	
Remarks	1. The deduction shall be done until the scores of each item are deducted completely. In case of major hidden dangers of personal, equipment and operational safety, the assessor may order the termination of the operation. 2. When equipment, working environment, safety belt, safety helmet, tool, shielding clothes, etc., do not conform to the operation condition, the assessor may order the termination of the operation	

S/N	Project name	Quality requirements	Score	Deduction standard	Reasons for deduction	Deduction	Scoring
1	Site re-survey	1) The Responsible Person shall go to the work site to check the line name, pole and tower number, on-site working conditions, defective parts and so on. 2) The site meteorological conditions should meet the operation requirements. 3) Check whether the work order is complete and unmodified, check whether the safety measures listed are consistent with the actual situation on site, and supplement it if necessary	3	1) Deduct 1 point for failure to check the double title. 2) Deduct 1 point for failure to verify on-site working conditions (meteorology), defective parts. 3) Deduct 0.5 points for any alteration in the work order, and deduct 1 point for incorrect work order number. Deduct 1.5 points for each incomplete work order			

Standard for Professional Training and Assessment for Operation Maintenance of UHV AC Transmission Line

Table (Cont'd)

S/N	Project name	Quality requirements	Score	Deduction standard	Reasons for deduction	Deduction	Scoring
2	Site layout	Install the security fence and hang the signboards correctly: 1) The security fence should take full account of falling objects from the heights and the influence on road traffic. 2) The entrance and exit of the security fence shall be set reasonably. 3) Signs such as "Access from Here", "Work Here", "Access from Here" shall be properly and well arranged	2	1) Deduct 0.5 points for failure to arrange the fence at the work site. 2) Deduct 0.5 points for failure to arrange the warning board. 3) Deduct 0.5 points for failure to hang the tower climbing operation sign			
3	Hold a pre-shift meeting	1) All staff and personnel shall wear safety helmets and work clothes correctly. 2) Responsible Person shall wear red vest and read out the work order and be clear with work task and division of personnel; explain safety measures and technical measures in work; check (inquire after) mental state of all working personnel; inform of hazards in work and precontrol measures. 3) All working personnel shall sign on the work order for confirmation.	3	1) Deduct 0.5 points for improper dress, and deduct 0.5 points for improper dress of each person. 2) Give no points to this item for no division of labor, and deduct 1 point for unclear division of labor. 3) Deduct 0.5 points for the site Responsible Person not wearing a safety monitoring vest. 4) Deduct 1 point for the work shift member failing to sign or signing incompletely on the work order.			

Part II Skill Module Training and Assessment Standards

Table (Cont'd)

S/N	Project name	Quality requirements	Score	Deduction standard	Reasons for deduction	Deduction	Scoring
4	Inspection of tools	1) The staff shall place the tools on the moisture-proof tarpaulin as required; the moisture-proof tarpaulin shall be clean and dry. 2) Inspection of personal tools: adjustable wrench, flat pliers, pin-pulling pliers and tool kits which shall meet the quality requirements. 3) Test-run the winching, and check whether the gear position is normal; Whether the lever hoist meets the quality requirements. 4) Check the wire rope and tackle, confirm the connection is reliable and the force meets the requirements. 5) Check whether safety belt and personal safety line meet the quality requirements within the test period. 6) Check whether the wire lifter, wire clamp and wire tumble tool meet the requirements within the test period. 7) Confirm the conductor type, length and appearance meet the requirements	10	1) Deduct 1 point for failure to use moisture-proof cloth and place tools to designed positions. 2) Deduct 1 points/item for failure to check qualified label of tool test and appearance inspection. 3) Deduct 1 point for each error or missing item 4) Deduct 0.5 points for each inspection result unreported			
5	Venue layout	1) Arrange winching at the traction field, complete the position arrangement of winching and angle tackle. 2) Arrange winching at tension field, correctly complete the position arrangement of winching, angle tackle and wire coil	4	Deduct 2 point for each error or missing item			

Table (Cont'd)

S/N	Project name	Quality requirements	Score	Deduction standard	Reasons for deduction	Deduction	Scoring
6	Climbing the tower	1) The electrician on tower shall check the double-designation, pole number and phase, check whether the tower meets the requirements for climbing and report the results to Responsible Person. 2) The electrician on tower shall fasten the safety belt, and properly conduct impulse test on safety belt, backup protection rope, and falling protector, and climb up the tower only after report to the Responsible Person. 3) When climbing the tower, the electrician shall fasten the anti-fall device, climb at constant speed, tread on shackles, and grasp the main materials. After reaching the working point of the cross arm, fasten the safety belt on the solid and reliable components, check whether the buckle is in correct position and select the appropriate position to arrange the tackle transmission rope	5	1) Deduct 2 points for failure to fasten the safety belt or for failure to perform the impulse test on the safety belt and the backup protection rope. 2) Deduct 2 points for grasping the shackles with hands. 3) Deduct 1 point for inconvenient suspension position of tackle transmission rope for taking tools. 4) Deduct 2 points for falling object at high place. 5) Deduct 2 points for tools colliding with the tower body during the transfer process. 6) Deduct 1 point for knotting and disordered rope in tool transfer. 7) Deduct 2 points for the Responsible Person failing to monitor the operation in place. 8) Deduct 2 points for incorrect operation of electrician on tower			
7	Inspection power and install ground wire	1) The ground electrician shall pass the test pen (electroscope) and other tools to the electrician on tower with a rope.	4	1) Deduct 4 points for failure to wear insulating gloves in electricity testing and installation of ground wire. 2) Deduct 4 points for installation of ground wire by winding the conductor			

Part II Skill Module Training and Assessment Standards

Table (Cont'd)

S/N	Project name	Quality requirements	Score	Deduction standard	Reasons for deduction	Deduction	Scoring
7	Inspection power and install ground wire	2) Electricity testing and setting ground wire. Use acceptable electroscope and wear acceptable insulating gloves to test the front and rear towers of the working section, install the ground wire after verifying that the circuit has no voltage, and install the ground wire with an insulating rope or an insulating handle. It is forbidden to install the ground wire by directly operating the metal part of the ground wire by hand. Make sure the collection of the ground wire is tightly connected to the conductor	4				
8	Operation on the conductor	1) The electrician on tower should inspect the rust of the hardware fittings, conduct impulse test on the insulator string, and let the insulator string enter into the conductor with the permission of the Responsible Person. 2) The electrician on tower should enter the sub-conductor, remove the spacer from the conductor and mark the removal position.	18	1) Deduct 3 points for failure to use double safety belt or loss of safety protection. 2) Deduct 2 points for failure to conduct impulse test for insulator string. 3) Deduct 1 point for failure to check armor clamps. 4) Deduct 2 points for transferring spacer without transmission rope. 5) Deduct 1 point for no mark. 6) Deduct 2 points for failure to protect conductor and tower material when lifting conductor.			

· 711 ·

Standard for Professional Training and Assessment for Operation Maintenance of UHV AC Transmission Line

Table (Cont'd)

S/N	Project name	Quality requirements	Score	Deduction standard	Reasons for deduction	Deduction	Scoring
8	Operation on the conductor	3) The electrician on tower should enter the conductor at the tangent tower along the rope ladder and install the single-sheave tackle, then remove the vibration damper, lift the sub-conductor with the wire lifter and 6T lever hoist, take off the armor clamps, and turn the sub-conductor into the single-sheave tackle. 4) After removing part armor clamps for connection, turn the ground wire into the single-sheave tackle fixed on the ground wire cross arm. 5) The operator should connect the 5T winching to the sub-conductor with wire clamp, and then tighten the sub-conductor with 9T lever hoist to take off the conductor. Loose the conductors at both ends to the ground for the re-sisting-tensile towers on both sides. 6) The operator shall remove the strain clamp at the conductor end, install the net cover on the old and new conductors, bind them firmly, and connect the old and new conductors with the rotary joint.	18	7) Deduct 2 points in case of nonstandard use of rope ladder in climbing up and down. 8) Deduct 2 points for failure to connect the winching and conductor first. 9) Deduct 2 points for incorrect position of lever hoist installed. 10) Deduct 1 point for failure to protect the tower material during installation of steering tackle at each place. 11) Deduct 3 points for non-standard layout of the new conductor 12) Deduct 3 points for no binding of net cover 13) Deduct 2 points for insufficient length of net cover 14) Deduct 1 point for incorrect use of rotating joint			

Part II Skill Module Training and Assessment Standards

Table (Cont'd)

S/N	Project name	Quality requirements	Score	Deduction standard	Reasons for deduction	Deduction	Scoring
9	Installation of crossover and accessories	1) Start 2 winchings on the traction field and tension field at the same time under unified command by specially-assigned person. 2) After the new conductor arrives at the traction field, the tension field shall be crimped with the strain clamp. After crimping, the crimped conductor shall be connected to the insulator armor clamp by winding and winching. 3) Tighten the ground wire at the traction field, and mark when the conductor sag is consistent with the original sag. 4) Loosen the conductor to the ground, crimp the strain clamp at the other end, then tighten the grinding rope. Connect the ground wire to the insulator string armor clamp and the new conductor to the cable clamp by the tangent tower, fasten the bolts, and install the vibration damper according to the previously marked position. 5) Remove tools in sequence, such as grinding rope, lever hoist, tackle, wire clamp and back-up protection rope.	35	1) Deduct 2 points for conductor falling to ground every time during crossover 2) Deduct 1 point for oversize strain during crossover 3) Deduct 2 points for thread jamming due to no one has been assigned to command 4) Deduct 2 points in case the old wires are not sorted out and recovered in time 5) Deduct 2 points in case the sag is not measured by a specially assigned person 6) Deduct 2 points for inconsistent sag with other conductor 7) Deduct 3 points in case the model of the strain clamp is not consistent with that of the conductor 8) Deduct 2 points for failure to conduct impulse test after connection 9) Deduct 2 points in case that the crimping process does not meet the requirements 10) Deduct 2 points for failure to check the bolts fastened properly 11) Deduct 1 point for each improper installation position of spacer			

Table (Cont'd)

S/N	Project name	Quality requirements	Score	Deduction standard	Reasons for deduction	Deduction	Scoring
9	Installation of crossover and accessories		35	12) Deduct 2 points for unacceptable installation and improper position of spacer 13) Deduct 3 points for failure to check the connection of hardware fittings 14) Deduct 2 points for each winding of transmission rope 15) Deduct 2 points for failure to check the conductor sag 16) Deduct 2 points for each damage to tower material			
10	Remove of ground wire	Remove the ground wire at the position where the ground wire is installed. The removal of the ground wire must be operated with an insulating rope or an insulating handle. It is forbidden to install the ground wire by directly operating the metal part of the ground wire by hand.	4	1) Deduct 4 points for failure to wear insulating gloves in removal of ground wire.			
11	Return to the ground	After checking that there is no object left on tower, the electrician on tower shall report to the Responsible Person and then climb down the tower with transmission rope after approval of the Responsible Person	5	1) Deduct 2 points for failure to use the falling protector when climbing down the tower. 2) Deduct 2 points for loosing the protection of the safety belt when moving on the tower. 3) Deduct 1 point for grasping the tower nail when climbing down the tower. 4) Deduct 2 points for any objects left on tower.			

Part II Skill Module Training and Assessment Standards

Table (Cont'd)

S/N	Project name	Quality requirements	Score	Deduction standard	Reasons for deduction	Deduction	Scoring
12	End of the work	1) The Responsible Person shall organize all working members to put working apparatus and materials in order and put them in a special kit (bag) after cleaning; clean the site to ensure that "the materials are removed and the site is cleaned after construction". 2) In the post-shift meeting, Responsible Person will give work summaries and comments. Comment on the construction quality of the work; Comment on the implementation of safety measures of all working members and report to the dispatcher to restore the power transmission status	5	1) Deduct 2 points for failure to clean the tools. 2) Deduct 2 points for missing tools. 3) Deduct 2 points for failure to hold the post-shift meeting. 4) Deduct 2 points for failure to remove the fence. 5) Deduct 2 points for failure to report to dispatcher			
	Total		100				

· 715 ·

Module 17　Training and Assessment Standard for Power-cut Replacement of 1000kV AC Transmission Line Overhead Ground Wire

Ⅰ. Training Standard

(Ⅰ) Training Requirements

Designation of module	Power-cut replacement of 1000kV AC transmission line overhead ground wire	Type of training	Operation
Training method	Practical operation training	Hours of training	21 training hours
Training objectives	1. Understand the type, structure and installation way of overhead ground wire. 2. Operators can test electricity and install ground wire on the ground wire of UHV 1000kV line. 3. Operators can complete power-cut replacement of 1000kV AC transmission line overhead ground wire		
Training venue	UHV AC training line		
Training content	Power-cut replacement of 1000kV AC transmission line overhead ground wire		
Scope of application	Maintenance personnel of UHV AC transmission line		

(Ⅱ) Referenced Rules and Specifications

(1) Electrotechnical Terminology - Overhead Line (GB/T 2900.51-1998).

(2) Code for Design of 1000kV Overhead Transmission Line (GB50665-2011).

(3) Maintenance Code for 1000kV AC Transmission Line (DL/T209-2008).

(4) Operation Code for 1000kV AC Transmission Line (DL/T307-2010).

(5) State Grid Corporation of China Working Regulations of Power Safety (Transmission Line Section) (Q/GDW1799.2-2013).

(6) Operating Code for Overhead Transmission Line (DL/T741-2010).

(Ⅲ) Teaching Design for Training

The design is to complete the work task of "power-cut replacement of 1000kV AC transmission line overhead ground wire". Each training stage shall be designed according to the standard operation procedure for work task completion. Each stage includes specific training objectives, training contents, hours of training, training methods (training resources), training environment, examination and evaluation, etc., as shown in Table 2-17-1.

Part II Skill Module Training and Assessment Standards

Table 2-17-1 Training Content Design for Power-cut Replacement of 1000kV AC Transmission Line Overhead Ground Wire

Training schedule	Training objectives	Training content	Hours of training	Training methods and resources	Preparation of training conditions	Assessment and evaluation
1. Theoretical teaching	1. Be familiar with the type, structure and installation of overhead ground wire. 2. Be familiar with the operation procedure of electricity testing and setting ground wire. 3. Be familiar with the operation procedure of power-cut replacement of 1000kV AC transmission line overhead ground wire	1. The type, structure and installation way of overhead ground wire. 2. The operation procedure of electricity testing and setting ground wire. 3. The operation procedure of power-cut replacement of 1000kV AC transmission line overhead ground wire	2	Training methods: Lecture. Training resources: PPT, relevant regulations and specifications	Multimedia classroom	Attendance, classroom questions and assignments
2. Preparations	Be able to complete the preparation before operation	1. Work site survey. 2. Preparation of the standardized operation card. 3. Filling of the work order. 4. Preparation of tools and materials for this operation	1	Training methods: 1. Site survey and cleaning of tools and materials shall be practiced at site. 2. Preparation of operation card and the filling of work order shall adopt lecture method. Training resources: 1. 1000kV practical training line. 2. UHV tools warehouse. 3. Blank work order	1. UHV training transmission line 2. Multi-Media classroom	
3. Work site preparation	Be able to complete the preparations of work site	1. Work site re-survey. 2. Work permit. 3. Work site layout. 4. Pre-shift meeting. 5. Inspection of tools and materials	2	Training methods: demonstration and role play. Training resource: 1000kV practical training line	1000kV training line	

· 717 ·

Standard for Professional Training and Assessment for Operation Maintenance of UHV AC Transmission Line

Table (Cont'd)

Training schedule	Training objectives	Training content	Hours of training	Training methods and resources	Preparation of training conditions	Assessment and evaluation
4. Trainer's demonstration	The trainees can preliminarily understand the operation process of the task through inspecting and learning from each other's work	1. Electricity testing and setting ground wire. 2. Tangent tower threadbare. 3. Resisting-tensile tower thread slack and connecting new and old ground wire. 4. Installation of crossover and accessories	8	Training methods: Demonstration. Training resource: 1000kV practical training line	1000kV training line	
5. Group training of trainees	1. Operators can test electricity and install ground wire. 2. Operators can complete power-cut replacement of 1000kV AC transmission line overhead ground wire	1. The trainees are grouped (22 persons per group) to train the skills of electricity testing, installing ground wire and replacing overhead ground wire. 2. Trainers guide the operation of trainees and conduct safety supervision	7	Training methods: Role play. Training resource: 1000kV practical training line	1000kV training line	Score the operation of trainees according to the detailed rules for skill assessment and scoring
6. End of the work	1. Enable the trainees to further distinguish the shortcomings of the operation process and facilitate the promotion in the later stage. 2. Train the trainees in the working style of safe and civilized production	1. Cleaning up the work site. 2. Report to dispatcher. 3. Comment and summarize the work task this time at post-shift meeting	1	Training method: Lecture and inductive method	1000kV training line	

(Ⅳ) Operation Flow

1. Work Task

Complete power-cut replacement of 1000kV AC transmission line overhead ground wire.

2. Requirements for Weather and Work Site

(1) Power-cut replacement of 1000kV AC transmission line overhead ground wire should be carried out in good weather.

In case of thunder (hear thunder, see lightning), snow, hail, rain, fog, etc., and the wind force is greater than level 5, any working shall not be conducted. When emergency power-cur repair is required in bad weather, relevant personnel shall be organized to fully discuss and prepare necessary safety measures, which can be implemented after being approved by the unit.

(2) The operators shall be in good mental state, the members of the working group shall carefully study the work order and safety technical measures, and all the personnel shall be conform to the "four clear" (i.e., clear to the operation tasks, the hazards, the operation procedures and the safety measures).

(3) The contact responsible for power-cut and transmission must fulfill the working licensing procedures with the dispatcher and it is strictly prohibited to make an appointment for stop or transfer electricity. Responsible Person must obtain the license work order of the licensor before checking the electricity on the line to be overhauled, setting up the ground wire and carrying out the maintenance work.

(4) After power-cut, the Responsible Person shall make records carefully.

(5) It is strictly prohibit to climb the tower if a honeycomb is found on tower by checking before climbing the pole.

(6) Double safety belt shall be used by the personnel who work on the tower.

3. Preparations

3.1 Hazards and precontrol measures

(1) Hazard- precontrol measures for climbing live line:

① Before climbing the pole and tower for operation, the Responsible Person and the members of the working team shall check whether the double-name and identification marks carefully (color code, distinguishing mark, etc.) are matched the name of the power-cut line.

② It shall check the root, base, etc. of the iron tower to keep them solid and reliable before climbing the tower.

③ It shall check the climbing tools, instruments and facilities such as safety belt, shackles, tower materials to keep them complete and firm before climbing the pole and tower.

④ Before climbing the pole, the intermediate operators who do not involved the installation of the grounding wire shall carefully verify that the phase sequence, color code, name, and number of the line are consistent with the power-cut line, and confirm that the line name is not error or in opposite order.

(2) Hazard - falling accident

Precontrol measures:

① During climbing, in order to prevent the operator climbing the pole from falling one after another, the space between them shall not be less than 1.6m.

② Before climbing the iron tower, the dirt on the shoe soles shall be removed, and it shall check whether toolkit is complete to make sure that persons will not be hurt by the falling objects during climbing.

③ Operator shall wear safety helmet and soft soles, keep in narrow movement range with uniform steps when climbing pole and tower.

④ During climbing, the safety belt shall be properly packed, the long tail rope shall be placed in the kit, and the main belt shall be hung on the shoulder to

prevent the safety belt from hooking shackles and tower materials during climbing, resulting in falling off.

⑤ Operators shall not lose the protection from the safety belt, and make firm step and clench when moving on the pole and tower.

⑥ When reaching the position of operation point, fasten the safety belt (rope), it shall be firm and reliable, and shall not be hung low for high use.

⑦ The safe distance between the human body, headless rope, etc. and the conductor must not be less than 9.5m before the power test, and special person shall be assigned to supervise the work.

(3) Hazard - injury caused by objects falling from heights.

Precontrol measures:

① The ground personnel shall not stand under the position vertical to the operation point. The personnel on tower shall prevent falling objects from injury, and the tools and materials used shall be transmitted by ropes.

② Tool bags shall be used when working at heights, relatively large tools shall be fixed on solid components and not allowed to be randomly placed.

③ In the process of lifting with a winching, special personnel shall be assigned to command and cooperate in a unified way. Impulse inspection shall be carried out immediately after the insulator string are lifted off the ground.

(4) Hazard - Prevent injury caused by induced electricity.

① In order to prevent injury caused by induced electricity, electricity testing shall be carried out and the ground wire shall be installed on the maintenance line.

② The overhead ground wire shall be reliably grounded before necessary contact.

(5) Hazard - Site operating safety supervision.

① From the beginning to the end of the operation, the safety Supervisor must always carry out uninterrupted safety supervision on the operators on the site.

② The Responsible Person or the Supervisor must wear a safety vest.

(6) Hazard - Transportation safety.

It is necessary to pay attention to the driving safety of the vehicle, drive carefully when driving, and illegal driving is not allowed.

3.2 Selection of tools and instruments and materials

Refer to Table 2-17-2 for tools, instruments and materials required by power-cut replacement of 1000kV AC transmission line overhead ground wire. Before delivering tools and instruments out of warehouse, application voltage class and test period shall be carefully checked and they shall be inspected to ensure that appearance is intact, connection is firm, rotation is flexible,and they meet the working task requirements. After delivering tools and instruments out of warehouse, they shall be stored in tools bag or tool kit for transportation to avoid contamination and damp. Metal tools and insulated tools shall be separately loaded and transported to avoid deformation, damage or other defects caused by mixed loading and transportation.

Table 2-17-2 Tools, Instruments and Materials Required for the Power-cut Replacement of 1000kV AC Transmission Line Overhead Ground Wire

S/N	Name	Specification	Unit	Qty.	Remarks
1	Operation hanging wheel		Nos.	1	Other tools
2	Grinding rope	φ16	Roll	1	Other tools
3	Lever blocks	6T	Nos.	3	Metal tool
4	Winching	5T	Set	3	Motor tool
5	Lifting rope tackle	1T	Set	3	Metal tool
6	Noose		Nos.	3	Other tools

Table (Cont'd)

S/N	Name	Specification	Unit	Qty.	Remarks
7	Wire clamp		Nos.	2	Metal tool
8	Wire lifter		Set	1	Metal tool
9	Single-wheel tackle		Nos.	3	Metal tool
10	Paying-off rack		Set	2	Metal tool
11	Paying-off disk		Set	2	Metal tool
12	Steel wire sleeve	$\varphi 18$	Nos.	8	Other tools
13	Iron tackle	10T	Nos.	8	Metal tool
14	Shackle	10T	Nos.	8	Metal tool
15	Shackle	8T	Nos.	6	Metal tool
16	Hydraulic press (including hydraulic pliers)		Set	2	Motor tool
17	Interphone		Nos.	10	Other tools
18	Full-covered safety belt	Including 20m rear safety rope with buffer bag	Set	7	PPE
19	Safety vest		Piece	3	Other tools
20	Security fence		Volume	5	Other tools
21	Sole timber		Piece	Several	Other tools
22	Ground Wire	Al-clad steel stranded wire JLB20A, 150	Roll	3	Material
23	Strain clamp	Same model	Set	2	Material

3.3 Division of labor for operators

Division of labor for operators of the task is shown in Table 2-17-3.

Table 2-17-3 Personnel allocation for power-cut replacement of 1000kV AC transmission line overhead ground wire

S/N	Post	Qty. (person)	Responsibilities
1	Responsible Person	1	Be responsible for the labor division of operating personnel of the work task, field investigation before work, preparation of the operation plan, work order filling, field re-investigation, performing work permit procedures, holding pre-shift meeting, implementation of the site safety measures, safety supervision in the operation process, dealing with emergency situations in work, quality surveillance of the work, and the summary after the work

Part II Skill Module Training and Assessment Standards

Table (Cont'd)

S/N	Post	Qty. (person)	Responsibilities
2	Safety Supervisor	1	Be responsible for the safety monitoring work during this operation
3	Workers working at height	7	Be responsible for the operation of power-cut replacement of 1000kV AC transmission line overhead ground wire
4	Ground auxiliary personnel	10	Be responsible for ground auxiliary works during the operation.
5	Winching operator	2	Be responsible for the winching work during this operation
6	Winching signal command worker	2	Be responsible for commanding the start and stop of 2 sets of winches

4. Work Procedure

The workflow of this task is shown in Table 2-17-4.

Table 2-17-4 Work Procedures for Power-cut Replacement of 1000kV AC Transmission Line Overhead Ground Wire

S/N	Work Content	Operation Steps and Standards	Safety Measures and Precautions	Responsible Person
1	Site re-survey	The Responsible Person shall complete the following work: (1) Check the line title, the number of the pole and tower and ensure the phases are correct; guarantee that the foundation and the pole and tower are intact and in normal condition; ensure that the cross and span distance meets the safety requirements; confirm the defect conditions and the specifications and models of earth wires. (2) Check that the site meteorological conditions such as wind speed should meet the operation requirements. (3) Check that the terrain and environment shall meet the operation requirements. (4) Check that the safety measures listed in the work order are in line with the actual situations on site, and the measures will be supplemented if necessary	(1) Correctly wear helmet, working clothes, work shoes and protective gloves. (2) Operation under meteorological conditions that may endanger the safety of operators is forbidden. (3) Non-operation personnel and vehicles are strictly prohibited from entering the working site.	

Table (Cont'd)

S/N	Work Content	Operation Steps and Standards	Safety Measures and Precautions	Responsible Person
2	Work Permit	The contact person for power off/on before the operation must contact the dispatcher to complete the work permit procedure.	No work shall be started without the Work Permit of approver	
3	Site layout	Install the security fence and hang the signboards correctly: (1) The security fence should take full account of falling objects from the heights and the influence on road traffic. (2) The entrance and exit of the security fence shall be set reasonably. (3) Signs such as "Access from Here", "Work Here", "Slow Down" or "Blocking" shall be properly arranged.	When the influence on road traffic safety is uncontrollable, the traffic management department should be contacted in time to strengthen the on-site control of traffic safety.	
4	Hold a pre-shift meeting	(1) All working personnel shall line up. (2) The Responsible Person shall wear red waistcoat and read out the work order and be clear with work task and division of personnel; explain safety measures and technical measures in work; check (inquire after) mental state of all working personnel; inform of hazards in work and precontrol measures. (3) All working personnel shall sign on the work order for confirmation.	(1) The work order shall be filled in, issued and approved in a standardized manner, and the signature shall be complete. (2) All working personnel shall be in good mental states. (3) All working personnel shall be clear with task division of works, safety measures and technical measures	
5	Inspection tool	(1) All necessary tools and instruments shall be prepared as per the operation requirements and placed on the moisture-proof tarpaulin regularly according to the category and location. The appearance and test certificate of tools and instruments shall be checked to ensure there is no omission. (2) The inspector shall report to the Responsible Person that all inspection results are in conformity with the operation requirements	(1) The waterproof tarpaulin shall be enough in quantity and reasonable in position, and be clean and dry. (2) The appearance of tools and instruments is acceptable after inspection, there is no damage, damp, deformation and malfunction, and the certificate is in its validity	

Part II Skill Module Training and Assessment Standards

Table (Cont'd)

S/N	Work Content	Operation Steps and Standards	Safety Measures and Precautions	Responsible Person
6	Climbing the tower	(1) Check the name and serial number of the line before climbing the pole and tower for operation. For multi-circuit lines on the same tower, the Responsible Person and the members of the working team shall check the double name and identification mark carefully (color code, distinguishing mark, etc.). (2) It shall check the root, base, etc. of the pole and tower to keep them solid and reliable before climbing the tower. (3) During climbing, in order to prevent the operator from falling, the distance between the operators shall not be less than 1.6m, the safety belt shall be properly packed, the long tail rope shall be placed in the kit, and the main belt shall be hung on the shoulder to prevent the safety belt from hooking shackles and tower materials during climbing, resulting in falling off. (4) When climbing the tower to the cross arm, the Supervisor and the operator shall check the identification mark and double-designation of the line outage again and can enter the cross arm on the line outage side only after confirming no error.	(1) Operator shall wear safety helmet and soft soles, and keep in narrow movement range with uniform steps when climbing pole and tower. (2) During climbing, the safety belt shall be properly packed, the long tail rope shall be placed in the kit, and the main belt shall be hung on the shoulder to prevent the safety belt from hooking shackles and tower materials during climbing, resulting in falling off. (3) Operators shall not lose the protection from the safety belt, and make firm step and clench when moving on the pole and tower. (4) When reaching the position of operation point, fasten the safety belt (rope), it shall be firm and reliable, and shall not be hung low for high use. (5) The safe distance between the human body, headless rope, etc. and the conductor must not be less than 9.5m before the verification of live part, and special person shall be assigned to perform supervision during the work	
7	Inspection power and install ground wire	(1) After the pole for climbing is in place, fasten the safety belt to a firm and reliable component or pole, and check whether the buckle is in place correctly. Appliances such as test pen (electroscope) must be transferred by rope.	(1) Before electricity test, test shall be carried out on charged equipment to verify that the electroscope is in good condition. When it's unable to test on charged equipment, the electroscope can be verified with high pressure generator, etc.	

Table (Cont'd)

S/N	Work Content	Operation Steps and Standards	Safety Measures and Precautions	Responsible Person
7	Inspection power and install ground wire	(2) Standing at the correct position during electricity testing: The position should be suitable for work and no electric shock hazard; Insulating gloves must be worn for electricity testing. The power inspection of the line shall be carried out phase by phase, according to the sequence of first near then far, prior lower layer and posterior upper layer. (3) Check whether the ground wire is intact, and install the ground wire according to the procedures (connect the ground terminal prior to the conductor terminal). (4) When ground wire is installed, it must be operated with insulating rope or insulated handle, and it is forbidden to install the ground wire by directly operating the metal part of the ground wire by hand. Make sure the collection of the ground wire is tightly connected to the conductor	(2) Insulating gloves must be worn during high voltage electricity test. The length of the telescopic insulation rod of the electroscope should be fully drawn. The hand should not exceed the guard ring at the handle during the electricity test. Safe distance should be kept between human body and electroscope. It is not allowed to conduct electricity test outdoor directly in rainy and snowy weather. (3) The ground wire must be first connected to the ground terminal, then conductor terminal, and must be in good contact; Demolition is done in reverse order. It is forbidden to install the ground wire by winding the conductor	
8	Tangent tower threadbare	(1) The operator shall install the single-sheave tackle in the suitable position of the tangent tower ground wire cross arm and remove the vibration damper on the ground wire. (2) The operator should lift the ground wire with wire lifter and 6T lever hoist. (3) After removing part hardware fittings for connection, turn the ground wire into the single-sheave tackle fixed on the ground wire cross arm		
9	Installation of wire clamp	(1) After measuring and recording the position of vibration damper on the ground wire, the operator shall remove the vibration damper (2) After the operator uses a hanging wheel to clamp the wire clamp to a suitable position on the ground wire, tighten the grinding rope. (3) Tighten the ground wire at both ends with a 6T lever hoist, and take off the ground wire insulator set fitting	Conduct impulse test after tightening the ground wire, and proceed to the next step after confirming there is no error.	

Part II Skill Module Training and Assessment Standards

Table (Cont'd)

S/N	Work Content	Operation Steps and Standards	Safety Measures and Precautions	Responsible Person
10	Resisting-tensile tower thread slack and connecting new and old ground wire	(1) The operator should loosen the 6T lever hoist at both ends to make the grinding rope stressed and then remove the lever hoist. (2) Start the winching and use the winching to loosen the ground wires at both ends to the ground. (3) After loosening to the ground, the ground personnel shall cut off the strain clamp at the ground wire end, polish and crimp the new and old ground wires	(1) During paying-off, the winding and winching operation should be stable. Maintain the balance of grounding traction, and prevent the ground wire from slipping. (2) When using the winching, the number of twining loops of the winching shall not be less than 5, and it shall be rolled from below and be neat, and the number of tail rope control personnel shall not be less than 2. (3) When laying off and stringing, personnel shall neither stand beside, cross the stressed hauling rope, the interior angle side of the ground wire, in the dispatched ground wire circles, nor below the vertical direction of the hauling rope and overhead line in case of any injury due to accidental deviation from lines. (4) Any part of human body must not be directly above the hydraulic press when ground wire is crimped. The crimping machine shall have fixed facilities and shall be placed stably during operation. The thread-holding personnel on both sides shall be aligned with the position and their fingers shall not extend into the moulding-die	
11	Installation of crossover and accessories	(1) Start 2 winchings on the traction field and tension field at the same time under unified command by specially-assigned person. (2) After the new ground wire arrives at the traction field, the tension field shall be crimped with the strain clamp. After crimping, the crimped ground wire shall be connected to the insulator armor clamp by winding and winching.	(1) During the paying-off construction, special personnel shall be assigned to watch over temporary stay wire, cross span, tangent tower and stressed tools, and check that the stress conditions are in good condition.	

Table (Cont'd)

S/N	Work Content	Operation Steps and Standards	Safety Measures and Precautions	Responsible Person
11	Installation of crossover and accessories.	(3) Tighten the ground wire at the traction field, and mark when the ground wire sag is consistent with the original sag. (4) Loosen the ground wire to the ground, crimp the strain clamp at the other end, then tighten the grinding rope. Connect the ground wire to the insulator string armor clamp and the new ground wire to the cable clamp by the tangent tower, fasten the bolts, and install the vibration damper according to the previously marked position	(2) When paying-off, communication should be kept smooth with unified signals and commands. (3) During construction, it is forbidden to cross or linger under the ground wire. (4) In the process of crossover, special personnel shall check the stress of ground anchor, steering tackle and grinding rope at any time and make timely adjustment. (5) During paying-off, the winding and winching operation should be stable. Maintain the balance of grounding traction, and prevent the ground wire from slipping.	
12	Removal of tools and instruments	(1) Remove tools in sequence, such as grinding rope, lever hoist, wire clamp and backup protection rope, etc. (2) Remove ground wire	(1) The removal sequence of ground wire: conductor terminal first and then the ground terminal	
13	End of the work	(1) Responsible Person shall organize the staff to put working apparatus and materials in order, and clean the site to ensure "materials are removed and the site is cleaned after completion of construction". (2) In the post-shift meeting, Responsible Person will give work summaries and comments. Comments include the construction quality of this work and the implementation of safety measures from all working personnel. (3) Responsible Person shall report the end of the work to the work approver, resume the power transmission of the power-cut line and terminate the work order		

II. Assessment Standard

Detailed Rules for Assessment and Scoring of Operation and Inspection Skills of UHV AC Transmission Line

Fill-in Column of Examinee	No.:		Name:		Position:		Date:	MM/DD/YYYY
Fill-in Column of Assessor	Grade:		Assessor:		Team Assessment Leader:		Starting time:	Closing time:	Operation Duration:
Assessment module	Power-cut replacement of 1000kV AC transmission line overhead ground wire		Assessee		Maintenance personnel of UHV AC transmission line		Assessment method	Operation	Assessment Time Limit	360min
Job Description	Power-cut replacement of the left overhead ground wire in the strain section on 1000kV UHV transmission line									
Work Specifications and Requirements	1. Given conditions: The left overhead wire of 003#-005# tower strain section of 1000kV AC practical training line needs to be replaced. The line has been cut off, tested, and grounded. The tools and materials used have been tested, the work order has been handled, and the safety measures have been taken. 2. The work shall be carried out in good weather. In case of lightning (hearing thunder or seeing lightning), snow, hail, rain, fog and so on with wind force greater than Level 5, working shall be prohibited. 3. This assignment needs 1 Responsible Person, 1 Safety Supervisor, 1 electrician on tower, 6 auxiliary workers on tower, 10 auxiliary workers on the ground, 2 winching operators to assist the personnel to finish the up and down transfer of tools, materials and other non-technical work. 4. Responsibilities of Responsible Person: Be responsible for the labor division of operating personnel of the work task, field investigation before work, preparation of the operation plan, work order filling, field re-investigation, performing work permit procedures, holding pre-shift meeting, implementation of the site safety measures, safety supervision in the operation process, dealing with emergency situations in work, quality surveillance of the work, and the summary after the work. 5. Special Supervisor: Be responsible for safety control of the work site. 6. Responsibilities of operators working at heights: Be responsible for power-cut replacement of 1000kV AC transmission line overhead ground wire.									

Table (Cont'd)

Work Specifications and Requirements	7. Responsibilities of ground auxiliary personnel: Be responsible for transferring tools and materials and cooperating with electricians on tower. 8. Responsibilities of winching operators: Be responsible for the winching work during the operation. 9. In case of thunder, rain, strong wind or any other circumstance threaten the safety of the staff during working, the Responsible Person or Supervisor may stop working temporarily according to the circumstances. Given conditions: 1. Training base: The left overhead wire of 003#-005# tower strain section of UHV 1000kV AC practical training line; Ground wire type: Al-clad steel stranded wire JLB20A, 150. 2. Work orders have been handled, safety measures have been completed, and oral application (dispatcher or assessor) shall be made at the beginning and end of the work. 3. The instrument shall be used safely and correctly to test the insulating tool. 4. The operation must be carried out according to the working procedures. The scores of the items to be carried out shall be deducted for the process error. In case of major hidden dangers of personal, equipment and operational safety, the assessor may order the termination of the operation (assessment)
Assessment scenario preparation	1. Line: The left overhead wire of 003#-005# tower strain section of UHV 1000kV AC line. Work content: power-cut replacement of 1000kV AC transmission line overhead ground wire; Model of ground wire: JLB20A-150. 2. Tools and instruments as required: 1 hanging wheel, a bundle of Φ16 grinding rope, 3 lever hoists (6T), 3 winchings (5T), 3 sets of lifting rope tackles, 3 nooses, 2 wire clamps, 1 wire lifter, 3 single-sheave tackles, 2 sets of pay-off racks, 2 sets of pay-off reels, 8 steel wire sleeves (Φ18), 8 iron tackles (10T), 8 shackles (10T), 6 shackles (8T), 2 sets of hydraulic presses (including hydraulic pliers), some walkie-talkies, 7 double-insurance safety belts, and several wood blocks. 3. Material: 1 reel of ground wire of the same model, 1 set of splicing sleeve, and 2 sets of strain clamps 4. The work site shall be monitored, and the safety measures (fence, etc.) on the work site have been fully implemented; non-operation personnel are prohibited from entering the site, and the staff must wear safety helmet when entering the work site
Remarks	1. The deduction shall be done until the scores of each item are deducted completely. In case of major hidden dangers of personal, equipment and operational safety, the assessor may order the termination of the operation. 2. When equipment, working environment, safety belt, safety helmet, tool, shielding clothes, etc., do not conform to the operation condition, the assessor may order the termination of the operation

Part II Skill Module Training and Assessment Standards

Table (Cont'd)

S/N	Project name	Quality requirements	Score	Deduction standard	Reasons for deduction	Deduction	Scoring
1	Site re-survey	1) The Responsible Person shall go to the work site to check the line name, pole and tower number, on-site working conditions, defective parts and so on. 2) Confirm that the site meteorological conditions meet the operational requirements. 3) Check whether the work order is complete and unmodified, check whether the safety measures listed are consistent with the actual situation on site, and supplement it if necessary	5	1) Deduct 1 point for failure to check the double title. 2) Deduct 1 point for failure to verify on-site working conditions (meteorology), defective parts. 3) Deduct 0.5 points/item for any alteration in the work order filling, and deduct 1 point for incorrect work order number. Deduct 1.5 points for each incomplete work order			
2	Work Permit	1) The Responsible Person shall contact the on-duty control personnel and fulfill the work permit procedures according to the contents of the work order. 2) Reporting content is standardized and complete	2	1) Deduct 2 points for failure to contact the dispatching department (referee) for deactivating the re-closing. 2) Deduct 0.5 points for non-standard or incomplete terminology reporting respectively			
3	Site layout	Install the security fence and hang the signboards correctly: 1) The security fence should take full account of falling objects from the heights and the influence on road traffic. 2) The entrance and exit of the security fence shall be set reasonably. 3) Signs such as "Access from Here", "Work Here", "Access from Here" shall be properly and well arranged	3	1) Deduct 0.5 points for failure to arrange the fence at the work site. 2) Deduct 0.5 points for failure to arrange the warning board. 3) Deduct 0.5 points for failure to hang the tower climbing operation sign			

Standard for Professional Training and Assessment for Operation Maintenance of UHV AC Transmission Line

Table (Cont'd)

S/N	Project name	Quality requirements	Score	Deduction standard	Reasons for deduction	Deduction	Scoring
4	Hold a pre-shift meeting	1) All staff and personnel shall wear safety helmets and work clothes correctly. 2) Responsible Person shall wear red vest and read out the work order and be clear with work task and division of personnel; explain safety measures and technical measures in work; check (inquire after) mental state of all working personnel; inform of hazards in work and precontrol measures. 3) All working personnel shall sign on the work order for confirmation.	3	1) Deduct 0.5 points/person for the staff not dressing uniformly. Deduct 0.5 points/person for the staff not dressing uniformly 2) Give no points to this item for no division of labor, and deduct 1 point for unclear division of labor. 3) Deduct 0.5 points for the on-site Responsible Person not wearing a safety monitoring vest. 4) Deduct 1 point for the work shift member failing to sign or signing incompletely on the work order.			
5	Inspection of tools	1) The staff shall place the tools on the moisture-proof tarpaulin as required; the moisture-proof tarpaulin shall be clean and dry. 2) The tools shall be placed according to the requirements of the fixed location management; Inspect the appearance of tools and test certificates without omission. 3) Adjustable wrench, flat pliers, pin-pulling pliers and tool kits shall meet the quality requirements. 4) Test-run the winching, and check whether the gear position is normal, and check lever hoist.	10	1) Deduct 1 point for failure to use moisture-proof cloth and place tools to designed positions. 2) Deduct 0.5 points/item for failure to check qualified label of tool test and appearance inspection. 3) Deduct 1 point for failure to test the tools. 4) Deduct 2 points for incorrectly testing the winching; deduct 2 points for improper check of lever hoist.			

Part II Skill Module Training and Assessment Standards

Table (Cont'd)

S/N	Project name	Quality requirements	Score	Deduction standard	Reasons for deduction	Deduction	Scoring
5	Inspection of tools	5) Check the wire rope and tackle, confirm the connection is reliable and the force meets the requirements. 6) The safety belt and personal safety line meet the quality requirements within the test period. 7) The wire lifter, wire clamp and wire tumble tool meet the requirements within the test period. 8) Confirm that the type, length and appearance of ground wire and connecting pipe meet the requirements. 9) Tower climbing personnel shall check the double name, pole number and phase again and report them	10	5) Deduct 1 point for each error and missing item. 6) Deduct 1 point for each improper inspection. 7) Deduct 0.5 points for each error and missing item. 8) Deduct 1 point for each error and missing item. 9) Deduct 2 points for failure to check the double-designation			
6	Climbing the tower	1) Check whether there is no abnormality at pole and tower base. 2) After the electricians on tower wear the safety belts and conduct the impulse test, fasten the safety belts and correctly carry the lifting rope (slinging over the shoulder with the lifting rope end folded double, fast knot) and climb up the tower one after another.	8	1) Deduct 1 point for no check of one item. 2) Deduct 2 points for failure to fasten the safety belt or for failure to perform the impulse test on the safety belt and the backup protection rope. Deduct 1 point for not carrying lifting rope, and deduct 1 point for nonstandard carrying method of lifting rope. 3) Deduct 2 points for grasping shackles with hands and deduct 1 point for not climbing the tower along the main material of the shackles side. 4) Deduct 2 points for failure to use safety belt properly.			

Table (Cont'd)

S/N	Project name	Quality requirements	Score	Deduction standard	Reasons for deduction	Deduction	Scoring
6	Climbing the tower	3) In the process of climbing the tower, fasten the anti-falling protection device, climb the tower at a uniform speed, grasp the main material, and hang the safety belt on the shoulder; The distance between the operators shall not be less than 1.6m, and the Responsible Person shall strengthen the operation supervision. 4) The safety belt protection shall not be lost from the tower body to the working point on the cross arm. 5) The transfer tackle shall be installed at the correct position for easy operation.	8	5) Deduct 1 point for inconvenient suspension position of tackle transmission rope for taking tools. 6) Deduct 2 points for metal tools that are difficult to ensure safe distance during transfer; deduct 2 points for tools that are not bound securely. 7) Deduct 2 points for falling object at high place. 8) Deduct 2 points for tools colliding with the tower body during the transfer process. 9) Deduct 1 point for knotting and disordered rope in tool transfer.			
7	Inspection power and install ground wire	1) After climbing the pole in place, fasten the safety belt to a firm and reliable component or pole, and check whether the buckle is in place correctly. Appliances such as test pen (electroscope) must be transferred by rope. 2) Check whether the ground wire is in good condition, and install the ground wire according to the procedures (first connect the ground terminal, then the conductor terminal). The ground collet shall be firmly connected on the cross arm.	11	1) Deduct 5 points for failure to test electricity and install ground wire. 2) Deduct 3 points for incorrect installation order of ground wire. 3) Deduct 3 points for insecure connection of ground wire. 4) Deduct 2 points for failure to wear insulating gloves.			

Part II Skill Module Training and Assessment Standards

Table (Cont'd)

S/N	Project name	Quality requirements	Score	Deduction standard	Reasons for deduction	Deduction	Scoring
7	Inspection power and install ground wire	(3) When ground wire is installed, it must be operated with insulating rope or insulated handle, and it is forbidden to install the ground wire by directly operating the metal part of the ground wire by hand. Make sure the collection of the ground wire is tightly connected to the conductor	11				
8	Tangent tower threadbare	1) The operator shall install the single-sheave tackle in the suitable position of the tangent tower ground wire cross arm and remove the vibration damper on the ground wire. 2) The operator should lift the ground wire with wire lifter and 6T lever hoist. 3) After removing part hardware fittings for connection, turn the ground wire into the single-sheave tackle fixed on the ground wire cross arm	6	1) Deduct 2 points for failure to protect ground wire and tower material when lifting ground wire. 2) Deduct 2 points for improper operation of lever hoist. 3) Deduct 2 points for nonstandard use of rope ladder in climbing up and down			

Standard for Professional Training and Assessment for Operation Maintenance of UHV AC Transmission Line

Table (Cont'd)

S/N	Project name	Quality requirements	Score	Deduction standard	Reasons for deduction	Deduction	Scoring
9	Installation of wire clamp	1) After measuring and recording the position of vibration damper on the ground wire, the operator shall remove the vibration damper. 2) After the operator uses a hanging wheel to clamp the wire clamp to a suitable position on the ground wire, tighten the grinding rope. 3) Tighten the ground wire at both ends with a 6T lever hoist, conduct impulse test and take off the ground wire insulator set fitting after confirmation	8	1) Deduct 3 points for failure to record the position of vibration damper. 2) Deduct 3 points for improper use of hanging wheel. 3) Deduct 2 points for failure to conduct impulse test before taking off the insulator set fitting			
10	Resisting-tensile tower thread slack and connecting new and old ground wire	1) The operator should loosen the 6T lever hoist at both ends to make the grinding rope stressed, impulse the grinding rope and then remove the lever hoist. 2) Start the winching and use the winching to loosen the ground wires at both ends to the ground. 3) Arrange the angle tackle of the new ground wire and its direction on the tension field tower. 4) The ground personnel shall cut off the strain clamp at the ground wire end, polish and crimp the new and old ground wires. The crimping process of splicing pipe shall meet the requirements.	15	1) Deduct 2 points for failure to carry out the impulse test. 2) Deduct 1 point for failure to protect the tower material during installation of steering tackle at each place. 3) Deduct 3 points for non-standard layout of the new ground wire. 4) Deduct 2 points for failure to polish the ground wire for each place. 5) Deduct 3 points in case that the crimping process does not meet the requirements			

· 736 ·

Part II Skill Module Training and Assessment Standards

Table (Cont'd)

S/N	Project name	Quality requirements	Score	Deduction standard	Reasons for deduction	Deduction	Scoring
11	Installation of crossover and accessories	1) Start 2 winchings on the traction field and tension field at the same time under unified command by specially-assigned person. 2) In the process of crossover, a special persons should be assigned to guard the mid-range center to prevent ground wire falling off. 3) The operators on tower should pay attention to on the joint to prevent wire jamming. 4) The winching operator shall control the crossover speed to prevent excessive tension. 5) After the new ground wire arrives at the traction field, the tension field shall be crimped with the strain clamp. After crimping, the crimped ground wire shall be connected to the insulator armor clamp by winding and winching. 6) Tighten the ground wire at the traction field, and mark when the ground wire sag is consistent with the original sag.	12	1) Deduct 2 points for ground wire falling to ground every time during crossover. 2) Deduct 1 point for oversize strain during crossover. 3) Deduct 2 points for thread jamming due to no one has been assigned to command. 4) Deduct 2 points in case the old ground wires are not sorted out and recovered in time. 5) Deduct 2 points in case the sag is not measured by a specially assigned person; 2 points for inconsistent sag with original sag. 6) Deduct 3 points in case the model of the strain clamp is not consistent with that of the ground wire.			

Table (Cont'd)

S/N	Project name	Quality requirements	Score	Deduction standard	Reasons for deduction	Deduction	Scoring
11	Installation of crossover and accessories	7) Loosen the ground wire to the ground, crimp the strain clamp at the other end, then tighten the grinding rope. Connect the ground wire to the insulator string armor clamp and the new ground wire to the cable clamp by the tangent tower, fasten the bolts, and install the vibration damper according to the previously marked position. 8) Connect the new ground wire to the cable clamp by the tangent tower, fasten bolts and install vibration damper according to the original requirements.		7) Deduct 2 points for failure to conduct impulse test after connection. 8) Deduct 2 points in case that the crimping process does not meet the requirements. 9) Deduct 2 points for failure to check the bolts fastened properly. 10) Deduct 1 point for each improper installation position of spacer, and 2 points for unacceptable installation and improper position of spacer			
12	Return to the ground	1) Check whether the armor clamps and accessories are complete and in place, whether the connecting part is firm. Loosen the grinding rope and check whether the tower material is damaged, check the sag of the ground wire, and check whether the connecting armor clamps meet the requirements. 2) Remove tools in sequence, such as grinding rope, lever hoist, ground wire, tackle, wire clamp and backup protection rope, etc. and transfer to the ground. When removing the grounding wire, the conductor terminal and ground terminal shall be removed successively. First remove the upper layer, then the lower layer, first the distant end, then the near end.	7	1) Deduct 3 points for failure to check the connection of armor clamp, and 2 points for failure to check the ground wire sag. 2) Deduct 2 points for each winding of transmission rope, 2 points for each damage to tower damage. 3) Deduct 2 points for incorrect removal order of ground wire. 4) Deduct 4 points for no lifting rope, deduct 2 points for nonstandard carrying method of lifting rope. 5) Deduct 1 point for grasping shackles with hands and 4 points for not climbing the tower along the main material of the shackles side. 6) Deduct 4 points for any object left			

Part II Skill Module Training and Assessment Standards

Table (Cont'd)

S/N	Project name	Quality requirements	Score	Deduction standard	Reasons for deduction	Deduction	Scoring
		3) After checking that there is no object left on tower, the equipotential electricians shall report it to the Responsible Person and then climb down the tower with insulated transmission rope after obtaining the consent of the Responsible Person. 4) Correctly climb down the main material along the shackles side, and carry the lifting rope (lifting rope head shall be double folded, fast knot, and hung over the shoulder). 5) Confirm no object left					
13	End of the work	1) The Responsible Person shall organize all working members to put working apparatus and materials in order and put them in a special kit (bag) after cleaning; clean the site to ensure that "the materials are removed and the site is cleaned after construction". 2) In the post-shift meeting, Responsible Person will give work summaries and comments. Comments include the construction quality of this work and the implementation of safety measures from all working personnel. 3) Responsible Person shall report the completion of the work to the work approver, resume the power transmission of the power-cut line and terminate the work order. 4) Complete operation as required within the specified time	10	1) Deduct 3 points for failure to clean the work site. 2) Deduct 2 points for failure to hold the post-shift meeting. 3) Deduct 2 points for failure to report to the work approver. 4) Deduct 1 point for every extension of 1 minute. After extension of 5 minutes, incomplete items will not be scored			
	Total		100